ANALYTICAL AND COMPUTER METHODS IN FOUNDATION ENGINEERING

McGRAW-HILL
BOOK COMPANY
New York
St. Louis
San Francisco
Düsseldorf
Johannesburg
Kuala Lumpur
London
Mexico
Montreal
New Delhi
Panama
Rio de Janeiro
Singapore
Sydney
Toronto

JOSEPH E. BOWLES
Professor of Civil Engineering
Bradley University

Analytical and Computer Methods in Foundation Engineering

This book was set in Times Roman.
The editors were B. J. Clark, Madelaine Eichberg, and M. E. Margolies;
the cover was designed by Jo Jones;
and the production supervisor was Thomas J. LoPinto.
The drawings were done by Oxford Illustrators Limited.
The printer and binder was The Maple Press Company.

Library of Congress Cataloging in Publication Data

Bowles, Joseph E
Analytical and computer methods in foundation engineering.

Includes bibliographies.
1. Electronic data processing—Foundations.
I. Title.
TA775.B62 624'.15'02854 73-6698
ISBN 0-07-006753-8

ANALYTICAL AND
COMPUTER METHODS IN
FOUNDATION ENGINEERING

1234567890MAMM79876543

CONTENTS

PREFACE

This textbook has been written to provide practicing soil engineers, foundation engineers, and engineering students with the finite-element and other computer-oriented computational tools to solve most soil-structure interaction problems encountered in practice. Computer-program listings are included for each major analytical procedure introduced so that the user can either punch a card deck or obtain a set of card decks from the author to immediately begin problem solving.

The book uses both fps (U.S. units) and the metric system: where metric units are given, the SI values are used unless common usage dictates otherwise. At least one problem in each problem-oriented chapter is solved using SI units. The computer programs are written in the widely used FORTRAN IV language and can be used for either system of units.

Chapter 2 includes enough material on soil mechanics to permit the text to be self-sufficient and to provide a ready source of soil-engineering information for the types of problems considered in this book.

Chapter 3 considers reinforced-concrete design as applicable to foundation engineering using the ACI 318-71 Code and extends the coverage to include spread and combined footings. This coverage makes it possible for this text to be used in a first course in foundation engineering. For the practicing engineer this chapter provides a handy reference.

Chapter 5 presents new methods of analysis for the beam on an elastic foundation and ring foundations developed by the author. The method for beams on elastic foundations is critically evaluated with test data and is compared with conventional design procedures. Chapter 7 includes coverage of mat foundations and eccentrically loaded footings. The finite-difference and finite-element methods of solution are used in the analysis of mat foundations. New material is included in this chapter on the analysis of footings with large eccentricity (including footing weight), footings with notches, and a simple finite-element method to solve irregular-shaped mat foundations. The author compares these methods with conventional practice.

Chapters 9 and 10 present new finite-element procedures developed by the author for lateral-pile and sheet-pile analysis and compare the analytical procedure with field and/or laboratory test results.

Chapter 11 contains the author's latest version of the wave equation together with pile-load test comparisons.

Chapter 12 contains a new method of static pile-stress analysis using a finite-element procedure developed by the author. This method is also compared to field pile-load test results. The method can include piles with batter, partial embedment, and the effect of lateral deformation ($P\Delta$) on bending stresses induced in the pile.

Chapter 13 contains the recently developed matrix analysis of three-dimensional pile groups, and Chapter 14 includes the Bishop method of slope stability, modified by the author to compute effective stresses directly and to include a very large number of different soil properties within the assumed failure arc.

This text uses matrix notation where applicable, but the notation has been kept as simple as possible and is used consistently throughout the text. This feature is of considerable value especially to those with limited familarity with the method.

Computer-program operations, input data, data units, and limitations are outlined prior to the program listing, and at least one set of data-card entries for a fps and a metric problem are completely listed in the proper sequence for use in the computer program. This feature gives the user immediate access to the programs with a minimum of uncertainty. Most chapters show a complete or slightly edited computer output for the problems solved. This program-documentation technique has been used successfully by the author both in the classroom and in correspondence with other engineers.

The anticipated method of classroom instruction is that the instructor will obtain the card decks from the author, load them on the computer, and assign problems to the class. This allows the student to observe the effect of varying soil and/or structure parameters rather than spending his limited time working a single (in many cases) complex problem by hand by outmoded and often highly approximate methods.

Furnishing the program listings also enables the student to spend his time profitably on problem solving rather than on program writing and debugging operations.

The practicing engineer will find this text easy to read, and the notation is such that comprehension develops quickly even if the reader is not familiar with matrix operations. The program listings and deck availability from the author enable nearly every design office to have access to highly efficient and powerful design tools for analyzing this class of problem. Even more important is that the foundation engineer can vary the applicable design parameters to achieve the best and most economical design (and in many cases with a much higher level of confidence).

The author wishes to express sincere appreciation to Marian J. Frobish, of the Bradley University Computer Center, for providing guidance in developing and helping to debug the included computer programs. Appreciation is also expressed to Dr. H. Y. Fang of Lehigh University and those others who reviewed the manuscript.

Especial thanks are due to my wife, Faye, for typing the entire manuscript several times and catching many errors. Without her cooperation and assistance it is doubtful if this text would have ever been completed.

JOSEPH E. BOWLES

ANALYTICAL AND
COMPUTER METHODS IN
FOUNDATION ENGINEERING

1

COMPUTER PROGRAMMING USING FORTRAN IV

1-1 ELEMENTS OF WRITING COMPUTER PROGRAMS

A computer program will be defined as a means of communicating a sequence of instructions to enable the computer to perform a desired set of operations. The discussion of this text limits these operations to various mathematical operations such as multiplying, dividing, raising the power, extracting roots, and obtaining trigonometric functions.

As the reader is assumed to have had a course in computer programming, this chapter is primarily a rapid reference source on computer methods. For a more complete discussion and other techniques, see Bradley (1969) and IBM (1968).

1-2 TYPES OF COMPUTATIONAL VARIABLES AVAILABLE

Computers use two types of numbers and variable identification in arithmetic computations; the variables are always identified alpha-numerically.

1 *Fixed-point* numbers, or integers, where no decimal point is used, for example, 1, 5, 6, 8, 200

2 *Floating-point*, or numbers with a decimal point, for example, 1.00, 1.41416, 3.1416, 500., 500.00, 500.001

Alpha-numeric identification means using alphabet letters as a prefix with or without numerals to provide variable identity. For example,

A	SLOPE2	STRESS	INERTA
K	CAT	$MIN	MODEL
ABLE	PHI1	MOM	G4

Fixed-point variables are always automatically identified by the computer as

I, J, K, L, M, N

or prefixed with these letters. Thus, K, MOM, INERTA, MODEL of the preceding paragraph refer to fixed-point numbers.

Floating-point variables use the remainder of the alphabet and the dollar symbol $, thus

A, ABLE, CAT, STRESS, $MIN

would be floating-point variables. It should be evident from the manner of writing variables that the programmer has considerable latitude in variable usage and will attempt to use the variable names for identification where possible. Section 1-5 puts a length limitation on a variable of six alpha-numeric characters.

Fixed- and floating-point numbers can be interchanged as part of the program specification instructions (see Sec. 1-8).

One should generally not intermix fixed and floating variables in any given computation although some computers do allow this. Identifying a floating-point operation (or number) or a fixed-point computation as an integer variable always effects a truncation of the decimal *down* to the nearest integer.

1-3 SINGLE AND DOUBLE PRECISION

Ordinarily IBM computers perform arithmetical operations in *single precision*. Single-precision computations use a maximum of seven digits (filled out with zeros or truncated if more than seven digits) plus space for the sign and decimal point, for example,

−9998887. or .0000123

Specifying *double-precision* computations allows the use of a maximum of 16 digits with sign and decimal point, for example,

−56789.83542155774

The single-precision number is a *word* consisting of two half-words, a half-word being 2 bytes. A byte consists of 8 bits. The byte is used to describe the computer's

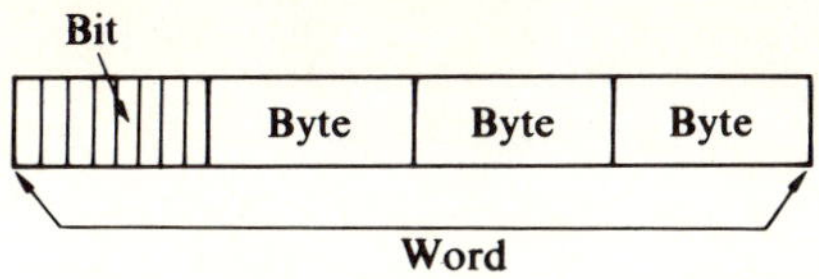

core capacity in blocks of 2^{10} bytes. To condense discussion, capacity of computer memory is described in terms of K, where $K = 2^{10}$ bytes. A computer of core memory size

$$16\ K = 16(1{,}024) = 16{,}384 \text{ bytes}$$

and

$$128\ K = 131{,}072 \text{ bytes}$$

A certain number of bytes are associated with the computer bookkeeping. For the 360/40 system with 128 K core storage this is about 14 K. Thus, in this system for all practical purposes the programmer would be able to use 114 K/4 ≈ 29,000 words (or seven-digit numbers) in a computer with this core capacity. One consumes computer core at a very rapid rate using double-precision numbers, since in all double precision the capacity is about 14,500 words. One should not arbitrarily assume that double precision provides more computational accuracy. Actually matrices of size 60 × 60 can be inverted in single precision on the IBM 360 with little loss of accuracy. In fact a 90 × 90 can be inverted in single precision to the precision of the input data.

1-4 ADDITION, SUBTRACTION, MULTIPLICATION, DIVISION (+, −, *, /)

Consider three variables, A, B, C. To add A and B one may write in computer language (FORTRAN IV) that the sum is a new variable D as follows:

D = A + B

To subtract C from A and from D, introduce new variables F and G:

F = A − C

G = D − C

To multiply A and B and the product AB = H by F

$$H = A * B$$

$$P = A * B * F$$

or

$$P = H * F$$

Obviously in this last case an error message would result if P = H * F were not preceded by H = A * B, as H does not have an identity until the product operation has been performed.

To divide F by P

$$Q = F/P \qquad \text{or} \qquad Q = (A - C)/P$$

and not Q = A − C/P since this would simply subtract the value of C/P from A, which was not wanted.

Note also that if A = 20., C = 4., and P = 2.,

$$(A - C)/P = (20. - 4.)/2. = 8.0$$

Further

$$(A - C)/P + 1.0 = 9.0$$

but

$$(A - C)/(P + 1.0) = 5.333$$

1-5 LENGTH OF VARIABLE; SUBSCRIPTED VARIABLES

No variable can be more than six alpha-numeric characters long, exclusive of subscript identification. For example,

AA3	3 characters long
ALINE	5 characters long
ASLOPE	6 characters (maximum length)
ASLOP1	6 characters (maximum length)
ASLOPE(I,J)	6 characters (maximum length)
ELASTIC	incorrect (7 characters long)

Variables may be subscripted so that they can be stored for later computational use or written in a particular output form (FORMAT). All subscripted variables *must be* DIMENSIONED, preferably at the beginning of the program (see Sec. 1-9).

The dimension specification (statement) may be bigger than the largest value of the subscript, or equal to it, but never smaller. The DIMENSION statement is an instruction to the computer to reserve or allocate in core the number of spaces specified on the DIMENSION statement. Variables may be subscripted single, double, or more. Examples:

A(I)	Single; A(20) reserves 20 spaces
A(I,J)	Double; A(20,20) reserves 400 spaces
A(I,J,K)	Triple; A(2,8,12) reserves 192 spaces

The spaces reserved are in single precision. If A(20) referred to a double-precision variable, 40 spaces would be effectively allocated.

The subscript counters (I,J), or whatever other variables are used, must be specified as fixed-point.

1-6 EXPONENTS AND ROOTS

To raise a variable to any power (subject, of course, to computer limitations of approximately ±75 on the IBM 360) consider the following:

Given: A, B, C

Find: $D = A^2 \quad F = AB^3 \quad G = A(BC)^4 \quad D = (ABC)^2$

Solution:

```
D = A**2          G = A*(B*C)**4
F = A*B**3        D = (A*B*C)**2
```

Note, however, that A*B*C**2 = ABC^2.

To obtain $\sqrt{A}$, $\sqrt[3]{C}$, $\sqrt[4]{D}$, and $\sqrt{A^2 + B^2}$, one may introduce variables and solve as follows:

G = A**.5	$\sqrt{A}$	Q = (A**2 + B**2)**.5
H = C**.333	$\sqrt[3]{C}$	
P = D**.25	$\sqrt[4]{D}$	

For the first and last root extraction, one may use a computer-furnished software system subroutine:

```
G = SQRT(A)
Q = SQRT(A**2 + B**2)
```

or

```
Q = DSQRT(A**2 + B**2)
```

where the prefix "D" in DSQRT is for double precision. The argument $A^2 + B^2$ *cannot be negative* for this particular subroutine, which may happen, for example, in solving roots of a quadratic equation of the general form

$$X = \frac{-b \pm \sqrt{b^2 - 4ac}}{2a}$$

Two alternate subroutines (CSQRT and DCSQRT) are used if the square root is negative or complex.

One may use either fixed or floating values of exponents in raising to powers; conversely, for all root operations it is necessary to use floating point.

To raise e to any exponent or obtain logarithms of numbers, computer software routines can be used.

$$e^X = \text{EXP(X) or DEXP(X) in double precision}$$

$$\log X = \text{ALOG10(X) or DLOG10(X)}$$

$$\ln X = \text{ALOG(X) or DLOG(X)}$$

1-7 TRIGONOMETRIC AND HYPERBOLIC FUNCTIONS

Trigonometric functions are obtained as computer software subroutines using:

Function	Computer single-precision subroutine
Sine X	SIN (X)
Cosine Y	COS (Y)
Tangent Z	TAN (Z)
Cotangent A	COTAN (A)
Arc sine B	ARSIN (B)
Arc cosine C	ARCOS (C)
Arc tangent D	ATAN (D)
Hyperbolic sine F	SINH (F)
Hyperbolic cosine G	COSH (G)
Hyperbolic tangent P	TANH (P)

For double precision, prefix the subroutine name with a D; for example, DSIN(X), DTANH(P).

The arguments used in the trigonometric subroutines must be in radians; angles computed from the inverse functions will be in radians. An angle is converted to radians as follows:

$$\text{Angle in radians} = \frac{\text{angle in degrees}}{57.2957795131}$$

$$1 \text{ rad} = \frac{180}{3.1415926535}$$

1-8 SIGNS; FINDING LARGEST OR SMALLEST VALUE IN AN ARRAY

Signs can be converted by either inserting a sign or multiplying by -1.0.

If a value is being compared (or tested) and it is not known in advance what the sign will be, one may use the absolute value as

ABS(arg) or DABS(arg)

depending on single- or double-precision (DABS) usage. In double precision both the argument and the number being compared must be of similar precision. Using this subroutine effectively converts any value or values in an array used as the argument to plus sign.

To find the largest or smallest value of a variable in a group or array, one may use still another computer software subroutine. The largest value of a list of values X, Y, Z, W, U, V is

```
AMAX0 (I,J,K,L,M,M1)
AMAX1 (-X,Y,-Z,W,A,...)*
MAX0  (...,F,G)
MAX1  (-...A,B,C,...)
DMAX1 (A1,B1,C2,F2,...)
```

using AMAX for the absolute maximum value when signs are included and MAX– when signs are not to be considered. Further, the zero (0) or (1) identifies whether the variables are fixed or floating point (AMAX1 and MAX1 are floating point). One may write a series of statements to do this as an alternative method.

The smallest value of a list or array can be obtained using

```
AMIN0 (N,N1,N2,NN)
AMIN1 (X,Y,Z,A)
MIN0  (N,N1,N2,...)
MIN1  (A,B,C,...)
DMIN1 (A,B,C,...)
```

* See the computer program in Chap. 5.

where A, 0, and 1 are for absolute minimum, fixed point, or floating point, respectively. In both maximum- and minimum-value subroutines the D prefix is double precision, and if the argument is not double precision, an error message will be displayed.

1-9 COMPUTER TECHNIQUES

This section summarizes a few of the many techniques the author has found advisable to have readily accessible when writing computer programs.

Referring to the program listing of Chap. 5, it can be said that any computer program consists of the following:

a SPECIFICATION statements listed in order and in proper sequence (observing that not all of the following five types of statements may be required in a given program) as follows:

1 DIMENSION TITLE (20), X(40,2), Y(40,40), E(30,30)
2 COMMON D,W
3 EQUIVALENCE (E(1,1), A(1,1))
4 IMPLICIT INTEGER (A,BX) (see Sec. 1-12)
IMPLICIT REAL*4 (N2,LK)
IMPLICIT REAL*8 (N3,LKK)
INTEGER C, CX
5 DOUBLE PRECISION X, Y

b ARITHMETIC statements (the program) interspersed with:
1 READ (1,200)K, Z1, Z2, D (input statements)
2 WRITE (3,1008) Z1, Z2, D (output statements)
3 FORMAT statements (input or output specifications)

The READ statements allow input of data for computations according to the specifications contained in the

```
200 FORMAT (I5,3F10.4)
```

and the WRITE statement causes the variable identified (for example, Z1, Z2, and D) to be written according to the specifications contained in

```
1008 FORMAT ('1', //, T5, 'Z1 =    ' F6.3, 5X, 'Z2 =    '
                    F6.3, 5X, 'D =    ' F8.2, '   LB', //)
```

Output is very important for ease of interpretation. The apostrophe is a very useful device in output. At the left parenthesis if it encloses a 1 ('1'), it starts a new page; enclosing a zero ('0') skips two lines, and leaving a blank (' ') skips one line.

Apostrophes are used elsewhere within the parentheses to enclose anything the printer is to write out. If a word contains an apostrophe, e.g., *don't*, it is written 'DON''T'; i.e., double interior apostrophes are used. Specifically referring to the input/output (I/O) of the example just given earlier under ARITHMETIC statements, the statement

200 FORMAT (I5,3F10.4)

says that K is a fixed-point number inserted in the first five data-card spaces *and right-justified.* A one-digit number uses the fifth space; two-digit, uses the fourth and fifth spaces, etc. The next 30 card spaces are used at 10 spaces each for the three floating-point variables Z1, Z2, D. The decimal point does not have to be used, but if it is not included, the right four spaces are automatically pointed off by the computer to satisfy the F10.4 field specification.

Now assume

Z1 = 200.02

Z2 = 25.0

D = 3255.6

The output specification

1008 FORMAT ('1', //, T5, 'Z1 = ',. . .)

is as follows:

'1', sets new page.
//, advances the paper two spaces from page top for first printed line.
T5, starts the printer to typing Z1 = in fifth column and is of general form *Tw*.
'Z1 = ' F6.3 causes to be written as shown: Z1 = 200.02.
5X, skips five spaces to start next printing on the same line and is of general form *wX*. Note that the computer will add zeros in the following formats.
'Z2 = ' F6.3 causes Z2 = 25.00, as shown on same line as Z1.
5X, skips five spaces.
'D = ' F8.2, ' LB', writes "D = 3255.60 LB" as shown. Note that the skip within the ' LB' is also written.
// advances the paper two spaces for next printed output.

The commas are to set off various commands and block information in sequence.

Other methods, including H (or Hollerith field specifications) and an elaboration of the T specifications, can be used to put output into almost any form. The use of *Tw* and *wX* as shown here, together with apostrophes, will take care of most output

FORMATS. For this reason it seems unnecessary to elaborate on alternative field specifications. Now consider

```
      WRITE (3,XYZ)X
XYZ FORMAT ('1', T5, 'THE X VALUES ARE', //, T5, F10.4)
```

Again the first material inside the parenthesis is '1' followed by a comma. Since the heading is to start in the fifth type space from the left margin, we have T5. The // advances the paper two spaces, and F10.4 is the maximum size of X to be written, one number to a line, starting no closer to the edge than five spaces and terminating in the fifteenth column. An alternate output format where a page start has already been made is as follows. If we want to write the footing

Length = EL

Width = B

Modulus of elasticity = EC

with units identified and the modulus of elasticity on the second line, this is done as

```
    WRITE (3,104) EL, B, EC
104 FORMAT (//,T5, 'FTG LENGTH =   ', F6.2, '  FT', 5X,
    'FTG WIDTH =   ', F6.2, '  FT', /, T5,' MODULUS OF
    ELAS =   ', F8.2, '  KSF', //)
```

interpreted as follows:

//, skips two lines.

T5, starts F of "FTG" in fifth column.

'FTG LENGTH =   ' is printed since it is enclosed in apostrophes.

F6.2 is the maximum size of footing length which can be written with two decimals (999.99); and no sign is included since the footing length is not likely to be negative.

' FT', writes units skipping the space enclosed within the apostrophes.

5X, skips five spaces to start next output information.

'FTG WIDTH =   ', F6.2, which is written with a maximum of field width of six spaces.

F6.2, as before, with units of ' FT'.

,/, advances the printer *one* line.

T5, starts writing of 'MODULUS OF ELAS =   ' in column 5.

F8.2, uses a maximum of eight digits with two decimals since E is usually larger than the footing length.

' KSF', is units of E with a space left before K of KSF so that number and units are separated by a space.

//) advances the printer two lines in anticipation of the next data to be written.

Sometimes, especially if the digit field size is not known in advance, it is desirable to specify the output in E-FORMAT; as an example,

xxxxx FORMAT (..., *aEw.d*,...)

where a = integer representing desired number of repeats of output

E = specification of output in single precision with exponent

w = integer specifying number of columns to be used

$.d$ = number of columns reserved for the decimal (or fraction)

Value	FORMAT	Printed
−0.004	E10.3	−0.400E − 02
250.2	E10.2	0.25E + 03

1-10 DATA CARDS

COMMENT Cards

Put C in column 1 and any desired alpha-numeric information in the remaining 79 spaces. Continuation cards may be used, but put a C in column 1 of each.

SPECIFICATION Cards

They are started in column 7 and can be used to include column 72.

Arithmetic, Input, Output, Format

Column	Information
1–5	Statement numbers
6	Blank unless a continuation; if continuation, put any alphabetic or numeric symbol, but this column must be filled
7–72	Statement
73–80	Data-processing identification (if desired)

Data Cards (Input or Output)

Spaces 1 to 80 may be used according to format specifications; for example,

3I5 uses first 15 spaces.

2I5, 6F10.4 uses 70 spaces with eight data entries; the first two are fixed-point.

8F10.4 uses all 80 spaces with eight data entries, each 10 spaces wide.

1-11 ALPHA-NUMERIC DATA

Occasionally it is desired to read (or write) alpha-numeric data into a program. This is done in the computer program displayed in Chap. 7.

This can be accomplished using an A-format specification, which in the general form is

A*w*

where *w* is a positive integer (either 4 or 8) designating the number of characters to be processed. If it is known how many characters are required (say it is desired to read and write the following, identified as K,

K = 4X + 6Y

which is 11 characters including blanks, =, +, etc.), then

```
      READ (1,1000)TITLE
1000  FORMAT (3A4)
      WRITE (3,1001)TITLE
1001  FORMAT (3A4)
```

Several of the programs included in this text use FORMAT (20A4), which reserves one data card of alpha-numeric data since $4 \times 20 = 80$ character spaces are reserved.

A DIMENSION statement must be included with this mode of output communication, otherwise only one character in the data will be saved for output. This statement might read

```
DIMENSION TITLE(20)
```

If 20 does not reserve enough space, of course, not all the alpha-numeric data will be stored.

1-12 FIXED POINT TO FLOATING POINT

When it is desirable to use certain of the alphabet segment normally reserved to identify fixed-point variables (I through N) as floating-point variables (say M2, N), one may use

```
REAL*4 M2, N     4 converts to single precision
REAL*8 M2, N     8 converts to double precision
```

or

```
DOUBLE PRECISION M2, N
```

From this it follows that DOUBLE PRECISION and REAL*8 do the same thing and are interchangeable.

To convert any variable beginning with, say, M, J, and L to floating point, one may use

IMPLICIT REAL*4 (M,J,L)

As an example, if variables MCOL, J, LCM are used in the program, they will be used as floating-point variables in single precision. If REAL*8 had been used, double precision would have resulted.

Floating Point to Fixed Point

To convert parts of the alphabet to fixed point other than (I through N), one may use

IMPLICIT INTEGER (A,G)

to convert any variable beginning with A or G (as A, AC, ABLE, G1, GMX, G) to fixed point. To convert a block of the alphabet, say A through G, to fixed point, use

IMPLICIT INTEGER (A−G)

To convert only certain variables use

INTEGER A, B, RAT

This converts only variables identified with A, B, and RAT to fixed point.

1-13 IF STATEMENTS

There are two types, logical and computed.

Logical IF Statements

IF(A.GE.B)X = Y	states that	if A ≥ B, then X = Y
IF(A.LT.B)L = 4	states that	if A < B, then L = 4
IF(A.EQ.C)W = 0.0	states that	if A = C, then W = 0.
IF(A.NE.B)GO TO 7	states that	if A ≠ B, transfer operations to statement labeled 7
IF(A.LE.D)W1 = W2	states that	if A ≤ D, then W1 = W2
IF(A.GT.D)Z = 6.	states that	if A > D, then Z = 6.

Other program statements may also be used, such as

READ (1, XY2)   or   WRITE (3, XXX)

instead of, for example, Z = 6., L = 4, X = Y.

Computed IF Statements

They are of the form

```
   IF(A−B) 67, 68, 69
67 DO ···
68 Z = K + 1
69 AA = B + C/D
```

which states that if A − B is

Sign	Operation
−	GO TO 67
0	GO TO 68
+	GO TO 69

and the *first* statement following this IF statement must have one of the three control-transfer identification numbers (or a statement number) to avoid an error message.

1-14 GO TO STATEMENTS

There are two types, routine and computed.

Routine GO TO Statements

When a routine statement like

```
GO TO 60
```

is encountered, it transfers control to statement numbered 60. Be careful, because if statement number 60 precedes this statement in a program, it can cause the computer to loop indefinitely.

Computed GO TO Statements

A computed GO TO statement like

```
GO TO (61, 62, 63,...),LL
```

sends control to statement

61 when LL = 1
62 when LL = 2
63 when LL = 3
....

Obviously there must be a means of incrementing the variable named LL (a fixed-point variable must be used) and a means of returning to this GO TO statement after LL is incremented.

1-15 SUBROUTINES AND DISK STORAGE

A program may use subroutines, e.g.,

```
SUBROUTINE INVERT
```

and

```
SUBROUTINE (B,IX)
```

which may be used at appropriate locations in the main program with the statements

```
CALL INVERT
```

and CALL (W,5) if the variable to be used is SUBROUTINE (B,IX). In this illustration the variables to use in the subroutine are W (for B) and 5 (for IX). In other words, the called subroutine has an array B and the identification IX. When the subroutine is called, all B variables become the W variables of the main program and IX will have the integer value 5.

A program may also use disk work space, a method varying somewhat at different computer centers. Work space is used as required with the typical statement within a DO loop as

```
WRITE (5) W(I,J)
```

to put information onto the disk. The statement

```
REWIND 5
```

returns the disk locator to the initial point where W(I,J) was begun. The statement

```
READ (5) (W(I,J), J = 1, M)
```

will recall the W(I,J) information stored in the disk. Care must be taken to WRITE (5), REWIND 5, and READ (5) in the correct sequence especially, for example, if after using

```
WRITE (5) (W(I,J), J = 1, M)
```

one immediately began to write a matrix A(I,J) as

```
WRITE (5) (A(I,J), J = 1, N)
```

Data of A(I,J) follow W(I,J). Now if the very next computation involved recalling the *A* matrix, the statement

```
REWIND 5
```

would still return the disk to the starting point of W(I,J), not to the start of A(I,J). If REWIND 5 were used just after the end of W(I,J), then A(I,J) would be written *over* W(I,J). Most computers have more than one work space, for example, 4, 5, 6. If A(I,J) is used prior to W(I,J), it would be better to use

```
WRITE (5) for A(I,J)
WRITE (6) for W(I,J)
```

and a sequential problem would not occur since use of A(I,J) would simply require

```
REWIND 5
READ (5) A(I,J)
```

It is absolutely necessary to WRITE to the disk in the same order that will be used on recall from the disk. For example, if the order of writing is

```
   DO 60 I = 1, L
60 WRITE (6) (A(I,J), J = 1, N)
```

one cannot

```
   DO 61 J = 1, N
61 READ (6) (A(I,J), I = 1, L)
```

but one can

```
      DO 61 K = 1, L
   61 READ (6) (A(K,KK), KK = 1, N)
```

If it is necessary to back up one record, the instruction BACKSPACE is used; for example,

```
      DO 62 K = 1, L
   62 READ (6) A(I,K)
      BACKSPACE 6
```

would backspace one record of A(I,K) located on work area 6 on disk or tape. This cannot be used for writing over or redefining records and using the same storage.

The COMMON statement is used to indicate that certain variables are common (the variable identity is used) to more than one program or subroutine. When COMMON statements are used, the order of variables must be maintained. When COMMON statements are used, the program map immediately following the program listing will contain a COMMON BLOCK. An inspection of the COMMON BLOCK of the programs using common variables will indicate whether the common storage area has been properly entered.

For example, COMMON I, J, W in two programs may map as:

	I	J	W
Main program	0	4	8
Subroutine	0	4	8

indicating that the variables are in common. Or if one has

	I	J	W
Main program	0	4	8
Subroutine	0	4	2C

the W variable is not in common (i.e., not in same core location since 2C and 8 are different core storage locations) in the subroutine, and peculiar events would undoubtedly occur when the subroutine started using W since whatever was stored at 2C and not at 8 would be used. One should routinely check the COMMON BLOCK map when using COMMON since an error message may not occur; the first warning is a miserable set of output. It is necessary to dimension COMMON subscripted variables in all common programs the same way.

1-16 CONSERVATION OF COMPUTER CORE

A very useful technique for saving computer storage, especially of matrices, is to scan the program for matrices which are used perhaps once. For example, in a program of interest Y (64,64) is used and inverted, and a matrix W(12,12) is computed. The Y matrix is no longer needed. As a matter of fact, the following sequence of events takes place (see computer program of Sec. 7-6).

Form Y(64,64) maximum size
Invert Y
Form W(I,J) using Y
Form W1(I,J) using W(I,J)
Compute XMX(I,J), XMY(I,J), using W(I,J)
Compute SOILR(I,J), SOILP(I,J), using W1(I,J)

Since SOILR(I,J) and SOILP(I,J) are the same size as XMX(I,J) and XMY(I,J) but do not depend on XMX(I,J) and XMY(I,J), we can compute XMX(I,J) and XMY(I,J), write them as output, then write over these matrices with new matrices, thus using the same core locations. The space reserved for XMX(I,J) and XMY(I,J) will now be used to store SOILR(I,J) and SOILP(I,J). This can be done using the statement

```
EQUIVALENCE (SOILP(1,1), XMX(1,1)), (SOILR(1,1), XMY(1,1))
```

to tell the computer this is our intention and that the starting points (1,1) are the same, although since they are the same size, there should be no problem.

What about the gigantic space Y(64,64) going to waste after W(I,J) is obtained? Let us be even cleverer. Since W, W1, XMX, XMY, SOILP, and SOILR all follow Y, we write

```
EQUIVALENCE (W1(1,1), Y(1,1)), (XMX(1,1), Y(3,17)),
            (XMY(1,1), Y(7,1))
```

thus effecting a saving of considerable space. The Y storage will be used as shown in Fig. 1-1. Here it is seen that W1, XMX, and XMY all use 144 spaces. To write 144 spaces into Y will use two and a fraction rows ($144/64 = 2.25$). Since there is plenty of room, let us start W1 in Y(1,1). Three rows later we start XMX(1,1) in Y(3,17). To save counting, we simply put XMY in Y(7,1), leaving a gap between the end of XMX(12,12) and XMY(1,1) of $64 \times 7 - 2 \times 144 = 160$ spaces still containing Y(I,J) values.

It should be noted that COMMON is an alternative statement which can be used to accomplish this superposition of storage. The COMMON statement also

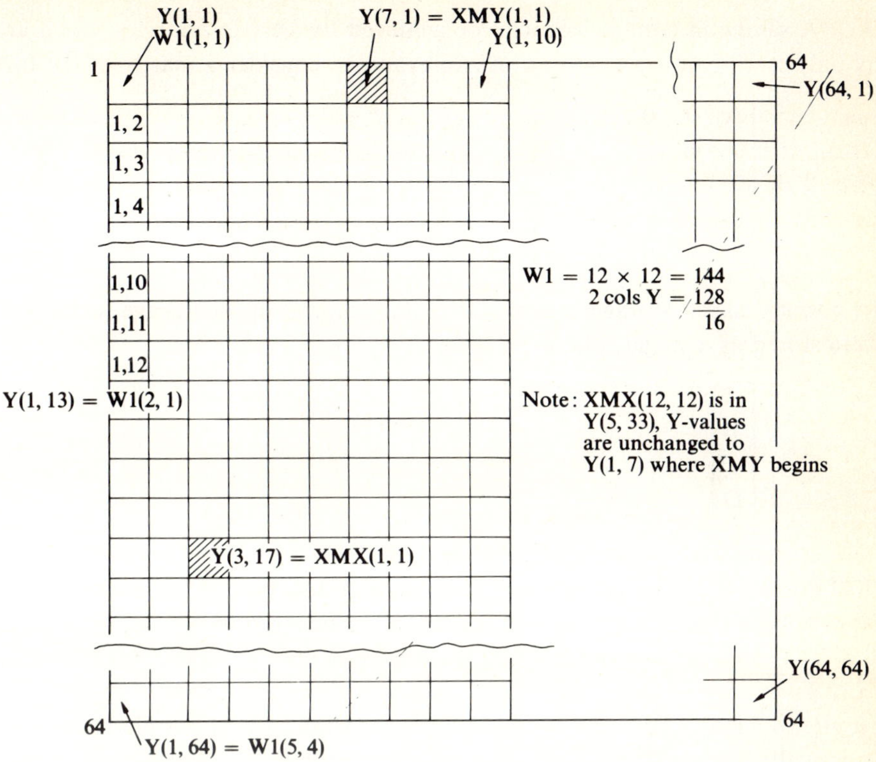

FIGURE 1-1
Computer core allocation for Y(64,64) and use of EQUIVALENCE to store three matrices W1, XMX, and XMY of size 12 × 12 in common core.

uses the same storage area between two subroutines. For example if W is a subscripted variable and

```
COMMON I, J, W
```

is used between two or more subroutines and/or the main program, all use the same common W(I,J) storage area, which results in a considerable saving in storage. Thus, in addition to avoiding the need to reidentify I, J and saving a space W(I,J) in each subroutine, the COMMON statement saves one or more W(I,J) storage areas and causes the computer to identify I and J as required. Note, too, that within a subroutine, one can further use

```
EQUIVALENCE (A(I, J), W(I, J))
```

to save storage. The use of the EQUIVALENCE statement is illustrated in Chaps. 5 and 7.

A saving in core storage can be achieved for certain matrices which are sparse by only storing nonzero values. As an example, consider a matrix of the form

$$S = \begin{bmatrix} A & 0 & 0 & 0 & 0 \\ 0 & B & & & \\ 0 & & C & & \\ 0 & & & D & \\ 0 & 0 & 0 & 0 & E \end{bmatrix}$$

In normal matrix storage this would require S(5,5), or 25 storage locations. If this were stored as a single column of S(5,1)

$$S = \begin{bmatrix} A \\ B \\ C \\ D \\ E \end{bmatrix}$$

The storage requirement is $5 \times 1 = 5$ storage locations. To store S as a column later to be multiplied by another matrix of dimension C(5,N) to obtain K as

$$K = SC$$

requires a slightly different method of computation. In the simple routine when S is square all that is required in the way of statements is

```
      DO  36  I = 1, 5
      DO  36  J = 1, 5
36 SAT = S(I,J) *C(J,I)
```

The computer program in Chap. 5 uses an S matrix in two columns.

A saving in core storage can also be achieved where one might multiply three matrices as follows:

$$P = ASA^TX$$

where

$$F = SA^TX$$

and

$$X = ASA^{T-1}P$$

The matrices would be A, S, A^T of sizes $N \times M$, $M \times M$, and $M \times N$ and a P matrix given of $N \times 1$. The X matrix of size $N \times 1$ is to be found. Since A^T is the transpose matrix of A, rather than transposing and storing, let us transpose A as we need it, thus storing only A. Also let us store S in two columns to save core. The SA^T matrix can be built as follows:

```
      DO 40 I = 1, M
      DO 40 J = 1, N
      KA = I
      IF(I/2*2.EQ.I)KA = KA-1
      SAT(I, J) = S(I, 1)*A(J, KA)+S(I, 2)*A(J, KA + 1)
   40 CONTINUE
```

This requires zeroing an A matrix of size $N \times M + 1$ rather than size $N \times M$ since any odd value of M will give $KA + 1$ as $M + 1$. Unpleasant results will be obtained regardless of the value of S(1,2) if this happens and A(J, M + 1) has not been defined.

Now let us save even more space. Suppose we no longer need A(I,J) after we build ASA^T. We will need it to build the ASA^T, but let us look at the routine.

```
      DO 104 I = 1, N
      DO 15 J = 1, N
      EE(J) = 0.
      DO 15 K = 1, M
   15 EE(J) = EE(J)+A(I, K)*SAT (K, J)
      DO 16 L = 1, N
   16 E(I, L) = EE(L)
  104 CONTINUE
```

What we will do is DIMENSION a new variable EE(J) with J = N. As noted, we build EE(J) one row (horizontally), using one row of A(I,K). The row of A(L,N) is not needed after building ASAT(1,N) = EE(N), and so *after* completing that row, the DO 16 loop is entered and E(I,L) is computed and placed over A(I,N). The statement

```
      EQUIVALENCE (E(1,1), A(1,1))
```

stores the new E matrix one row at a time over A. Care must be exercised to double-precision A if E is double precision, or very unwanted results may occur.

1-17 LOCATING ERRORS AND DEBUGGING PROGRAMS

Keypunch errors such as omitting commas, parentheses, etc., are printed on the program listing where they occur. Unclosed DO loops or references to statement numbers which have not been punched are indicated at the beginning of the

PROGRAM MAP. Errors in arithmetic statements, such as dividing by zero, are of the general form

ILF225I PROGRAM INTERRUPT OLD PSW IS xxxxxxxixxxxABCD
ILF226I PROGRAM INTERRUPT OLD PSW IS xxxxxxxixxxxABCD

where i refers to the type of error [see a list of error messages; e.g., IBM (1968)]. The last four terms (ABCD) are useful for finding which statement produced this interrupt.

We use this error message as follows:

1 Look at the PHASE MAP to see the entry point under any of the headings

LOCORE, LOADED, or REL-FR

on the IBM/360 system this number *may be* 3800 or 4000.

2 Obtain the last four digits and/or characters of the PSW and subtract the entry value (3800 or 4000) from these digits using binary techniques. In binary arithmetic, the alpha characters are understood to be

A = 10 B = 11 C = 12 D = 13 E = 14 F = 15

For example, suppose ABCD was 8CD8, as

ILF225I PROGRAM INTERRUPT OLD PSW IS
FF15000D82008CD8

Subtracting 3800, we have

8CD8 =	8	12	13	8
	3	8	0	0
	5	4	D	8

Note that 13−0 = 13, which is a D.

As another example, take ABCD as 96CA, which converted is

9	6	12	10
3	8	0	0
5	14	12	10
5	E	C	A

In this case in column 3 from the right, 6−8 is not possible, and so we borrow but 16 instead of 10, thus 16 + 6 − 8 = 14 = E.

3 Now one enters the MAP with the value of 54D8 or 5ECA and finds the statement line closest to the map location. This pinpoints the location within one or two lines, so that one can concentrate on locating the error here. The i will indicate the type of error, e.g., dividing by zero, intermixing fixed- and floating-point numbers, or unidentified variable usage.

This technique will not find subscript or address errors. Subscript errors can be found by using debugging subroutines such as

DEBUG SUBCHK $(n_1,n_2,\ldots,n)$

and

CALL PDUMP (a_1,a_2,f_n)

DEBUG SUBCHK is used to force the program to continue using the incorrect subscript(s) after writing it out. The variables n_1, n_2, . . . are only those in subscripted arrays one wants to check, but by omitting the n identification all the subscripts are checked. Obviously, if everything is to be checked, the statement

DEBUG SUBCHK

would go in order at the end of the program as follows

```
STOP
DEBUG SUBCHK
END
```

The statement

CALL PDUMP (a_1,a_2,f)

causes variables from a_1 through a_2 to be written according to format f. Ordinarily f would be

4 integer
5 single precision
6 double precision

Looking at the SCALAR MAP helps determine what to have written with the PDUMP statement since this gives the order of storage/usage. In many cases one might not know exactly what to have written out at any time, i.e., how to make a good assignment of a_2 so that the error will be included; therefore one inspects the SCALAR MAP.

REFERENCES

BRADLEY, JOHN H. (1969): "Programmer's Guide to the IBM System/360," McGraw-Hill, New York, 336 pp.

IBM SYSTEMS REFERENCE LIBRARY (1968): "IBM System/360 FORTRAN IV Language," 8th ed., International Business Machines Corporation, New York.

LEE, R. M. (1967): "A Short Course in FORTRAN IV Programming," McGraw-Hill, New York, 233 pp.

SCHEID, FRANCIS (1970): "Introduction to Computer Science," Schaum's Outline Series, McGraw-Hill, New York, 281 pp.

2

SOIL MECHANICS, EXPLORATION, BEARING CAPACITY, AND SETTLEMENT

2-1 INTRODUCTION

It is presumed that the reader is already familiar with the principles of soil mechanics. This chapter provides a reference for the topics in soil mechanics to be considered or pertinent to the chapters that follow. A brief discussion of current soil tests and exploration procedures, including selected references, is provided so that the reader can evaluate the soil parameters needed in the foundation problems that follow. This information should enable one to estimate probable errors and in any case put the computer I/O soil information into proper perspective. An inspection of computer output data can indicate the possibility or impossibility of a solution, whether the correctness of the solution can be determined by inspection or not.

For an enlarged discussion on soil mechanics the reader should consult texts such as Lambe and Whitman (1969), Sowers and Sowers (1970), Terzaghi and Peck (1967), Bowles (1968, 1970), Leonards (1961), Richart et al. (1970), or Tomlinson (1969).

This chapter uses the metric system primarily, and Sec. 2-13 discusses the applicable SI (Système International d'Unités) conversion units.

At this point it will be useful for the reader to remember that

$$
\begin{aligned}
62.5 \times 1 \text{ g/cu cm} &= 62.5 \text{ lb/cu ft} \\
1 \text{ g/cu cm} &= 9.807 \text{ kN/cu m} \\
1 \text{ kg/sq cm} &= 98.07 \text{ kN/sq m} \\
1 \text{ ton (metric)} &= 1{,}000 \text{ kg} = 2{,}204 \text{ lb} \\
1 \text{ lb} &= 453.5924 \text{ g} \\
1 \text{ kg} &= 1{,}000 \text{ g} = 2.204 \text{ lb}
\end{aligned}
$$

2-2 BASIC DEFINITIONS

Certain volumetric-gravimetric relationships in soil mechanics are defined. From these basic relationships any additional needed relationships can be derived if sufficient data are available or assumptions are made.

The tabulation on page 27 gives the common definitions and includes several additional equations frequently used.

In addition to these basic definitions it is necessary to consider the effect of water in the soil mass on the weight relationships. Water which is trapped in the soil voids above the water table and which does not move (because of discontinuities in the soil pores or surface-tension effects), increases the weight of the dry soil mass by the amount of water entrapped. Soil below the water table is buoyed up (like any object submerged in water) by the displaced volume of water. For obtaining the buoyant unit weight γ' of the submerged portion of the soil mass this becomes

$$\gamma' = \gamma_{sat} - \gamma_w$$

where γ_{sat} is the saturated unit weight of soil.

Pore pressure is another term of considerable importance in soil mechanics. It is the static water pressure in any soil pore that one would obtain from a column of water in a piezometer (calibrated tube) above that point for positive pore pressure (the usual case) or below the point considered (negative pore pressure). This concept is important because the fluid-pressure effect on the soil grains surrounding the point is exactly the same as if the water table were coincident with the water level in the piezometer. If the water table is below the level of water in a piezometer tube, the pore pressure is often termed *excess pore pressure*. The effective (intergranular or grain-to-grain contact) pressure $\bar{\sigma}$ is

$$\bar{\sigma} = \sigma - \Delta u$$

where σ = total pressure to point under consideration

Δu = pressure due to piezometer column height of water

Definition	Symbol	Equation	Range
Water content	w	$\frac{W_w}{W_s} = \frac{\text{weight of water}}{\text{weight of dry soil}}$	$0 \leq w < \infty$
Void ratio	e	$\frac{V_v}{V_s} = \frac{\text{volume of voids}}{\text{volume of solids}}$ $e = \begin{cases} \frac{n}{1-n} & \text{in terms of porosity} \\ wG_s & \text{when } S = 100\% \end{cases}$	$0 < e < \infty$
Porosity	n	$\frac{V_v}{V_t} = \frac{\text{volume of voids}}{\text{total sample volume}}$ $n = \frac{e}{1+e}$	$0 < n < 1$
Degree of saturation	S	$\frac{V_w}{V_v} = \frac{\text{volume of water}}{\text{volume of voids}}$	$0 \leq S \leq 1$
Specific gravity	G_s	$\frac{\gamma}{\gamma_w} = \frac{\text{unit weight of material}}{\text{unit weight of water (4°C)}}$	2.40–3.00 2.60–2.70*
Unit weight	γ	$\frac{W}{V} = \frac{\text{weight of material}}{\text{corresponding volume}}$ $\gamma_{dry} = \begin{cases} \frac{G_s\gamma_w}{1+e} & \\ \frac{G_s\gamma_w}{1+wG_s} & \text{when } S = 100\% \end{cases}$	

* Most common range.

2-3 LABORATORY AND FIELD TESTING FOR FOUNDATION EVALUATION

Laboratory and field soil-testing programs enable the foundation engineer to establish the foundation design criteria, and to establish the probable foundation behavior, based on experience with foundation behavior of similar constructions in similar soils. The judgment factor is based on a realization of the heterogenous nature of soils and soil deposits and coupled with current findings on the character and physical properties of the site material from field and laboratory investigation.

Blind faith in laboratory tests—especially if they are limited in number—is the essence of lawsuits. The test data should be analyzed together with sample inspection, boring records, and site inspection. The type of structure cost and structural loads should be considered. If possible, the worst as well as ideal site conditions should be analyzed to bracket the actual situation.

The laboratory tests should be chosen to yield the desired and necessary information as economically as possible. Elaborate and refined tests are justified only if the small increase (generally) in data accuracy will yield worthwhile savings in design or eliminate risk of a costly failure.

Soil tests of interest to the foundation engineer in order of increasing costs (approximately) are:

1 Visual examination
2 Natural moisture content w_N
3 Liquid and plastic limits w_L, w_P
4 Grain-size analysis (mechanical)
5 Unconfined compression q_u
6 Laboratory vane
7 Moisture-density or relative density
8 Permeability
9 Direct shear
10 Triaxial compression (ϕ and c)
11 Consolidation
12 Chemical analysis

Tests 2, 3, and 4 are in any laboratory text, e.g., Bowles (1970), Lambe (1951). The hydrometer test for grain sizes smaller than the no. 200 sieve is seldom used in foundation work. The complete sieve analysis is rarely needed for building construction. On occasion the no. 4 and no. 200 sieves may be used to refine the classification of the soil. The no. 40 sieve is used to obtain soil for the liquid- and plastic-limit test.

The natural-moisture plot with depth tends to indicate the softer cohesive materials, alerts the engineer to abrupt changes, and may indicate preconsolidated soils.

The unconfined-compression test q_u is rather routine on cohesive samples. Many organizations also use pocket penetrometer values as a check on the unconfined-compression value.

The laboratory vane test is simple to perform in the laboratory on sensitive or fine-grained samples. A small vane is inserted into the sample, and the torque to shear a known volume of soil is measured.

Permeability tests are useful in dam studies, but find little use in building construction unless the construction site must be dewatered.

Direct shear and triaxial testing are necessary to evaluate the soil-strength parameters, angle of internal friction ϕ, and cohesion c. The direct-shear test is also termed a plane-strain test. Angles of internal friction from direct-shear (or plane-strain) tests tend to be from 2 to 4° [Lee (1970), with a summary from several sources] higher than in triaxial tests. Little difference is obtained with fine sands. For retaining-wall or slope-stability problems, the plane-strain tests may be more realistic than triaxial tests since these failures are unidirectional.

Soil parameters from either direct-shear or triaxial tests depend heavily on the degree of saturation and resulting excess pore-water pressures developed during the test. Values range in clays and silts from $\phi = 0°$ for unconsolidated undrained (UU) tests to $\phi = 30°$ or more for consolidated drained (CD) tests. Pore pressure in granular materials with large coefficients of permeability k (order of 10^{-2} to 10^2 cm/sec) dissipates rapidly, and thus there is little effect on the angle of internal friction. The water may, however, provide a small lubrication effect to reduce ϕ perhaps 1 to 2°. Nonsaturated granular material may exhibit small amounts of cohesion due to surface-tension effects. Work of Schmertman and Osterberg (1960) indicates that cohesion tends to be mobilized in tests before the friction component, i.e., the two strength components are not acting at peak values simultaneously. Other aspects of this test will be considered in Sec. 2-6.

The consolidation test is used to predict the time-dependent settlements of structures situated on saturated, fine-grained deposits where the coefficient of permeability is so low that it takes long periods of time for the pore water to move from the points beneath the structure which are subjected to stress (and resulting pore-pressure increases) to a new equilibrium. The length of time t is proportional to the length of the drainage path squared, or

$$t = \frac{TH^2}{C_v} \tag{2-1}$$

where T = time factor

H = length of drainage path (L)

C_v = coefficient of consolidation (L^2/Time)

For a pore-pressure increase of Δu = constant through the depth H (the usual case considered) the time factor T for a given percent consolidation U is:

$U\%$	0	10	20	30	40	50	60	70	80	90	100
T	0.00	0.008	0.031	0.071	0.126	0.197	0.287	0.403	0.567	0.848	∞

The coefficient of consolidation C_v can be obtained by solving Eq. (2-1) for C_v using laboratory time values for t. Also

$$C_v = \frac{k(1 + e)}{a_v \gamma_w} \tag{2-2}$$

where k is the coefficient of permeability and the term a_v is the slope of the void-ratio-versus-pressure curve (natural scale)

$$a_v = \frac{\Delta e}{\Delta p}$$

without regard to the negative sign. The coefficient of volume compressibility

$$m_v = \frac{a_v}{1 + e} \tag{2-3}$$

is useful in consolidation settlement computations (refer to Fig. 2-1) as follows. By proportion

$$\frac{S}{H} = \frac{\Delta e}{1 + e} \tag{2-4}$$

but

$$\Delta e = a_v \, \Delta p$$

and

$$S = \frac{a_v \, \Delta p H}{1 + e}$$

or

$$S = \Delta p m_v H \tag{2-5}$$

Some people have interpreted Eq. (2-5) as being applicable to soils other than saturated, fine-grained materials (such as fine to medium saturated or nonsaturated sands). By inspection of m_v the units of $1/m_v$ are those of the modulus of elasticity; thus, Eq. (2-5) could be written

$$S = \frac{\Delta p H}{1/m_v} = \frac{\Delta p H}{E_s}$$

which is the conventional expression for axial deformation in any text on mechanics of materials.

More conventionally, however, the consolidation settlement is computed using

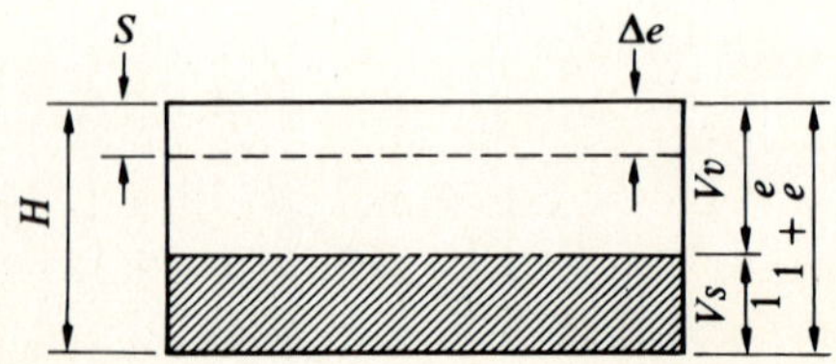

FIGURE 2-1
Settlement due to change in void ratio.

the compression index C_c, obtained as the slope of the void ratio versus log pressure curve as

$$C_c = \frac{\Delta e}{\log(p_2/p_1)} \tag{2-6}$$

Solving for Δe and substituting into Eq. (2-4) gives

$$S = \frac{C_c H}{1 + e} \log \frac{p_2}{p_1} \tag{2-7}$$

where p_2 = new in situ pressure = present p_1 + increase Δp

p_1 = present value of overburden pressure (generally termed p_0)

The compression index C_c as given in Eq. (2-7) has the disadvantage of requiring a consolidation test, which may require several days. An approximate relationship exists between the liquid limit and C_c:

$$C_c \approx a(w_L - b) \tag{2-8}$$

where a, b are constants with typical values as follows:

Soil type	a	b
Certain remolded clays (primarily British)	0.007	10
Certain clays of medium sensitivity	0.009	10
Black cotton soil, India	0.007	10
Brazil clay	0.005	10
East Pakistan	0.0078	14

One must apply rules of thumb when the soil is overconsolidated (or preconsolidated) to reduce the computed value from Eq. (2-7) to a more likely value. Two methods can be used: (1) compare the increase in pressure Δp to the difference between present overburden pressure p_0 and the preconsolidation pressure p_0' as

$$\frac{\Delta p}{p_0' - p_0}$$

and guess (using experience) at the reduction, or (2) use the preconsolidation pressure p_0' value in the denominator of Eq. (2-7) instead of p_0. The second method is somewhat more realistic, but it has not yet been used enough to permit valid conclusions to be drawn. A saturated cohesive soil is preconsolidated if the natural moisture content w_N is closer to the plastic than liquid limit.

Chemical analysis is rarely used in foundation engineering. Occasionally it may be necessary to chemically stabilize a soil, in which case a knowledge of the soil's mineral composition may be desirable.

Table 2-1 lists some representative soil properties of soils from widely distributed geographical locations. This tabulation is from several sources not referenced since values are for illustrative purposes rather than quantitative use.

Table 2-1 REPRESENTATIVE SOIL PROPERTIES FROM DIFFERENT GEOGRAPHICAL LOCATIONS*

Type of soil and/or description and location	G_s	γ, g/cu cm	w_L, %	w_P, %	Angle of internal friction ϕ, deg†	Cohesion c, kg/sq cm†
Quartz sand	2.65–2.67					
Dense		1.68–1.92			34–45	
Loose		1.44–1.76			30–35	
Fine sand	2.65–2.67					
Dense		1.44–1.76			32–38	
Loose		1.36–1.52			28–32	
Gravelly or coarse sand	2.65–2.67					
Dense		1.76–2.08			34–50	
Loose		1.52–1.76			30–35	
Fine sand (E. Pakistan)	2.65	1.35–1.54				
Micaceous silt (Georgia)	2.69	1.27	30–45	20–4	25	
Loess, silty (Kans., Nebr.)		1.28–1.3	25–35	5–10		$0.07w$
Clayey (Kans., Nebr.)		1.28–1.50	35–40	15–20	25–30	0.3–0.7
Iowa (Wisconsin age)	2.68–2.72		34–58	23–27		
Russian	2.63–2.68	1.32–1.83	28–33	10–20	32–33	0.02–0.17
Redeposited (Argentina)	2.63	1.30–1.46	34–40	26–36	20	0.1–0.50
Rhineland silt (Aachen, Germany)			29	21	32	0.12
Clayey silt (S. Melbourne, Australia)			30	14	5	2.0
Chalk (London)	2.70–2.72	1.52–1.62			10–15UU 20CD	8–20
Glacial till						
Otterbrook till (New England)	2.77	1.85–2.00	15–28	6–16	32CD	
Silty clay, stiff (Central Ill.)	2.55	1.68–1.92	NP–40	NP–20		
(Wis.)	2.62	1.68–1.92	25–50	15–25		

Clay, fat (Israel)	2.77	1.38	64	32		
Dark gray (Bankok)	2.6–2.85	1.44–1.84	70–90	27–50	19	
Brazil	2.69–2.76	1.47–1.8	26–31	12–14		
Chicago	2.80		28	18	26	0.1R
Overconsolidated (Norway)		1.90	39	21	27–34	0.2
Leda (Canada)	2.78–2.83	1.04–1.08	40–80	18–30	17	0–1.5
London (overconsolidated)	2.74–2.84	1.6	70–90	24–30	18–19	0.9–3.5
						1.8UU
Weald	2.76		43	18	22–23	0.1UU
(Aswan, Egypt)	2.65	1.5–1.8	71	33		0.12–0.5
Expansive (Tex.)		1.29	88	71	11.3	1.0
Organic (Tex.)		1.38	61	47	9	0.3
Tertiary (Denmark)	2.77		127	36	10	
Silty (Vienna)	2.76		47	22	$17\frac{1}{2}$	
Residual clayey granite (Puerto Rico)	2.73	1.70	NP		43	0.17
Residual soil (Hong Kong)		1.47–1.61	40–60	30–50	37	0.90
Kaolin (Singapore)		1.44	56	23	$25\frac{1}{2}$	0.17–0.2
Granite (Singapore)		1.50	35–50	10–15	35	0.10
Holloysite clay (Gaum)		1.05	< 50		19	0.7–2.0
Residual clay (Calif.)		1.30	55	42	25	1–1.2
Black cotton soil (India)			46–97	22–49		$0.14q_u$
Shale						
Nebraska		1.78	> 50	47–161	33	0.9
Bearpaw, weathered (Canada)		1.36–1.53			6–20	0.2–0.4
S. Dak.		1.53–1.77			8–25	0.1–2
N. Dak.		1.53–1.85			20	7
Trinity, Tex.		1.85–2.14			26	0–0.5

* This list is not complete and is intended to be merely suggestive. It displays similarities in soil properties and consistency indexes w_L and w_P from different locations. Local soil or the same soil within any geographical location listed may vary considerably from this table.

† KEY: UU, undrained test; CD, consolidated drained test; R, remolded soil; q_u, unconfined compression test; w, wet; NP, nonplastic.

Table 2-2 SOIL COMPONENTS AND FRACTIONS, INCLUDING PARTIAL USC SYMBOLS

Soil	Soil component	USC symbol	Grain-size range and description	Significant properties
Coarse-grained components no. 4 sieve	Boulder		Rounded to angular, bulky, hard, rock particle, average diam > 30 cm	Boulders and cobbles are very stable components, used for fills, ballast, and to stabilize slopes (riprap); because of size and weight, their occurrence in natural deposits tends to improve the stability of foundations; angularity of particles increases stability
	Cobble		Rounded to angular, bulky, hard, rock particle, average diam between 15 and 30 cm	
	Gravel	G	Rounded to angular bulky, hard, rock particle, passing 3-in sieve (76.2 mm), retained on no. 4 sieve (4.76 mm)	Gravel and sand have essentially same engineering properties, differing mainly in degree (no. 4 sieve is an arbitrary division and does not correspond to significant change in properties); they are easy to compact, little affected by moisture, not subject to frost action; gravels generally more pervious, stable, and resistant to erosion and piping than sands; well-graded sands and gravels generally less pervious and more stable than poorly graded and uniform gradation; irregularity of particles increases stability slightly; finer, uniform sand approaches characteristics of silt, i.e., decrease in permeability and reduction in stability with increase in moisture
	Coarse		5 to 2 cm	
	Fine		2 cm to no. 4	
	Sand	S	Rounded to angular, bulky, hard, rock particle, passing no. 4 sieve (4.76 mm), retained on no. 200 sieve (0.74 mm)	
	Coarse		No. 4 to 10 sieves	
	Medium		No. 10 to 40 sieves	
	Fine		No. 40 to 200 sieves	

Fine-grained components no. 200 sieve	Silt	M	Particles smaller than no. 200 sieve (0.74 mm); identified by behavior, i.e., slightly or nonplastic regardless of moisture; little or no strength when air-dried	Silt is inherently unstable, particularly when moisture is increased, with a tendency to become quick when saturated; it is relatively impervious, difficult to compact, highly susceptible to frost heave, easily erodible, and subject to piping and boiling; bulky grains reduce compressibility; flaky grains, i.e., mica, diatoms, increase compressibility and produce an elastic silt
	Clay	C	Particles smaller than no. 200 sieve (0.74 mm); identified by behavior, i.e., can be made to exhibit plastic properties within a certain range of moisture; considerable strength when air-dried	Distinguishing character of clay is cohesion or cohesive strength, which increases with decrease in moisture; permeability of clay is very low; difficult to compact when wet and impossible to drain by ordinary means; when compacted, is resistant to erosion and piping, is not susceptible to frost heave, is subject to expansion and shrinkage with changes in moisture; properties are influenced not only by size and shape (flat, platelike particles) but also by mineral composition, i.e., type of clay mineral, and chemical environment or base-exchange capacity; in general, the montmorillonite clay mineral has greatest and illite and kaolinite the least adverse effect on the properties
	Organic matter	O	Organic matter in various sizes and stages of decomposition	Organic matter even in moderate amounts increases the compressibility and reduces the stability of the fine-grained components; it may decay, causing voids, or by chemical alteration change the properties of a soil; hence organic soils are not suitable for engineering uses

2-4 SOIL CLASSIFICATION AND IDENTIFICATION

Soil classification as used for most foundation work consists in establishing the basic type of soil, e.g., rock, gravel, sand, silt, or clay (listed in decreasing value as a foundation material) as shown in Table 2-2.

The basic material shown in Table 2-2 is modified to obtain a soil classification. The following is typical of soils classified in this manner:

Color (wet state)	Description and soil classification	Unified classification
Brown	Sand, very fine, silty; nonplastic; contains a few angular gravel particles	SM
Tan	Sandy clay; sand particles are coarse, rounded; moderately tough at plastic limit	CL
Tan	Sand, coarse to fine, well graded, clean; sand is subrounded, gravel is angular	SW
Gray	Clayey sand; medium to fine sand; contains gravel-size shale fragments; clay portion moderately plastic	SC
Tan	Silt; contains slight amount of very fine sand	ML
Tan	Silty clay; slightly sandy; slight to moderate plasticity when wet	CL
Tan	Well-graded sand (approximately 10% of size 7 to 25 cm)	SW

In this system the classification specialist may use a certain amount of imagination, but it is imperative that he understand the classification method to obtain the correct interpretation of the gravel, sand, silt, or clay particles.

It is usual practice to use the Unified Soil Classification (USC) system directly or somewhat modified to obtain these terms. The USC, proposed by A. Casagrande (1948) as a method of classifying soils for airfield construction, considers the soil as:

Coarse-grained More than 50 percent of soil larger than 0.074 mm (no. 200 sieve)

Fine-grained More than 50 percent of soil smaller than 0.74 mm

The coarse-grained soil is:

Gravel More than 50 percent of the soil is coarser than 4.76 mm (no. 4 sieve)

Sand More than 50 percent of the soil is between 4.76 and 0.074 mm

If the soil is fine-grained, the fine-grained fraction (passing a no. 200 sieve) of the total mass is classified as an organic or inorganic silt or clay. This can be estimated using the plasticity chart of Fig. 2-2. An organic soil can be determined by visual inspection (it is usually dark in color) and odor. If a considerable reduction, say 20 to 30 percent, occurs in the liquid limit over that of a fresh sample upon oven drying, the soil is probably organic.

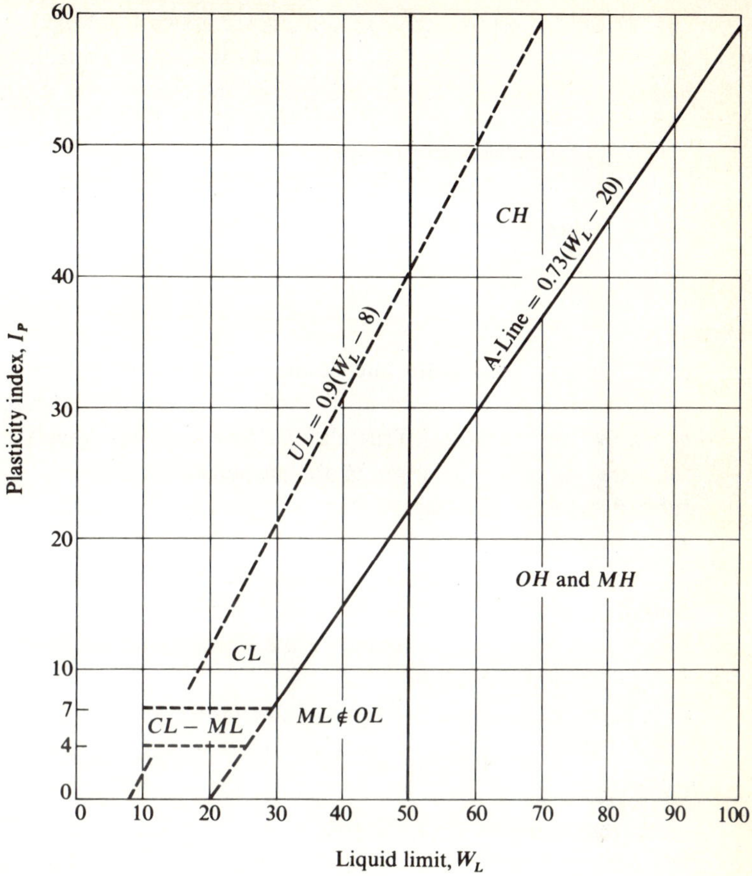

FIGURE 2-2
The Casagrande *A* chart for use in classifying soils. The upper-limit line is reported by the U.S. Corps of Engineers as the upper limit of soils encountered.

Visual Classification

Generally the soil for foundation work is classified visually, supplemented with as few laboratory tests as possible (see Table 2-2). The liquid- w_L and plastic- w_P limit tests are often routine along with natural-moisture tests. With experience one can classify the soil rather accurately by the following:

1 For sand and gravel by an inspection. This may be supplemented by comparing laboratory standards (small jars of classified materials as fine, medium, and coarse sand).

2 For sands and fine-grained soils, place a small sample in water in a test tube or small jar and observe the rate of sedimentation and amounts in the strata making up the sediments at the end of the test. Typical times are as follows:

Approximate time to settle through 12 cm of water	Grain diam, mm	Differentiates
2 sec	>0.4	Coarse and fine sand
30 sec	>0.06	Sand and silt
10 min	>0.03	Coarse and fine silt
1 h	>0.01	Silt and clay

Since clay will flocculate and settle out rapidly as larger lumps, a dispersing agent using 8 to 10 cu cm of 4% sodium hexametaphosphate can be used to neutralize the particles. True clay particle sizes will keep the suspension turbid for a day or more. If some of the suspension is reduced in water content and rubbed between the fingers, clay will feel slippery, whereas silts and very fine sands are gritty.

3 Wet a spot on a lump of soil and rub it. If the finger or a spatula leaves a smooth slippery surface when rubbed across the spot, the material is clay. If the spot streaks, this is related to the number of grains larger than clay in the soil (silt or fine sand sizes).

4 Check the crushing strength of dry lumps (approximate minus no. 40 sieve size material). Clay lumps generally have higher crushing strength than silts as follows:

$$\begin{array}{ll} \text{CH} > \text{MH} & w_L > 50\% \\ \text{CL} > \text{ML} & w_L < 50\% \\ \text{CH} > \text{CL} & \\ \text{MH} > \text{ML} & \end{array}$$

However, an MH silt *may* have a higher crushing strength than a CL clay. In general, the dry crushing strength increases with increasing liquid limit.

5 Check the ease or difficulty of making a plastic-limit thread. Silts require more effort at higher water content than clays. It is nearly impossible to form threads with rock flours and fine sands.

2-5 SOIL EXPLORATION

All construction projects require knowledge of the surface and subsurface site conditions. How extensive this knowledge must be depends in part on the magnitude of the project, but in any case it must be adequate to provide structural stability and general construction and public safety.

Surface conditions can be obtained by a visual on-site inspection. Subsurface conditions can be obtained only by some method of soil exploration. As a minimum the subsurface exploration should determine the stratification of the deposit, the kinds of materials making up the various strata, and the location of the water table. The location of the water table may be impracticable; however, if it is in a zone which will influence construction or affect the design in any way, it must be located accurately. In other cases ascertaining its nonexistence in this zone is adequate.

In order to identify the soils encountered in the various strata, it is necessary to recover samples. The cheapest and most popular means of exploration is boring. A current and popular method is to use hollow auger boring, which is simply a continuous-flight, hollow-stem auger, truck-mounted, which augers a hole at the desired location. This system commonly uses:

Hole (approx OD), in	Auger OD, in	Hollow-stem ID, in
8	7	$2\frac{3}{4}$
$8\frac{3}{4}$	$7\frac{3}{4}$	$3\frac{1}{4}$
$9\frac{1}{4}$	$8\frac{1}{4}$	$3\frac{3}{4}$

Material is continually discharged at the ground surface and is intermixed with that from varying locations. To obtain samples, the drill is halted at intervals of $2\frac{1}{2}$ to 5 ft, the drive disconnected from the auger, and a tube sampler inserted through the hollow stem to recover soil at the drill-tip location.

An alternative method of boring, termed *wash boring*, is also popular. This method utilizes a chopping bit, which is raised, rotated, and dropped onto the soil in the hole, thus chopping it up. Water is circulated through the hollow end of the bit to bring the cuttings to the surface. Drilling mud (a bentonite clay) may be used where the soil being penetrated will cave and may provide enough cohesion for the boring to stay open. This method of drilling can be used in rock, but the rate of hole advance is greatly reduced. Obviously, this method does not produce samples at the ground surface representative of the material at the drill point in either soil or rock.

Rock samples are obtained with some type of core drill, and the sample is a rock core. Soil samples must be obtained with some type of tube sampler.

Standard Penetration Test (SPT)

The common tube sampler is the standard split sampler (also called split spoon), a device 24 in long by 2 in OD by $1\frac{3}{8}$ in ID, which accomplishes two things:

1 A disturbed soil sample is recovered, which can be visually inspected for classification and stored in containers for later laboratory verification and analysis.

2 An SPT datum (a number) is obtained. The test consists in:

a Seating the standard split sampler 6 in into the soil at the bottom of the borehole.

b Driving the sampler 12 in (additional) into the soil and recording the blows for each 6 in of penetration. The sum of the blows to advance the sampler 12 in is N (the penetration number).

The SPT consists of driving the standard split sampler using a 140-lb weight dropping 30 in onto the end of the drill rod, to the far end of which the sampler is attached. A guide is used to align the drive weight. The test has many shortcomings [Fletcher (1965)], but since about 1927 it has been widely used both in the United States and abroad.

Some effort has been made to modify and supplant the test because of its shortcomings [Palmer and Stuart (1957); Schmertman (1967, 1970)], but the SPT is so widely used and so many people have built up considerable and successful experience with it that it is doubtful that it will ever become obsolete.

Undisturbed Samples

When relatively undisturbed cohesive soil samples are needed for strength and settlement tests in the laboratory, thin-walled tubes may be inserted into the borehole (or through the hollow stem of the auger) and pushed into the soil to recover samples. These thin-walled tubes should have an area ratio defined as

$$A_r = \frac{\text{OD}^2 - \text{ID}^2}{\text{ID}^2} \tag{2-9}$$

of 10 to 15 percent compared to 112 percent of the standard split sampler.

Suitable precautions must be taken to make sure the collected sample reaches the laboratory in as undisturbed a condition as possible.

It is nearly impossible to obtain an undisturbed soil sample due (at the very least) to loss of overburden pressure. There is always the loss of static water pressure from samples below the water table.

Cohesionless soils represent an even more formidable problem since the smallest disturbance may destroy their structure. Even if the sample is recovered "undisturbed," there is the problem of transporting and handling to get it to the laboratory and into the testing apparatus intact.

In gravelly soil the blow count may be either too high or too low and requires considerable judgment to arrive at a penetration number for use in Eq. (2-28). The efforts of Palmer and Stuart (1957) in placing a cone on the end of the split sampler are aimed at reducing the judgment factor (see Fig. 2-3).

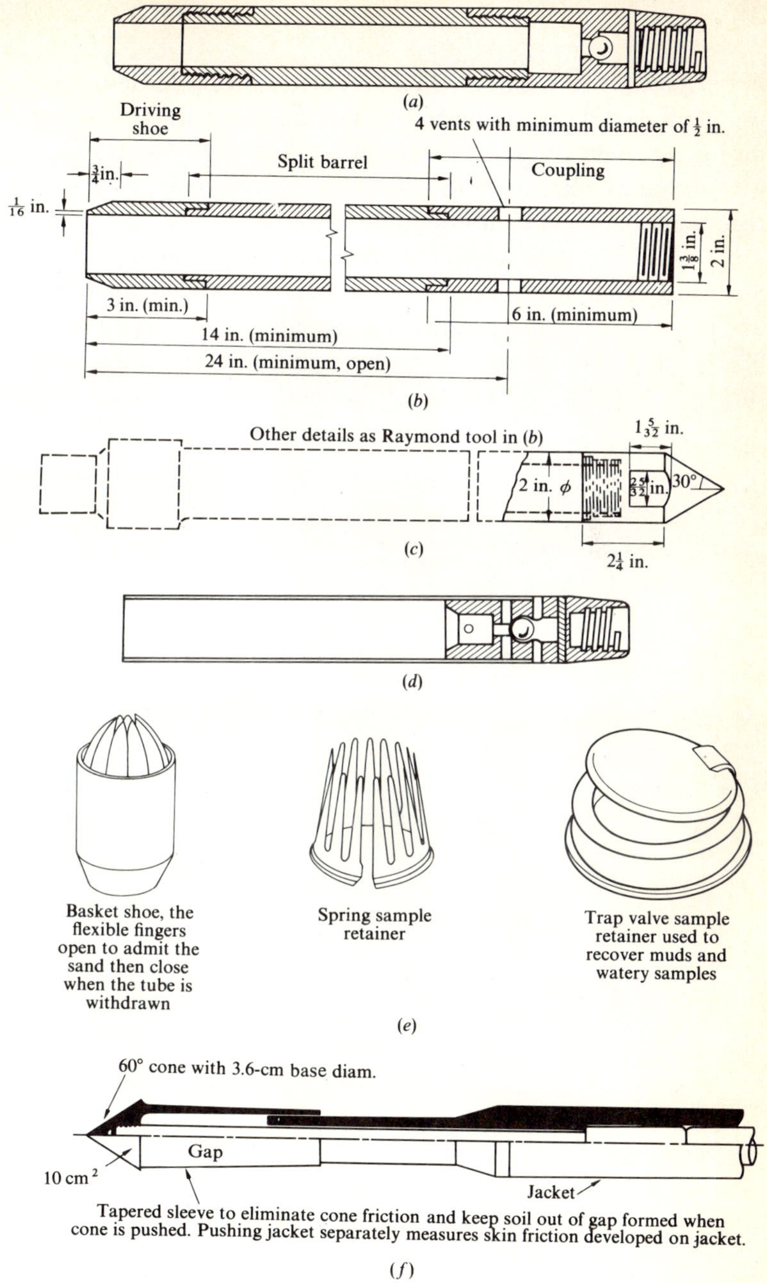

FIGURE 2-3
Penetration and sampling devices: (*a*), (*b*) standard split spoon as widely used; (*c*) modification of standard split spoon with 60° cone for use in gravelly soils [*Palmer and Stuart* (1957)]; (*d*) thin-walled tube sampler for cohesive soils; (*e*) inserts for the standard split-spoon sampler for recovering cohesionless soil and muds; (*f*) Dutch cone modified to measure both point resistance C_R and skin friction [*Vermeiden* (1948)].

Shockley and Garber (1953) reported a series of tests which indicated that the sampling disturbance is related to the relative density. Changes in relative density during sampling operations within a range of ± 15 percent can easily occur.

It seems reasonable to assume that if the in-place density can be duplicated, the in-place grain structure should be approximately duplicated.

Because of the formidable problem of recovering cohesionless samples "undisturbed" and intact, it has become common practice to relate the SPT blow count and the experience factor to the in situ relative density and predicted soil performance. The initial proposal was made by Terzaghi and Peck (1948), as shown by the solid line of Fig. 2-4. Gibbs and Holtz (1957) found that blow counts near the ground surface

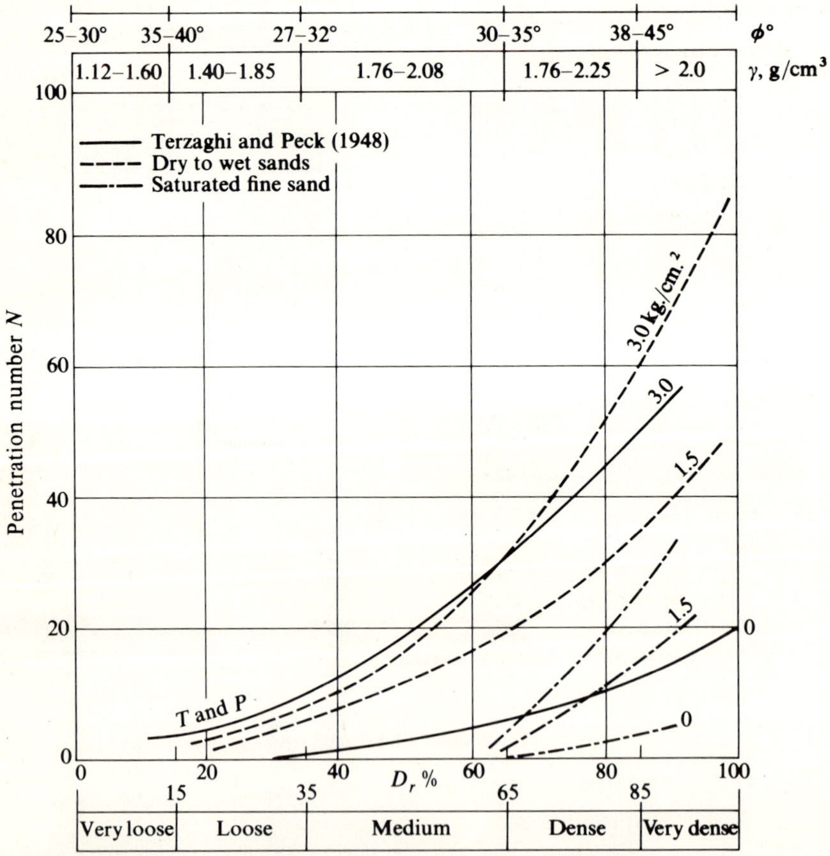

FIGURE 2-4
Chart for correcting SPT blow count for effect of overburden pressure. [*After Gibbs and Holtz* (1957).]

are smaller for the same relative density than deeper in the ground. In order to be able to use the Terzaghi and Peck curve without being overly conservative one may correct the blow count for SPT near the ground surface. In order to make the correction it is necessary to know the in situ relative density and overburden pressure. As an example, we are given

$$
\begin{aligned}
D_r &= 0.60 \\
\gamma &= 1.8 \text{ g/cu cm} \\
\text{Depth} &= 2 \text{ m} \approx 6 \text{ ft} \\
N' &= 12 \text{ blows}
\end{aligned}
$$

Find the corrected blow count, i.e., enter at N', project horizontally to approximately 1.08 kg/sq cm and $D_r = 60$ percent, then project vertically to the Terzaghi and Peck curve. By inspection N should be approximately 26 blows.

This chart also indicates that blow-count corrections may be necessary in saturated fine sands.

Other data shown on Fig. 2-4 are for preliminary estimation and are not intended to replace laboratory testing.

Vane Testing

When the soil is likely to be disturbed in recovery, so that the resulting laboratory information would be misleading, as in saturated, fine-grained soils, in-place *vane* testing should be considered *if gravel is not present* [Gray (1957), Gibbs et al. (1960)]. Soils of this type are considered to be *sensitive* clays (or silts). Sensitivity of a soil is defined as

$$S_t = \frac{\text{undisturbed strength}}{\text{remolded strength}} \tag{2-10}$$

and the range of values is approximately as follows:

S_t	Degree of sensitivity
2–4	Normal
4–8	Sensitive
>8	Extra sensitive

It will be very difficult to recover "undisturbed" samples in soils with sensitivities greater than about 4.

Dutch Cone

The Dutch cone is a cone-shaped device (Fig. 2-3) developed in the Netherlands in the early 1930s. It has since been refined somewhat but is in essence a cone with a pointed tip 3 cm long and a 60° angle (60° with horizontal at base of cone). The cone projection encompasses a base diameter of 3.57 cm and an area of 10 sq cm. It is attached to a 1.90-cm rod. The resistance of pushing the cone at a velocity of 1 to 2 cm/sec for 10 cm is measured. The resistance is related to bearing capacity. Later versions [Vermeiden (1948)] include a calibrated length of pipe sleeve for the push rod of the same OD as the cone base (3.6 cm). By pushing the sleeve the shear resistance of the soil can be measured separately. The Dutch cone is also used with considerable success in silt and fine-sand deposits.

The Dutch cone is widely used in Europe and appears to be making considerable inroads in other parts of the world.

Schmertman (1970) has made some comparisons of the Dutch cone, the SPT, and a relatively new device, termed a *screw plate*, claimed to provide reasonably good in situ data.

2-6 SHEAR STRENGTH

This section considers the essentials of soil shear strength needed in order to attack the problems given herein. There is a massive amount of literature on this topic, including two conferences devoted solely to the subject, one at Boulder, Colorado, in 1960 (Research Conference on Shear Strength of Cohesive Soils, ASCE) and the other at Ottawa, Canada, in 1963 (ASTM STP 361).

Cohesive Soils

The Mohr-Coulomb strength theory is

$$\tau = c + \sigma \tan \phi \qquad (2\text{-}11)$$

where τ = shear strength at failure

c = cohesion of soil

σ = normal stress on failure surface

$\tan \phi$ = friction coefficient

The correct values of c and ϕ are evaluated on the basis of effective stresses on the failure plane, i.e., using the intergranular or effective stress $\bar{\sigma}$ for the σ term in

Eq. (2-11). The statement of what should be done is easy; how to do it is the difficulty.

For fully saturated soils the effective stress can be obtained with a fair degree of accuracy (probably ± 10 percent). In partially saturated soils the effective stress is an educated guess. The principal reasons for discrepancies between measured and actual values are (1) the lack of means for accurately obtaining the pore pressure Δu to compute

$$\bar{\sigma} = \sigma - \Delta u$$

and (2) for partially saturated soils the location of the pore water. Usual methods of measuring pore pressure are to attach waterlines to the ends of the sample in a triaxial cell or insert a large needle piezometer into the sample near midheight and connect the line to a pressure transducer or other pore-pressure measuring device. Unless the pore pressure is the same at the pressure takeoff point as on the failure surface, errors result. In the direct-shear test there is currently no means of measuring the pore pressure. In this test the only method for obtaining the true soil parameters is to load the specimen, wait until consolidation halts, then test so slowly that pore pressures do not build up again as the grains on the failure surface move about.

Fortunately, many problems do not require precise evaluation of the parameters.

TYPES OF SHEAR TESTS

A Unconsolidated-Undrained (UU, or Quick) Test

No drainage during application of confining cell pressure σ_3 or normal load in direct shear. The sample obviously consolidates somewhat, depending on initial degree of saturation and how soon testing starts after load application. No drainage is allowed during the test; hence the descriptive term *undrained.* The unconfined-compression test is considered to be a UU test with $\sigma_3 = 0$. This test yields minimum apparent values of ϕ and maximum cohesion.

B Consolidated-Undrained (CU) Test

Drainage is allowed during application of the σ_3 or normal load. The sample consolidates with respect to the applied pressure as observed via drainage (or vertical settlement in the direct-shear test). No drainage is allowed during the test. Larger values of ϕ are obtained, and cohesion is somewhat less than in the UU test.

C Consolidated-Drainage (CD or CS) Test

Only difference between this test and the CU test is that drainage takes place during the test and the test rate is slow enough to ensure that pore pressures do not build up. True or effective values of ϕ and c are obtained.

Approximately true values of ϕ and c are obtained, however, in the UU and CU tests if pore pressures are measured and the normal stress corrected for pore pressure. Where construction or final loads occur so rapidly that pore pressures cannot dissipate, the UU shear-test parameters should be used. Examples are rapidly constructed embankments on clay substrata, strip loading rapidly placed on a clay deposit, or a rapidly constructed clay dam core. An embankment constructed very slowly or a strip slowly loaded over a long time represents a CD analysis.

COMMON METHODS OF DETERMINING SHEAR STRENGTH

One method of evaluating the approximate unconfined compression strength of a soil is to use the SPT of Sec. 2-5. A relationship between N and q_u is shown in Fig. 2-5.

The field vane test may be used. The vane is embedded 45 to 50 cm into the soil to be tested and is rotated at about 0.1°/sec, so that failure occurs in 3 to 10 min. The vane is 5 to 10 cm in diameter with a height-to-diameter (h/d) ratio of at least 2.

$$\tau = \frac{T}{\pi(d^2h/2 + d^3/6)} \tag{2-12}$$

The reader should consult ASTM STP 193 (1956), ASTM STP 399 (1965), and Aas (1965) for additional vane-shear information.

UNCONFINED-COMPRESSION TEST (COHESIVE SOILS)

This is the most commonly used test of all. Much of the time, however, it is used on SPT samples which are highly disturbed due to the large tube-area ratio. The disturbed values tend to be conservative, but they may be too conservative and should be supplemented by "undisturbed" thin-wall-tube samples if the shear strength is coming in low, say, less than 0.5 kg/sq cm.

DIRECT-SHEAR AND TRIAXIAL TESTS

The direct-shear and triaxial tests are relatively simple (if pore pressure is not measured) and should be used with undisturbed samples if there is any question about the results obtained by the unconfined-compression or SPT method. These tests do not improve on vane-test data, in general, as the vane is ordinarily used for soils where sample recovery is difficult due to high sensitivity of the soil.

Cohesionless Soils

In cohesionless soils

$$\tau = \sigma \tan \phi \tag{2-13}$$

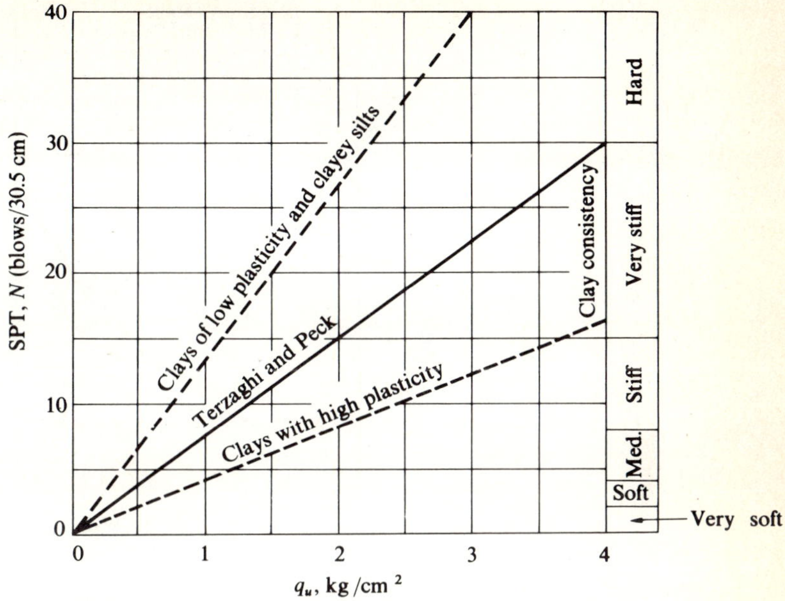

FIGURE 2-5
Approximate relationship between SPT blow count N and the unconfined compressive strength of cohesive soils.

Generally pore pressure is not a problem. The angle of internal friction is very sensitive to the soil density, however (Fig. 2-6), and a small increase in density may change ϕ by 8 to 10°.

It is difficult to obtain undisturbed samples for direct-shear or triaxial tests. One usual practice is to duplicate the in situ density with a laboratory sample and obtain ϕ.

2-7 POISSON'S RATIO

As this material property is of considerable value in the solution of settlement problems, is indirectly of value in lateral-stress problems, may be used to evaluate the modulus of subgrade reaction, and has general use in three-dimensional stresses, it warrants a brief discussion here.

Poisson's ratio is defined as the ratio of lateral (ε_3) to longitudinal strain (ε_1) as

$$\mu = \frac{\varepsilon_1}{\varepsilon_3} \tag{2-14}$$

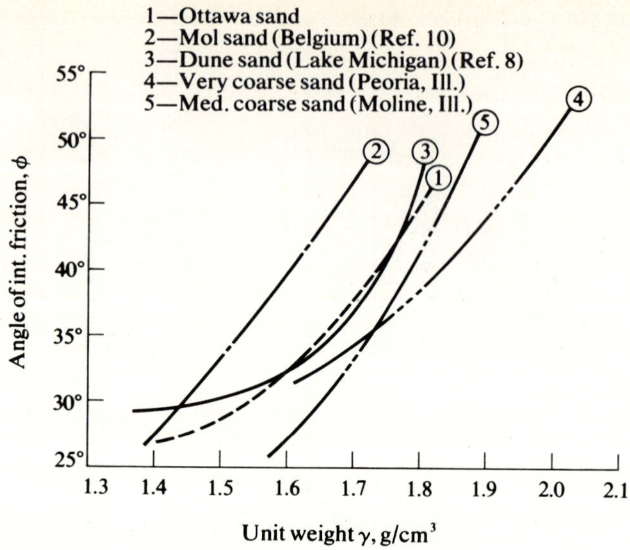

FIGURE 2-6
Relationship between density and angle of internal friction ϕ for several cohesionless soils.

This elastic constant is particularly difficult to evaluate in the laboratory. It appears that it may be determined by measuring the volumetric strain in a triaxial test. If we let ΔV be the change in volume of the triaxial specimen and V the initial volume, then

$$\varepsilon_v = \frac{\Delta V}{V}$$

This value of ε_v equals $\varepsilon_1 + \varepsilon_2 + \varepsilon_3$, the sum of the three principal strains. In the triaxial test $\varepsilon_2 = \varepsilon_3$; therefore,

$$\varepsilon_v = \frac{\Delta V}{V} = \varepsilon_1 + 2\varepsilon_3$$

Since

$$\varepsilon_1 = \frac{1}{E}(\sigma_1 - \mu\sigma_2 - \mu\sigma_3)$$

in general for a triaxial test with $\sigma_2 = \sigma_3$

$$\varepsilon_1 = \frac{1}{E}(\varepsilon_1 - 2\mu\sigma_3)$$

Similarly, in general

$$\varepsilon_3 = \frac{1}{E}(\sigma_3 - \mu\sigma_1 - \mu\sigma_2)$$

and for the triaxial test

$$\varepsilon_3 = \frac{1}{E}[\sigma_3 - \mu(\sigma_1 + \sigma_2)]$$

Solving and substitution gives

$$\varepsilon_v = \varepsilon_1(1 - 2\mu) \tag{2-15}$$

from which

$$\mu = \frac{1}{2}\left(1 - \frac{\varepsilon_v}{\varepsilon_1}\right) \tag{2-16}$$

Lee (1970) has presented data indicating that Poisson's ratio will be affected by the confining pressure, i.e., reduces as pressure increases. Jakobson (1957) has shown that μ also depends on soil density and in general increases as

$$\mu = \mu_0 \sin \phi \tag{2-17}$$

with ϕ the angle of internal friction and μ_0 is a constant to be determined for the soil. It has been common [Terzaghi (1943), for example] to evaluate Poisson's ratio as a condition of no lateral strain and (in the plane-strain case) to obtain

$$\mu = \frac{K}{1 + K} \tag{2-17a}$$

where K is the coefficient of lateral pressure (K_a or K_0 is often used).

From Eq. (2-16) it is evident that if $\varepsilon_v = 0$, that is, no volume change, then Poisson's ratio is 0.5. This is the situation for incompressible materials such as water; therefore, fully saturated soils with low coefficients of permeability would initially experience $\mu = 0.5$. It should be evident also that if the soil structure collapsed so that a *decrease* in volume occurred, Poisson's ratio would be larger than 0.5 since the ratio $\varepsilon_v/\varepsilon_1$ now becomes additive to 1.00.

It is possible to obtain Poisson's ratio in the field by two other methods: (1) by seismic techniques [Maxwell and Fry (1967)] or (2) by the use of a borehole pressure meter [Calhoon (1969), Dixon and Jones (1968), Livneh et al. (1971)]. The pressure meter operates by expanding a cylinder in the borehole. By observing the amount of expansion and the pressure to obtain this deformation, one may use the theory of an infinitely thick cylinder subjected to an internal pressure to obtain the desired elastic constants. Users of this device are able to make reasonably good estimates of at-rest earth pressure (K_0 condition) as additional information.

It can be shown [Seely and Smith (1952)] that the radial displacement of a thick-walled cylinder of radii r_1, r_2 subjected to an internal pressure p is

$$\Delta r = \frac{pr_1}{E_s}\left(\frac{r_2^2 + r_1^2}{r_2^2 - r_1^2} + \mu\right)$$

which for $r_2 \to \infty$ becomes

$$\Delta r = \frac{pr_1}{E_s}(1 + \mu) \tag{2-18}$$

With two values of internal pressure p and two measured values of Δr one can compute Poisson's ratio and the in situ stress-strain modulus E_s.

The seismic method utilizes the velocity of either the shear or Rayleigh waves through the soil mass. These waves are induced by applying energy to the soil mass, generally in a vibratory mode, and the resulting shock-wave velocity is measured. When vibration energy is applied to the soil, three types of energy waves are produced:

P Compression waves
S Shear waves
R Rayleigh waves

The Rayleigh wave travels near the ground surface and is relatively easy to detect, thus obtaining its velocity. The shear wave should be used, but since it is somewhat more difficult to obtain and since there is little difference in the two velocities, the Rayleigh wave velocity is often used.

From seismic principles

$$G = v_s^2 \rho \tag{2-19}$$

where G = shear modulus

v_s = velocity of shear-wave velocity $\approx$ Rayleigh wave velocity v_r

ρ = mass density of soil, γ/g

γ = wet unit weight of soil

g = acceleration of gravity

The shear modulus is also

$$G = \frac{E_s}{2(1 + \mu)}$$

Thus if E_s is known, one can obtain μ as

$$\mu = \frac{E_s}{2v_s^2 \rho} - 1 \tag{2-20}$$

As this is not very precise because of problems in evaluating E_s, one may also measure the compression wave (P wave), which, being the fastest, gets to the pickup unit first. If this value is obtained, the ratio of the two velocities is

$$M = \frac{v_c}{v_s} \approx \frac{v_c}{v_r}$$

and

$$\mu = \frac{M^2 - 2}{2(M^2 - 1)} \qquad (2\text{-}21)$$

We can now obtain the soil modulus of elasticity using Eqs. (2-19) and (2-21) as

$$E_s = 2(1 + \mu)G$$

Table 2-3 lists typical values of Poisson's ratio μ as compiled from several sources.

Table 2-3 POISSON'S RATIO FOR SELECTED MATERIALS

Material	Poisson's ratio μ
Sand:	
Dense	0.3–0.4
Loose	0.2–0.35
Fine (e = 0.4–0.7)	0.25
Coarse (e = 0.4–0.7)	0.15
Rock (basalt, granite, limestone, sandstone, schist, shale)	0.1–0.4 depending on rock type, density, and quality; commonly 0.15–0.25
Clay	
Wet	0.1–0.3
Sandy	0.2–0.35
Silt	0.3–0.35
Saturated clay or silt	0.45–0.50
Glacial till (wet)	0.2–0.4
Loess	0.1–0.3
Ice	0.36
Concrete	0.15–0.25
Steel	0.28–0.31

2-8 STRESS-STRAIN MODULUS (MODULUS OF ELASTICITY)

It would appear on initial examination that determining the stress-strain modulus E_s of a soil merely involves plotting a stress-strain curve as for other materials such as steel, concrete, etc. Unfortunately this is not the case, several major factors complicating the procedure.

1 The stress-strain curve is not linear (Fig. 2-7) over any significant portion of strain.

2 With a nonlinear curve, does one use:

a Initial tangent modulus?

b Any tangent modulus?

c A secant modulus? If a secant modulus is used, where are the curve intercepts to be taken?

3 The stress-strain curve is sensitive to the confining pressure σ_3. In this context, the unconfined-compression test has the confining pressure $\sigma_3 = 0$.

4 The stress-strain curve is sensitive to sample disturbance.

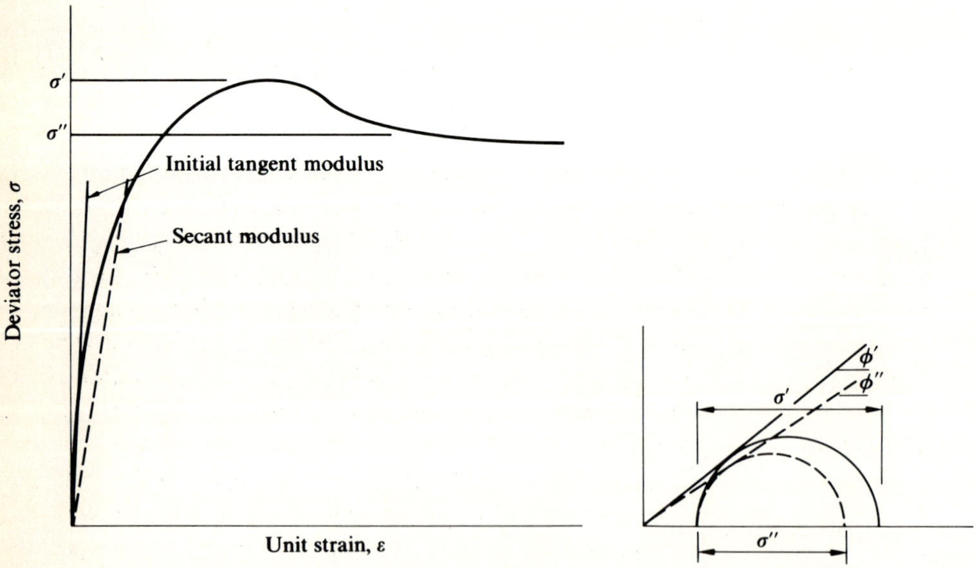

FIGURE 2-7
Effect of using the tangent modulus or secant modulus noting that other locations for both these moduli could have been used. Also shown is the effect on ϕ of using the peak stress to plot Mohr's circle or using the stress value at a larger strain.

These factors will now be considered individually. It is not universal practice to use either the initial tangent modulus or a secant modulus (a secant modulus is often obtained using the origin and the estimated structure contact pressure as intercepts). The tendency seems to be toward the initial tangent modulus, but this is probably because that location is not as sensitive to cell pressure (see Fig. 2-7).

The confining cell pressure affects E_s rather markedly. It appears appropriate to use as a cell pressure the in situ lateral pressure, which is

$$\sigma_h = \gamma z K_0$$

where K_0 is the at-rest lateral earth-pressure coefficient. It also appears, based on considerable research work [Brooker and Ireland (1965), Noorany and Seed (1965)], that the equation proposed by Jaky (1948)

$$K_0 = 1 - \sin\phi \tag{2-22}$$

is reasonably valid both for normally consolidated clay and cohesionless soils (although Brooker and Ireland prefer $0.95 - \sin\phi$ for normally consolidated clay). In this equation ϕ is the *effective* angle of internal friction.

Using data presented by Sherif and Koch (1970), we find the at-rest earth-pressure coefficient K_0 for a soil with an overconsolidation ratio OCR $= p_0'/p_0$ to be approximately

$$K_0 = 0.70 + 0.10(\text{OCR} - 1.2) \tag{2-23}$$

Values of K_0 for normally consolidated clays are approximately 0.60 ± 0.10 and for cohesionless materials 0.50 ± 0.10.

It appears from inspection of many stress-strain curves that using an arbitrary σ_3 value of 0.5 to 1.0 kg/sq cm and the initial tangent modulus is reasonably adequate.

A major problem, however, is that laboratory values of stress-strain modulus are generally too low [Ladd (1964), Klohn (1965), Soderman et al. (1968)] compared to in situ plate tests or full-scale structure performance. This is probably an accumulation of sample disturbance, problems of preparing samples with perpendicular ends when any gravel is present, extracting the samples from the collection tube, nonduplication of in situ stress and pore-water conditions, and interpretation of curve coordinates. Unconfined-compression values tend to be 4 to 10 times too low. Triaxial values range from approximately correct to 5 or 6 times too low.

Because of these shortcomings the borehole pressure-meter device cited earlier may hold considerable promise, although one should realize that it measures horizontal properties and may not provide the desired values unless the soil is isotropic.

De Beer (1967) cites the use of the cone (Dutch cone) penetrometer to obtain the stress-strain modulus of a soil. The relationship is simple

$$E_s = 1.5C_R$$

where C_R is the cone resistance in kilograms per square centimeter.

Typical ranges of *stress-strain moduli* for several soils are given in Table 2-4.

Table 2-4 TYPICAL VALUES OF STRESS-STRAIN MODULUS E_s*

Soil	E_s, kg/sq cm	Comments
Sandy gravel	800–3,000	
Sand:		Depends on Poisson's ratio, test method, and confining pressure in triaxial tests
Loose	100–250	
Dense	500–1,000	
Silty	50–200	
Fine to silty fine	50–180	
Shale	1,400–14,000	If under about 1,500, may be troublesome
Silt	20–200	
Clay:		Depends heavily on triaxial cell pressure, sample disturbance, and water content
Soft	3–30	
Medium	45–90	
Stiff	70–200	
Leda clay	650–1,100	
Norwegian clay	250–500q_u	q_u = unconfined-compression strength
Marine clay	14–70	
Glacial till	100–1,600	
Loess	150–600	Depends heavily on porosity n and water content

* Values are based on static tests and are not recommended for use in dynamic analysis. Note that with the wide range of values shown, these values are of use only for estimation and guidance of probable magnitude.

2-9 MODULUS-OF-SUBGRADE REACTION

This book makes considerable use of the concept of the modulus-of-subgrade reaction (refer to idealized stress-deformation curve Fig. 2-8), designed as

$$k_s = \frac{q}{\delta}$$

where k_s = modulus of subgrade reaction (FL^{-3})

q = intensity of soil pressure (FL^{-2})

δ = deformation at q pressure intensity (L)

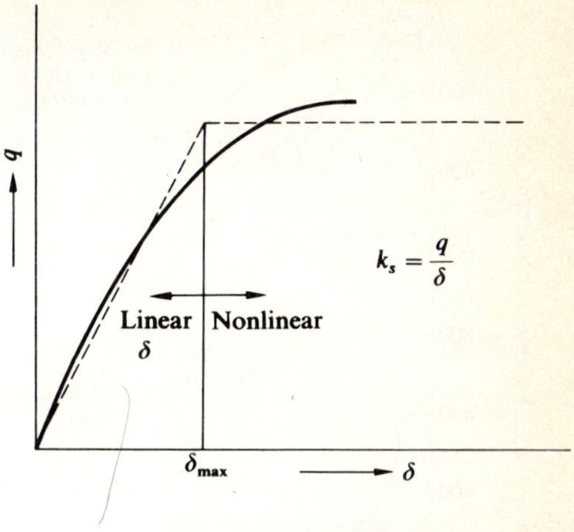

FIGURE 2-8
Qualitative load-deformation data for subgrade modulus k_s. Curve is divided into linear and nonlinear zones. If soil deformation exceeds the maximum deflection δ_{max}, the equation shown on the figure does not apply; i.e., soil pressure = const.

The concept of subgrade modulus has been widely used for rigid pavement design and with considerable success for beam-on-elastic-foundation problems.

There has been reluctance to use the subgrade-modulus concept for many foundation engineering problems, due in part to the apparent difficulty of obtaining a value to use. The most popular method has been to use plate-load tests.

Field load-test procedures have used square plates commonly 30.5 × 30.5 cm and round plates from 30.5 to 76 cm in increments of 15 cm. The plate thickness is commonly 2.5 cm to reduce bending, although the plates are machined so that the smaller plates can be stacked on the larger-diameter ones before loading to reduce bending effects. It should be realized that the subgrade-modulus concept as given here is valid only for elastic (flexible) plates such that the deformation is sufficient to obtain approximately uniform contact pressure over the plate area. Since plate-load tests for subgrade modulus are the same as for other plate-load tests, the reader should consult ASTM Standards part 11 for general load-test procedures.

Field load testing requires some means of providing a large load to the plate. This can be accomplished by jacking against a crawler tractor or loaded flat-bed equipment trailers or driving tension piles to be used with a crossbeam to provide the load reaction.

Plate-load test data are presented as shown in Fig. 2-9. From such plots the modulus-of-subgrade reaction is obtained. Since the curve is seldom straight over any appreciable range of deformation, one must arbitrarily choose coordinates. The Road Research Laboratory (1952) uses the pressure corresponding to 0.13 cm (0.05 in), obtaining

$$k_s = \frac{q}{0.13} \qquad \text{kg/cu cm}$$

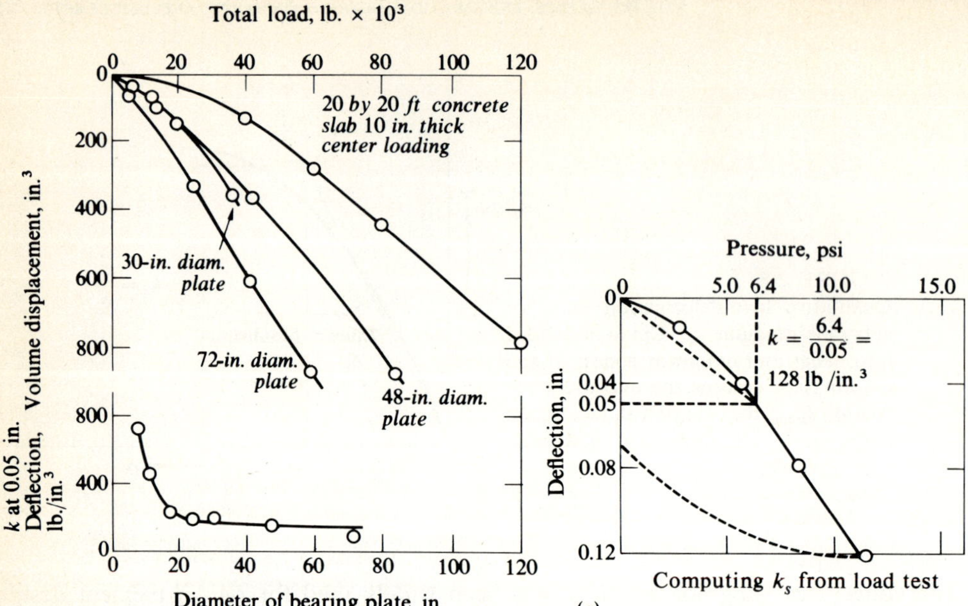

(a)

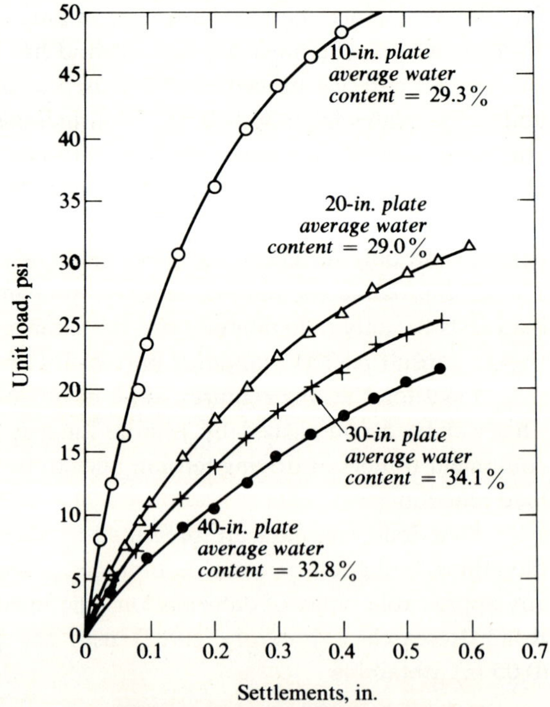

Summary of results of field bearing tests—buckshot clay, natural condition.

(b)

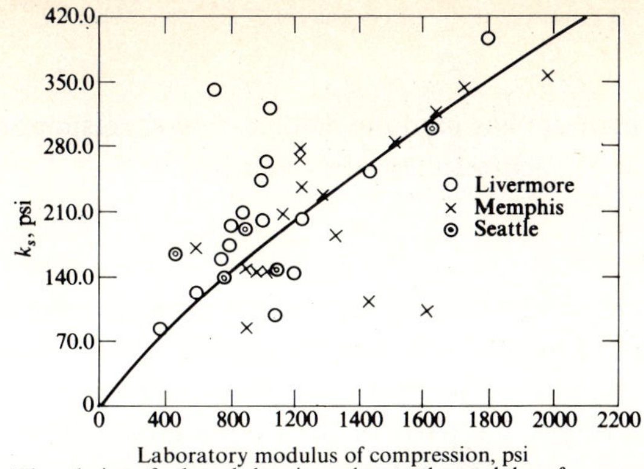

The relation of subgrade bearing values to the modulus of compression of the top 9 in. of subgrade as determined by laboratory triaxial tests.

(*c*)

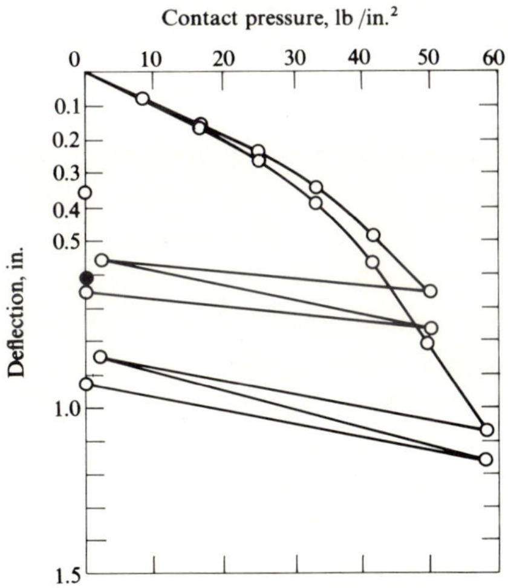

Plate load results (circular plate, 17.5-in. diameter)

(*d*)

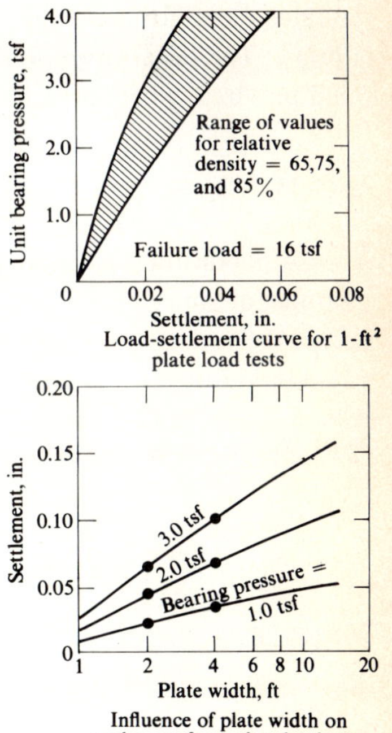

Influence of plate width on settlement from plate load tests

(*e*)

FIGURE 2-9
Typical load-test data accumulated from the literature cited. (*a*) Load-test data from Phillipe (1947), including a typical computation of k_s. Note that volume displacement = area of plate × average deflection. (*b*) Load-test data from Osterberg (1947). (*c*) Comparison of E_s from laboratory triaxial-test data and k_s. The author has converted the original ordinates to k_s since the original paper gave data based on $S = 0.2$ in, 30-in plate, and total load. (*Palmer*, 1947.) (*d*) Load-test data from Vesić and Johnson (1963). (*e*) Load-test data from D'Appolonia et al. (1968). Ton units here are 2,000 lb.

The U.S. Corps of Engineers has used the deformation corresponding to 0.7 kg/sq cm (10 psi) for evaluation of subgrades for airfields as

$$k_s = \frac{0.7}{\delta} \qquad \text{kg/cu cm}$$

The Navdocks Design Manual (1961) uses the pressure and corresponding deflection at one-half the yield pressure. The data can be plotted to a log-log scale, and where tangents to the two straight-line parts of the resulting curve intersect taken as yield pressure. If a zero correction occurs due to imperfect plate seating as determined from a natural arithmetic plot, the deformation data should be corrected prior to making the log plot for yield stress.

California bearing (CBR) tests have also been used to obtain the subgrade modulus [see, for example, Nascimento and Simoes (1957), Black (1961), Barata (1967)]. The penetration load at 0.25-cm (0.10-in) penetration converted to pressure (load per area, where area is computed based on a piston diameter of 4.95 cm) yields

$$k_s = \frac{q_{\text{CBR}}}{0.25}$$

A laboratory method proposed by Vesić (1961, 1963) uses the stress-strain modulus obtained from triaxial tests. To accept the validity of this method one must accept the validity of the concept of the soil stress-strain modulus E_s. Since the stress-strain modulus is widely used in spite of limitations noted in Sec. 2-8, there is no valid reason not to accept the concept of modulus-of-subgrade reaction.

The Vesić equation for subgrade reaction is

$$k_s' = 0.65 \sqrt[12]{\frac{E_s B^4}{E_b I_b}} \frac{E_s}{1 - \mu^2} \tag{2-25}$$

where $k_s' = k_s B \; (FL^{-2})$

E_s = stress-strain modulus (FL^{-2})

μ = Poisson's ratio

B = width of footing

$E_b I_b$ = flexural rigidity of footing

It should be noted that this value of subgrade modulus includes the footing width.

Inspection of Eq. (2-25) indicates that

$$0.65 \sqrt[12]{\frac{E_s B^4}{E_b I}}$$

will be approximately 0.9 to 1.5 with an average value for footings of modest proportions of about 1.2; thus

$$k_s' \approx 1.2 \frac{E_s}{1 - \mu^2} \tag{2-25a}$$

and is accurate to the degree of precision of laboratory determination of E_s.

Figure 2-9*c* indicates some discrepancy between the E_s versus field values of k_s based on a 76.2-cm (30-in) diameter plate and the Vesić method of extrapolation of E_s. For example, for the following data, plate = 1 in thick (assumed)

$$B = 30 \text{ in} \qquad E_p = 30 \times 10^6 \text{ psi}$$

$$E_s = 800 \text{ psi} \qquad \mu = 0.25$$

$$I_w = 0.88 \text{ rigid}^1 \qquad I_p = \frac{\pi d^4}{32} \qquad I_w = 0.82 \quad \text{square}$$

$$k_s = 0.65 \frac{0.88}{0.82} \sqrt[12]{\frac{800(30)^4(32)}{30 \times 10^6 \pi (30)^4}} \frac{800}{1 - 0.25^2}$$

$$= 0.70 \sqrt[12]{\frac{800(32)}{30 \times 10^6 \pi}} (853)$$

$$= 0.70(0.554)(853) = \frac{330}{30} = 110 \text{ pci} < 140 \qquad \text{in Fig. 2-9}c$$

This discrepancy is of the same order of magnitude as shown in Chap. 5 using the Vesić plate-load test data. Much of the discrepancy can be attributed to unreliable E_s values determined by triaxial tests in the laboratory, as discussed in the preceding section. Noting both here and in Chap. 5 that the Vesić values of k_s are lower and also noting that laboratory values of E_s generally are lower, it would appear that the Vesić equation is satisfactory if one can obtain the correct value of E_s.

Terzaghi (1955) and Vesić (1961) proposed correcting k_s for length-to-width effects. The Terzaghi proposal was

$$k_m = k_s \frac{m + 0.5}{1.5m}$$

where k_m is the square plate k_s corrected for $m = L/B$ effect.

From the data given on the only large-scale model tests in the literature [Vesić and Johnson (1963)] it appears that corrections for $L/B > 1$ are not necessary. This is illustrated in Chap. 5, where the Vesić data are used to compare the effect of k_s on beams on an elastic foundation.

[1] See Bowles (1968), p. 87.

Vesić and Johnson (1963) provided both plate-load test data (Fig. 2-9*d*) and triaxial test data. Comparing these data (with $\mu = 0.25$) by several methods and including the beam widths (20 cm) yields:

Method	k_s, kg/sq cm
Vesić (by Vesić, see Table 5-1)	71.9, 90.3, 109.8
Using δ at $q = 0.7$ kg/sq cm (10 psi)	123.2
Using q at $\delta = 0.127$ cm (0.05 in)	123.2
Using q at 1.75 kg/sq cm $\approx$ one-half yield	134.7

The stress-strain modulus used by Vesić was 83.4 kg/sq cm. Other data from these experiments are given in Table 5-1. It can be seen from the above illustrations, however, that k_s is not highly sensitive to curve coordinates.

A reasonable approximation of k_s can be obtained from the allowable soil pressure. This method is presented on the assumption that the allowable soil pressure is based on some maximum amount of deformation S, including a safety factor (SF); thus

$$k_s = \frac{(\text{SF})q_a}{S} \tag{2-26}$$

where S and q_a are in consistent units.

From a readily available reference [D'Appolonia et al. (1968) in Fig. 2-9] the following data are available:

$$q = 3 \text{ tsf} \approx 3 \text{ kg/sq cm}$$

$$\text{SF} = 10 \text{ (bearing)}$$

$$\text{SF} = 50 \text{ (settlement for 2.54 cm based on plate-load tests and the writer's interpretation)}$$

$$k_s = \frac{3}{0.0508} = 59 \text{ kg/cu cm}$$

By Eq. (2-26) the subgrade modulus is computed as

$$k_s = \frac{50(3)}{2.54} = 59 \text{ kg/cu cm}$$

These computations are based on the load test on sand of density approximately $D_r = 0.85$.

For cohesive soil the unconfined-compression test may be used in a similar manner. Neglecting the overburden pressure (see Sec. 2-10), we have

$$q_a = \frac{1.2cN_c}{\text{SF}} \approx q_u$$

Therefore (q_u in kilograms per square centimeter)

$$k_s = 1.2q_u \qquad \text{kg/cu cm}$$

For piles where the soil surrounds the structural element, k_s as determined by the methods cited herein should be doubled; thus we obtain for cohesive material

$$k_s = 2.4q_u \qquad \text{kg/cu cm}$$

Terzaghi (1955) has cited the value of

$$k_s = 2.2q_u \qquad \text{kg/cu cm}$$

as a good approximation, or a difference of about 7 percent. Table 2-5 gives some typical values of k_s.

It is important to obtain the correct value of k_s to compute deflections. This is equally true for E_s in any elastic analysis of deflections. To obtain a soil pressure which

Table 2-5 TYPICAL VALUES FOR SUBGRADE MODULUS k_s FOR SURFACE MEMBERS*

		k_s, kg/sq cm	
Soil type	Unified classification	Dense	Loose
Gravel, gravelly	GW	15–20	5–10
	GP	10–20	5–10
	GC	8–15	
	GM	5–15	
Sand, sandy	SW	6–15	1–3
	SP	5–8	1–3
	SC	6–15	
	SM	3–8	

Cohesive soils

	Consistency q_u, kg/sq cm		
	Soft 0.1–1.0	Medium 1.5–4.0	Hard 4.0 or more
Clays and silts	3–5q_u	3–5q_u	3–5q_u

* For lateral piles k_s is approximately 1.5 to 2.0 times the values shown.

can then be compared to a probable maximum or to a reasonable value of soil pressure which the soil can carry, it is recommended that k_s be used. The true value may be off by as much as a factor of 4 with little effect on computations, as illustrated in subsequent chapters. What is important is to use a method which reflects the soil-structure interaction so that actual soil pressures can be obtained. One can also obtain deflections, but unfortunately with the same limitations as other elastic methods. In the author's opinion, therefore, the subgrade modulus is as valid as the concepts of shear strength or stress-strain modulus and is obtained with about the same degree of precision and difficulty.

2-10 BEARING CAPACITY

The bearing capacity of soils can be determined analytically using Terzaghi's (1943), Hansen's (1970), Meyerhof's (1951), or Balla's (1961) methods, all of which require the soil parameters c and ϕ. The methods of Terzaghi, Meyerhof, and Hansen are quite similar, and only the Hansen equation (1970 version) will be presented here:

$$q_{\text{ult}} = cN_c s_c i_c d_c g_c b_c + qN_q s_q i_q d_q g_q b_q + \tfrac{1}{2}\gamma B N_\gamma s_\gamma i_\gamma d_\gamma g_\gamma b_\gamma \qquad (2\text{-}27)$$

where q_{ult} = ultimate soil pressure

N_c = bearing-capacity factors depending on angle of internal friction ϕ (Table 2-6)

s = shape factors

i = inclination factors when foundation loads have horizontal and vertical components

d = depth factors

$q = \gamma D_f$ = *effective* overburden-pressure effects of soil surrounding footing

B = footing width

γ = unit weight of soil; use submerged unit weight as applicable

Table 2-6 gives the means of computing and representative values of N_c, N_q, and N_γ for selected ϕ values.

Table 2-7 gives values of shape, depth, and inclination factors as proposed by J. B. Hansen (1970).

There is some question of which bearing capacity equations to use. Generally the Hansen equations compute close to Terzaghi's values, with Meyerhof's values higher. A comparison of values by Milović (1965) shows that Hansen's equations appear best for cohesive soils and that Balla's method is best for soils with little or no cohesion. Balla's method will not be presented here, as it is readily available elsewhere [Bowles (1968)]. The Terzaghi, Meyerhof, and Hansen equations have generally been preferred to the Balla method because they are easier to use. The author does not wish to deemphasize the analytical solution of bearing capacity; however, settlement considerations prevail much of the time.

Through wide usage and satisfactory performance but considered much too conservative, the bearing capacity of cohesionless soils is often based on the SPT value N, as in the following equation, originally proposed in chart form [Terzaghi and Peck (1948)] and revised by the author to a 50 percent increase over the original values:

$$q_a = 0.6(N - 3)\left(\frac{B + 0.305}{2B}\right)^2 F_d \tag{2-28}$$

where q_a = allowable bearing pressure, kg/sq cm

B = footing width, m

F_d = depth factor = $1 + D/B < 2$

N = penetration number from a depth approximately $B/2$ below footing base and corrected for overburden pressure using Fig. 2-4

The original equation recommended correction for the water table, but current evidence indicates that this is unnecessary as N includes the effect of water.

Table 2-6 BEARING-CAPACITY FACTORS

Equations for use in Eq. (2-27) together with selected values for check purposes

$N_q = K_p \exp(\pi \tan\phi)$

where $K_p = \tan^2(45 + \phi/2)$

$N_c = (N_q - 1)\cot\phi$

$N_\gamma = 1.80(N_q - 1)\tan\phi$

Typical values

ϕ	N_c	N_q	N_γ
0	5.14	1.00	0
10	8.34	2.47	0.47
30	30.14	18.40	18.08
40	75.32	64.18	95.14

Table 2-7 BEARING-CAPACITY FACTORS FOR EQ. (2-27)

Shape factors

$$s_c = 1 + 0.2 i_c \frac{B}{L}$$

$$s_q = 1 + \frac{B i_q}{L} \sin \phi$$

$$s_\gamma = 1 - \frac{0.4 i_\gamma B}{L} \geq 0.6$$

Depth factors

$$d_c = \begin{cases} 1 + 0.35 \dfrac{D}{B} & D \leq B \\ 1.4 \tan^{-1} \dfrac{D}{B} & D > B \end{cases}$$

$$d_q = \begin{cases} 1 + 2 \tan \phi (1 - \sin \phi)^2 \dfrac{D}{B} & D \leq B \\ 1 + 2 \tan \phi (1 - \sin \phi)^2 \tan^{-1} \dfrac{D}{B} & D > B \end{cases}$$

$d_\gamma = 1.00$ all cases

Inclination factors*

$$i_c = \begin{cases} 0.5 - 0.5 \sqrt{1 - \dfrac{H}{A_f c}} & \phi = 0 \\ i_q - \dfrac{1 - i_q}{N_q - 1} & \phi > 0 \end{cases}$$

$$i_q{}^\dagger = \left(1 - \frac{0.5H}{V + A_f c \cot \phi}\right)^5$$

$$i_\gamma{}^\dagger = \begin{cases} \left(1 - \dfrac{0.7H}{V + A_f c \cot \phi}\right)^5 & \text{horizontal ground} \\ \left(1 - \dfrac{0.7H - \eta^\circ/450^\circ}{V + A_f c \cot \phi}\right)^5 & \text{sloping ground} \end{cases}$$

Ground-slope factors‡

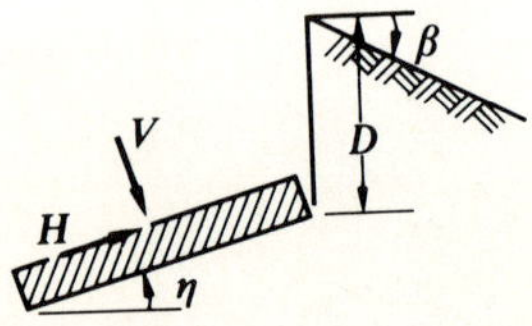

$$g_c = 1 - \frac{\beta^\circ}{147^\circ}$$

$$g_q = (1 - 0.5 \tan \beta)^5 = g_\gamma$$

Table 2-7 **BEARING-CAPACITY FACTORS FOR EQ. (2-27)** (*Continued*)

Base factors‡

$$b_c = 1 - \frac{\eta^\circ}{147^\circ}$$

$$b_q = \exp\,(-2\eta \tan\,\phi)$$

$$b_\gamma = \exp\,(-2.7\eta \tan\,\phi)$$

where A_f = effective footing contact area $B'L'$
L' = effective footing length = $L - 2e_L$
B' = effective footing width = $B - 2e_B$
e_B, e_L = eccentricity of resultant soil pressure with respect to center of footing area
c = cohesion of base soil
H, V = load components parallel and perpendicular to base, respectively
$\tan\,\delta$ = coefficient of friction between footing and underlying soil
ϕ = angle of internal friction of soil, deg
η = slope of footing, deg
β = slope of ground surface, deg

SOURCE: Hansen (1970).

NOTE: L and B are interchangeable (in computing the effective footing width) depending on location of load eccentricity.

* Limitation $H \leq V \tan\,\delta + cA_f$.

† Note i_γ and $i_q > 0.0$.

‡ Limitations for g and b factors: η and β (+) as shown; $\eta + \beta \leq 90^\circ$.

In cohesive soils the use of unconfined-compression test data is widespread to obtain the bearing capacity as follows. Assume a square footing

$$N_c = 5.14 \qquad N_q = 1.00 \qquad N_\gamma = 0.0 \qquad \text{(Table 2-6)}$$

also

$$s_c = 1.2 \qquad d_c = 1 + \frac{0.35D}{B}$$

and take $D/B = 1.0$

$$c = \frac{q_u}{2}$$

Now substituting into Eq. (2-27) and neglecting the $\frac{1}{2}\gamma N_\gamma$ term

$$q_{\text{ult}} = \frac{1.2(5.14)(1.35)q_u}{2} + qD_f$$

we obtain

$$q_{\text{ult}} \approx 3q_u + qD_f$$

It is usual practice with cohesive (clay) soils to take the allowable soil pressure as one-third of ultimate (SF = 3) soil pressure as computed by the various methods, e.g., Eq. (2-27), obtaining the allowable soil bearing pressure as approximately

$$q_a = \frac{q_{\text{ult}}}{3} = q_u$$

and neglecting $qD_f/3$.

2-11 SAFETY FACTORS IN BEARING CAPACITY AND SUBGRADE MODULUS

It is considered normal practice to compute the allowable bearing pressure from the ultimate bearing capacity using a safety factor as follows:

Soil or load condition	SF
Cohesionless soils	2.0
Cohesive soils	3.0
For transient loads such as wind, earthquake, certain live loads	2.0
Dead loads or long-time live loads	2 or 3, depending on soil type
Settlements	1.5–3 designer prerogative

It may be questioned whether applying safety factors to the cohesion and angle of internal friction separately might be more appropriate. Current favored practice is to compute the ultimate soil pressure and divide by the selected safety factor.

The equation given for allowable soil pressure based on the penetration number [Eq. (2-28)] tacitly assumes a SF > 1 and a maximum settlement of 2.5-cm (1 in).

It is inappropriate to apply a safety factor to the soil modulus. The soil modulus is essentially a "spring" concept, and you either have one or not. Dividing by 1.5, 2.0, etc., simply makes the deflection response different. A safety factor is applied in problems using this concept by inspection of the resulting soil pressures and comparing to allowable values or on some other rational engineering basis.

2-12 ELASTIC (OR IMMEDIATE) SETTLEMENTS

Soil being a pseudo-elastic[1] material, it has some movements associated with stress increases. All the deformation is elastic-plastic, but most people compute (or estimate) the settlement using the theory of elasticity of a body on the surface of a homogeneous

[1] In the sense that it deforms somewhat proportionally under load but little of the deformation is recoverable.

isotropic elastic body of finite thickness as

$$S = \mu_1 \mu_2 q B \frac{1 - \mu^2}{E_s} \tag{2-29}$$

where S = settlement (L)

q = contact pressure (FL^{-2})

B = width of surface body (L)

μ = Poisson's ratio (Table 2-2)

E_s = stress-strain modulus of soil (FL^{-2})

μ_1, μ_2 = influence factors based on L/B, stratum thickness, and footing depth (Fig. 2-10)

The major limitations of Eq. (2-29) are determinations of E_s and the fact that soils tend to be anisotropic more often than isotropic. Equation (2-29) tends to yield the correct order of magnitude of settlement in spite of these limitations.

Lambe (1964) proposed that the settlement could be analyzed simply as

$$S = \int_0^H \varepsilon \, dz \tag{2-30}$$

The procedure is as follows. Replace the integral with a summation across several increments of the total stratum thickness (as ΔH) and compute the average strain in each thickness increment as

$$\varepsilon = \frac{\text{average stress increase in increment}}{\text{average } E_s \text{ in increment}}$$

Next compute the incremental deformation as

$$S_i = H_i \varepsilon_i$$

and sum the number of increments taken.

De Beer (1967) indicates that settlements can be calculated which average (mean) about twice as high as observed settlements using the cone-penetration-test data. This is done by obtaining a coefficient C as

$$C = \frac{3}{2} \frac{C_R}{p_0} \tag{2-31}$$

where p_0 is the overburden pressure at cone-point elevation in kilograms per square centimeter, and using a modification of the consolidation equation (2-7):

$$S = \int_0^H \frac{1}{C} 2.3 \log \frac{p_0 + \Delta p}{p_0} \tag{2-32}$$

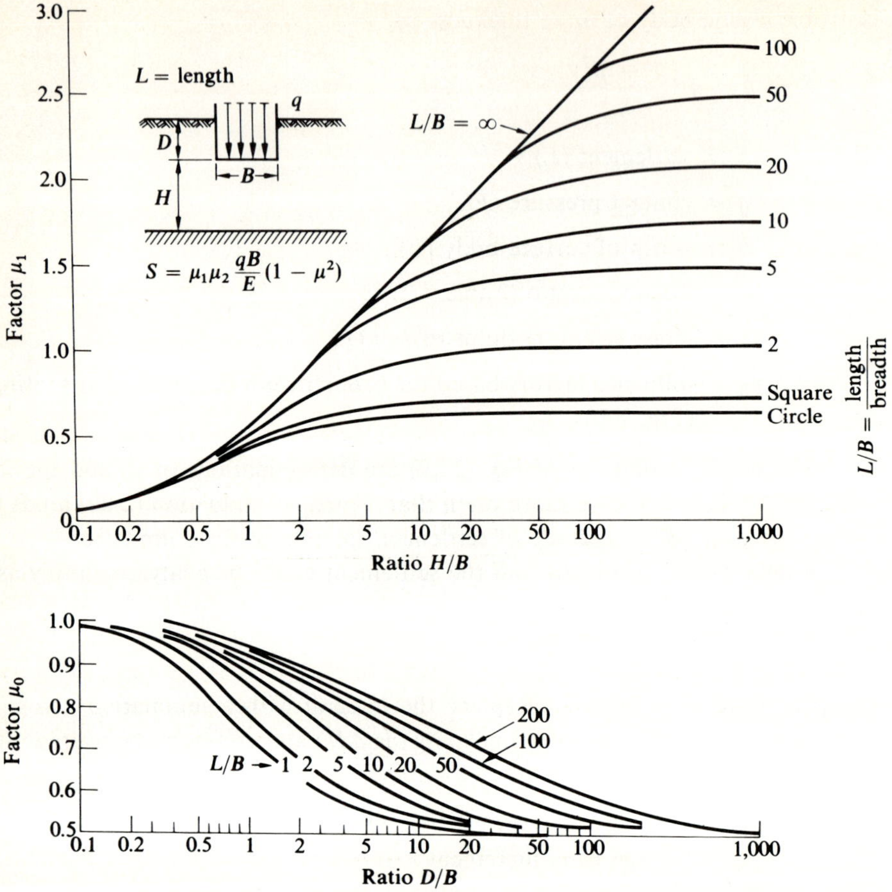

FIGURE 2-10
Influence factors for use in the elastic (immediate) settlement equation. [*After Janbu et al.* (1966).]

Several factors complicate settlement computations:

1 The stress-strain modulus varies with depth for nearly all soils. For any homogeneous isotropic soil layer

$$E_s \approx E_0 + Az^n$$

where z is depth and n is the exponential power.

2 The depth of increased stresses depends on the size of the footing and loads and can be taken as mB, where m is somewhere between 1 and 2.5.

3 It is difficult to evaluate the average increase in stress in a soil stratum due to an applied load on a footing. The Boussinesq and Westergaard equations have been used, as well as a 2 vertical to 1 horizontal spread of pressure [see Bowles (1968), chap. 2]. Influence charts have been made for selected cases to obtain pressure with depth.

Now considering a case of 2 : 1 pressure, a square footing, and the other cited factors (refer to Fig. 2-11), the average pressure within a depth mB is

$$\Delta q = \frac{1}{mB}\int_0^{mB} \frac{\sigma_0 B^2}{(B+z)^2}\,dz$$

Integration yields

$$\Delta q = \frac{\sigma_0}{1+m}$$

From Eq. (2-30) the average strain is the average stress divided by the average E_s. The average stress-strain modulus is

$$E_s = \frac{1}{mB}\int_0^{mB} (E_0 + Az^n)\,dz = \frac{(n+1)E_0 + A(mB)^n}{n+1}$$

and the settlement is

$$S = \Delta q \frac{mB}{E_s}$$

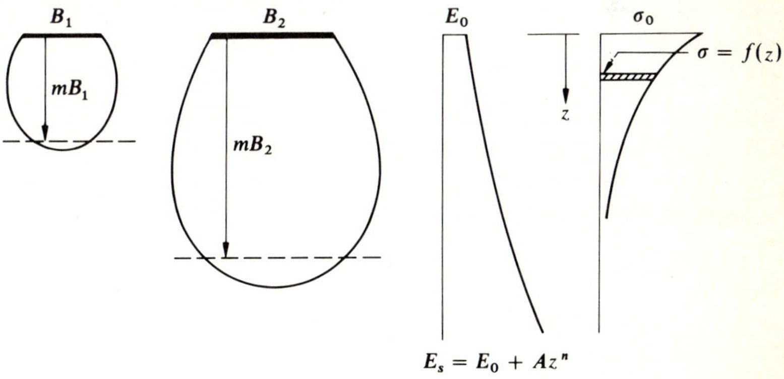

FIGURE 2-11
Relationship of influence of footing size on depth of stress increase. If poor soil exists below depth mB_1, plate-load-test data would be in error. Also shown is probable variation of stress-strain modulus E_s with depth. The exponent n may be greater or less than 1.0. Variation of stress due to footing load can be analyzed using a Boussinesq or 2:1 slope.

Substituting gives

$$S = \frac{n+1}{m+1} \frac{\sigma_0 mB}{(n+1)E_0 + A(mB)^n}$$

The settlement of two footings of widths B_1, B_2 and the same contact pressure σ_0 would be in the ratio of

$$\frac{S_1}{S_2} = \frac{(n+1)E_0 + A(mB_2)^n}{(n+1)E_0 + A(mB_1)^n} \frac{mB_1}{mB_2} \tag{2-33}$$

An approximation used in lieu of an equation of the form of Eq. (2-33) for $B_1 = 1$ ft is

$$\frac{S_1}{S_2} \approx \left(\frac{B+1}{2B}\right)^2$$

Considerable evidence [D'Appolonia et al. (1968); de Beer (1967) with references; Meyerhof (1965)] indicates that this latter approximation is just that and may be unconservative in many cases. Equations of the form of Eq. (2-33), however, are not overly conservative, but E_0, A, m, and n present considerable difficulty.

2-13 THE INTERNATIONAL SYSTEM OF METRIC UNITS

At the time of writing (1972) all the countries in the world except the United States and Canada have already converted to the metric system of engineering units (or are in the process). This Système International d'Unités (SI) will be referred to herein as the *metric system.* Units and conversion factors are included as it is expected that the United States and Canada will be at least started on conversion to the SI units during the useful life of the text. The engineering system now in use in the United States and Canada will be referred to as the fps (foot-pound-second) system.

The metric standard of length is the meter, 1,000 millimeters (mm). The conversion to fps length is

1 in = 25.4 mm   exact value

Since the metric units are multiples of 10, one obtains

Unit	Number of millimeters
Millimeter (mm)	1
Centimeter (cm) = 10 mm	10
Decimeter (dm) = 10 cm	100
Meter = 10 dm = 100 cm	1,000
Kilometer = 1,000 m	10^6

The cgs units of force is the *dyne*. *A dyne is that force acting on a body of 1 gram mass which will accelerate it 1 centimeter per second per second.*

$$1 \text{ dyne} = 1 \text{ g-cm/s}^2$$

since

$$\text{Weight} = mg$$

and $g = 32.17$ ft/sec^2 or 980.7 cm/sec^2, one obtains

$$1 \text{ gram weight} = 1 \text{ g} \times 980.7 \text{ cm/sec}^2 = 980.7 \text{ dyn}$$

A *newton* (N) is the force which acting on a body of mass of 1 kilogram will accelerate it 1 m/sec^2.

$$1 \text{ newton} = 1 \text{ kg-m/sec}^2 = 1{,}000 \text{ g} \times 100 \text{ cm/sec}^2 = 10^5 \text{ dyn}$$

$$1 \text{ kilonewton (kN)} = 1{,}000 \text{ N}$$

$$1 \text{ meganewton (MN)} = 1{,}000 \text{ kN}$$

Critical metric units used in this text are:

fps	Metric
square inch (sq in)	sq cm
square foot (sq ft)	sq cm or sq m
kips/square inch (ksi)	kg/sq cm or kN/sq m
kips/square foot (ksf)	kg/sq cm or kN/sq m
kips/cubic foot (kcf)	kg/cu cm or kN/cu m

Table 2-8 presents useful factors to convert from fps to metric units and shows values of fps, cgs, and SI units. Commonly used units are kg/sq cm and kN/sq m for pressure and g/cu cm and kN/cu m for unit weights. The context or usage generally indicates whether force or mass units are to be used.

Several useful conversions are as follows:

$$29{,}600 \text{ ksi} = 2{,}081{,}200 \text{ kg/sq cm} = 204{,}103{,}300 \text{ kN/sq m}$$

$$3{,}250 \text{ ksi} = 2{,}240.6 \text{ kN/sq cm} = 2{,}240{,}610 \text{ kN/sq m}$$

$$3{,}000 \text{ psi} = 211 \text{ kg/sq cm} = 20{,}700 \text{ kN/sq m}$$

$$62.5 \text{ pcf} = 1 \text{ g/cu cm} = 9.807 \text{ kN/cu m}$$

$$0.144 \text{ kcf} = 22.62 \text{ kN/cu m}$$

$$1 \text{ kip} = 4.4475 \text{ kN}$$

Table 2-8 CONVERSION FACTORS*

To convert from	To	Multiply by
	Length	
inch	centimeter (cm)	2.54
foot	centimeter (cm)	30.48
	meter (m)	0.3048
mile	kilometer (km)	1.609
	Area	
sq inch	sq cm	6.451
sq foot	sq cm	929.03
	sq m	0.092903
	Volume	
cu inch	cu cm	16.38706
cu foot	cu m	0.0283169
gal (U.S.)	cu m	0.003785
	liter	3.785
	Force	
gram force	dyne	980.7
kilogram force (kg_f)	newton (N)	9.807
pound force (lb_f)	kilogram force (kg_f)	0.4535924
kips (1,000 lb_f)	newton	4,447.4735
	kilonewton (kN)	4.44747
kips/ft	kN/m	14.59136
kips/in	kN/cm	1.751
	Pressure or stress (force area)	
kg/sq m	N/sq m	9.807
kg/sq cm	ton/sq m	10.0
	kN/sq cm	0.009807
	kN/sq m	98.07
kips/sq in (ksi)	kN/sq cm	0.689428
	kg/sq cm	70.31
	kN/sq m	6,894.28
kips/sq ft (ksf)	kg/sq cm	0.4882
	ton/sq m	4.882
	kN/sq m	47.87777
lb/sq in (psi)	N/sq m	6,894.28
	kg/sq cm	0.07031
	Bending moment, or torque	
inch-pound	meter-kilogram force (m-kg_f)	0.0152
	newton-meter (N-m)	0.1130
foot-kips	meter-ton (M-T)	0.138255
	kN-m	1.3556
foot-pound	(M-kg_f)	0.138255
	N-m	1.3556

Table 2-8 CONVERSION FACTORS (*Continued*)

	Mass per unit volume	
lb/cu ft	kg/cu m	16.01846
	g/cu cm	0.01601846
	ton/cu m	0.01601846
kips/cu ft	ton/cu m	16.01846
	kN/cu m	157.09304
g/cu cm	lb/cu ft	62.4279
	kN/cu m	9.807
lb/gal (U.S.)	kg/cu m	119.8
	Inertia	
in^4	cm^4	41.62
cm^4	m^4	1×10^{-8}
ft^4	m^4	0.00836097

* Foot-pound-second (U.S. customary) to metric units; 1 kilonewton = 1×10^3 N; 1 meganewton = 1×10^6 N; 1 ton = 1,000 kg.

PROBLEMS

2-1 Some people advocate that

$$C_c = c(e_0 - d)$$

Make a search of the literature and find as many values of c and d as possible for various geographical locations. Be sure to cite the reference source.

2-2 Perform a laboratory test to obtain triaxial-test data using a cell pressure of 10 psi. Find E_s and use the method proposed by Lambe (1964) to compute the expected settlement of a footing 10 × 10 ft loaded with 300 kips on a sand layer 15 ft thick. Make a comparison of settlements using your E_s, a reasonable value of Poisson's ratio, and Fig. 2-10.

2-3 From references cited, make a study of the validity of an equation of the form of Eq. (2-33).

2-4 Make a table of bearing-capacity factors for each degree of angle of internal friction from 0 to 50°.

2-5 It may be possible to make curves for shape, depth, and inclination factors once and for all; if so make the appropriate plots.

2-6 Make a table listing the advantages and disadvantages of the SPT, cone penetrometers, and the vane shear test.

2-7 Using the plate-load-test data herein or other data, refer to Terzaghi and Peck (1948, p. 422) for the settlement equation

$$S = S_1 \left(\frac{2B}{B + 1} \right)^2$$

where B = footing width

S_1 = settlement of 30.5 × 30.5 cm plate

S = settlement of footing of width B

and with reference to Douglas Bond, The Influence of Foundation Size on Settlement, *Geotech.*, vol. 11, no. 2, June, 1961, pp. 121–143, and D'Appolonia et al. (1968), make an attempt to obtain an equation which will better describe $S = f(B)$. Note that you may have to use more terms $S = f(B,E,?,?,?,)$.

REFERENCES

AAS, G. (1965): A Study of the Effect of Vane Shape and Rate of Strain on the Measured Values of In-situ Shear Strength of Clays, *Proc. 6th Int. Conf. Soil Mech. Found. Eng., Montreal*, vol. 1, pp. 141–145.

BALLA, A. (1961): Bearing Capacity of Foundations, *J. Soil Mech. Found. Div. ASCE*, vol. 89, SM5, pp. 13–34.

BARATA, F. E. (1967): Contribution to a Better Application and More Correct Analysis of Bearing Plate Tests, *Proc. 3rd Panam. Conf. Soil Mech. Found. Eng., Caracas*, vol. 1, pp. 591–612.

BLACK, W. P. M. (1961): The Calculation of Laboratory and In-situ Values of California Bearing Ratio from Bearing Capacity Data, *Geotech.* (*Lond.*), vol. 11, no. 1, March, pp. 14–21.

BOWLES, J. E. (1968): "Foundation Analysis and Design," McGraw-Hill, New York.

——— (1970): "Engineering Properties of Soils and Their Measurement," McGraw-Hill, New York.

BROOKER, E. W., and H. O. IRELAND (1965): Earth Pressures at Rest Related to Stress History, *Can. Geotech. J.*, vol. 2, no. 1, February, pp. 1–15.

CALHOON, MAX L. (1969): Pressure Meter Field Testing of Soils, *Civ. Eng., ASCE*, vol. 39, no. 7, July, pp. 71–74.

CASAGRANDE, ARTHUR (1948): Classification of Soils, *Trans. ASCE*, vol. 113, pp. 901–991.

D'APPOLONIA, D. J., E. D'APPOLONIA, and R. F. BRISSETTE (1968): Settlement of Spread Footing on Sand, *J. Soil Mech. Found. Div., ASCE*, vol. 94, SM3, May, pp. 735–759.

DE BEER, E. E. (1967): Bearing Capacity and Settlement of Shallow Foundations on Sand, *Proc. Conf. Bear. Capacity Settlement Found., Duke Univ.*, pp. 15–33.

——— (1970): Experimental Determination of the Shape Factors and the Bearing Capacity Factors of Sand, *Geotech.* (*Lond.*), vol. 20, no. 4, December, pp. 387–411.

DIXON, S. J., and W. V. JONES (1968): Soft Rock Exploration with Pressure Equipment, *Civ. Eng., ASCE*, vol. 38, no. 10, October, pp. 34–36.

FLETCHER, G. F. A. (1965): Standard Penetration Test: Its Uses and Abuses, *J. Soil Mech. Found. Div., ASCE*, vol. 82, SM1, pp. 67–75.

GIBBS, H. J., and W. G. HOLTZ (1957): Research on Determining the Density of Sands by Spoon Penetration Testing, *Proc. 4th Int. Conf. Soil Mech. Found. Eng., London*, vol. 1, pp. 35–39.

———, J. W. HILF, W. G. HOLTZ, and F. C. WALKER (1960): Shear Strength of Cohesive Soils, *Res. Conf. Shear Strength Cohesive Soils, Boulder, Colo.*, pp. 133–162.

GRAY, HAMILTON (1957): Field Vane Shear Tests of Sensitive Cohesive Soils, *Trans., ASCE*, vol. 122, pp. 844–863.

HANSEN, J. B. (1970): A Revised and Extended Formula for Bearing Capacity, *Dan. Geotech. Inst. Bull.* 28, Copenhagen [see also *Bull.* 11 (1961) and *Bull.* 13 (1966)].

JAKOBSON, B. (1957): Some Fundamental Properties of Sand, *Proc. 4th Int. Conf. Soil Mech. Found. Eng., Lond.*, vol. 1, pp. 167–171.

JAKY, J. (1948): Pressure in Silos, *Proc. 2nd Int. Conf. Soil Mech. Found. Eng., Rotterdam*, vol. 1, pp. 103–107.

JANBU, N., L. BJERRUM, and B. KJAERNSLI (1966): *Norw. Geotech. Inst. Pub.* 16, Oslo, pp. 30–32.

KLOHN, E. J. (1965): The Elastic Properties of a Dense Glacial Till Deposit, *Can. Geotech. J.*, vol. 2, no. 2, May, pp. 116–140 (includes discussion).

LADD, C. C. (1964): Stress-Strain Modulus of Clay in Undrained Shear, *J. Soil Mech. Found. Div., ASCE*, vol. 90, SM5, September, pp. 103–132.

LAMBE, T. W. (1951): "Soil Testing for Engineers," Wiley, New York.

——— (1964): Methods of Estimating Settlement, *J. Soil Mech. Found. Div., ASCE*, vol. 90, SM5, September, pp. 43–67.

———, and R. V. WHITMAN (1969): "Soil Mechanics," Wiley, New York.

LEE, K. L. (1970): Comparison of Plane Strain and Triaxial Tests on Sand, *J. Soil Mech. Found. Div., ASCE*, vol. 96, SM3, May, pp. 901–923.

LEONARDS, G. A. (ed.) (1961): "Foundation Engineering," McGraw-Hill, New York.

LIVNEH, M., M. GELLERT, and J. UZAN (1971): Determination of the Elastic Modulus of Soil by the Pressure Meter Test: Theoretical Background, *J. Mater., JMLSA*, vol. 6, no. 2, June, pp. 348–355.

MAXWELL, A. A., and Z. B. FRY (1967): A Procedure for Determining Elastic Moduli of In-situ Soils by Dynamic Techniques, *Proc. Int. Symp. Wave Propag. Dyn. Prop. Earth Mater., Univ. N. Mex.*, pp. 913–919.

MEYERHOF, G. G. (1951): The Ultimate Bearing Capacity of Foundations, *Geotech. (Lond.)*, vol. 2, no. 4, pp. 301–331.

——— (1965): Shallow Foundations, *J. Soil Mech. Found. Div., ASCE*, vol. 91, SM2, March, pp. 21–31.

MILOVIĆ, D. M. (1965): Comparison between the Calculated and Experimental Values of the Ultimate Bearing Capacity, *Proc. 6th Int. Conf. Soil Mech. Found. Eng., Montreal*, vol, 2, pp. 142–144.

NASCIMENTO, V., and A. SIMOES (1957): Relation between CBR and Modulus of Strength, *Proc. 4th Int. Conf. Soil Mech. Found. Eng., Lond.*, vol. 2, pp. 166–168.

NAVDOCKS DESIGN MANUAL DM-7 (1961): Soil Mechanics, Foundations and Earth Structures, Dept. of the Navy, Bureau of Yards and Docks, p. 7-4-6.

NOORANY, TRAJ, and H. B. SEED (1965): In-situ Strength Characteristics of Soft Clays, *J. Soil Mech. Found. Div., ASCE*, vol. 91, SM2, March, pp. 49–80.

OSTERBERG, J. O. (1947): General Discussion, *Symp. Load Tests Bear. Capacity Soils*, ASTM STP 79, pp. 128–139.

PALMER, D. J., and J. G. STUART (1957): Some Observations on the Standard Penetration Test and the Correlation of the Test and a New Penetrometer, *Proc. 4th Int. Conf. Soil Mech. Found. Eng., London*, vol. 1, pp. 231–236.

PALMER, L. A. (1947): Field Loading Tests for the Evaluation of the Wheelload Capacities of Airport Pavements, *Symp. Load Tests Bear. Capacity Soils*, ASTM STP 79, pp. 9–30.

PHILIPPE, R. R. (1947): Field Bearing Tests Applied to Pavement Design, *Symp. Load Tests Bear. Capacity Soils*, ASTM STP 79, pp. 65–70.

RICHART, F. E., JR., J. R. HALL, JR., and R. D. WOODS (1970): "Vibrations of Soils and Foundations," Prentice-Hall, Englewood Cliffs, N.J.

ROAD RESEARCH LABORATORY (1952): "Soil Mechanics for Road Engineers," H.M.'s Stationery Office, London, pp. 376–382.

SCHMERTMAN, JOHN H. (1967): Static Cone Penetrometers for Soil Exploration, *Civ. Eng.*, vol. 37, no. 6, June, pp. 71–73.

——— (1970): Static Cone to Compute Static Settlement over Sand, *Soil Mech. Found. Div., ASCE*, vol. 96, SM3, May, pp. 1011–1043.

———, and J. O. OSTERBERG (1960): An Experimental Study of the Development of Cohesion and Friction with Axial Strain in Saturated Cohesive Soils, *Res. Conf. Shear Strength Cohesive Soils, Boulder, Colo.*, pp. 643–694.

SEELY, F. B., and J. O. SMITH (1952): "Advanced Mechanics of Materials," p. 305, Wiley, New York.

SHERIF, M. A., and D. E. KOCH (1970): Coefficient of Earth Pressure at Rest as Related to Soil Precompression Ratio and Liquid Limit, *High. Res. Rec.* 323, HRB, Washington, pp. 39–48.

SHOCKLEY, W. G., and P. K. GARBER (1953): Correlation of Some Physical Properties of Sand, *Proc. 3rd Int. Conf., Zurich*, pp. 203–206.

SODERMAN, L. G., Y. D. KIM, and V. MILLIGAN (1968): Field and Laboratory Studies of Modulus of Elasticity of Clay Till, *Highw. Res. Rec.* 243, HRB, Washington, pp. 1–11.

SOWERS, G. B., and G. F. SOWERS (1970): "Introductory Soil Mechanics and Foundations," 3d ed., Macmillan, New York.

TERZAGHI, K. (1943): "Theoretical Soil Mechanics," arts. 135 and 161, Wiley, New York.

——— (1955): Evaluation of Coefficient of Subgrade Reaction, *Geotech.* (*Lond.*), vol. 5, no. 4, pp. 297–326.

———, and R. B. PECK (1948, 1967): "Soil Mechanics in Engineering Practice," 1st and 2d eds., Wiley, New York.

TOMLINSON, M. J. (1969): "Foundation Design and Construction," 2d ed., Wiley, New York.

VERMEIDEN, J. (1948): Improved Sounding Apparatus as Developed in Holland since 1936, *Proc. 2d Int. Conf. Soil Mech. Found. Eng., Rotterdam*, vol. 2, pp. 280–287.

VESIĆ, A. S. (1961): Bending of Beams Resting on Isotropic Elastic Solid, *J. Eng. Mech. Div., ASCE*, vol. 87, EM2, April, pp. 35–53.

———, and W. H. JOHNSON (1963): Model Studies of Beams Resting on a Silt Subgrade, *J. Soil Mech. Found. Div., ASCE*, vol. 89, SM1, February, pp. 1–31.

3

ELEMENTS OF STRUCTURAL DESIGN: SPREAD FOOTINGS, COMBINED FOOTINGS

3-1 INTRODUCTION

This chapter presents the elements of reinforced-concrete design of simple spread footings and the conventional design methods of combined footings. Concrete design is based on strength (or ultimate-strength) design of the ACI 318-71 Building Code Requirements for Reinforced Concrete. Discussion will be brief since it is assumed that the reader has had a course in reinforced-concrete design.

Elements of reinforced concrete applicable to footings will be introduced: Sec. 3-3 considers simple spread footings, Sec. 3-7 combined footings, and Sec. 3-9 trapezoid footings. Spread footings with eccentricity will be considered in Chap. 7 along with mat foundations.

Notation

To assist the reader, the following notation, generally complying with the ACI 318-71 Code, will be used throughout. Any other notation will be identified where used.

A = area, BL
A_b = area of any steel reinforcing bar (rebar)
A_g = gross area of concrete, bd
A_s = area of steel/unit of width (also total amount required)
b = width of concrete element
B = least lateral dimension of footing, usually width
d = effective depth of concrete
D = total footing depth (also used to indicate bar diameters)
DL = dead load (working-design value)
E_c = modulus of elasticity of concrete
E_s = modulus of elasticity of steel
f_c' = 28-day compressive concrete strength (F1C in computer programs)
f_y = yield strength of steel reinforcement
L = footing length
L_d = required embedment depth for rebars, tension or compression
LL = live load (working-design value)
M_u = bending moment (strength design)
m = small-end width of trapezoid footing
n = modular ratio E_s/E_c; large-end width of trapezoid footing
p = percent reinforcement steel = A_s/bd
p_b = percent reinforcement steel at balanced design
q_a = allowable soil pressure (also used as q in text for certain equations)
q_{ult} = product of allowable soil pressure and load factor
T = tensile steel force = $A_s f_y$
u = concrete bond stress
v = concrete shear stress (may be subscripted)
w = column width (or diameter if round column)
ϕ = concrete workmanship factors (also angle of internal friction)
ψ = bond stress factor

3-2 REINFORCED-CONCRETE DESIGN FOR FOOTINGS (ACI 318-71)

The latest (1971) revision of the ACI Standard Building Code Requirements for Reinforced Concrete places major emphasis on strength design, the only method considered in this text.

Strength design entails converting design working loads to design ultimate loads through the use of *load factors* as:

$$P_u = 1.4DL + 1.7LL \qquad (a)$$

$$P_u = 1.25(DL + LL + WL) \qquad \text{with wind} \tag{b}$$

$$P_u = 0.9DL + 1.1WL \qquad \text{alternative with wind} \tag{c}$$

where P_u = ultimate strength-design load

WL = wind loading

For earthquake loading substitute EL for WL as applicable.

Strength design also considers workmanship and other uncertainties by use of ϕ factors, which are applied to increase the design requirements as follows:

Design quantity	ϕ factor
Moment	0.90
Diagonal tension, bond, and anchorage	0.85
Compression members, spiral	0.75
Tied	0.70
Unreinforced footings (bending)	0.65

The concrete strain at ultimate stress (strength) is taken as 0.003 in/in. Generally the yield strength f_y of the reinforcing steel is not to exceed 80,000 psi without additional design consideration (ACI, art. 9.4). Reinforcing steel of $f_y = 60{,}000$ psi appears to be the most popular grade at present (1973).

Basic Design Elements

Considering a section of a concrete flexural member (Fig. 3-1) and summing horizontal forces ($\sum F_h = 0$), we have $C = T$; also let the working concrete stress be $0.85f'_c$. But $C = f_c ab$, and from the figure, $T = A_s f_y$. Substituting and solving for a gives

$$a = \frac{A_s f_y}{0.85 f'_c b} \tag{3-1}$$

Equating the internal resisting moment $M = Ty = Cy$ to the external applied ultimate moment M_u

$$M_u = Ty = Cy$$

and substituting $T = A_s f_y$ and the value of $y = d - a/2$, we obtain

$$M_u = A_s f_y \left(d - \frac{a}{2}\right)$$

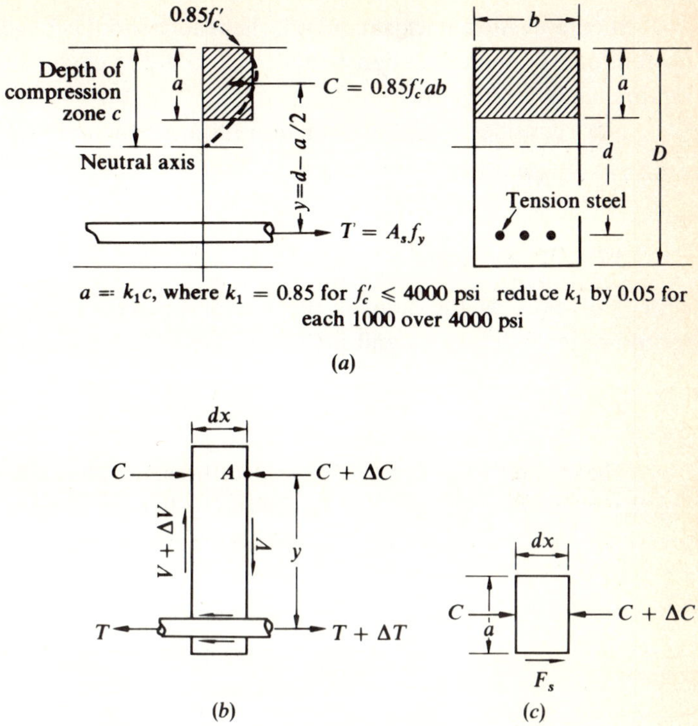

FIGURE 3-1
Derivation of strength-design equations for reinforced concrete: (*a*) flexural stress condition; use rectangular stress block; (*b*) development of bond; (*c*) development of shear.

Inserting the recommended workmanship factor for uncertainties gives the ACI Code design equation of

$$M_u = \phi A_s f_y \left(d - \frac{a}{2} \right) \tag{3-2}$$

The percentage of steel at a cross section of a flexural member is $p = A_s/bd$, and the percent at *balanced* design is p_b. To ensure a tensile failure rather than a sudden compression-zone failure the design percentage of steel p_d is taken as *not more than* $0.75p_b$, where

$$p_b = \frac{0.85k_1 f'_c}{f_y} \frac{87{,}000}{f_y + 87{,}000} \tag{d}$$

Table 3-1 provides maximum percent of steel to use in design ($0.75p_b$) for various combinations of f'_c and f_y.

If compression steel is required, the reader should consult any standard textbook on reinforced-concrete design or the ACI (318-71) Code, art. 10.3, as several design requirements must be met.

Next consider Fig. 3-1*b* and sum moments at point *A* as a convenience

$$(V_u + \Delta V_u)\,dx - \Delta T y = 0 \tag{e}$$

but

$$\Delta T = V_u\,dx \sum o$$

with $\sum o$ taken as the sum of the bar perimeters of the tension steel. Dropping second-order differentials and including the ϕ factor, we obtain for bond stresses

$$u = \frac{V_u}{\phi y \sum o} \tag{3-3}$$

The current ACI Code, however, considers an alternative to Eq. (3-3) by specifying the embedment length L_d of reinforcing bars in tension (art. 12-5) and in compression (art. 12.6) rather than in terms of bond stress.

In tension the basic development length L_d is:

Bar size		
11 or smaller	$\dfrac{0.04A_b f_y}{\sqrt{f'_c}}$	but not less than $0.0004Df_y$
14	$\dfrac{0.085f_y}{\sqrt{f'_c}}$	
18	$\dfrac{0.11f_y}{\sqrt{f'_c}}$	

Table 3-1 MAXIMUM ALLOWABLE PERCENT OF STEEL*

		f_y, psi (kg/sq cm)		
f'_c, psi (kg/sq cm)	k_1	45,000 (3,164)	50,000 (3,515)	60,000 (4,219)
3,000(211)	0.850	0.0278	0.0206	0.0160
3,500(246)	0.850	0.0325	0.0241	0.0187
4,000(281)	0.850	0.0371	0.0275	0.0214
4,500(316)	0.800	0.0393	0.0291	0.0226
5,000(352)	0.800	0.0437	0.0324	0.0252
5,500(387)	0.750	0.0450	0.0334	0.0259
6,000(422)	0.750	0.0491	0.0364	0.0283

* Table includes 25 percent reduction for bending using strength design ACI 318-71, art. 8.1. Note that k_1 is reduced for $f'_c > 4$ ksi.

For top reinforcement (if there is more than 12 in of concrete beneath the bar) multiply L_d from above as follows:

Conditions	Factor
For bars of f_y < 60,000 psi	1.4
For bars of f_y > 60,000 psi	$2 - \frac{60{,}000}{f_y}$
In all cases (top or bottom)	L > 12 in

This operation allows for some consolidation of the fresh concrete away from beneath the top rebars, with the resulting loss of bonding.

For no. 11 and smaller rebars the expression for L_d tabulated above is obtained (see Fig. 3-2) as follows. Equate tension force to bond resistance

$$A_b f_y = \phi u \pi D L_d$$

but the bond stress

$$u = \psi \frac{\sqrt{f'_c}}{D}$$

For top bars $\psi = 6.7$, and for other bars $\psi = 9.5$. Substituting $\psi = 9.5$ and solving for L_d, we obtain

$$L_d \approx \frac{0.04 A_b f_y}{\sqrt{f'_c}}$$

Standard hooks can be used to reduce the required value of L_d but are not required in most footing problems. Consult ACI, art. 12.8 if L_d must be reduced.

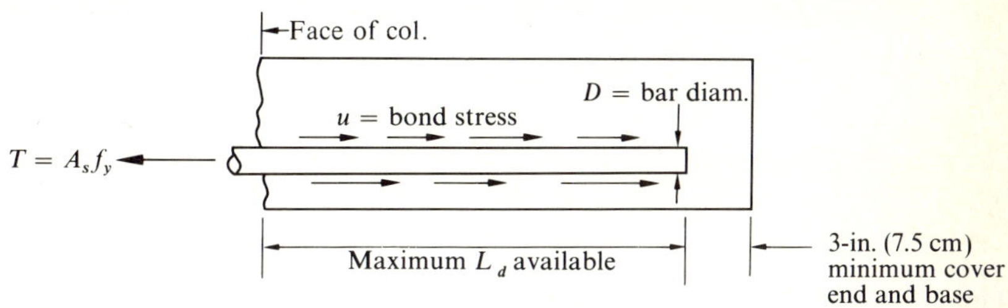

FIGURE 3-2
Derivation of embedment depth L_d for steel reinforcement, top and bottom bars. Bottom bar shown.

For compression bars the development length is

$$L_d = \frac{0.02 f_y D}{\sqrt{f'_c}}$$

but not less than $0.0003 f_y D$ or 8 in. The development length may be reduced in certain cases, for example, 25 percent when the bars are enclosed in spirals of not less than no. 2 rebar size and the spiral pitch is not over 4 in (see ACI, art. 12.6).

For shear, consider that the compression stresses are zero at the neutral axis; therefore, inspection of Fig. 3-1*c* indicates that the maximum shear stress can be evaluated from summing horizontal forces

$$C + F_s - C - \Delta C = 0$$

but

$$F_s = b(dx)v_u$$

and from Eq. (*e*), neglecting the second-order differential $dx\ \Delta V_u$,

$$\Delta T y = V_u\, dx$$

From Fig. 3-1

$$\Delta C = \Delta T$$

and equating $F_s = \Delta C$, we obtain

$$b(dx)v_u = V_u \frac{dx}{y}$$

Solving for the shear stress and including the ϕ factor, we have

$$v_u = \frac{V_u}{\phi y b}$$

Research indicates [Report ACI-ASCE Committee (1962)] that enough uncertainty exists to replace the distance y with the effective concrete depth to obtain the design equation

$$v_u = \frac{V_u}{\phi b d} \tag{3-4}$$

The allowable design shear stress without the use of shear reinforcement (the usual case for footings) is

$$v_d \leq 2\phi\sqrt{f'_c}$$

(see also Table 3-2). This stress is usually termed *beam* shear, or in the case of footings *wide-beam shear*. The controlling condition for square footings and often other types of footings is *diagonal-tension shear*. The design value of diagonal-tension stress (ACI Code, art. 11.10.3) is

$$v_d \leq 4\phi\sqrt{f'_c}$$

The bearing pressure exerted by the column on the footing may be a critical factor in controlling the depth, especially if the column concrete has a much higher compressive stress f'_c than the footing. The column contact stresses are not to exceed $0.85f'_c$ of the footing unless the supporting surface (footing or pedestal) is larger than the column (ACI, art. 10.14). If the supporting surface is larger than the column (see Fig. 3-3), the allowable contact stress can be computed as

$$f_c \leq 0.85\phi f'_c \sqrt{\frac{A_2}{A_1}}$$

but the ratio of $\sqrt{A_2/A_1}$ cannot exceed 2. Use 0.70 for the ϕ factor.

A footing may be designed with no tensile reinforcement if the flexural stresses do not exceed

$$f_t \leq 5\phi\sqrt{f'_c} \qquad \text{strength design}$$

and the shear stresses meet the requirements already given (ACI, art. 15.7). Note, however, that shrinkage and temperature reinforcement (ACI, art. 7.13) should always be used.

Metric conversion factors for equations presented in this chapter are listed in Sec. 3-11.

Table 3-2 ALLOWABLE WIDE-BEAM AND DIAGONAL-TENSION SHEAR BY ACI 318-71 CODE

	f_c, psi (kg/sq cm)			
	3,000 (211)	3,500 (246)	4,000 (281)	5,000 (352)
Wide beam $2\phi\sqrt{f'_c}$:				
psi (kg/sq cm)	93.1(6.5)	100.6(7.1)	107.5(7.6)	120.2(8.5)
ksf (kN/sq m)	13.41(642)	14.49(694)	15.48(741)	17.31(829)
Diagonal tension $4\phi\sqrt{f'_c}$:				
psi (kg/sq cm)	186.2(13.1)	201.1(14.1)	214.0(15.0)	240.4(17.0)
ksf (kN/sq m)	26.81(1,283.7)	28.96(1,386.4)	30.96(1,475.4)	34.62(1,657.4)

$\phi = 0.85$

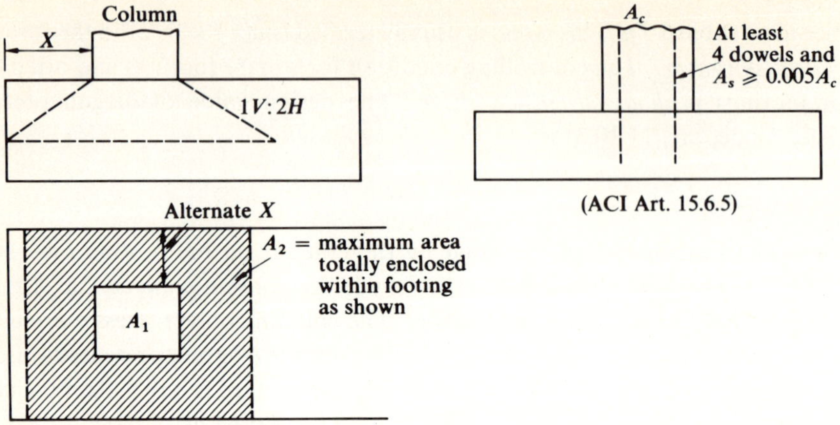

FIGURE 3-3
Method of determining allowable column pressure on footing and minimum column-dowel requirements.

3-3 SPREAD-FOOTING DESIGN

Table 3-3 summarizes the pertinent sections of the ACI 318-71 Code most generally applicable to footing (and foundation) design.

The first step is to obtain the lateral dimensions of the footing, one of which may occasionally be given if the footing cannot be made square. For square footings, the lateral dimension B is

$$B = \frac{\sum \text{(working loads)}}{\text{allowable soil pressure}}$$

This value of B is usually rounded to the nearest larger 0.25 ft (say 7.5 cm) as a practical field construction practice.

With the footing dimension established, the ultimate soil pressure (for computational purposes only) is found as

$$q_{\text{ult}} = \frac{P_{\text{ult}}}{B^2}$$

This value of soil pressure will be used in Eqs. (3-5) and (3-6) to obtain the footing thickness. Note that P_{ult} does not include the estimated weight of the footing for three reasons: (1) the side dimension B is generally rounded upward slightly; (2) the footing will displace its own volume of soil, resulting in a net increase in the difference in soil and concrete unit weights multiplied by the footing depth, which will rarely amount to

Table 3-3 SUMMARY OF MOST COMMON FOOTING REQUIREMENTS ACI 318-71

Design factor	ACI Code article	General requirements
Spacing of reinforcement	7.4	Not less than D or 1 in or 1.33 (max aggregate size); not more than 3 × depth of footing or 18 in
Lap splices	7.5.2	Not for bars > no. 11
In tension	7.6	See section in Code
In compression	7.7	See section in Code
Temperature Shrinkage	7.13 10.5.2	$p = \begin{cases} 0.002 & \text{for } f_y = 40 \text{ to } 50 \text{ ksi} \\ 0.0018 & \text{for } f_y = 60 \text{ ksi or welded wire fabric} \end{cases}$
Minimum reinforcement cover	7.14	3 in against earth
Design methods flexure	8.1	$M_u = \phi A_s f_y[(d - a)/2]$
		$a = A_s f_y/0.85 f'_c b$
Maximum reinforcement	10.3.2	$p_a = 0.75 \times$ Eq. (d)
		$p = A_s/bd \leq p_a$
Minimum reinforcement	10.5.1	$p \geq 200/f_y$ if footing of variable thickness; for slabs of uniform thickness use shrinkage and temperature percentage
k_1 factor	10.2.7	$k_1 = \begin{cases} 0.85 & \text{for } f'_c \leq 4{,}000 \text{ psi} \\ 0.85 - 0.05 & \text{for each } 1{,}000 \text{ psi over } 4{,}000 \text{ psi} \end{cases}$
Limits of compression reinforcement	10.9	$0.01 \leq A_{st}/A_g \leq 0.08$
Modulus of elasticity	8.3	$E = w^{1.5} 33\sqrt{f'_c}$ psi for w between 90 and 155 pcf
		$E_c = 57{,}000\sqrt{f'_c}$ psi for $w = 140$ to 150 pcf
		$E_s = 29{,}000{,}000$ psi
		Take $n = E_s/E_c$ to nearest integer > 6
Load factors ϕ	9.2	Flexure = 0.90; shear = 0.85; bearing = 0.70; flexure plain concrete = 0.65

(Continued)

Table 3-3 SUMMARY OF MOST COMMON FOOTING REQUIREMENTS ACI 318-71 (*Continued*)

Design factor	ACI Code article	General requirements
Load	9.3.1	1.4 × dead load; 1.7 × live load
Bearing	10.14	$q_{brg} \leq R0.85\phi f_c'$ (see Fig. 3-3) $R \leq 2$
Shear, wide-beam	11.10.1*a* 11.2.1 11.4.1	$v_u = V_u/bd \qquad v_c = 2\phi\sqrt{f_c'}$
Diagonal (punch) tension	11.10.1*b*	$v_u = V_u/bd \qquad v_c = 4\phi\sqrt{f_c'}$
Shear reinforcement	11.11.1	For footings only 50% effective
Development of reinforcement	12.5 12.6	See general requirements and values given in text earlier
Grade beams	14.3–15.10	
Footings	15.1	General footing considerations
Location of bending moments	15.4.2	See Fig. 3-5
Distribution of reinforcing in rectangular footings	15.4.4	Percent in zone of width $B = 2/[L/(B+1)]$
Shear	15.5 11.10.1*a* 11.10.1*b*	See Fig. 3-4
Transfer of stress at base of column	15.6.5 10.14.1	At least 4 dowels with total $A_s \geq 0.005A_g$
Unreinforced pedestals	15.7	$f_c = 0.85\phi f_c' \qquad \phi = 0.70$
		$f_t = 5\phi\sqrt{f_c'} \qquad \phi = 0.65$
Round columns	15.8	Equivalent square column side, $a = \sqrt{A_c}$
Minimum edge thickness	15.9	8 in for unreinforced footing; 6 in above reinforcement
Maximum tensile stress in unreinforced footings	15.7.2	$f_t \leq 5\phi\sqrt{f_c'} \qquad \phi = 0.65$

over 100 lb/sq ft; (3) the allowable bearing pressure is not known to a precision of more than about ± 300 lb/sq ft.

Since shear or diagonal-tension strength is relatively low, this usually controls the design of spread footings. That is, one finds the depth to satisfy shear without using shear reinforcement,[1] then computes the steel requirements for bending and checks column bearing if required. If dowels are required to transfer part of the column stresses to the footing, the footing depth is checked for adequacy of dowel transfer action.

Since the design is controlled by the footing depth and the depth usually depends on diagonal tension, it is desirable to obtain an expression for footing depth once and for all. Referring to Fig. 3-4*b*, the perimeter for a square column is $4(w + d)$, and the area enclosed by this perimeter is $(w + d)^2$. With the soil pressure on the base of the footing as

$$q = \frac{P_u}{BL}$$

the desired expression can be obtained by summing vertical forces on the diagonal-tension zone shown in Fig. 3-4*b*

$$P_u - P' - dv_c\,(\text{perimeter}) = 0$$

Substituting and rearranging gives

$$4v_v(w + d) + q(w + d)^2 = P_u$$

Simplifying, we obtain the desired expression for square columns as

$$d^2\left(v_c + \frac{q}{4}\right) + d\left(v_c + \frac{q}{2}\right)w = \frac{(BL - w^2)q}{4} \tag{3-5}$$

For round columns (use diam $= w$) the applicable equation becomes

$$d^2\left(v_c + \frac{q}{4}\right) + d\left(v_c + \frac{q}{2}\right)w = \frac{(BL - A_{\text{col}})q}{\pi} \tag{3-6}$$

These two quadratic equations for footing depth are used in the computer program included in this chapter.

One should not convert a round column to an equivalent square column to use Eq. (3-5) instead of Eq. (3-6). The use of an equivalent square column instead of the actual column diameter will give an unsafe difference in computed depth, which can be considerable, depending on the column dimensions and loads. The included

[1] It is usual practice, both as an economy and to ensure reasonable footing rigidity, not to use shear reinforcement in footings.

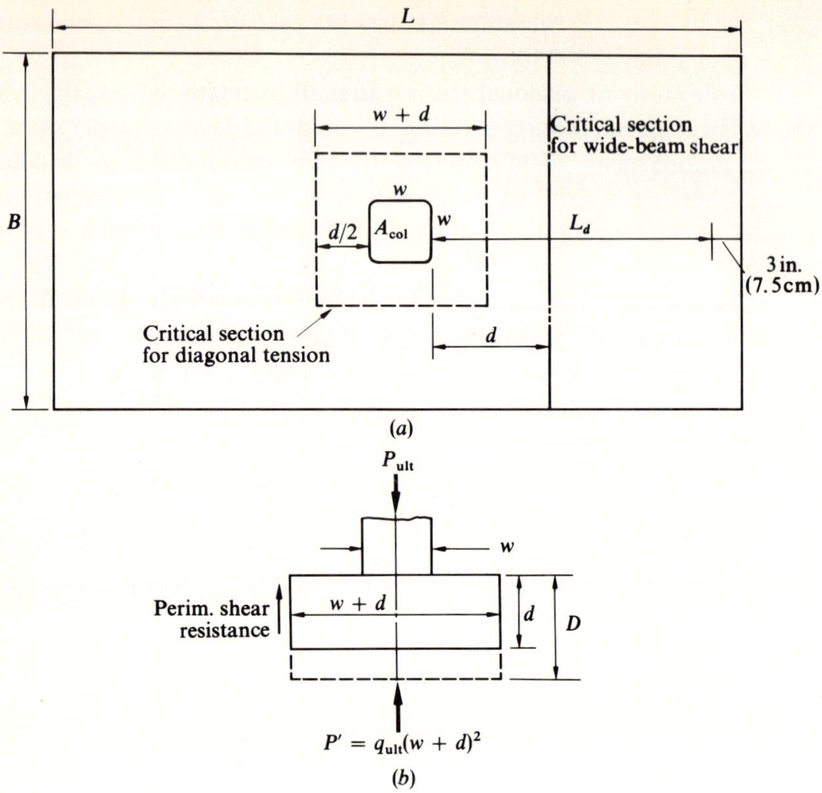

FIGURE 3-4
Critical sections for shear. For round columns use w = diameter. (*a*) Furnished L_d and critical sections for shear computations. (*b*) Obtaining depth for diagonal tension.

computer program considers either round or square columns to compute the correct footing depth.

For baseplates, which may be rectangular, the computer program is not programmed to work for a steel column on a rectangular baseplate directly. One may use the program, however, and obtain a conservative footing depth by taking the smaller of the two baseplate dimensions corrected by the distance X of Fig. 3-5 as the column width w. The depth will be somewhat less conservative using

$$w = \sqrt{\text{area of baseplate}}$$

Once the effective footing depth d is established, the next step is to compute the required area of steel to satisfy bending requirements. Rearranging Eq. (3-2) results

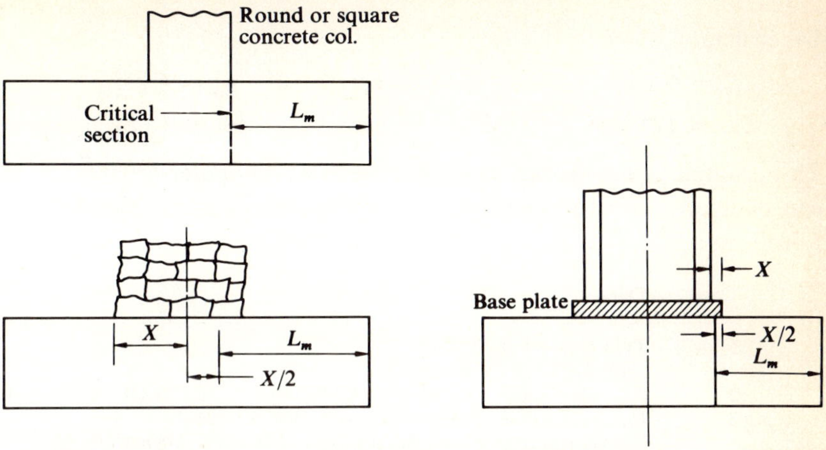

FIGURE 3-5
Critical sections for computing bending moments (ACI, art. 15.4).

in a quadratic equation in A_s. Figure 3-5 indicates the critical sections for bending-moment computations for several types of footing load. Taking L_m as an equivalent cantilever beam, the ultimate moment is computed as $M_u = q_{\text{ult}} L_m^2/2$. Bond stresses are indirectly checked by computing the required embedment length L_d and comparing this value to the length actually available (see Fig. 3-2 for maximum available length).

Column bearing stresses are checked according to ACI, art. 10.14 (see Fig. 3-3); however, in any case a minimum of four dowels with a total area of not less than $0.005A_c$ must be provided.

EXAMPLE 3-1 Design a square spread footing for the conditions shown in Fig. E3-1.1.

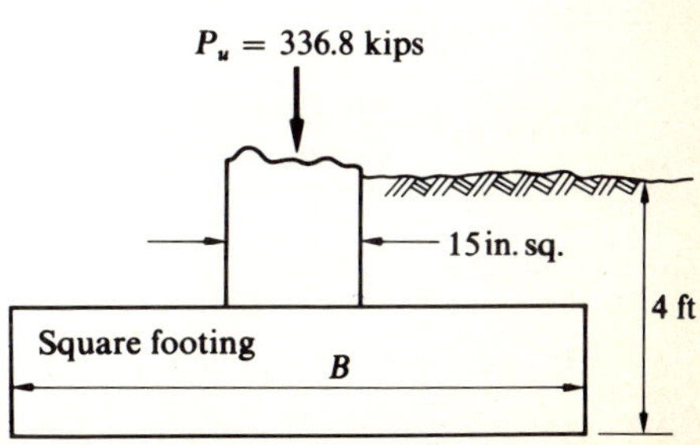

FIGURE E3-1.1

Other data:

$$f'_c = 4 \text{ ksi} \qquad f_y = 60 \text{ ksi} \qquad q_a = 4 \text{ ksf}$$

$$DL = 124 \text{ kips} \qquad LL = 96 \text{ kips} \qquad \text{col} = 15 \text{ in}$$

This example is also worked using the included computer program so that the reader can compare the two solutions. The computer input consists of four data cards, as follows:

Card	Data
1	TITLE (see Fig. E3-1.2)
	UT1–UT8
2	FT IN KIPS FT-KIPS KIPS/SQ FT KIPS/CU FT LB/SQ IN SQ IN
3	15. 0. 60000. 4000. 0.0 4.0 124.0 196.0
	NBAR (number of reinforcing bars on DATA USBAR/···/)
4	9

The output is shown in Fig. E3-1.2 along with a design sketch.

```
J E BOWLES   EXAMPLE 3-1   SQUARE SPREAD FOOTING DESIGN   (USD)

    *****SQUARE FOOTING DESIGN BY ACI 318-71-****

GIVEN DESIGN DATA AND CONDITIONS
   COL SIDE = 15.000 IN
   COL DIAM =   0.0    IN
   YIELD STR OF STEEL = 60000.0 LB/SQ IN
   28-DAY CONCRETE STR = 4000.0 LB/SQ IN
   LENGTH OF RECT FTG =   0.0     FT
   ALLOWABLE SOIL PRESS =    4.000 K/SQ FT

   COL LOADS:   DEAD =   124.00 KIPS
                LIVE =    96.00 KIPS

****FOOTING WIDTH, B = 7.500 FT

      EFFECTIVE FOOT DEPTH = 12.762 IN<----
         AREA OF STEEL REQD =   0.5250 SQ IN/FT
            ACTUAL % STEEL = 0.003
         MAX ALLOW % STEEL = 0.021
ACTUAL SOIL PRESS =    3.911 K/SQ FT
ULT SOIL PRESS =    5.988 K/SQ FT
LD FURN =   34.500 IN<----

 BAR    BARS REQD & SPAC CEN-TO-CEN        AS FURN        AS REQD  LD REQD
  IN             NO          IN             SQ IN          SQ IN     IN
0.5000           21        4.286            4.123          3.937    12.000
0.6250           13  ---/--6.923            3.988          3.937    15.000
0.7500            9       10.000            3.976          3.937    18.000

DOWELS TO SATISFY ACI ART 15.6.5 OF 1/2% OF COL AREA AND AT LEAST 4 BARS IS:
 BAR DIA    NO REQD   LD, IN      1/2 % REQD      AS FURN, SQ IN
    0.3750    11        8.000        1.125          1.215
    0.5000     6        9.487        1.125          1.178
    0.6250     4       11.859<----   1.125          1.227
```

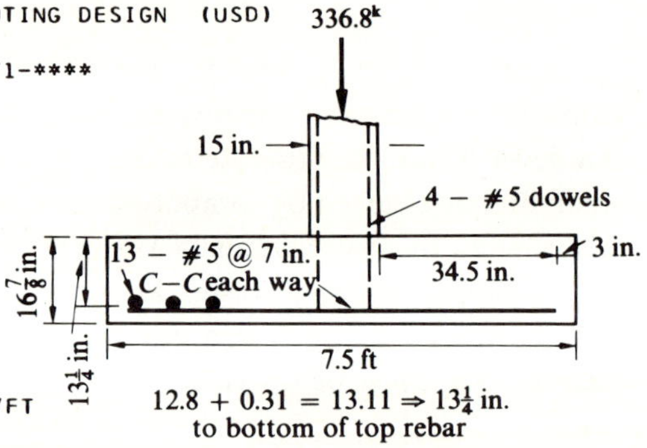

FIGURE E3-1.2
Computer output and final-design sketch using output data.

SOLUTION (partial, by hand):

$$B^2 = \frac{220}{4} = 55 \qquad B = 7.415 \approx 7.5 \text{ ft}$$

$$P_u = 1.4(124) + 1.7(96) = 5.988 \text{ kips} \qquad q_{\text{ult}} = \frac{336.8}{7.5^2} = 5.988 \text{ ksf}$$

Find the depth for diagonal tension [Eq. (3-5)]:

$$w^2 = 1.25^2 = 1.563$$

$$BL = B^2 = 7.5^2 = 56.25$$

$$v_c = 215 \text{ psi} = 30.96 \text{ ksf} \qquad \text{Table 3-2}$$

$$\frac{q}{2} = 2.994 \text{ ksf} \qquad \frac{q}{4} = 1.497 \text{ ksf}$$

$$v_c + \frac{q}{4} = 32.457$$

$$\left(v_c + \frac{q}{2}\right) w = 42.4425$$

$$\frac{(B^2 - w^2)q}{4} = 81.87$$

Substituting into Eq. (3-5) gives

$$32.457d^2 + 42.4425d = 81.87$$

$$d^2 + 1.31d = 2.52$$

from which

$$d = 1.064 \text{ ft} = 12.76 \text{ in}$$

Find the required steel area A_s for bending

$$A_s > 0.002 \text{ shrinkage}$$

$$< 0.021 \text{ maximum code} \qquad \text{Table 3-1}$$

Find the ultimate moment; the "cantilever" is

$$L = \frac{7.5 - 1.25}{2} = 3.125 \text{ ft}$$

$$M_u = \frac{5.988(3.125)^2}{2} = 29.239 \text{ ft-kips} = 350.87 \text{ in-kips}$$

$$M_u = \phi A_s f_y \left(d - \frac{a}{2}\right) \qquad \phi = 0.90 \qquad \text{[Eq. (3-2)]}$$

$$a = \frac{A_s f_y}{0.85 f'_c b} = \frac{60A_s}{40.8} = 1.47A_s \qquad \text{taking 12-in width for } b$$

Substituting into Eq. (3-2) and dividing by 0.9 gives

$$60A_s(12.76 - 0.735A_s) = 389.86$$

$$A_s^2 - 17.36A_s = 8.841$$

Completing the square, we have

$$A_s = 0.525 \text{ sq in/ft}$$

Check the actual percent of steel:

$$p = \frac{0.525}{12.76 \times 12} = 0.0034 > 0.002 < 0.021 \qquad \text{O.K.}$$

Check the bond:

$$L_d \text{ furnished} = 3.125(12) - 3 = 34.5 \text{ in}$$

(using 3 in of concrete cover on sides as well as bottom).

Select rebars: use thirteen no. 5 bars at 7 in center to center

$$A_s = 13(0.31) = 4.00 \text{ sq in}$$

$$\text{Required } A_s = 7.5(0.525) = 3.94 \text{ sq in} \qquad \text{O.K.}$$

$$\text{Required } L_d = \frac{0.04(0.31)(60{,}000)}{\sqrt{4{,}000}} = 11.6 \text{ in} < 12 \text{ in}$$

$$\text{Alternate } L_d = 0.0004(0.625)(60{,}000) = 15.0 \text{ in} \qquad \text{controls}$$

Since neither required value of L_d is as large as the furnished value, no further check need be made.

Check column bearing on footing (refer to Fig. 3-3); by inspection

$$A_2 = (15 + 48)^2 = 63^2 \qquad A_1 = 15^2$$

$$\text{Ratio} = \sqrt{\left(\frac{63}{15}\right)^2} = 4.2 > 2 \qquad \text{use 2}$$

$$P_{\text{allow}} = \text{ratio } (0.85f'_c)\phi A = 1.7(225)(4)(0.7) = 1{,}071 \text{ kips} \gg 336.8 \text{ kips} \quad \text{O.K.}$$

To satisfy ACI, art. 15.6.5, we must use four dowels of $0.005A_c$:

$$A_{\text{reqd}} = 0.005(225) = 1.125 \text{ sq in}$$

Use four no. 5 bars:

$$A_s = 4(0.31) = 1.25 \text{ sq in} > 1.125 \text{ required}$$

Check depth of footing for dowels:

$$L_d = 0.02 f_y D/\sqrt{f'_c} = 11.86 \text{ in} > 8 \text{ in}$$

$$L_d = 0.0003 f_y D = 11.25 \text{ in} < 11.86 \text{ in}$$

$$< 12.76 \text{ in (footing depth)}$$

Both values of L_d are workable; use 11.875-in dowels. ////

3-4 RECTANGULAR SPREAD FOOTINGS

Rectangular spread footings are designed in a manner similar to square spread footings. In fact, the reader will note that Eqs. (3-5) and (3-6) are derived for either type of footing.

There are, however, two differences in design: (1) longitudinal steel requirements (steel parallel to long side) are larger than transverse steel requirements, and (2) the ACI Code requires that a certain percent of the total short-side steel requirement be placed in a zone of width B centered on the column and parallel to the transverse direction (Fig. 3-6). This percentage of the total short-side steel requirement is

$$S = \frac{2}{L/B + 1} \tag{3-7}$$

If this percent of the total required steel for the short side does not leave enough steel for the two end zones to satisfy minimum shrinkage requirements, the end-zone steel should be increased to satisfy that amount.

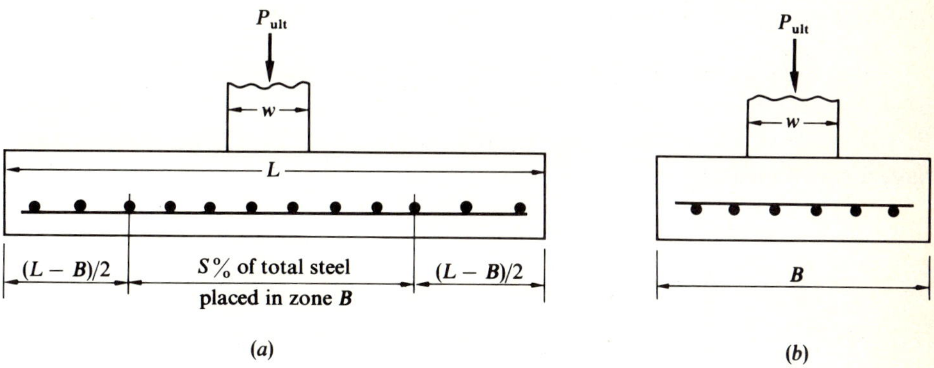

FIGURE 3-6
Rectangular-footing steel requirements for short side (ACI, art. 15.4.4). (*a*) Place half of remaining steel in each zone; if inadequate for shrinkage, use the amount to satisfy shrinkage requirements. (*b*) Note longitudinal steel on bottom for greater efficiency.

The computer program for square spread footings will also design a rectangular footing. It is programmed to make a footing square if it computes almost square by rounding up on the computed (not the required) dimension.

The program checks wide-beam shear for rectangular footings, as wide-beam shear may control the design depth for footings with L/B ratios larger than about 1.1. It has been shown [Furlong (1965)] that wide-beam shear will not control the design of square footings; however, this statement has not been proved mathematically.

EXAMPLE 3-2 Design a rectangular footing using metric units. Given data are as follows (partially shown in Fig. E3-2.1): length = 4.6 m, $f_c' = 281$ kg/sq cm, $f_c = 3{,}515$ kg/sq cm, $q_a = 168$ kN/sq m, column diameter = 46 cm, $DL = 1{,}015$, and $LL = 1{,}010$ kN.

This problem is also worked using the computer program with part of the output displayed. Data for the computer-solution input are as follows:

Card	Data
1	TITLE (see Fig. E3-2.1)
	UT1–UT8
2	M CM KN KN–M KN/SQ M KN/CU M KG/SQ CM SQ CM
3	0.0 46.0 3515. 281. 4.6 168. 1025. 1010.
	NBAR (number of bars on CARD DATA SIBAR/· · ·/)
4	9

These four data cards represent the I/O shown following.

Note that since most of the solution here is similar to the spread-footing solution except for units (the reader may verify this), only the significantly different computations will be shown. Other computation data will be on the computer output sheets.

SOLUTION (partial): We will need to find the equivalent "cantilever" beams of the footing for bending moments in both directions. Since the column is round, we will compute an equivalent square column of side w'. A column of equal area will have a side dimension of

$$w' = \sqrt{1{,}661.9} = 40.77 \text{ cm} = 0.4077 \text{ m}$$

$$L' = \frac{4.600 - 0.408}{2} = 2.096 \text{ m} = 209.6 \text{ cm}$$

$$L_d = 209.6 - 7.5 = 202.1 \text{ cm} \qquad \text{checks computer output}$$

$$M_u = \frac{253.8(2.096)^2}{2} = 557.60 \text{ kN-m} = 5{,}685{,}735.7 \text{ kg-cm}$$

$$a = 0.1472A_s \qquad \frac{M_u}{\phi f_y} = 17.973$$

```
J E BOWLES   EXAMPLE 3-2   RECTANGULAR FOOTING DESIGN USING METRIC UNITS

     ***RECTANGULAR FOOTING DESIGN BY ACI 318-71****

GIVEN DESIGN DATA AND CONDITIONS
   COL SIDE =   0.0    CM
   COL DIAM = 46.000 CM
   YIELD STR OF STEEL =   3515.0 KG/SQ CM
   28-DAY CONCRETE STR =   281.0 KG/SQ CM
   LENGTH OF RECT FTG =   4.600 M
   ALLOWABLE SOIL PRESS = 168.000 KN/SQ M

   COL LOADS:   DEAD = 1025.00 KN
                LIVE = 1010.00 KN

DEPTH DIA TEN = 59.613 CM    DEPTH WIDE BEAM = 53.498 CM
****FOOTING LENGTH, L =   4.600 M
    FOOTING WIDTH, B =   2.700 M

      EFFECTIVE FOOT DEPTH = 59.613 CM
         AREA OF STEEL REQD = 31.3611 SQ CM/M
             ACTUAL % STEEL = 0.005
         MAX ALLOW % STEEL = 0.028
ACTUAL SOIL PRESS = 163.849 KN/SQ M
ULT SOIL PRESS = 253.785 KN/SQ M
LD FURN = 202.117 CM

 BAR    BARS REQD & SPAC CEN-TO-CEN      AS FURN      AS REQD   LD REQD
  CM              NO         CM           SQ CM        SQ CM      CM
 2.2220           22       12.273        85.310       84.675     48.299
 2.5400           17       15.882        86.140       84.675     63.113
 2.8650           14 <--|-->19.286       90.254       84.675     80.297
 3.2260           11       24.545        89.911       84.675    101.807
 3.5810            9       30.000        90.645       84.675    125.446

 STEEL REQD SHORT SIDE = 11.9225 SQ CM/M
                ACTUAL % STEEL = 0.002
TOTAL STEEL SHORT SIDE = 54.8436 SQ CM
      STEEL IN WIDTH B = 40.5692 SQ CM
LD IN SHORT SIDE = 107.117 CM

BAR DIA   BARS REQD AND SPAC CEN-TO-CEN   AS FURN  AS REQD    LD REQD
  CM           ZONE B          END ZONE    SQ CM    SQ CM       CM

 0.9520       57    4.737      11    8.636    56.233   54.844     19.040
 1.2700       33    8.182       6   15.833    57.005   54.844     25.400
 1.5880       21   12.857       4   23.750    57.437   54.844     31.761
 1.9050       15   18.000       3   31.667    59.855   54.844     38.101
 2.2220<--|-->11   24.545       2   47.500    58.166   54.844     48.299
 2.5400        9   30.000       2   47.500    65.872   54.844     63.113
 2.8650        7   38.571       2   47.500    70.914   54.844     80.297

DOWELS TO SATISFY ACI ART 15.6.5 OF 1/2% OF COL AREA AND AT LEAST 4 BARS IS:
 BAR DIA    NO REQD   LD, CM      1/2 % REQD    AS FURN, SQ CM
    0.9520    12       20.300        8.310         8.542
    1.2700     7       20.300        8.310         8.867
    1.5880     5       25.140        8.310         9.903
    1.9050<----4------>30.159        8.310        11.401
```

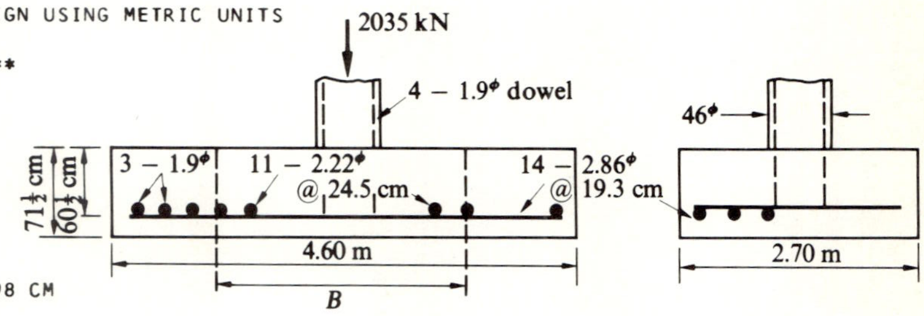

FIGURE E3-2.1
Computer output and final-design sketch. (Sketch is not to scale, and not all rebars are shown.)

Substituting into Eq. (3-2) gives

$$A_s(59.61 - 0.0736A_s) = 1{,}797.3 \qquad \text{depth from computer output}$$

$$A_s = 31.36 \text{ sq cm/m}$$

Use fourteen 2.87ϕ (no. 9) bars at 19.3 cm center to center (also from computer output)

$$L_d = 80.3 < 202.1 \text{ cm} \qquad \text{O.K.}$$

The short-side moment and steel requirements are now computed:

$$L' = \frac{2.700 - 0.408}{2} = 1.146 \text{ m}$$

$$L_d = 114.6 - 7.5 = 107.1 \text{ cm} \qquad \text{output check}$$

$$M_u = \frac{253.8(1.146)^2}{2} = 166.7 \ k\text{N-m} = 1{,}699{,}806.6 \text{ kg-cm}$$

Substituting into Eq. (3-2) gives

$$0.0736A_s^2 - 59.61A_s = -537.32$$

$$A_s = 9.12 \text{ sq cm/m}$$

The percent steel furnished is

$$p = \frac{9.12}{59.6(100)} = 0.0015 < 0.002$$

Using 0.002, the required A_s is

$$5{,}960(0.002) = 11.92 \text{ sq cm/m}$$

The total A_s for the short side is 11.92(4.6) = 54.8 sq cm. The L/B is 1.70.

$$\text{Percent steel in zone } B = S = \frac{2}{1.70 + 1} = 0.741$$

$$\text{Steel in width } B = 54.8(0.741) = 40.6 \text{ sq cm}$$

Use eleven 2.22ϕ (no. 7) bars at 24.5 cm center to center in zone B (actual A_s = 42.6). Check end zone:

$$\text{Width} = \frac{4.6 - 2.7}{2} = 0.95 \text{ m}$$

$$\text{Minimum } A_s = 11.92(0.95)(0.002) = 11.32 \text{ sq cm} \qquad \text{requires three 2.22}\phi \text{ bars}$$

Use 11 + 3 + 3 = 17—2.22ϕ (no. 7) bars at uniform spacing. Use 4—1.9ϕ (no. 6) dowels to tie column to footing. See final sketch in Fig. E3-2.1. ////

3-5 COMPUTER PROGRAM TO DESIGN SQUARE AND RECTANGULAR SPREAD FOOTINGS

This computer program will proportion a spread footing for minimum depth and list a selection of bar sizes which can be used. The design engineer must inspect the output and select the final bar sizes and footing depth to complete an economical and adequate design considering practical limitations. This program does not intermix bar sizes.

This program will design either a square or rectangular two-way spread footing. Design is based on strength (or ultimate-strength design) and ACI 318-71. Either round or square columns may be used. The column must be centrally located on the footing, as eccentric footing loads are not considered. The program is valid for any strength concrete. Steel yield strength is limited to 60 ksi. This program will solve metric-unit problems, but only United States rebar sizes in fps and metric units are currently included (no. 3 to no. 11 bars). Conversion factors are used (21) on card DATA FFU/. . ./, alternate (2,4,6,. . .) being for metric units so that the user does not have to punch this as part of the data. The United States rebar sizes are on cards DATA USBAR/. . ./ and DATA SIBAR/. . ./ (for metric equivalents). Users of foreign rebars should change the card DATA SIBAR/. . ./.

Lines	Operation
1–2	DIMENSION
3–5	DATA Conversion factors for computations in either fps (IUNIT = 1) or metric. Cards 4 and 5 are rebar diameters in fps and metric units. Only nine bars are used. If more bars are included, increase the numbers and change the appropriate DIMENSION statement
7	READ TITLE (any alpha-numeric description of problem and UNITS (UT1 - UT8, see Examples 3-1 and 3-2 for entries)
9	READ A = column side; DIA = diameter (if round); FY, F1C = steel and concrete stresses; EL = length of rectangular footing; QALL = allowable soil pressure; PD, PL = dead and live working loads. Read A and DIA in inches (or centimeters); f_y, f'_c in psi or kg/sq cm; L in feet or meters; q_a in ksf or kN/sq m and loads in kips or kilonewtons. Punch 0.0 for column type not used and EL = 0.0 if footing is square. NBAR = number of rebars to use in design (currently nine); this is a separate card
11–17	Identifies computation constants and use of rebar diameters in inches or centimeters
26–51	Computes footing size and adjusts for rectangular footings so that narrow side = B
52–87	Finds depth d for diagonal-tension and wide-beam shear
88–115	Finds required reinforcing steel
124–145	Finds bar sizes and L_d
147–176	Finds bar sizes and L_d for short side of rectangular footing
177–217	Checks column dowel-bar requirements

```
C      J E BOWLES  SPREAD FOOTING DESIGN (ACI 318-71) SQUARE OR RECTANGUL
C      FPS AND METRIC CONVERSIONS AND US REBARS & METRIC DIAM OF US SIZES
C       ARE ON CARDS "DATA XXX/........./ BELOW
C      USE UNITS CONSISTENT WITH UT1 THRU UT8  SEE EXAMPLE 3-1 AND 3-2 OF
C      NBAR = NO OF BARS ON DATA CARD--US BARS = #3 TO #11--IF MORE IN ME
C       SYSTEM INCREASE NUMBER ON "DATA SIBAR/.../" & CHANGE DIMENSION ST
       DIMENSION AREA(10),PERM(10),COMLD(10),BARL(10),TITLE(20)
       DIMENSION FFU(2,21),FU(21),BDIA(10),USBAR(10),SIBAR(10)
       DATA FFU/12.,100.,.144,98.07, 2.,.53,4.0,1.06, 87000.,6117.,200.,1
      14.06,.001, .009807,.25,.100, .50,.15,3.,7.5,4000.,281.2, 1000.,70.
      231, 1000., 101.968,.50,1.2,.02,.0755,.0003,.00427,8.0, 20.3,.04,
      3.0594, .0004,.00569,12.,30.,4.,10./
       DATA USBAR/.375,.500,.625,.750,.875,1.000,1.128,1.270,1.410/
       DATA SIBAR/.952,1.270,1.588,1.905,2.222,2.540,2.865,3.226,3.581/
       DATA Z/'M'/
       DOUBLE PRECISION UT4,UT5,UT6,UT7,UT8
  6000 READ(1,1000,END=150)TITLE,UT1,UT2,UT3,UT4,UT5,UT6,UT7,UT8
  1000 FORMAT(20A4,/,3(A4,6X),5(A8,2X))
       READ(1,1003)A,DIA,FY,F1C,EL,QALL,PD,PL,NBAR
  1003 FORMAT(8F10.4,/,2I5)
       IUNIT = 1
       IF(UT1.EQ.Z)IUNIT = 2
       DO 2500 J = 1,21
  2500 FU(J) = FFU(IUNIT,J)
       DO 2600  J = 1,NBAR
       BDIA(J) = USBAR(J)
  2600 IF(UT1.EQ.Z)BDIA(J) = SIBAR(J)
       WRITE(3,1001)TITLE
  1001 FORMAT('1',//,T5,20A4,//)
       IF(EL.LE.0.)WRITE(3,1007)
  1007 FORMAT(T10,'*****SQUARE FOOTING DESIGN BY ACI 318-71-****',//)
       IF(EL.GT.0.1)WRITE(3,1008)
  1008 FORMAT(T10,'***RECTANGULAR FOOTING DESIGN BY ACI 318-71****',//)
    15 WRITE(3,1009)A,UT2,DIA,UT2,FY,UT7,F1C,UT7,EL,UT1,QALL,UT5,PD,UT3,P
      1L,UT3
  1009 FORMAT(T5,'GIVEN DESIGN DATA AND CONDITIONS',/,T8,'COL SIDE =',F7.
      13,1X,A2,/,T8,'COL DIAM =',F7.3,1X,A2,/,T8, 'YIELD STR OF STEEL =',
      2F8.1,1X,A8      ,/,T8,'28-DAY CONCRETE STR =',F7.1,1X,A8      ,/,T8,
      3 'LENGTH OF RECT FTG =',F7.3,1X,A2,/,T8,'ALLOWABLE SOIL PRESS =',F
      48.3,1X,A7,//,T8,'COL LOADS:',T20, 'DEAD =', F8.2,1X,A4,/,T20, 'LIV
      5E =',F8.2,1X,A4,//)
       PTOT = PD + PL
       IF(EL.LT.1.)GO TO 16
       B = PTOT/(EL*QALL)
       GO TO 17
    16 B = SQRT (PTOT/QALL)
    17 IB = B
       BNEW = 0.
       AB = IB
       IF(IB.EQ.B)GO TO 19
       DO 18  K = 1,11
       AB = AB + FU(8)
       IF (AB.GT.B)GO TO 19
    18 CONTINUE
C      FOOTING MADE SQUARE IF REQ'D DIMENSION IS WITHIN -0.5 FT
    19 B = AB
       IF(EL.LT.1.OR.EL.GT.(B+FU(9)))GO TO 9
       BDIF = EL - B
       IF(ABS (BDIF).GT.FU(9))GO TO 7
       BNEW = EL
       B = BNEW
       EL = 0.
       WRITE (3,1015)
  1015 FORMAT(T5,'***** FOOTING MADE SQUARE ***')
       GO TO 9
     7 BNEW = EL
       EL = B
       B = BNEW
C      FIND DEPTH TO SATISFY 'PUNCHING' SHEAR (OR DIAGONAL TENSION)
     9 PULT = 1.4*PD + 1.7*PL
       IF(EL.LE.0.01)GO TO 21
       QULT = PULT/(B*EL)
       AF = B*EL
       GO TO 22
    21 QULT = PULT/(B**2)
       QACT = PTOT/(B**2)
       AF = B**2
    22 IF(DIA.GT.0.01)GO TO 23
       ACOL1 = A**2
       A = A/FU(1)
       GO TO 24
C      CONVERT DIAM TO EFFECTIVE SQUARE
    23 A = SQRT (0.7854*(DIA**2))/FU(1)
       ACOL1 = 0.7854*DIA**2
    24 UVC = FU(4)*.85*SQRT(F1C)
       EEL = (B-A)/2.
       V = UVC*FU(2)
       IF(DIA.GT..015)GO TO 38
       C1 = V + QULT/4.
       C2 =(V + QULT/2.)*A
       C3 =-(AF-A**2)*QULT/4.
       DE = (-C2+ SQRT (C2**2-4.*C1*C3))/(2.*C1)
       DE = DE*FU(1)
       GO TO 41
    38 CP1 = V + QULT/4.
```

```
          CP2 = (V + QULT/2.)*(DIA/FU(1))
          ACOL = 0.7854*((DIA/FU(1))**2)
          CP3 = -(AF - ACOL)*QULT/3.1416
          DE = (-CP2 + SQRT (CP2**2 - 4.*CP1*CP3))/(2.*CP1)
          DE = DE*FU(1)
C         CHECK WIDE-BEAM SHEAR FOR RECTANGULAR FOOTING
       41 IF(EL.LT..1)GO TO 20
          VCWB = FU(3)*.85*SQRT(F1C)
          V1 = VCWB*FU(2)
          AA = V1/QULT
          DWB =((EL-A)/(2.*(AA+1)))*FU(1)
          IF(DWB.GE.DE)DE = DWB
          WRITE(3,997)DE,UT2,DWB,UT2
      997 FORMAT(//,T5,'DEPTH DIA TEN =',F7.3,1X,A2,3X,'DEPTH WIDE BEAM =',F
         17.3,1X,A2)
C         FIND REINF STEEL FOR FOOTINGS--USE 3-INCH OR 7.5 CM COVER
          EEL =(EL-A)/2.
          QACT = PTOT/(B*EL)
          SLD = (B-A)*FU(1)/2. - FU(10)
       20 CONTINUE
          IF(F1C-FU(11))52,52,51
       51 DEL = (F1C-FU(11))/FU(12)
          IDEL = DEL
          ADEL = IDEL
          AKI = 0.85 -0.05*ADEL
          GO TO 53
       52 AKI = 0.85
       53 PB = 0.85*AKI*F1C*(FU(5)/(FY*(FU(5)+FY)))*.75
          G = 0.5*FY/(0.85*F1C*FU(1))
          KK = 1
       25 ULMOM = QULT*(EEL**2)/2.
          AEL = EEL*FU(1) - FU(10)
          F = ULMOM*FU(1)*FU(13)/(.9*FY)
       26 AS = (DE- SQRT(DE**2-4.*G*F))/(2.*G)
          PER = AS/(DE*FU(1))
          IF(PER-0.002)27,27,28
       27 PER = 0.002
          AS = PER*DE*FU(1)
C         IF PERCENT STEEL REQ'D EXCEEDS PB, FTG DEPTH INCR .5 (1.2 CM) INCH
       28 IF(PER-PB)29,29,30
       29 IF(KK.GT.1)GO TO 36
          GO TO 55
       30 DE = DE + FU(14)
          GO TO 26
       55 CONTINUE
          IF(EL.GT.B)GO TO 6
          WRITE(3,1025)B,UT1
     1025 FORMAT(T5, '****FOOTING WIDTH, B =',F6.3,1X,A2/)
          GO TO 31
        6 WRITE(3,1013)EL,UT1,B,UT1
     1013 FORMAT(T5, '****FOOTING LENGTH, L =',F7.3,1X,A2,/,T10,'FOOTING WID
         1TH, B =',F7.3,1X,A2//)
       31 WRITE(3,1010)DE,UT2,AS,UT8,UT1,PER,PB,QACT,UT5,QULT,UT5,AEL,UT2
     1010 FORMAT(T11,'EFFECTIVE FOOT DEPTH =',F7.3,1X,A2,/,T14,'AREA OF STEE
         1L REQD =',F8.4,1X,A5,'/',A2,/,T18, 'ACTUAL % STEEL =',F6.3,/, T14,
         2 'MAX ALLOW % STEEL =',F6.3,/,T5,'ACTUAL SOIL PRESS =',F8.3,1X,A7,
         3/,T5,'ULT SOIL PRESS =',F8.3,1X,A7,/,T5,'LD FURN =',F8.3,1X,A2,//)
C         ROUTINE FOR BAR PROPERTIES AND TO SIZE REBARS
          DO 32  I = 1,NBAR
          PERM(I) = 3.1416*BDIA(I)
          AREA(I) = 0.7854*(BDIA(I)**2)
          ASTOT = AS*B
          BARL(I) = FY*AREA(I)/FU(7)
          COMLD(I) = FU(15)*FY*BDIA(I)/SQRT(F1C)
          ABMLD = FU(16)*FY*BDIA(I)
          IF(COMLD(I).LT.ABMLD)COMLD(I)=ABMLD
       32 IF(COMLD(I).LT.FU(17))COMLD(I) = FU(17)
          WRITE(3,1012)UT2,UT2,UT8,UT8,UT2
     1012 FORMAT(T6,'BAR',T12,'BARS REQD & SPAC CEN-TO-CEN',T44,'AS FURN',T5
         16,'AS REQD', T65,'LD REQD',/,T7,A2,T20,'NO',T32,A2,T45,A5,T57,A5,T
         267,A2)
          DO 33  J = 1,NBAR
          BARS = ASTOT/AREA(J)
          NN = BARS
          IF(NN.LT.BARS)NN = NN+1
          AC = NN
          SPAC = B*FU(1)/AC
          TOTAS = AC*AREA(J)
          TENLD = FU(18)*AREA(J)*FY/SQRT(F1C)
          ABT = FU(19)*BDIA(J)*FY
          IF(TENLD.LT.ABT)TENLD = ABT
       33 IF(SPAC.GE.FU(21).AND.SPAC.LE.FU(20))WRITE(3,1014)BDIA(J),NN,SPAC,
         1TOTAS,ASTOT,TENLD
     1014 FORMAT(T5,F7.4,T20,I2,T29,F7.3,T44,F7.3,T55,F7.3,T65,F7.3)
C         FIND REQD STEEL IN SHORT DIRECTION FOR RECTANGULAR FOOTING
          IF(EL.LT.0.01)GO TO 42
          EEL = (B-A)/2.
          KK = KK+1
          GO TO 25
       36 TOTSD = AS*EL
          SZONE = TOTSD*2./(EL/B+1.)
          WRITE(3,1016)AS,UT8,UT1,PER,TOTSD,UT8,SZONE,UT8,SLD,UT2
     1016 FORMAT(//,T6,'STEEL REQD SHORT SIDE =',F8.4,1X,A5,'/'A2,/,T19,'ACT
         1UAL % STEEL =',F6.3,/,T5,'TOTAL STEEL SHORT SIDE =',F8.4,1X,A5,/,
         3T11,'STEEL IN WIDTH B =',F8.4,1X,A5,/,T5,'LD IN SHORT SIDE =',F8.3
         2,1X,A2,//)
```

```
          DAS = (TOTSD-SZONE)/2.
          WRITE(3,1018)UT2,UT8,UT8,UT2
     1018 FORMAT(T5,'BAR DIA',T14,'BARS REQD AND SPAC CEN-TO-CEN',T46,'AS FU
         1RN',T55,'AS REQD',T65, 'LD REQD',/,T8,A2,T20,'ZONE B', T34,'END ZO
         2NE',T47,A5, T56,A5,T67,A2,/)
          DP = .75*DE
          DO 37  J = 1,NBAR
          SLL = FU(18)*AREA(J)*FY/SQRT(F1C)
          SLDA = FU(19)*BDIA(J)*FY
          IF(SLL.LT.SLDA)SLL = SLDA
          BAR = SZONE/AREA(J)
          BARR = DAS/AREA(J)
          NB = BAR
          NT = BARR
          IF(NB.LT.BAR)NB = NB+1
          IF(NT.LT.BARR)NT=NT+1
          SPACS = B*FU(1)/NB
          SPACE = FU(1)*(EL-B)/(2.*NT)
          ASC = NB*AREA(J)
          ASE = NT*AREA(J)
          ASSDT = ASC +2.*ASE
          IF((SPACE.GT.DP.AND.SPACS.GT.DP).OR.SLL.GT.SLD)GO TO 42
       37 WRITE(3,1019)BDIA(J),NB,SPACS,NT,SPACE,ASSDT,TOTSD,SLL
     1019 FORMAT(T5,F7.4,T17,I2,T20,F7.3,T32,I2,T35,F7.3,T47,F7.3,T56,F7.3,
         1T66,F7.3)
C         COMPUTATIONS FOR REQUIRED COLUMN DOWELS
       42 DAMIN = 0.005*ACOL1
          FBRG = 0.85*0.70*F1C*FU(7)
          A = A*FU(1)
          APRIM = A + 4.*DE
          RATIO = SQRT((APRIM/A)**2)
          IF(RATIO.GT.2.)RATIO = 2.
          PMAX = RATIO*FBRG*(A**2)
          IF(PMAX-PULT)40,39,39
       40 DIFF = PULT-PMAX
          WRITE(3,1021)DIFF,UT3
     1021 FORMAT(//,T5,'DOWELS REQD ACI ART 12.6 FOR LOAD DIFF =',F8.3,1X,A4
         1,//)
          KK = 0
          WRITE(3,1022)UT1,UT3,UT8
     1022 FORMAT(T5,'BARS FOR ACI ART 22.6,(MIN AREA ART 15.6)',/,T7,'BAR DI
         1A',T17,'LD,'1X,A2,T25,'LOAD/BAR,'1X,A4,T42,'NO BARS REQD',3X,'AREA
         2 FURN,'1X,A5,/)
          DO 46  I = 1,NBAR
          BAREQ = DIFF/BARL(I)
          JJ = BAREQ
          IF(JJ.LT.BAREQ)JJ = JJ + 1
          AA = JJ
          AFURN = AREA(I)*AA
          IF(AFURN.LT.DAMIN.OR.JJ.LT.4)GO TO 2000
          WRITE(3,1023)BDIA(I),COMLD(I),BARL(I),JJ,AFURN,DAMIN
     1023 FORMAT(T7,F7.4,T18,F7.3,T27,F8.3,T45,I3,7X,F8.3,2X,'(',F8.3,')')
          GO TO 46
     2000 WRITE(3,1035)BDIA(I)
     1035 FORMAT(T8,F7.4,3X,'  THIS BAR DOES NOT SATISFY ACI ART 15.6**')
       46 CONTINUE
       39 WRITE(3,1026)UT2,UT8
     1026 FORMAT(//,T5,'DOWELS TO SATISFY ACI ART 15.6.5 OF 1/2% OF COL AREA
         1 AND AT LEAST 4 BARS IS:',/,T6,'BAR DIA',T16,'NO REQD', T25, 'LD,
         2',A2, T35, '1/2 % REQD',3X,'AS FURN, ',A5)
          DO 48  J = 1,NBAR
          BARCO = DAMIN/AREA(J)
          NT = BARCO
          IF(NT.NE.BARCO)NT = NT+1
          IF(NT.LE.3)NT = 4
          BARCO = NT
          AFURN = AREA(J)*BARCO
          IF(AFURN.LT.DAMIN)GO TO 48
          WRITE(3,1024)BDIA(J),NT, COMLD(J),DAMIN,AFURN
     1024 FORMAT(T9,F7.4,T18,I3,T25,F7.3,T36,F7.3,4X,F7.3)
          IF(NT.LE.4)GO TO 6000
       48 CONTINUE
      150 STOP
          END
```

3-6 DESIGN LIMITATIONS

The design principle utilized has been that the spread footing is absolutely *rigid*, i.e.,

$$q = \frac{P}{A_{\text{footing}}}$$

Nothing has been said about what footing proportions (depth, width, cantilevered length from column face, modulus of elasticity) or material properties are necessary to achieve rigidity. Studies [Schultze (1961)] and analysis based on the theory of elasticity [Borowicka (1936, 1938)] have shown that the actual soil pressure is generally not uniform but is somewhat as shown in Fig. 3-7. From this it is apparent that footings on cohesionless materials and depending on the flexural rigidity EI may approach a uniform soil pressure. Footings on soils with cohesion tend to higher edge pressures, depending again on the flexural rigidity. The work of Borowicka has been used [Bowles (1973)] to obtain approximate design edge pressures, as shown in Fig. 3-8, if the designer feels this is appropriate.

The application of edge pressures with reduced average contact pressures does not change the design as much as might be expected since the effect is somewhat nullified through use of larger contact pressures in the conventional design. Another mitigating factor is that footing rigidity is significantly increased by obtaining the footing depth to satisfy concrete shear strength without the use of shear reinforcement.

3-7 COMBINED FOOTINGS (RECTANGULAR)

Column footings adjacent to property lines may be eccentrically loaded if the column is placed as close as possible to the property boundary, as shown in Fig. 3-9*a*. A problem may exist where rectangular footings may interfere, as Fig. 3-9*b*. The situation of Fig. 3-9*c* may arise if the soil is of such low allowable bearing capacity that the resulting footing dimensions conflict. In these three cases a possible solution is to put more than one column on a *combined footing*. When more than one line of columns is on the footing, it is termed a *mat* or *raft*, with a solution to be considered later (Chap. 7). Some engineers are of the opinion, however, that it may be more economical, even if the foundation site is saturated with footings, to use spread footings, if at all possible, to avoid placing both positive- and negative-bending steel. A small side benefit in formwork savings may be obtained where footings are very close by pouring alternate footings so that the poured footing can be used as a form by inserting a thin spacer board between footings.

A combined footing may have more than two columns in a line; however, this chapter and its computer program consider only a two-column case. The reason will be apparent later.

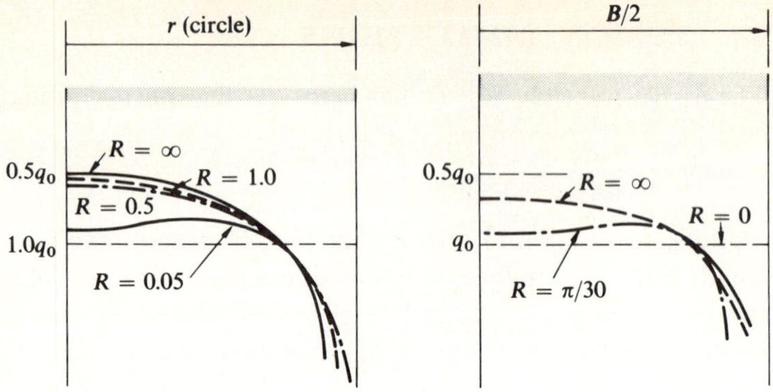

$$R = \frac{1-\mu_s^2}{1-\mu_f^2}\left(\frac{E_f}{6E_s}\right)\left(\frac{t_f}{r}\right)^2$$

FIGURE 3-7
Footing pressures based on theoretical considerations. [*After Borowicka* (1936, 1938).]

Combined footings may be of any shape; this section considers only rectangular shapes, and Sec. 3-9 considers a trapezoid.

Combined footings may more properly be considered slabs or mats if the L/B ratio is over about 2 to 2.25. The ACI Code does not provide much guidance on their design except to leave it largely to the designer's judgment (art. 15.10). The design

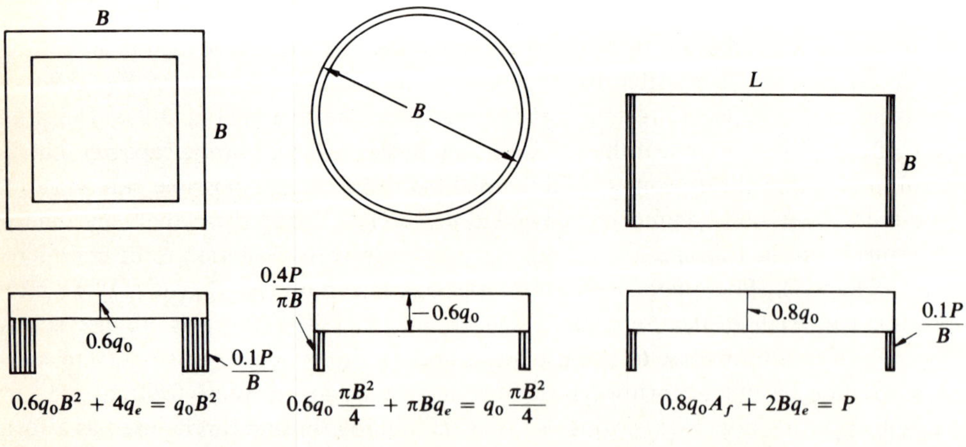

$0.6q_0B^2 + 4q_e = q_0B^2$

(*a*)

$0.6q_0\frac{\pi B^2}{4} + \pi Bq_e = q_0\frac{\pi B^2}{4}$

(*b*)

$0.8q_0A_f + 2Bq_e = P$

(*c*)

FIGURE 3-8
Alternative pressure distribution for footings on cohesive soil: (*a*) square; (*b*) circle; (*c*) rectangle. [*After Bowles* (1973).]

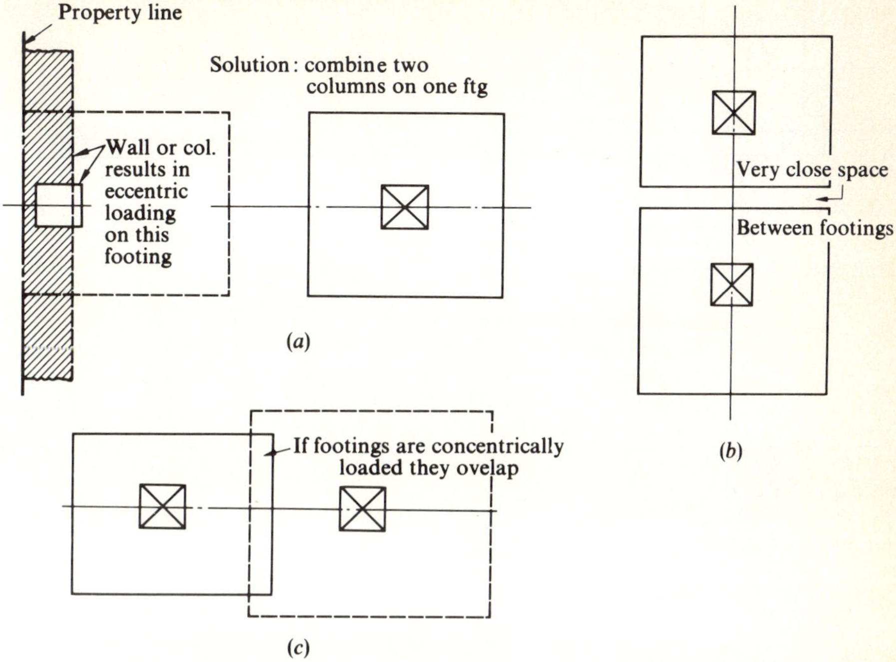

FIGURE 3-9
Some conditions where combined footing may be the most practical solution: (*a*) column or load-bearing wall is so close to the property line that a footing would be eccentrically loaded; (*b*) footings are so close together that it may be more practical to combine them; (*c*) column loads are such that the resulting spread footings interfere.

steps outlined in the following paragraphs are those considered to be accepted practice. It is not unreasonable to consider a combined footing as a slab. The reader may make this comparison between the combined footing and the same footing analyzed as a mat in Chap. 7.

The following steps constitute proportioning and finding the required steel area as done in the computer program (included in this chapter).

Referring to Fig. 3-10, the assumption is one of uniform soil pressure beneath the footing. If one can utilize the distance AK shown, then either B or L may be specified and the resultant of column loads can be made to coincide with the center of the area. By making the assumption of uniform soil pressure the load resultant *must coincide with the center of area.*

Some engineers have proposed two approaches to converting to USD from working loads: (1) convert the loads and soil pressure to "ultimate" values and proceed

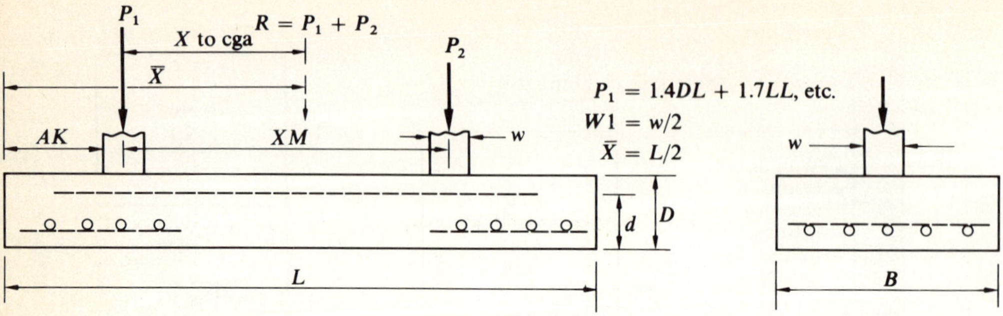

FIGURE 3-10
Conventional analysis of combined footing. The load resultant R coincides with the center of area (cga) for a uniform soil distribution beneath the footing. Note location of both positive and negative reinforcing steel.

or (2) use the working-load values then multiply the resulting shear and moment values by a ratio

$$UR = \frac{\sum P_{\text{ult}}}{\sum P_{\text{work}}}$$

The author recommends using method (1) to ensure a better computed closure of the moment diagram. It is found that method (2) will result in a small eccentricity between the resultant location and the center of area. The difference is of no practical significance, but is annoying in trying to obtain a moment-diagram closure as an arithmetical check.

The first step is to find the footing dimensions. It will be necessary as a part of the early computations to obtain the "ultimate" soil pressure as

$$q_{\text{ult}} = q_a(UR)$$

$$RX = P_2(XM)$$

from which X is found. Now $X + W1 + AK = L/2$ by inspection of Fig. 3-10. If the footing width B is not given (and note that the end distance AK may be zero),

$$L = 2(X + W1 + AK)$$

$$B = \frac{R}{Lq_{\text{ult}}}$$

If B is given (AK must not be limited),

$$L = \frac{R}{Bq_{\text{ult}}}$$

and one must compute AK.

Once footing dimensions are established, the total soil pressure per linear foot of footing (total load per foot, TLPF) is

$$\text{TLPF} = Bq_{\text{ult}}$$

as shown in Fig. 3-11.

Now the shear and moment diagrams can be constructed using conventional mechanics-of-materials methods. This is a tedious operation and ideally suited for the computer program, which finds the moment at each 1 ft (0.3 m) along the member, the maximum value at zero shear, and the values at the faces of the columns. Shears are likewise found at the critical locations of column faces, and the point of zero shear is located.

Footing depth is computed considering both wide-beam and diagonal-tension possibilities. Note that this can be a formidable endeavor since Fig. 3-11 indicates four possible cases of wide-beam shear (all but one can be eliminated by visual inspection and with only minor difficulty on the computer) and four cases (two at each end) of diagonal tension. These four cases depend on the value of C and AK producing a perimeter, which may be three- or four-sided. Remember that wide-beam allowable stress is $2\phi\sqrt{f_c'}$ and diagonal tension is $4\phi\sqrt{f_c'}$.

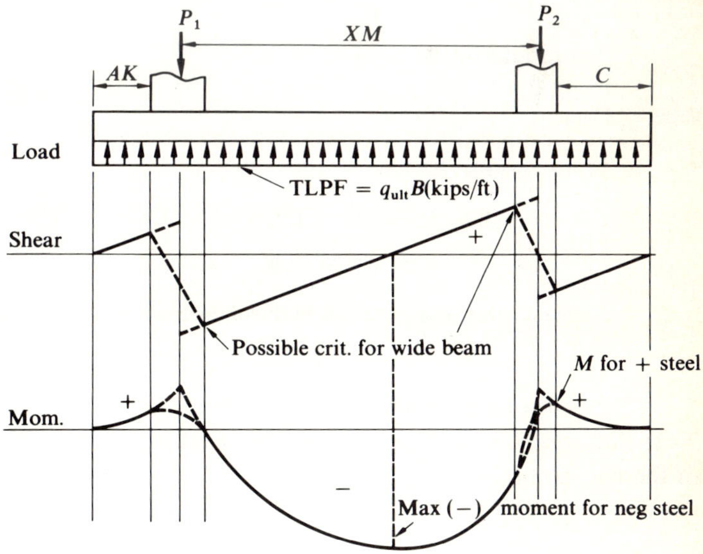

FIGURE 3-11
Soil-pressure assumptions and qualitative shear and moment diagram. Note that critical design locations are the same whether column loads are treated as distributed or point loads.

After obtaining the depth to satisfy shear without using shear reinforcement, the next operation is to determine longitudinal steel requirements both positive (or bottom steel) at the ends and negative (or top steel) between columns.

Steel for bending in the transverse direction must also be provided. Column bearing stresses and dowels are checked and provided in the same manner as for spread footings.

EXAMPLE 3-3 Design a combined rectangular footing for the conditions and data shown in Fig. E3-3.1. Use the computer output of Figs. E3-3.3 and E3-3.4 to check and/or obtain remaining design information.

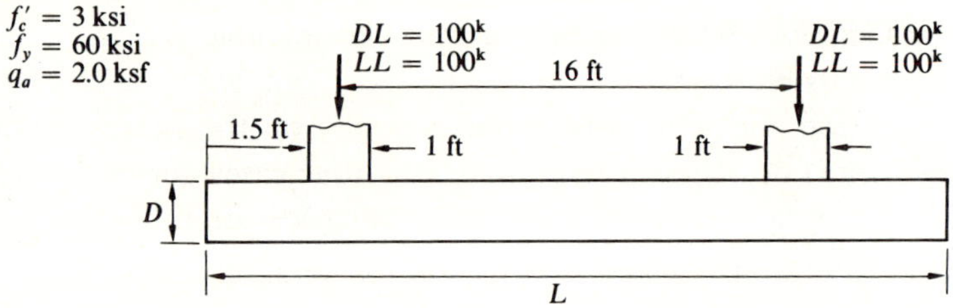

FIGURE E3-3.1

The computer input is as follows:

Card	Data
1	TITLE (see Fig. E3-3.3)
	UT1–UT7
2	FT IN KIPS FT-KIPS KIPS/SQ FT KIPS/LIN FT LB/SQ IN
	FU1–FU8
3	12. .144 2.00 4.00 87000. 200. .001 1.0
4	.50 .50 100. 100. 100. 100. 16.0 2.0
	(Column half widths, DL, LL both columns, column spacing and allowable soil pressure)
5	3000. 60000. 1.50 0.0

These five cards represent input data. Output is on Figs. E3-3.3 and E3-3.4. Note that card UT1-UT7 and card FU1-FU8 are also used in the trapezoid footing of the next section.

SOLUTION (partial): Find "ultimate" values for loads and soil pressure

$$P_{ult} = 140 + 170 = 310 \text{ kips} \qquad \text{and} \qquad UR = \frac{310}{200} = 1.55$$

$$q_{ult} = 2 \times 1.55 = 3.10 \text{ ksf}$$

Summing moments about column 1, we find the location of the load resultant and center of footing area as

$$X = \frac{16(310)}{620} = 8.00 \text{ ft}$$

Find the footing length L since it is now fixed:

$$L = 2(8 + 0.5 + 1.5) = 20.00 \text{ ft}$$

Find B:

$$B = \frac{\sum P_{\text{ult}}}{Lq_{\text{ult}}} = \frac{620}{20(3.10)} = 10.00 \text{ ft}$$

$$\text{TLPF} = 3.1 \times 10 = 31.00 \text{ kips/lin ft}$$

Based on the computed dimensions and using conventional methods the shear and moment diagrams (author uses the computer printout) are completed as shown on Fig. E3-3.4.

Wide-beam shear from the shear diagram by inspection of values of $232.5 - 31d$ (kips) is used to find the effective footing depth as 16.901 in. Diagonal-tension shear similar to that for simple spread footings is investigated for two cases.

Case 1 four-sided perimeter.

Case 2 three-sided perimeter. It should be evident that depending on the value of AK, the punch-out zone could be three-sided.

Check the computer output for case 1 (either column from symmetry) and refer to Fig. E3-3.2.

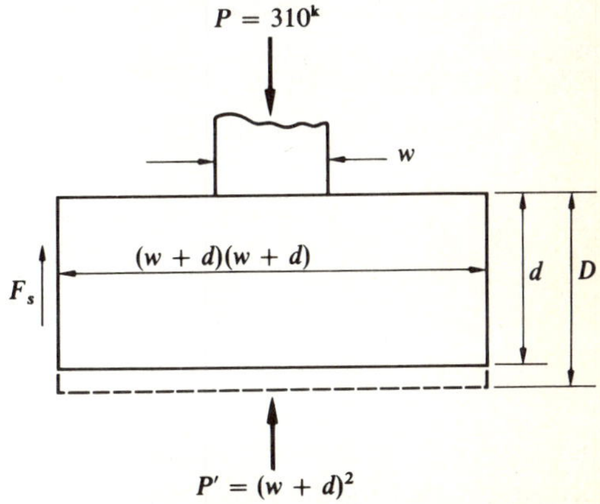

FIGURE E3-3.2

The perimeter shear resistance is

$$4(w + d)v_c d = F_s$$

The upward soil pressure is

$$(w + d)^2 q_{\text{ult}} = P' \qquad \text{case 1}$$

Summing vertical forces gives

$$F_s + P' - P_{\text{ult}} = 0$$

By substitution we obtain

$$110.34d^2 + 113.44d = 306.9$$

$$d = 1.231 \text{ ft} = 14.77 \text{ in} \qquad \text{checks computer output}$$

Checking a case 2 possibility is left as an exercise for the reader.

The computation for area of steel is similar to that for a spread footing except that here the computed moment is for the entire footing width of 10 ft; thus

$$a = \frac{A_s f_y}{0.85 f'_c(120)}$$

and the resulting computation for *top* or negative A_s is

$$A_s = 13.246 \text{ sq in per 10-ft width}$$

The designer may select a series of bars to satisfy this requirement, e.g., seventeen no. 8 bars at 7 in center to center:

$$A_s \text{ furnished} = 17(0.785) = 13.35 \text{ sq in} > 13.25$$

The bars may be run all the way or terminated at the exterior column faces, as shown.

Positive steel in the two cantilevered end zones is based on a minimum of $200/f_y = 0.0033$. This gives

$$A_s = 0.0033(120)(17) = 6.76 \text{ sq in}$$

which is greater than the approximately 5.04 sq in shown in Fig. E3-3.3 as required to satisfy the bending moment.

```
J E BOWLES   EXAMPLE 3-3   COMBINED FOOTING DESIGN USING USD

                      FOOTING DESIGN INPUT DATA IS AS FOLLOWS:
COL NO    1/2 WIDTH COL, FT    DL, KIPS   L. LOAD, KIPS
 1          0.500               100.0       100.0
 2          0.500               100.0       100.0
     DIST BETWEEN COLS =16.000 FT
     DIST END FTG TO LT FACE COL 1 = 1.50 FT
     FOOTING WIDTH, B =  0.0  FT
          THE ALLOWABLE SOIL PRESSURE =  2.00 K/SQ FT

COLUMN TOTAL LOADS ARE:   COL 1 = 200.0 KIPS   COL 2 = 200.0 KIPS
CONCRETE AND STEEL STRESSES:   F1C =  3000. LB/SQ IN   FY = 60000. LB/SQ IN

     PULT COL 1 = 310.000 KIPS
     PULT COL 2 = 310.000 KIPS
     ULT LOAD/FT OF FTG =  31.000 K/LIN FT
COMPUTED FOOT DIMENSIONS:
     WIDTH =10.000 FT
    LENGTH =20.000 FT
   THE L/B RATIO =2.000

THE SHEAR AT LT FACE COL 1 =      46.50 KIPS
THE SHEAR AT RT FACE COL 1 =    -232.50 KIPS
THE SHEAR AT LT FACE COL 2 =     232.50 KIPS
THE SHEAR AT RT FACE COL 2 =     -46.50 KIPS
THE MOMENT AT LT FACE COL 1 =      34.87 FT-K
THE MOMENT AT RT FACE COL 1 =     -58.12 FT-K
THE MOMENT AT LT FACE COL 2 =     -58.13 FT-K
THE MOMENT AT RT FACE COL 2 =      34.87 FT-K
THE MAX MOMENT =     -930.00 FT-K AT A DIST = 10.000 FROM COL 1 END

MAX SHEAR USED FOR WIDE BEAM =  232.500 KIPS
  DEPTH OF CONCRETE FOR WIDE BEAM = 16.901 IN

     DEPTH OF CONCRETE FOR CASE 1 @ COL 1 = 14.771 IN

     DEPTH OF CONCRETE FOR CASE 1 @ COL 2 = 14.771 IN

     DEPTH OF CONCRETE FOR CASE 2 @ COL 1 = 15.127 IN

     DEPTH OF CONCRETE FOR CASE 2 @ COL 2 = 15.127 IN

     ***** DEPTH OF CONCRETE USED FOR DESIGN = 16.901 IN

*** AS = TOTAL STEEL AREA FOR FTG WIDTH OF B
DISTANCE
FROM END          SHEAR            MOMENT,FT-K        AS,SQ-IN

   0.0             0.0                0.0             0.000
   1.00           31.0000            15.5000          0.204
   2.00          -92.9999            23.2500          0.306
   3.00        -216.9998           -170.4999          2.272
   4.00        -185.9999           -371.9998          5.039
   5.00        -154.9999           -542.4998          7.456
   6.00        -123.9999           -681.9998          9.490
   7.00         -93.0000           -790.4998         11.110
   8.00         -62.0000           -867.9995         12.289
   9.00         -31.0002           -914.4998         13.006
  10.00          -0.0002           -929.9998         13.246
  11.00          30.9998           -914.4998         13.006
  12.00          61.9998           -867.9998         12.289
  13.00          92.9998           -790.5000         11.110
  14.00         123.9998           -682.0002          9.490
  15.00         154.9998           -542.5005          7.456
  16.00         185.9998           -372.0000          5.039
  17.00         216.9995           -170.5039          2.272
  18.00          92.9996             23.2461          0.306
  19.00         -31.0002             15.5002          0.204
  20.00          -0.0002              0.0005          0.000

MAX % STEEL =0.016 %      MAX STEEL AREA =   32.520 SQ IN
MIN % STEEL =0.003 %      MIN STEEL AREA =    6.760 SQ IN
```

FIGURE E3-3.3
Computer output for Example 3-3.

Compute transverse steel requirements. Note that the computer program does not perform this step. Take an equivalent (see Chap. 7) beam[1] of

$$w + 3d = 1.00 + 4.23 = 5.23 \text{ ft} \qquad \text{say } 5.0 \text{ ft}$$

$$q'_{\text{ult}} = \frac{310}{50} = 6.20 \text{ ksf}$$

$$M'_u = \frac{6.2}{2}(4.5)^2 = 62.8 \text{ ft-kips} = 753.6 \text{ in-kips}$$

Take an effective depth d of approximately $d - 1$ in to allow for bar diameters and place this steel on top of the longitudinal bars:

$$d' = d - 1 \text{ in} = 17 - 1 = 16 \text{ in}$$

Substituting into Eq. (3-2) and rearranging, we have

$$A_s(16 - 0.98A_s) = \frac{753.6}{0.9(60)}$$

$$0.98A_s - 16A_s = -13.96$$

$$A_s = 8.17 - 7.25 = 0.92 \text{ sq in/ft}$$

$$A_{s,\text{total}} = 5(0.92) = 4.60 \text{ sq in}$$

Use eight no. 7 bars at 7.5 in center to center. The steel area furnished is

$$A_s = 8(0.60) = 4.8 \text{ sq in} > 4.60$$

For remainder of footing use minimum for flexure (ACI, art. 10.5.1)

$$A_s = 0.0033(17)(12) = 0.68 \text{ sq in/ft}$$

$$\text{Total} = 0.68 \times 10 = 6.8 \text{ sq in}$$

Use twelve no. 7 bars at 10 in center to center

$$A_s \text{ furnished} = 7.20 \text{ sq in} > 6.80 \text{ sq in} \qquad \text{O.K.}$$

Figures E3-3.3 and E3-3.4 illustrate typical computer output and final design sketch. Note that no design has been made of column dowels to attach the columns to footing. This part of the design is omitted because it is identical to that of spread footings.

////

[1] Keep in mind that by making this zone "stiffer" it will tend to "attract" moment into a narrower width.

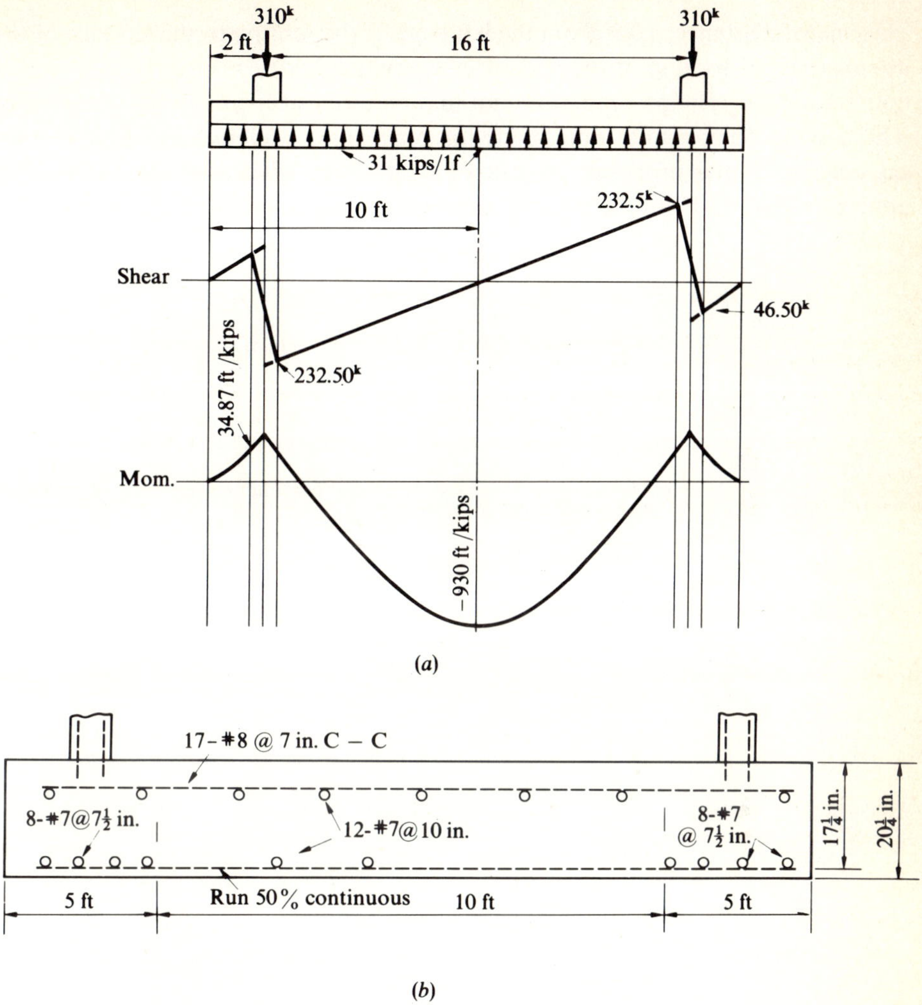

FIGURE E3-3.4
(*a*) Shear and moment diagrams and (*b*) final-design sketch (not to scale). Generally run negative steel and 50 percent positive steel all the way unless end overhangs are large.

3-8 COMPUTER PROGRAM TO DESIGN COMBINED FOOTING (CONVENTIONAL)

This program will design a continuous rectangular footing with two square columns loaded with axial loads (no moments) at any location based on USD and ACI 318-71. Column 1 is the left column, and footing orientation should be such as to achieve this placement. Distance *AK* is from the left edge of the footing to the left face of the first column. Final footing dimensions are symmetrical with respect to the resultant of column widths. The program does not find required steel areas in the short direction. The designer may round dimensions for practical considerations. Metric units can be used with this program via use of cards UT1-UT7 and FU1-FU8. The fps equivalents are in Example 3-3, and the metric entries are listed in Example 3-4.

Line	Operation
3	READ TITLE AND UT1–UT7 (two cards)
7	READ FU1–FU8 (see Examples 3-3 and 3-4 for numerical values)
8	READ W1, W2 = half widths of columns 1 and 2, ft; DL1, XLL1, DL2, XLL2 = column dead and line loads, kips; XM = center to center column spacing, ft; QALL = allowable soil pressure, ksf (for metric problems use meter and kN units)
9	READ F1C, FY = f'_c, f_y in psi; AK = distance from left end to left face of column 1, ft; B = footing width if controlling. If B > 0, then AK = 0
17–29	Finds footing dimensions
44–53	Computes critical shear and moment values at column faces
61–83	Finds footing depth to satisfy wide beam and two cases of diagonal tension at each end
86–120	Computes shear and moments at 1-ft (or 0.3-m) intervals and corresponding reinforcing steel required

```
           C       J E BOWLES  COMBINED FOOTING BY USD (ACI 318-71) DESIGN--LIMIT = 2
           C       W1,W2 = COL 1/2 WIDTHS, AK ALWAYS LEFT END
0001               DIMENSION TITLE(20)
0002               DOUBLE PRECISION UT5,UT6,UT7,UT8
0003          6000 READ(1,1000,END=150)TITLE,UT1,UT2,UT3,UT4,UT5,UT6,UT7,UT8
0004          1000 FORMAT(20A4/4(A4,6X),4(A8,2X))
0005               WRITE(3,1001)TITLE
0006          1001 FORMAT('1',//,T5,20A4,//)
0007               READ(1,5)FU1,FU2,FU3,FU4,FU5,FU6,FU7,FU8
0008               READ(1, 5)W1,W2,DL1,XLL1,DL2,XLL2,XM,QALL
0009               READ(1, 5) F1C, FY, AK, B
0010             5 FORMAT(8F10.4)
0011               P1=DL1+XLL1
0012               P2=DL2+XLL2
0013               WRITE(3,500)UT1,UT3,UT3,W1,DL1,XLL1,W2,DL2,XLL2,XM,UT1,AK,UT1,
                  1 B,UT1,QALL,UT5
0014           500 FORMAT(T25,'FOOTING DESIGN INPUT DATA IS AS FOLLOWS:'/T5,'COL NO',
                  13X,'1/2 WIDTH COL, ',A2,3X,'DL, ',A4,3X,'L. LOAD, ',A4,/ T6,'1',
                  2T15,F5.3,T35,F6.1,T46,F6.1/T6,'2',T15,F5.3,T35,F6.1,T46,F6.1/
                  3T10,'DIST BETWEEN COLS =',F6.3,1X,A2/T10,'DIST END FTG TO LT FACE
                  4COL 1 =',F5.2,1X,A2/T10,'FOOTING WIDTH, B =',F6.2,1X,A2/T15,
                  5'THE ALLOWABLE SOIL PRESSURE =',F6.2,1X,A7/)
0015               WRITE(3,501)P1,UT3,P2,UT3,F1C,UT7,FY,UT7
0016           501 FORMAT(T5,'COLUMN TOTAL LOADS ARE:    COL 1 =',F6.1,1X,A4,3X,'COL 2
                  1 =',F6.1,1X,A4/T5,'CONCRETE AND STEEL STRESSES:',3X,'F1C =',
                  2F7.0,1X,A8,3X, 'FY =',F7.0,1X,A8,//)
0017               UP1=1.4*DL1+1.7*XLL1
0018               UP2=1.4*DL2+1.7*XLL2
0019               XBAR = UP2*XM/(UP1+UP2)
0020               SOIRA = (UP1+UP2)/(P1+P2)
0021               QULT = SOIRA*QALL
0022               AKC = 0.
0023               IF(B.LE.0.)GO TO 6
0024               TLEN = (UP1+UP2)/(B*QULT)
0025               AKC = TLEN/2. - XBAR - W1
0026               AK = AKC
0027               GO TO 7
0028             6 TLEN = 2.*(AK + W1 +XBAR)
0029               B = (UP1+UP2)/(TLEN*QULT)
0030             7 RATIO = TLEN/B
0031               TLPF=(UP1+UP2)/TLEN
0032               WRITE(3,503)UP1,UT3,UP2,UT3,TLPF,UT6,B,UT1,TLEN,UT1,RATIO
0033           503 FORMAT(/T10,'PULT COL 1 =',F8.3,1X,A4/T10,'PULT COL 2 =',F8.3,1X,
                  1A4/T10,'ULT LOAD/FT OF FTG =',F8.3,1X,A8      ,/T5,'COMPUTED FOOT
                  2DIMENSIONS:'/T10,'WIDTH =',F6.3,1X,A2/T9,'LENGTH =',F6.3,1X,A2/
                  3T7,'THE L/B RATIO =',F5.3//)
0034               ANO = TLEN/FU8
0035               INO = ANO
0036               IF(INO.EQ.ANO)N = INO+1
0037               IF(INO.NE.ANO)N = INO + 2
0038               A = AK + XM + W1 - W2
```

```
      C=2.0*W1
      D=2.0*W2
      E = AK + XM + W1 + W2
      DP1=UP1/(2.*W1)
      DP2=UP2/(2.*W2)
C     COMPUTE SHEAR AND MOMENT AT CRITICAL LOCATIONS ALONG FOOTING
      V1L = TLPF*AK
      V1R = V1L + (TLPF-DP1)*C
      V2L = TLPF*A-UP1
      V2R = TLPF*E-UP1-UP2
      DIST = AK + C - V1R/TLPF
      XMAXM = .5*TLPF*(DIST**2) - UP1*(DIST-(AK+W1))
      EMOL1 = .5*TLPF*AK**2
      EMOR1 = .5*TLPF*(AK+C)**2 - UP1*W1
      EMOL2 = .5*TLPF*A**2 - UP1*(XM-W2)
      EMOR2 = .5*TLPF*(TLEN-E)**2
      WRITE(3,505)V1L,UT3,V1R,UT3,V2L,UT3,V2R,UT3,EMOL1,UT4,EMOR1,UT4,
     1 EMOL2,UT4,EMOR2,UT4,XMAXM,UT4,DIST
  505 FORMAT(//T5,'THE SHEAR AT LT FACE COL 1 =',F10.2,1X,A4/T5,'THE SHE
     1AR AT RT FACE COL 1 =',F10.2,1X,A4/T5,'THE SHEAR AT LT FACE COL 2
     3=',F10.2,1X,A4/T5,'THE SHEAR AT RT FACE COL 2 =',F10.2,1X,A4/
     3T5,'THE MOMENT AT LT FACE COL 1 =',F10.2,1X,A4/T5,'THE MOMENT AT R
     4T FACE COL 1 =',F10.2,1X,A4/T5,'THE MOMENT AT LT FACE COL 2 =',F10
     5.2,1X,A4/T5,'THE MOMENT AT RT FACE COL 2 =',F10.2,1X,A4/T5,'THE MA
     6X MOMENT =',F10.2,1X,A4,' AT A DIST =',F7.3,' FROM COL 1 END'//)
C     COMPUTE MAX AND MIN STEEL REQUIREMENTS
C     NOTE CONCRETE LIMITED TO F1C NOT MORE THAN 4000 PSI DUE TO PB
      PB = ((.85**2)*F1C/FY)*(FU5/(FU5 + FY))*.75
      SMIN = FU6/FY
      FYS = FY*FU7
      F1CP = F1C*FU7
      GG = 0.5*FYS/(0.85*F1CP*B*FU1)
C     WIDE BEAM SHEAR = WBS---DIAGONAL TENSION SHEAR = PUNS
  102 PUNS = FU4*.85*SQRT(F1C)
      WBS = FU3*.85*SQRT(F1C)
      VMAX = AMAX1(ABS(V1L),ABS(V1R),ABS(V2L),ABS(V2R))
  104 VCO = WBS*B*FU2
      DWB = FU1*VMAX/(TLPF + VCO)
C     DEPTH FOR 2-CASES:  CASE 1 = PERIM OF 4 SIDES; CASE 2 PERIM HAS 3
  106 DDB = 2.*B
      AA = PUNS*FU2 + QULT/4.
      BB1 = (PUNS*FU2 + QULT/2.)*C
      CC1 = -(UP1 - QULT*C**2)/4.
      DPS1 =(.5*(-BB1 + SQRT (BB1**2 - 4.*AA*CC1))/AA)*FU1
      BB2 = (PUNS*FU2 + QULT/2.)*D
      CC2 = -(UP2 - QULT*D**2)/4.
      DPS2 =(.5*(-BB2 + SQRT (BB2**2 - 4.*AA*CC2))/AA)*FU1
C      *** CASE 2 COMPUTATIONS--3-SIDED ZONE
  108 CV = PUNS*FU2/QULT
      AAA = 2.*CV + .5
      BB3 = 2.*AK*CV + 3.*C*CV + AK + 1.5*C
      CC3 = - UP1/QULT + AK*C + C**2
      DPS3 =(.5*(-BB3 + SQRT (BB3**2 - 4.*AAA*CC3))/AAA)*FU1
  110 EXT = TLEN-E
      BB4 = 2.*EXT*CV + 3.*D*CV + EXT + 1.5*D
      CC4 = - UP2/QULT + EXT*D + D**2
      DPS4 =(.5*(-BB4 + SQRT (BB4**2 - 4.*AAA*CC4))/AAA)*FU1
      DC = AMAX1(DWB,DPS1,DPS2,DPS3,DPS4)
      WRITE(3,300)VMAX,UT3,DWB,UT2,DPS1,UT2,DPS2,UT2,DPS3,UT2,DPS4,UT2,
     1 DC,UT2
  300 FORMAT(T5,'MAX SHEAR USED FOR WIDE BEAM =',F9.3,1X,A4/T7,'DEPTH OF
     1 CONCRETE FOR WIDE BEAM =',F7.3,1X,A2//T10,'DEPTH OF CONCRETE FOR
     2CASE 1 @ COL 1 =',F7.3,1X,A2//T10,'DEPTH OF CONCRETE FOR CASE 1 @
     3COL 2 =',F7.3,1X,A2//T10,'DEPTH OF CONCRETE FOR CASE 2 @ COL 1 =',
     4F7.3,1X,A2//T10,'DEPTH OF CONCRETE FOR CASE 2 @ COL 2 =',F7.3,1X,A
     52//T10,'***** DEPTH OF CONCRETE USED FOR DESIGN =',F7.3,1X,A2//)
C     COMPUTE MOMENTS @ 1-FT OR .3 METER INCREMENTS--FIND REQ'D STEEL AR
      WRITE(3,44) UT4,UT2
   44 FORMAT(T5,'*** AS = TOTAL STEEL AREA FOR FTG WIDTH OF B'/T5,'DISTA
     1NCE'/T5,'FROM END',T21,'SHEAR',T34,'MOMENT,',A4,T51,'AS,SQ-',A2/)
      X = 0.
      DO 100  I = 1,N
      IF(I.EQ.N)X = TLEN
      IF(AK)9,9,8
    8 IF(X-AK)10,10,9
   10 V = TLPF*X
      TM = V*X/2.
      GO TO 60
    9 ADIST = C + AK
      IF(X-ADIST)12,12,20
   12 V = TLPF*AK+(TLPF-DP1)*(X-AK)
      TM = TLPF*(X**2)/2. - DP1*((X-AK)**2)/2.
      GO TO 60
   20 IF(X-A)25,25,30
   25 V = TLPF*X - UP1
      TM = .5*TLPF*X**2 - UP1*(X-W1-AK)
      GO TO 60
   30 IF(X-E)35,35,40
   35 V = TLPF*X - UP1 -DP2*(X-A)
      TM = .5*TLPF*X**2 - UP1*(X-W1-AK) - .5*DP2*(X-A)**2
      GO TO 60
   40 V = TLPF*X - UP1 - UP2
      TM = .5*TLPF*X**2 - UP1*(X-W1-AK) - UP2*(X-E+W2)
   60 ULM = TM
      IF(TM.LT..00001)ULM = -TM
      F1 = ULM*FU1/(.9*FYS)
```

```
          AS = (DC-SQRT(DC**2-4.*F1*GG))/(2.*GG)
          WRITE(3,50)X,V,TM,AS
      100 X = X+FU8
       50 FORMAT(T8,F5.2, T18, F9.4, T36,F10.4, T51, F7.3)
          ASMAX = PB*B*FU1*DC
          ASMIN = SMIN*B*FU1*DC
          WRITE(3,55)PB,ASMAX,UT2,SMIN,ASMIN,UT2
       55 FORMAT(/T5,'MAX % STEEL =',F5.3,' %',T30,'MAX STEEL AREA =',F8.3,
         1' SQ 'A2/T5,'MIN % STEEL =',F5.3,' %',T30,'MIN STEEL AREA =',F8.3,
         2' SQ ',A2)
          GO TO 6000
      150 STOP
          END
```

3-9 TRAPEZOID FOOTING DESIGN

When it is necessary to use a combined footing and the distance AK of Sec. 3-7 is limited or the column loads are unequal, the resultant of the column loads may fall at a distance less than $L/2$ from one end of the footing (Fig. 3-12). If the distance $\bar{X}$ is

$$\frac{L}{2} > \bar{X} < \frac{L}{3}$$

one may use a trapezoid-shaped footing to obtain the assumed condition of load resultant coincident with the center of the area.[1] As with the rectangular combined footing of Sec. 3-7, the assumption is made that the footing is rigid and soil pressure uniform.

Equations can be derived (see any text on plane geometry) for a trapezoid to give

$$\bar{X} = \frac{L}{3}\frac{2m + n}{m + n} \qquad \text{locate center of gravity of area} \tag{3-8}$$

$$A = (m + n)\frac{L}{2} \qquad \text{area of trapezoid} \tag{3-9}$$

The proportioning and design of a trapezoidal footing follow.

1 Convert loads to ultimate and find the ultimate load ratio (UR of Sec. 3-7) and the ultimate soil pressure

$$q_{\text{ult}} = q_a UR$$

2 Locate the load resultant on the footing to obtain $\bar{X}$.

3 Solve Eqs. (3-8) and (3-9) simultaneously for m and n.

[1] It should be evident that if $X = L/2$, the solution of Sec. 3-7 is obtained; if $X = L/3$, the resulting footing is triangular.

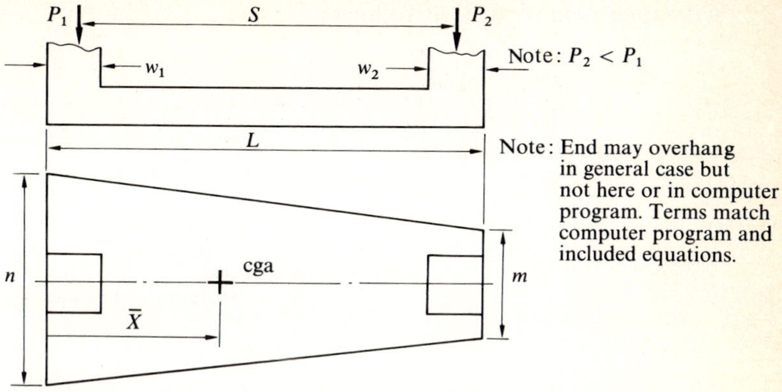

FIGURE 3-12
Trapezoid footing when one column load is larger than the other and L is limited.

4 Find the load per foot of footing at each end as

$$Q_{\text{big end}} = nq_{\text{ult}}$$

$$Q_{\text{small end}} = mq_{\text{ult}}$$

The load per foot of footing is linear from one end to the other.

5 Draw shear and moment diagrams noting

Shear = second-degree curve

Moment = third-degree curve

6 Find depth for wide-beam and diagonal tension.

7 Find steel for bending considering the footing width variable.

EXAMPLE 3-4 Make a partial design using metric units of a trapezoidal footing for the conditions shown in Fig. E3-4.1. Compare and use computer output of Fig. E3-4.4.

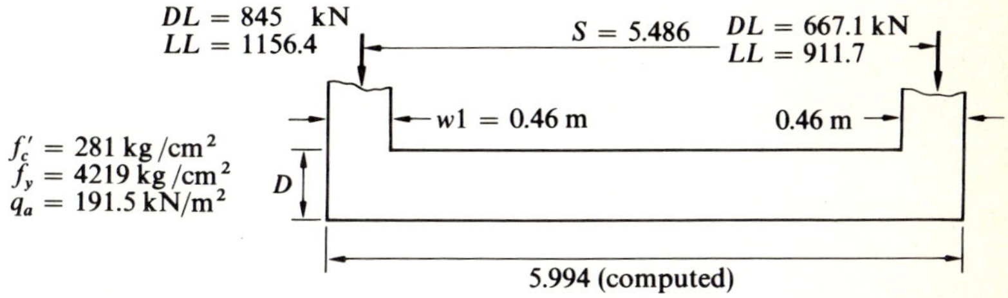

FIGURE E3-4.1

Computer data input is as follows:

Card	Data
1	TITLE (see Fig. E3-4.4)
	UT1 – UT7
2	M CM KN KN–M KN/SQ M KN/LN M KG/S QCM
	FU1 – FU8
3	100. 98.07 .530 1.06 6117. 14.06 .009807 .30
4	845. 1156.4 667.1 911.7 281. 4219. 191.51
	(loads, material properties)
5	.4572 .4572 5.4864 (column widths and spacing)

These five cards represent the input. The output is shown in Fig. E3-4.4.

SOLUTION The ultimate soil pressure is found as follows:

$$P_{\text{work}} = 2{,}001.4 + 1{,}578.8 = 3{,}580.2 \text{ kN}$$

$$P_{\text{ult}} = 3{,}148.88 + 2{,}483.83 = 5{,}632.71 \text{ kN}$$

$$UR = \frac{5{,}632.7}{3{,}580.2} = 1.573 \qquad q_{\text{ult}} = 191.51(1.573) = 301.25 \text{ kN/sq m}$$

$$\text{Required footing area} = \frac{5{,}632.7}{301.25} = 18.69 \text{ sq m}$$

$$(m + n)\frac{L}{2} = 18.69$$

$$m + n = \frac{18.69}{2.972} = 6.289 \text{ m} \tag{a}$$

Also the location of the center of area (cga) is

$$X = \frac{2{,}483.8(5.486)}{5{,}632.7} = 2.42 \text{ m} \qquad \text{and} \qquad \bar{X} = 2.42 + 0.23 = 2.65 \text{ m}$$

From Eq. (3-8)

$$\frac{2m + n}{m + n}\frac{L}{3} = \bar{X}$$

or

$$\frac{2m + n}{m + n} = 1.338 \tag{b}$$

Solving Eqs. (*a*) and (*b*) for m and n gives

$$m = 2.12 \qquad \text{and} \qquad n = 4.17 \text{ m}$$

The soil-pressure distribution is

$$Q_{\text{be}} = 4.17(301.3) = 1{,}257 \text{ kN/lin m}$$

$$Q_{\text{se}} = 2.12(301.3) = 638.0 \text{ kN/lin m}$$

from which the shear and moment diagrams of Fig. E3-4.2 can be constructed.

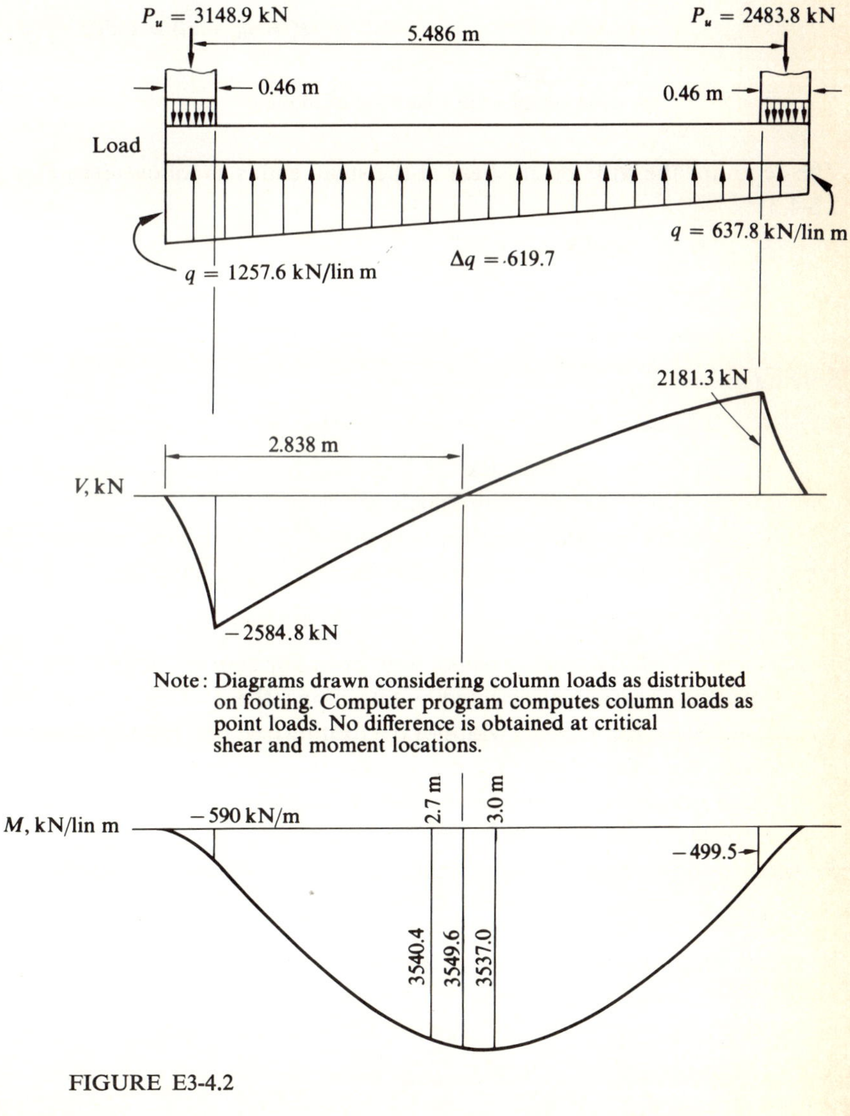

FIGURE E3-4.2

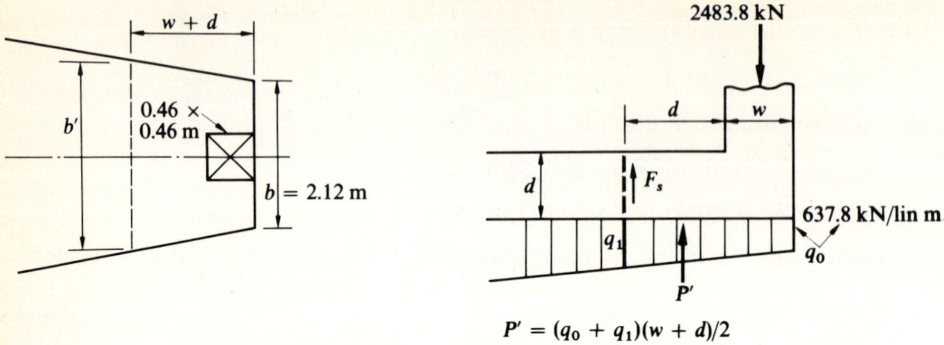

FIGURE E3-4.3
Wide-beam shear at small end of footing.

The depth for the wide-beam shear at the small end is as follows (see Figs. E3-4.2 and E3-4.3):

$$F_s + P' - 2{,}483.83 = 0$$

$$F_s = 1{,}687.9d + 255.5d^2$$

$$P' = 304.23 + 684.91d + 51.17d^2$$

Solving gives

$$d = 0.83 \text{ m} = 83 \text{ cm} \qquad \text{checks} \qquad (D \geq 83 + 9 \cong 93 \text{ cm})$$

It is also necessary to check the large end for wide-beam shear since it is usually not possible to tell by inspection which end is critical.

A check for diagonal tension is required, but it is only necessary to check the large end. It should be evident that the footing width dimension m should be

$$m > \text{column width} + d \qquad (2.127\ (.46 + .83))$$

so that a modified diagonal-tension zone does not control at small end.

The area of steel at selected points along the footing is computed with due consideration for the variable width of the footing (see Fig. E3-4.3) and using minimum percent as $200/f_y$ (or $14.06/f_y$).

Since no new concepts are needed for the selection for steel bars, this exercise will be left for the reader. The same procedure used for the rectangular footing may be used for transverse-steel computations, i.e., two equivalent beams of width

Column width + 3*d*

Since the footing width is varying, use the *average* footing width and soil pressure in the zone 3*d* for bending moment.

Checking for dowels (ACI, art. 15.62), since the distance *X* of Fig. 3-3 is zero,

$$f_c = 0.85(0.70)f_c' = 167.2 \text{ kg/sq cm}$$

```
J E BOWLES  EXAMPLE 3-4  TRAPEZOID FOOTING  METRIC UNITS

GENERAL INPUT DATA IS AS FOLLOWS

     COL 1                              COL 2
     DEAD LOAD =  845.00 KN             DEAD LOAD =  667.10 KN
     LIVE LOAD = 1156.40 KN             LIVE LOAD =  911.70 KN
     COL WIDTH = 0.46 M                 COL WIDTH = 0.46 M
       COL SPACING CEN TO CEN = 5.486 M
          28-DAY CONCRETE STRESS =  281.0 KG/SQ CM
          YIELD STRESS OF STEEL =  4219.0 KG/SQ CM
          THE ALLOWABLE SOIL PRESSURE = 191.51 KN/SQ M

THE COMPUTED QUANTITIES ARE AS FOLLOWS
TOTAL LENGTH =  5.94 M          TOTAL FTG AREA =  18.69 SQ-M
XBAR  =  2.42 M           DIST TO C.G. OF FTG AREA =  2.65 M
WIDTH @ BIG END =  4.17 M       WIDTH @ SMALL END =  2.12 M


ULT LOAD COL 1 =3148.88 KN         ULT LOAD COL 2 =2483.83 KN
THE ULTIMATE LOAD RATIO =1.57


ULT SOIL PRESSURE = 301.301 KN/SQ M  SOIL PRESS BIG END =1257.552 KN/LIN M
                                     SOIL PRESS SMALL END = 637.833 KN/LIN M
                         DIFF OF END PRESS, DELQ =619.719


THE SHEAR LEFT COL 1 =  284.75 KN
THE SHEAR RIGHT COL 1 =-2864.13 KN
THE SHEAR FACE COL 1 =-2584.82 KN
THE SHEAR LEFT COL 2 = 2335.29 KN
THE SHEAR RIGHT COL 2 = -148.54 KN
THE SHEAR FACE COL 2 = 2181.31 KN
THE MOM AT COL 1 =   32.65 KN-M
THE MOM AT FACE COL 1 = -590.06 KN-M
THE MOM AT COL 2 =    16.86 KN-M
THE MOM AT FACE COL 2 = -499.49 KN-M

THE SHEAR IS ZERO AT   2.838 M  FROM BIG END
 THE COMPUTED SHEAR AT THIS POINT IS  0.0  KN
THE MAX MOMENT @ ZERO SHEAR LOCATION = -3549.59 KN-M

DEPTH OF CONC FOR WIDE BEAM SHEAR SMALL END = 83.047 CM
DEPTH OF CONC FOR WIDE BEAM SHEAR LARGE END = 64.886 CM
DEPTH OF CONC FOR PUNCH SHEAR AT LARGE END = 69.905 CM
***** DEPTH OF CONC USED FOR DESIGN = 83.047 CM

MAX % STEEL =0.021 %      THE MAX ALLOW STEEL AREA =177.380 SQ CM/M
 MIN % STEEL =0.003 %      THE MIN. ALLOW STEEL AREA = 27.676 SQ CM/M
```

DIST. M	SHEAR, KN	MOMENT, KN-M	FOOT. WIDTH, M	AS, SQ CM/M
0.0	0.0	0.0	4.174	0.000
0.300	-2776.306	168.709	4.070	1.343
0.600	-2413.116	946.888	3.966	7.785
0.900	-2059.311	1617.517	3.862	13.743
1.200	-1714.889	2183.413	3.758	19.176
1.500	-1379.852	2647.386	3.655	24.039
1.800	-1054.199	3012.258	3.551	28.282
2.100	-737.930	3280.841	3.447	31.857
2.400	-431.045	3455.952	3.343	34.708
2.700	-133.544	3540.410	3.239	36.780
3.000	154.573	3537.020	3.136	38.013
3.300	433.306	3448.605	3.032	38.346
3.600	702.652	3277.973	2.928	37.715
3.900	962.617	3027.953	2.824	36.052
4.200	1213.199	2701.348	2.720	33.290
4.500	1454.398	2300.973	2.616	29.353
4.800	1686.207	1829.652	2.513	24.167
5.100	1908.637	1290.188	2.409	17.651
5.400	2121.684	685.402	2.305	9.716
5.700	2325.348	18.121	2.201	0.266
5.944	-0.005	0.012	2.117	0.000

FIGURE E3-4.4
Computer output for trapezoid footing using metric units.

and the maximum allowable column load is

$$P_b = \frac{167.2(46)^2(9.807)}{1{,}000} = 3{,}470 > 3{,}149 \text{ kN} \qquad \text{ultimate load column 1}$$

Therefore, dowels are not required. To satisfy ACI, art. 15.6.5, use four bars of A_s at least

$$A_s = 0.005(2{,}116) = 10.58 \text{ sq cm} \quad \text{say four no. 6 bars at } 4(2.84) = 11.36 \text{ sq cm}$$

$$L_d = \frac{0.0755 f_y D}{\sqrt{f_c'}}$$

$$= \frac{0.755(4{,}219)(1.905)}{\sqrt{281}} = 36.1 \text{ cm} > 20.3 \text{ cm} \qquad \text{O.K.}$$

Check:

$$L_d = 0.00427 f_y D = 0.00427(4{,}219)(1.905) = 34.3 \text{ cm} < 36.1 \text{ cm}$$

Use

$$L_d \approx 40 \text{ cm} \qquad \text{can use up to about 83 cm}$$

////

3-10 COMPUTER PROGRAM FOR TRAPEZOID FOOTING

The included trapezoid-footing-design computer program is based on ACI 318-71 (ultimate-strength design). The program is only for a footing with no overhang and axial loads (no column moments) at either end; i.e., column faces are flush with the ends of the footing. The column with the largest load is column 1. Concrete stresses are limited to 4,500 psi because of p_b. Steel yield stress is limited to 60 ksi. Column loads are treated as point loads acting at the center of columns. The program does not find required steel areas in the short direction. This program will solve metric problems by use of data cards UT1-UT7 and FU1-FU8 (see Examples 3-3 and 3-4 for entries).

Line	Operation
1–2	Bookkeeping
2	READ TITLE, UT1–UT7 (two cards)
7	READ FU1–FU8
8	READ PD1, PL1, PD2, PL2 = columns 1 and 2 dead and live loads; F1C, FY = concrete and steel stresses; QALL = allowable soil pressure; use consistent units
9	READ W1, W2 = column widths; S = column spacing (ft or m)
13–27	Finds ultimate soil pressure and footing dimensions
30–52	Finds shear and moments at 1-ft (or 0.3-m) increments and at critical locations
53–80	Finds depth to satisfy both wide beam and diagonal tension and total area of steel for bending at each increment along footing as specified by FU8 (1 ft and 0.3 m)

```
C       J E BOWLES  PROG TO SOLVE A TRAPEZOID FTG W/O END OVERHANG
        DOUBLE PRECISION UT5,UT6,UT7,UT8
        DIMENSION TITLE(20)
   6000 READ(1,1000,END=150)TITLE,UT1,UT2,UT3,UT4,UT5,UT6,UT7
   1000 FORMAT(20A4/4(A4,6X),4(A8,2X))
        WRITE (3,1001)TITLE
   1001 FORMAT('1',//,T5,20A4,//)
        READ(1,6)FU1,FU2,FU3,FU4,FU5,FU6,FU7,FU8
        READ(1,6)PD1,PL1,PD2,PL2,F1C,FY,QALL
        READ(1,6)W1,W2,S
      6 FORMAT(8F10.3)
        WRITE(3,300)PD1,UT3,PD2,UT3,PL1,UT3,PL2,UT3,W1,UT1,W2,UT1,S,UT1,
       1F1C,UT7,FY,UT7,QALL,UT5
    300 FORMAT(T5,'GENERAL INPUT DATA IS AS FOLLOWS'//T10,'COL 1',T40,'COL
       1 2',/ T10,'DEAD LOAD =',F8.2,1X,A4,T40,'DEAD LOAD =',F8.2,1X,A4/
       2T10,'LIVE LOAD =',F8.2,1X,A4,T40,'LIVE LOAD =',F8.2,1X,A4/ T10,'CO
       3L WIDTH =',F5.2,1X,A2,T40,'COL WIDTH =',F5.2,1X,A2/ T12,'COL SPACI
       4NG CEN TO CEN =',F6.3,1X,A2/T15,'28-DAY CONCRETE STRESS =',F7.1,
       51X,A8,   /,T15,'YIELD STRESS OF STEEL =',F8.1,1X,A8,   /,T15,'THE
       6ALLOWABLE SOIL PRESSURE =',F7.2,1X,A7///)
C       *******  FIND AREA OF FOOTING AND END DIMENSIONS
        TOTL = PD1+PL1+PD2+PL2
        U1 = 1.4*PD1 + 1.7*PL1
        U2 = 1.4*PD2 + 1.7*PL2
        RATIO = (U1 + U2)/TOTL
        QULT = RATIO*QALL
        OTM = S*U2
        XBAR = OTM/(U1+U2)
        X = XBAR + W1/2.
     70 AREA = (U1+U2)/QULT
        ELTOT = W1/2. + S + W2/2.
     71 APB = AREA*2.0/ELTOT
        TAPB = X*3.0*APB/ELTOT
        A = TAPB - APB
        B = APB - A
        D = S+W1/2.
        WRITE(3,301)ELTOT,UT1,AREA,UT1,XBAR,UT1,X,UT1,B,UT1,A,UT1
    301 FORMAT(T5,'THE COMPUTED QUANTITIES ARE AS FOLLOWS'/ T5,'TOTAL LENG
       1TH =',F6.2,1X,A2,T35,'TOTAL FTG AREA =',F7.2,' SQ-',A2/ T5,'XBAR
       2=',F6.2,1X,A2,T30,'DIST TO C.G. OF FTG AREA =',F6.2,1X,A2/ T5,'WID
       3TH @ BIG END =',F6.2,1X,A2,T36,'WIDTH @ SMALL END =',F6.2,1X,A2//)
C       ***FIND ULTIMATE SOIL PRESSURE, SHEAR AND MOMENTS IN FOOTING
     73 Q1 = QULT*B
        Q2 = QULT*A
        DELQ = Q1 - Q2
        G = DELQ/(2.*ELTOT)
        AJ1 = ELTOT/FU8
        J1 = AJ1
        J = J1
        IF(J1.EQ.AJ1)J = J1 + 1
        IF(J1.NE.AJ1)J = J1 + 2
        E = S + W1/2. - W2/2.
        VC1L = Q1*W1/2. - G*(W1/2.)**2
        VC1R = VC1L - U1
        VC2L = Q1*D - (G*D**2) - U1
        VC2R = VC2L - U2
        VCF1 = Q1*W1 - (G*W1**2) - U1
        VCF2 = Q1*E - (G*E**2) - U1
     77 EMC1 = Q1*(W1/2.)**2/2. - (DELQ*(W1/2.)**3/(6.*ELTOT))
        EMC2 = Q1*D**2/2. -(DELQ*D**3/(6.*ELTOT)) - U1*(D-W1/2.)
        EMCF1 = Q1*W1**2/2. - (DELQ*W1**3/(6.*ELTOT)) - U1*W1/2.
        EMCF2 = Q1*E**2/2. -(DELQ*E**3/(6.*ELTOT)) - U1*(E-W1/2.)
        ZX = (Q1 - SQRT (Q1**2 -4.*G*U1))/(2.*G)
        VO = Q1*ZX -(DELQ*ZX**2/(2.*ELTOT)) - U1
        EMMAX = Q1*ZX**2/2. - (DELQ*ZX**3/(6.*ELTOT))- U1*(ZX-W1/2.)
C       ***FIND DEPTH OF FOOTING TO SATISFY WIDE BEAM SHEAR
     78 VC = FU3*FU2*SQRT(F1C)*.85
C       **WIDE BEAM SHEAR DEPTH AT SMALL END
        B3 = (B-A)/ELTOT
        A1 = B3*(VC + QULT/2.)
        B1 = (VC + QULT)*(A + B3*W2)
        C1 = -U2+ A*W2*QULT + (B3*QULT/2.)*W2**2
        DC1 = (-B1 + SQRT (B1**2 - 4.*A1*C1))/(2.*A1)
C    ***WIDE BEAM SHEAR AT  LARGE END
        A4 = -B3*(VC + QULT/2.)
        B4 = (VC + QULT)*(B-B3*W1)
        C4 = -U1 + B*W1*QULT - (B3*QULT/2.)*W1**2
        DC2 = (-B4 + SQRT (B4**2 - 4.*A4*C4))/(2.*A4)
C       ** CHECK PUNCHING SHEAR AT LARGEST COLUMN LOAD (COLUMN 1)
     79 VCP = FU4*FU2*SQRT(F1C)*.85
        AI = VCP/QULT
        A8 = 2.*AI + 0.5
        B8 = 3.*W1*AI + 1.5*W1
        C8 = -U1/QULT + W1**2
     80 DC3 = (-B8+SQRT (B8**2-4.*A8*C8))/(2.*A8)
        DC = AMAX1(DC1, DC2, DC3)
C       **  COMPUTE AREA OF STEEL REQ'D AS SQ IN OR SQ CM PER UNIT WIDTH
        PB = (.85**2*F1C/FY)*(FU5/(FU5+FY))*.75
        ASMAX = PB*DC*FU1*FU1
        SMIN = FU6/FY
        ASMIN = SMIN*DC*FU1*FU1
        FYS = FY*FU7
        F1C = F1C*FU7
        DC = DC*FU1
        GG = .5*FYS/(.85*F1C*FU1)
```

```
      DC1 = DC1*FU1
      DC2 = DC2*FU1
      DC3 = DC3*FU1
      WRITE(3,302)U1,UT3,U2,UT3,RATIO
  302 FORMAT(T5,'ULT LOAD COL 1 ='F7.2,1X,A4,T38,'ULT LOAD COL 2 =',F7.2
     1,1X,A4/ T5,'THE ULTIMATE LOAD RATIO =',F4.2//)
      WRITE(3,303)QULT,UT5,Q1,UT6,Q2,UT6,DELQ
  303 FORMAT(T5,'ULT SOIL PRESSURE =',F8.3,1X,A7,2X,'SOIL PRESS BIG END
     1=',F8.3,1X,A8,/,T36,'SOIL PRESS SMALL END =',F8.3,1X,A8,/,T26,'DIF
     2F OF END PRESS, DELQ =',F7.3,//)
      WRITE(3,306)VC1L,UT3,VC1R,UT3,VCF1,UT3,VC2L,UT3,VC2R,UT3,VCF2,UT3,
     1 EMC1,UT4,EMCF1,UT4,EMC2,UT4,EMCF2,UT4
  306 FORMAT(T5,'THE SHEAR LEFT COL 1 =',F8.2,1X,A4/ T5,'THE SHEAR RIGHT
     1 COL 1 =',F8.2,1X,A4/ T5,'THE SHEAR FACE COL 1 =',F8.2,1X,A4/
     2T5,'THE SHEAR LEFT COL 2 =',F8.2,1X,A4/ T5,'THE SHEAR RIGHT COL 2
     3=',F8.2,1X,A4/ T5,'THE SHEAR FACE COL 2 =',F8.2,1X,A4/ T5,'THE MOM
     4 AT COL 1 =',F8.2,1X,A4/ T5,'THE MOM AT FACE COL 1 =',F8.2,1X,A4/
     5T5,'THE MOM AT COL 2 =',F8.2,1X,A4/ T5,'THE MOM AT FACE COL 2 =',F
     68.2,1X,A4/////)
      WRITE(3,307)ZX,UT1,VO,UT3,EMMAX,UT4
  307 FORMAT(T5,'THE SHEAR IS ZERO AT',F8.3,1X,A2,' FROM BIG END'/T6,'TH
     1E COMPUTED SHEAR AT THIS POINT IS',F6.2,1X,A4/ T5,'THE MAX MOMENT
     2@ ZERO SHEAR LOCATION =',F9.2,1X,A4/)
      WRITE(3,309)DC1,UT2,DC2,UT2,DC3,UT2,DC,UT2
  309 FORMAT(T5,'DEPTH OF CONC FOR WIDE BEAM SHEAR SMALL END =',F7.3,1X,
     1A2/ T5,'DEPTH OF CONC FOR WIDE BEAM SHEAR LARGE END =',F7.3,1X,A2/
     2 T5,'DEPTH OF CONC FOR PUNCH SHEAR AT LARGE END =',F7.3,1X,A2/
     3 T5,'***** DEPTH OF CONC USED FOR DESIGN =',F7.3,1X,A2)
      WRITE(3,316)PB,ASMAX,UT2,UT1,SMIN,ASMIN,UT2,UT1
  316 FORMAT(/T5,'MAX % STEEL =',F5.3,' %',5X, 'THE MAX ALLOW STEEL AREA
     1 =',F7.3,' SQ ',A2,'/',A2/ T6,'MIN % STEEL =',F5.3,' %',5X,'THE MI
     2N. ALLOW STEEL AREA =',F7.3,' SQ ',A2,'/',A2)
      WRITE(3,312) UT1,UT3,UT4,UT1,UT2,UT1
  312 FORMAT(//T5,'DIST. ',A2,T17,'SHEAR, ',A4,T31,'MOMENT, ',A4,T45,'FO
     1OT. WIDTH, ',A2,T64,'AS, SQ ',A2,'/',A2)
      F=0.
      DO 45 K=1,J
      IF(K.EQ.J)F = ELTOT
      IF(F-W1/2.)15,15,20
   15 V = Q1*F - G*F**2
      EM = Q1*F**2/2. - (DELQ*F**3/(6.*ELTOT))
      GO TO 31
   20 IF(F-D)25,25,30
   25 V = Q1*F - G*F**2 - U1
      EM = Q1*F**2/2. - (DELQ*F**3/(6.*ELTOT)) - U1*(F-W1/2.)
      GO TO 31
   30 V = Q1*F - G*F**2 - U1 - U2
      EM = Q1*F**2/2.-(DELQ*F**3/(6.*ELTOT))-U1*(F-W1/2.)-U2*(F-S-W1/2.)
   31 WIDTH = B - ((B-A)/ELTOT)*F
      IF(EM.LT..00001)EM = -EM
      F2 = EM*FU1/(.9*FYS*WIDTH)
      AS = (DC-SQRT(DC**2 - 4.*GG*F2))/(2.*GG)
      WRITE(3,74)F,V,EM,WIDTH,AS
   45 F = F+FU8
   74 FORMAT(T5,F8.3,T18,F10.3,T32,F10.3,T49,F6.3,T67,F7.3)
      GO TO 6000
  150 STOP
      END
```

3-11 SELECTED METRIC CONVERSION FACTORS FOR THE ACI CODE

The following are the ACI (318-71) Code metric conversion factors applicable to those design equations used in this text containing numerical factors.

fps*	Metric†
$2\phi\sqrt{f'_c}$	$0.530\phi\sqrt{f'_c}$
$4\phi\sqrt{f'_c}$	$1.06\phi\sqrt{f'_c}$
$5\phi\sqrt{f'_c}$	$1.33\phi\sqrt{f'_c}$
$\frac{200}{f_y}$	$\frac{14.06}{f_y}$
$\frac{0.04A_b f_y}{\sqrt{f'_c}}$	$\frac{0.0594A_b f_y}{\sqrt{f'_c}}$
$0.0004Df_y$	$0.00569Df_y$
$\frac{0.02Df_y}{\sqrt{f'_c}}$	$\frac{0.0755Df_y}{\sqrt{f'_c}}$
$0.0003Df_y$	$0.00427Df_y$
$\frac{87{,}000}{87{,}000 + f_y}$	$\frac{6{,}117}{6{,}117 + f_y}$

* f'_c and f_y = psi; A_b and D in inch units.

† f'_c and f_y = kg/sq cm; A_b and D in centimeter units.

PROBLEMS

Note: In all problems the given loads are working-design values.

3-1 Design a square footing as assigned from the following table by ultimate-strength design; f_y = 60,000 psi in all cases.

	w, in	f'_c, psi	DL, kips	LL, kips	q_{allow}, ksf
(*a*)	12	3,000	100	90	4.0
(*b*)	14	3,000	150	180	4.0
(*c*)	16	3,000	200	200	4.0
(*d*)	16	4,000	200	200	3.0
(*e*)	18	5,000	250	250	5.0
(*f*)	20	3,000	200	200	5.0

3-2 Design a rectangular footing by ultimate-strength design with $f_y = 60{,}000$ psi.

	w, in	f'_c, psi	DL, kips	LL, kips	q_a, ksf	L or B, ft
(*a*)	12	3,000	100	100	3.0	6.0
(*b*)	15	3,000	150	150	3.0	8.0
(*c*)	18	4,000	240	280	4.0	20.0
(*d*)	12	3,000	100	100	6.0	6.0

3-3 Repeat Prob. 3-1 using metric units. Refer to Example 3-4 for units.

3-4 (*a*) Plot a graph of d versus f'_c for part (*b*) of Prob. 3-1, varying f'_c from 3 to 6,000 psi.
(*b*) Plot a graph of A_s versus f_y for part (*b*) of Prob. 3-1, varying f_y from 40 to 60,000 psi.
(*c*) What is the optimum f'_c and f_y values for part (*b*) of Prob. 3-1 based on material quantities?

3-5 Design a combined rectangular footing for the figure shown below. Take $f_y = 60$ ksi, $f'_c = 4$ ksi. *Note:* *a*1 is *a* with $AK = 2.0$ ft, etc.

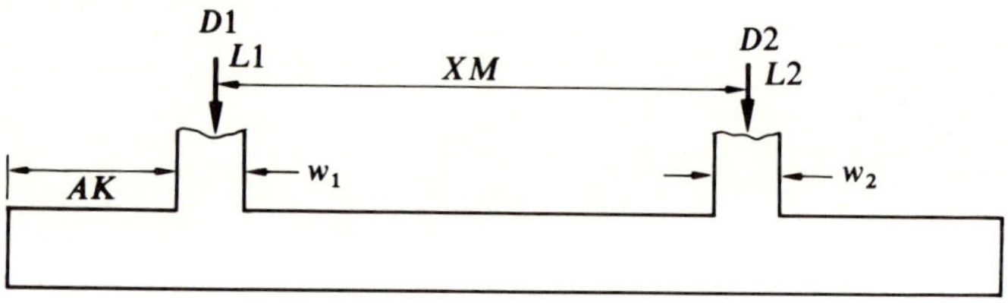

FIGURE P3-1

Key	Loads, kips		Column, in	AK, ft	XM, ft	q_a, ksf
a	D1 = 50	D2 = 75	$w_1 = 12$	0.0	16.	2.0
*a*1	L1 = 70	L2 = 75	$w_2 = 12$	2.0	16.	2.0
b	D1 = 80	D2 = 100	$w_1 = 12$	2.0	16.	2.5
*b*1	L1 = 90	L2 = 100	$w_2 = 12$	0.0	16.	2.5
c	D1 = 120	D2 = 130	$w_1 = 15$	0.0	18.	3.0
*c*1	L1 = 100	L2 = 150	$w_2 = 15$	1.5	18.	3.0

3-6 Repeat the assigned part of Prob. 3-5 if $B = 6$ ft.

3-7 Repeat the assigned part of Prob. 3-5 if $B = 8$ ft.

3-8 Repeat the assigned part of Prob. 3-5 using metric units.

3-9 Repeat the assigned part of Prob. 3-5 using $f'_c = 211$ kg/sq cm.

3-10 Design and detail a trapezoid footing for the given conditions. Take $f'_c = 4$ ksi; $f_y = 60$ ksi. Refer to the figure.

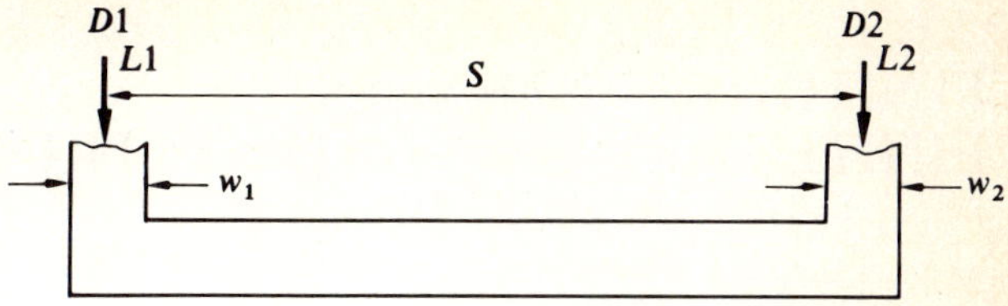

FIGURE P3-2

Key	Loads, kips		Column, in	S, ft	q_a, ksf
a	D1 = 200	D2 = 160	w_1 = 18	20.0	4.0
	L1 = 240	L2 = 150	w_2 = 18		
b	D1 = 120	D2 = 120	w_1 = 16	16.0	3.0
	L1 = 130	L2 = 140	w_2 = 16		
c	D1 = 300	D2 = 200	w_1 = 24	16.0	4.0
	L1 = 320	L2 = 210	w_2 = 18		

3-11 Repeat the assigned part of Prob. 3-10 using $f_c' = 3$ ksi; $f_c' = 5$ ksi.

3-12 Repeat the assigned part of Prob. 3-10 using $f_y = 3{,}516$ kg/sq cm and metric units.

REFERENCES

AMERICAN CONCRETE INSTITUTE (1971): ACI Standard Building Code Requirements for Reinforced Concrete, ACI 318-71.

BOROWICKA, H. (1936): Influence of Rigidity of a Circular Foundation Slab on the Distribution of Pressures over a Contact Surface, *Proc. 1st Int. Conf. Soil Mech. Found. Eng., Cambridge, Mass.*, vol. 2, pp. 144–149.

——— (1938): The Distribution of Pressure under a Uniformly Loaded Elastic Strip Resting on Elasto-isotropic Ground, *Proc. 2d Congr., Int. Assoc. Bridge Struct. Eng.*, Final Report, Berlin.

BOWLES, J. E. (1973): Spread Footings, chap. 15 in "Foundation Engineering Handbook," Van Nostrand, Princeton, N.J.

FURLONG, R. W. (1965): Design Aids for Square Footings, *J. Am. Concr. Inst.*, vol. 62, no. 3, March, pp. 363–371.

REPORT ACI-ASCE COMMITTEE 326 (1962): Pt. 2, Shear and Diagonal Tension, *Proc. ACI*, vol. 59, p. 277.

SCHULTZE, E. (1961): Distribution of Stress beneath a Rigid Foundation, *Proc. 5th Int. Conf. Soil Mech. Found. Eng., Paris*, vol. 1, pp. 807–813.

4

FINITE DIFFERENCES, FINITE-ELEMENT AND MATRIX ANALYSIS

4-1 INTRODUCTION

Considerable effort has been and is being made in formulating mathematical models to define the response of soil to loads and the interaction of the interface elements of structures with the loaded soil mass. The mathematical model which has had the most success in recent years has been the finite-difference method. Basically this method replaces differential equations with difference equations, i.e., replacing the differential

$$dy_{\lim\to 0} \quad \text{with} \quad \Delta y = \text{finite value}$$

and

$$dx_{\lim\to 0} \quad \text{with} \quad \Delta x = \text{finite value}$$

More recently the literature has begun identifying the method of using finite or discrete member elements to make up the mathematical model of a system as the *finite-element method.* For many two-dimensional problems, such as combined footings or piles, the methods are very similar; for three-dimensional soil-stress problems or seepage problems, the methods may differ considerably and often do.

The author introduces herein still another concept termed *matrix analysis* to model mathematically many of the problems which formerly used finite-difference methods. Actually, the finite-element method and the author's matrix analysis can be considered as the same technique; however, strictly speaking, all three are finite-element methods.

The author prefers the matrix method as the simplest of the methods. It is the easiest of the methods to account for nonlinear soil behavior. The model is not hard to visualize or formulate, and it is the simplest method to use when a large number of different load conditions must be considered.

Both the finite-difference and matrix (or finite-element) method will be considered in this chapter.

4-2 FINITE-DIFFERENCE MATHEMATICS

Basically the finite-difference method analytically analyzes simplified geometrica changes in discrete lengths. Consider Fig. 4-1*a*, which illustrates a simple beam loade with a uniform load as shown. From mechanics of materials, one may write th following differential equations:

$$EIy\frac{d^4y}{dx^4} = -q \qquad \text{intensity of load}$$

$$EI\frac{d^3y}{dx^3} = -qx + \frac{qL}{2} \qquad \text{shear}$$

$$EI\frac{d^2y}{dx^2} = -\frac{qx^2}{2} + \frac{qLx}{2} \qquad \text{bending moment}$$

$$EI\frac{dy}{dx} = \frac{qx^2}{6} + \frac{wLx^2}{4} - \frac{qL^3}{24} \qquad \text{slope}$$

$$EIy = -\frac{qx^4}{24} + \frac{qLx^3}{12} - \frac{qL^3x}{24} \qquad \text{deflection}$$

and substituting $x = L/2$ gives the deflection and moment at midspan as

$$y = -\frac{5qL^4}{384EI} = -\frac{qL^4}{76.8EI}$$

and

$$M = +\frac{qL^2}{8}$$

which are the expected values for this loading condition.

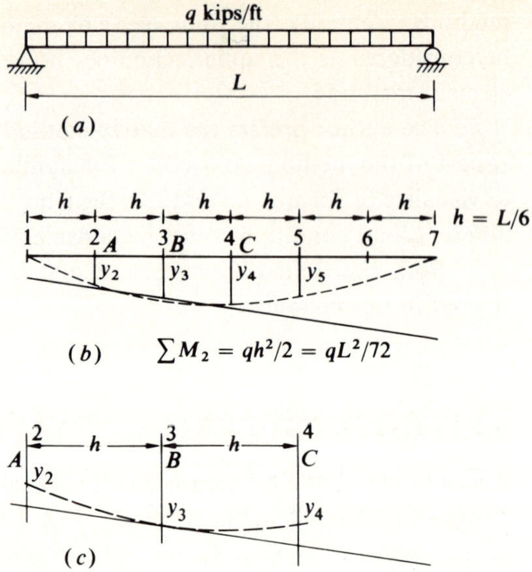

FIGURE 4-1
Beam and equivalent finite-difference approximation.

Referring to Fig. 4-1*b*, which is the elastic curve for the beam, and removing a segment *ABC* as in Fig. 4-1*c*, we see that the average slope at point 3 can be approximated as

$$\frac{dy}{dx} \approx \left.\frac{\Delta y}{\Delta x}\right|_3 = \frac{y_4 - y_3}{h} \qquad \text{first forward difference}$$

or

$$\left.\frac{\Delta y}{\Delta x}\right|_3 = \frac{y_3 - y_2}{h} \qquad \text{first backward difference}$$

Adding these two equations for a better value gives

$$2\left.\frac{\Delta y}{\Delta x}\right|_3 = \frac{y_4}{h} - \frac{y_3}{h} + \frac{y_3}{h} - \frac{y_2}{h} = \frac{y_4 - y_2}{h}$$

or

$$\left.\frac{\Delta y}{\Delta x}\right|_3 = \frac{y_4 - y_2}{2h} \qquad \text{first central difference} \tag{4-1}$$

The second derivative $d^2y/dx^2 \approx \Delta^2y/\Delta x^2$ can be formed as

$$\frac{\Delta^2 y}{\Delta x^2} = \frac{\Delta(\Delta y/\Delta x)}{\Delta x}$$

which is equivalent to

$$\frac{1}{h}\left(\frac{y_4 - y_3}{h} - \frac{y_3 - y_2}{h}\right)$$

and simplifying, we obtain

$$\frac{\Delta^2 y}{\Delta x^2} = \frac{y_4 - 2y_3 + y_2}{h^2} \tag{4-2}$$

In a similar manner we obtain

$$\frac{\Delta^3 y}{\Delta x^2} = \frac{y_5 - 2y_4 + 2y_2 - y_1}{2h^3} \tag{4-3}$$

Equations (4-1) to (4-3) are *central-difference* finite-difference equations. Table 4-1 gives several forms of the difference equations which the reader may find useful.

Let us compare the finite-difference solution of the beam to the exact solution given earlier of

$$y_4 = -\frac{qL^4}{76.8EI}$$

Recognizing that $EI\, d^2y/dx^2 \approx EI\, \Delta^2y/\Delta x^2 = M$, we have at point 2 of Fig. 4-1*b*

$$y_1 - 2y_2 + y_3 = M_2 = \frac{5qL^2}{72EI}\frac{L^2}{36}$$

at point 3

$$y_2 - 2y_3 + y_4 = M_8 = \frac{qL^2}{9EI}\frac{L^2}{36}$$

at point 4

$$y_3 - 2y_4 + y_5 = M_4 = \frac{qL^2}{8EI}\frac{L^2}{36}$$

Table 4-1 TABLE OF FINITE DIFFERENCES

First central differences	$y_n' = \dfrac{y_{n+1} - y_{n-1}}{2(\Delta x)}$
	$y_n'' = \dfrac{y_{n+1} - 2y_n + y_{n-1}}{(\Delta x)^2}$
	$y_n''' = \dfrac{y_{n+2} - 2y_{n+1} + 2y_{n-1} - y_{n-2}}{2(\Delta x)^3}$
	$y_u'''' = \dfrac{y_{n+2} - 4y_{n+1} + 6y_n - 4y_{n-1} + y_{n-2}}{(\Delta x)^4}$
Second central differences	$y_n' = \dfrac{-y_{n+2} + 8y_{n+1} - 8y_{n-1} + y_{n-2}}{12(\Delta x)}$
	$y_n'' = \dfrac{-y_{n+2} + 16y_{n+1} - 30y_n + 16y_{n-1} - y_{n-2}}{12(\Delta x)^2}$
	$y_n''' = \dfrac{-y_{n+3} + 8y_{n+2} - 13y_{n+1} + 13y_{n-1} - 8y_{n-2} + y_{n-3}}{8(\Delta x)^3}$
	$y_n'''' = \dfrac{-y_{n-3} + 12y_{n+2} - 39y_{n+1} + 56y_n - 39y_{n-1} + 12y_{n-2} - y_{n-3}}{6(\Delta x)^4}$
First forward differences	$y_n' = \dfrac{y_{n+1} - y_n}{\Delta x}$
	$y_n'' = \dfrac{y_{n+2} - 2y_{n+1} + y_n}{(\Delta x)^2}$
	$y''' = \dfrac{y_{n+3} - 3y_{n+2} + 3y_{n+1} - y_n}{(\Delta x)^3}$
	$y_n'''' = \dfrac{y_{n+4} - 4y_{n+3} + 6y_{n+2} - 4y_{n+1} + y_n}{(\Delta x)^4}$

Also, from symmetry, $y_3 = y_5$. At the reaction $y_1 = 0$; therefore, arranging terms, we have

$$-2y_2 + y_3 = \frac{5qL^4}{2{,}592EI}$$

$$y_2 - 2y_3 + y_4 = \frac{qL^4}{324EI}$$

$$2y_3 - 2y_4 = \frac{qL^4}{288EI}$$

Table 4-1 TABLE OF FINITE DIFFERENCES *(Continued)*

Second forward differences	$y_n' = \dfrac{-y_{n+2} + 4y_{n+1} - 3y_n}{2(\Delta x)}$
	$y_n'' = \dfrac{-y_{n+3} + 4y_{n+2} - 5y_{n+1} + 2y_n}{(\Delta x)^2}$
	$y_n''' = \dfrac{-3y_{n+4} + 14y_{n+3} - 24y_{n+2} + 18y_{n+1} - 5y_n}{2(\Delta x)^3}$
	$y_n'''' = \dfrac{-2y_{n+5} + 11_{n+4} - 24y_{n+3} + 26y_{n+2} - 14y_{n+1} + 3y_n}{(\Delta x)^4}$
First backward differences	$y_n' = \dfrac{y_n - y_{n-1}}{\Delta x}$
	$y_n'' = \dfrac{y_n - 2y_{n-1} + y_{n-2}}{(\Delta x)^2}$
	$y_n''' = \dfrac{y_n - 3y_{n-1} + 3y_{n-2} - y_{n-3}}{(\Delta x)^3}$
Second backward differences	$y_n' = \dfrac{3y_n - 4y_{n-1} + y_{n-2}}{2(\Delta x)}$
	$y_n'' = \dfrac{2y_n - 5y_{n-1} + 4y_{n-2} - y_{n-3}}{(\Delta x)^2}$
	$y''' = \dfrac{5y_n - 18y_{n-1} + 24y_{n-2} - 14y_{n-3} + 3y_{n-4}}{2(\Delta x)^3}$

Solving, we find

$$y_4 = \frac{qL^4}{75.13EI}$$

This is about 2 percent error too large. Fewer divisions would, of course, increase the percent error.

The finite-difference approach provides a means of solving both beam and plate problems. From the foundation engineering standpoint, the plate problem is a concrete slab on an elastic medium. Timoshenko et al. (1959) [see also Bowles (1968)] expands the differential equation of a plate

$$\frac{\partial^4 w}{\partial x^4} + \frac{2\,\partial^4 w}{\partial x^2\,\partial y^2} + \frac{\partial^4 w}{\partial y^4} = \frac{q}{D} + \frac{P}{D(\partial x\,\partial y)}$$

where $D = [Et^3/12(1 - \mu^2)]$

E = modulus of elasticity of plate material

t = thickness of plate

μ = Poisson's ratio

Making a direct substitution of the y'''' central-difference expressions and a product of y'' for $[2\,\partial^4 w/\partial x^2\,\partial y^2]$ from Table 4-1, and using $\partial x = \partial y = h$, we see that the finite-difference equation in terms of deflections at any point (node) within a plate using a square grid (see Fig. 4-2 for identification of subscripts) is

$$20w_0 - 8(w_T + w_B + w_R + w_L) + 2(w_{TL} + w_{TR} + w_{BL} + w_{BR}) + (w_{TT} + w_{BB} + w_{LL} + w_{RR}) = \frac{qh^4}{D} + \frac{Ph^2}{D} \qquad (4\text{-}4)$$

The sign convention is based on $+q$ and $+P$ in the downward direction. The q term may be upward soil pressure or downward plate loading. The soil pressure is based on the concept of subgrade reaction k_s resulting in

$$-q = k_s w$$

Carrying this deflection term to the left side of Eq. (4-4) results in an increase of the term $20w_0$ to

$$20w_0 + \frac{k_s h^4}{D}$$

Obviously dimensions must be consistent; e.g., if qh^4/D is in feet, then the deflection w is in feet.

Figure 4-2 illustrates the method of applying Eq. (4-4) to a point in a plate. If Eq. (4-4) is applied at a point within two nodes of an edge or at a corner, some of the deflections will fall off the plate. One of two approaches may be utilized: (1) use backward- or forward-difference expressions (partial reason for including in Table 4-1) or (2) consider the fictitious points off the plate and use

$$\frac{\partial^2 w}{\partial x^2} + \mu \frac{\partial^2 w}{\partial^2 y} = 0 \qquad \text{moment perpendicular to edge} = 0$$

$$\frac{\partial^3 w}{\partial x^3} + (2 - \mu)\frac{\partial^3 w}{\partial x\,\partial y^2} = 0 \qquad \text{shear perpendicular to edge} = 0$$

with appropriate interchange of ∂x, ∂y as required to provide enough extra equations to solve the problem. The U.S. Bureau of Reclamation (1954) solved this problem using a rectangular grid, which has been slightly modified by the author for computer programming (Fig. 4-3 and Table 4-2).

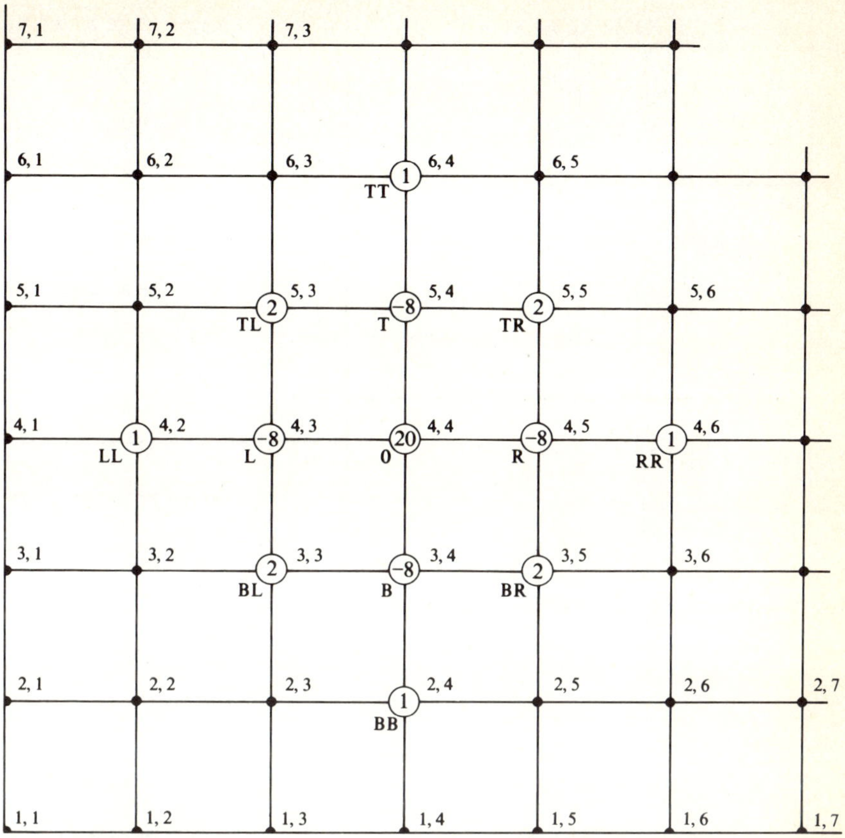

FIGURE 4-2
Method of applying Eq. (4-4) to a node more than two nodes from the edge of the plate. The node used above is located at (4,4) and is used 20 times. The node at (2,4) is used only once. Letters adjacent to nodes correspond to subscripts of Eq. (4-4).

A computer program to generate the coefficient matrix for any size mat *using a square grid* has been available [Bowles (1968)] for some time. The author has revised this program to include the rectangular grid and included it in Chap. 7 with results to be presented later.

4-3 MATRIX FORMULATION

Matrix-analysis procedures are powerful tools in solving many of the problems in foundation engineering, as the reader is about to discover. Since it is assumed that he has some knowledge of matrix operations, only a brief discussion is given here.

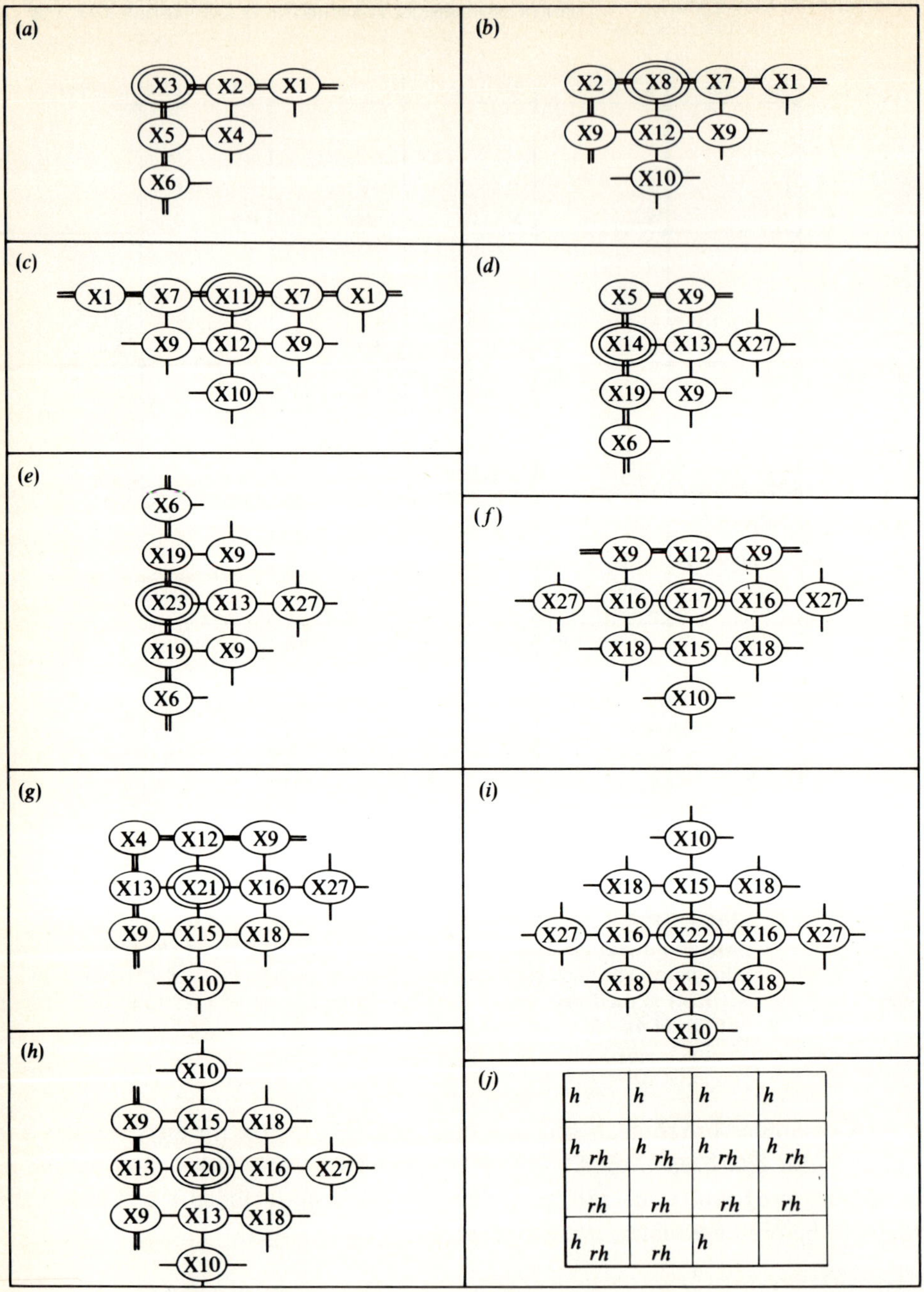

FIGURE 4-3
Deflection-coefficient matrix for indicated nodes. All horizontal grid dimensions are rh; all vertical values are h. Any set of grid coefficients is equated to nodal loads of $Ph^4/Drh^2 - (k_s rh^2/D)w_{ij}$. See Table 4-2 for identifications of coefficients.

Table 4-2 IDENTIFICATION OF VARIABLES IN FIG. 4-3 (USED IN COMPUTER PROGRAM OF CHAP. 7)

$X1 = \frac{1}{2r^4}(1 - \mu_2)$	$X2 = -\frac{1}{r^4}(1 - \mu^2) - \frac{2}{r^2}(1 - \mu)$
$X3 = \frac{1}{2r^4}(1 - \mu^2) + \frac{2}{r^2}(1 - \mu) + \frac{1}{2}(1 - \mu^2)$	$X4 = \frac{2}{r^2}(1 - \mu)$
$X5 = -\frac{2}{r^2}(1 - \mu) - (1 - \mu^2)$	$X6 = \frac{1}{2}(1 - \mu^2)$
$X7 = \frac{2}{r^4}(1 - \mu^2) - \frac{2}{r^2}(1 - \mu)$	$X8 = \frac{5}{2r^4}(1 - \mu^2) + \frac{4}{r^2}(1 - \mu) + 1.0$
$X9 = \frac{1}{r^2}(2 - \mu)$	$X10 = 1.0$
$X11 = \frac{3}{r^4}(1 - \mu^2) + \frac{4}{r^2}(1 - \mu) + 1.0$	$X12 = -\frac{2}{r^2}(2 - \mu) - 2.0$
$X13 = -\frac{2}{r^4} - \frac{2}{r^2}(2 - \mu)$	$X14 = \frac{1}{r^4} + \frac{4}{r^2}(1 - \mu) + \frac{5}{2}(1 - \mu^2)$
$X15 = -\frac{4}{r^2} - 4$	$X16 = -\frac{4}{r^4} - \frac{4}{r^2}$
$X17 = \frac{6}{r^4} + \frac{8}{r^2} + 5$	$X18 = \frac{2}{r^2}$
$X19 = -\frac{2}{r^2}(1 - \mu) - 2(1 - \mu^2)$	$X20 = \frac{5}{r^4} + \frac{8}{r^2} + 6$
$X21 = \frac{5}{r^4} + \frac{8}{r^2} + 5$	$X22 = \frac{6}{r^4} + \frac{8}{r^2} + 6$
$X23 = \frac{1}{r^4} + \frac{4}{r^2}(1 - \mu) + 3(1 - \mu^2)$	$X27 = \frac{1}{r^4}$

A matrix as used in this book is the group of factored coefficients from a series of *independent* simultaneous equations; e.g.,

$$3x_1 + 4x_2 + 5x_3 = -5 \qquad (a)$$

$$7x_1 + 9x_2 = 34 \qquad (b)$$

$$8x_1 + 10x_2 + 5x_3 = 18 \qquad (c)$$

is a set of three simultaneous equations which one can solve by elimination to obtain

$$x_1 = 1 \qquad x_2 = 3 \qquad x_3 = -4$$

In matrix notation one may write

$$\begin{bmatrix} 3 & 4 & 5 \\ 7 & 9 & 0 \\ 8 & 10 & 5 \end{bmatrix} \begin{bmatrix} x_1 \\ x_2 \\ x_3 \end{bmatrix} = \begin{bmatrix} -5 \\ 34 \\ 18 \end{bmatrix} \tag{4-5}$$

where it can be readily seen that column 1 represents the coefficients of x_1 as

$3x_1$ for Eq. (*a*)

$7x_1$ for Eq. (*b*)

$8x_1$ for Eq. (*c*)

Also noted is the use of a zero in Eq. (*b*) as the coefficient of x_3. This does not add or subtract from the set of equations.[1] Note that there are exactly three independent equations (none are multiples of another equation in the set or its duplicate) with three unknowns. Solution of systems problems with more or fewer unknowns than equations, such as four equations with three unknowns or four unknowns and three equations is beyond the scope of this discussion. If there are exactly as many independent equations as unknowns, one always has a *square* matrix, which is a necessary condition for the following discussion. The matrix illustrated is 3 × 3. Writing the full matrix is cumbersome, and a shorthand technique has developed so that the set of Eqs. (4-5) can also be written

$$A_{3\times 3}X_{3\times 1} = C_{3\times 1} \tag{4-6}$$

indicating that the coefficient matrix is of size 3 × 3, the X matrix is a column of 3 × 1, and the constant matrix is a column of 3 × 1. This form is useful for the beginner as it is easy to see whether the system is dimensionally (in size not units) correct. A method of checking the matrix dimensions is to cancel like *interior* dimensions, as illustrated in rewriting Eq. (4-6) below with canceled values shown

$$A_{3\times \not{3}}X_{\not{3}\times 1} = C_{3\times 1} \tag{4-6a}$$

obtaining an equality of matrix size of

$$3 \times 1 = 3 \times 1$$

[1] Only because there are some nonzero x_3 values. It would not be legal to add an x_4 and an extra column and row of zeros.

which is of course a necessary condition of equality. With practice one can write Eq. (4-6) as

$$AX = C \tag{4-7}$$

it being understood that A is a 3×3 and X and C are 3×1 matrices.

One can solve Eq. (4-7) for X to obtain

$$X = \frac{C}{A}$$

However, A is a matrix and a solution is not obtained without further operation on A as

$$X = A^{-1}C \tag{4-8}$$

where A^{-1} is the inverse (or invert) of the matrix A and must be placed with respect to C as shown. The reason is that in matrix usage

$$CA^{-1} \neq A^{-1}C$$

The *transpose* of a matrix $A = A^T$ is a rotation of the given matrix 90° and interchanging. The transposed A matrix of Eq. (4-5) is

$$A^T = \begin{bmatrix} 3 & 7 & 8 \\ 4 & 9 & 10 \\ 5 & 0 & 5 \end{bmatrix}$$

Thus, the first column of the A^T matrix is the first row of the A matrix, the second column of the A^T is the second row of A, etc.

At this point several other peculiarities of matrix operations will be presented. Given two matrices A, B of *equal size*,

$$A + B = B + A$$

$$A - B = B - A$$

$$AB \neq BA$$

The sum of two (or more matrices) of equal size is a new matrix of the same size, as follows. Find $A + A^T = F$, where A is the matrix of Eq. (4-5)

$$F = \begin{bmatrix} 3+3 & 7+4 & 8+5 \\ 4+7 & 9+9 & 10+0 \\ 5+8 & 0+10 & 5+5 \end{bmatrix} = \begin{bmatrix} 6 & 11 & 13 \\ 11 & 18 & 10 \\ 13 & 10 & 10 \end{bmatrix}$$

Matrix F above illustrates why only matrices of equal size can be added or subtracted.

Also given A, B and the matrix of Eq. (4-5),

$$BAX = BC$$

i.e., both sides of a matrix can be multiplied by a new matrix without changing the equality. This is an especially powerful tool when it is difficult to determine whether equations are, in fact, independent. For example, a ring-foundation problem the author worked had 20 unknown deflections, but 41 apparently independent equations could be written (20 moment, 20 shear, and $\sum F_v = 0$); it was obvious that some of the equations were dependent, but one could not tell by looking at the problem which were. Nor did trying various combinations of equations work with predictable results. In matrix notation, one had the following system, for this ring problem

$$A_{41 \times 20} X_{20 \times 1} = C_{41 \times 1}$$

solving

$$X_{20 \times 1} = A^{-1}_{41 \times 20} C_{41 \times 1}$$

but a nonsquare matrix *cannot be inverted.*[1] Therefore, premultiply both sides of the equation by A^T

$$A^T_{20 \times 41} A_{41 \times 20} X_{20 \times 1} = A^T_{20 \times 41} C_{41 \times 1}$$

Solving for X gives

$$X_{20 \times 1} = [A^T A]^{-1}_{20 \times 20} A^T C_{20 \times 1}$$

The matrix A^T was used for convenience as a premultiplier since it was known, although any other nonzero matrix of size 20 × 41 could have been used.

An identity matrix I is defined as a unit diagonal matrix. An identity matrix of size 3 × 3 is

$$I = \begin{bmatrix} 1 & 0 & 0 \\ 0 & 1 & 0 \\ 0 & 0 & 1 \end{bmatrix}$$

A unique property of matrices is

$$AA^{-1} = A^{-1}A = I \tag{4-9}$$

[1] A matrix with a zero on the main diagonal or with a determinant of zero is a *nonsingular* matrix (one with more than one solution) and cannot be inverted. Use of dependent equations to build a square matrix will result in a determinant of zero. In many problems changing the order of forming the equations will remove the zero from the diagonal. Many computer-inversion routines check for zeros on the main diagonal, interchange the equations to move the zero, then invert the matrix. None of the problems in this text form directly with a zero on the main diagonal.

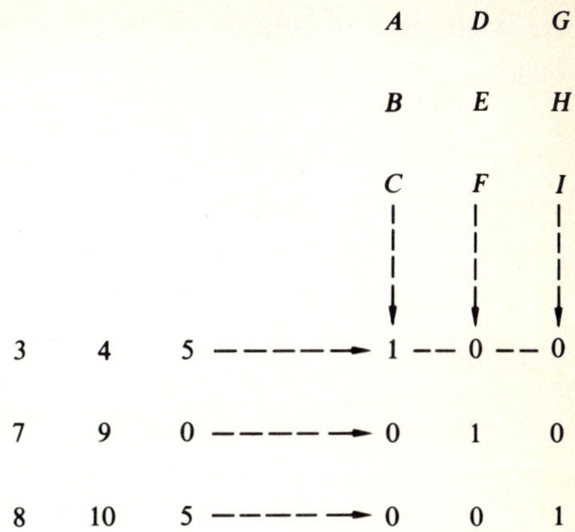

FIGURE 4-4
Convenient formation of matrices to yield their product, in this case, an identity matrix I.

or the product of a matrix and its inverse regardless of order (only exception to order of multiplication) is an identity matrix. Also, let us multiply Eq. (4-9) by A^{-1}:

$$A^{-1}AA^{-1} = A^{-1}I$$

but

$$AA^{-1} = I \qquad \text{also } A^{-1}A$$

therefore

$$A^{-1}I = A^{-1}I$$

and, dividing both sides by A^{-1},

$$I = I$$

The use of the identity matrix is useful in longhand inverting of a matrix. Referring to Fig. 4-4, the following three simultaneous equations can be written using matrix methods as a first step in inverting the given matrix. The constants 1, 0, 0 are, of course, the first column of the identity matrix.

$$3A + 4B + 5C = 1$$

$$7A + 9B + 0C = 0$$

$$8A + 10B + 5C = 0$$

Solving simultaneously, we have

$$A = -3.000 \qquad B = 2.333 \qquad C = 0.133$$

Next, the second column of the inverse matrix multiplied by each row in turn of the original matrix gives the second column of the identity matrix of 0, 1, 0; thus

$$3D + 4E + 5F = 0$$

$$7D + 9E + 0F = 1$$

$$8D + 10E + 5F = 0$$

Solving gives

$$D = -2.000 \qquad E = 1.667 \qquad F = 0.133$$

Repeat the procedure for the third column of the inverse matrix and equate to the corresponding terms in the third column of the identity matrix to obtain

$$3G + 4H + 5I = 0$$

$$7G + 9H + 0I = 0$$

$$8G + 10H + 5I = 1$$

and, solving,

$$G = 3.000 \qquad H = -2.333 \qquad I = 0.067$$

Re-forming these values into the A^{-1} matrix, we obtain for the original A matrix of Eq. (4-5) the inverse matrix

$$A^{-1} = \begin{bmatrix} -3.000 & -2.000 & 3.000 \\ 2.333 & 1.667 & -2.333 \\ 0.133 & -0.133 & 0.067 \end{bmatrix} \tag{4-10}$$

Figure 4-5 illustrates solving for values of X_i using Eq. (4-10) and the C matrix of Eq. (4-5).

The longhand process of inverting a matrix is difficult except for the special case of a 2×2 matrix. A 2×2 matrix is inverted as follows: Given, 2×2 matrix

$$A = \begin{bmatrix} 2 & 3 \\ 4 & 5 \end{bmatrix}$$

find the A inverse so that $A^{-1}A = I$.

The solution proceeds as follows:

1 Compute value of the determinant $= 2 \times 5 - 4 \times 3 = -2$.
2 Reverse numbers on main diagonal (left to right down) and reverse signs on

$$A^{-1}C = X \qquad [C] = \begin{bmatrix} -5.0 \\ +34.0 \\ +18.0 \end{bmatrix}$$

$$A^{-1} = \begin{bmatrix} -3.000 & -2.000 & 3.000 \\ 2.333 & 1.667 & -2.333 \\ 0.133 & -0.133 & 0.067 \end{bmatrix} \begin{matrix} \rightarrow 15.00 - 68.00 + 54.0 = +1 = X_1 \\ \rightarrow -11.67 + 56.67 - 42.0 = +3 = X_2 \\ \rightarrow -0.67 - 4.53 + 1.20 = -4 = X_3 \end{matrix}$$

FIGURE 4-5
Finding the solution of the system of equations using the product of the A^{-1} and the C matrices.

minor diagonal (left to right up) and divide by the value of the determinant, to obtain

$$A^{-1} = \frac{1}{-2}\begin{bmatrix} 5 & -3 \\ -4 & 2 \end{bmatrix}$$

3 Check results:

$$\begin{matrix} -\frac{5}{2} & \frac{3}{2} \\ 2 & -1 \\ \downarrow & \downarrow \end{matrix}$$

$$\begin{bmatrix} 2 & 3 \\ 4 & 5 \end{bmatrix} \begin{matrix} \rightarrow \\ \rightarrow \end{matrix} \begin{bmatrix} 1 & 0 \\ 0 & 1 \end{bmatrix} \qquad \begin{aligned} -\tfrac{5}{2} \times 2 + 3 \times 2 &= +1 \\ -\tfrac{5}{2} \times 4 + 5 \times 2 &= 0 \\ \tfrac{3}{2} \times 2 + 3(-1) &= 0 \\ \tfrac{3}{2} \times 4 + 5(-1) &= +1 \end{aligned}$$

Since an identity matrix is obtained, the inverse is correct.

Aside from this special case, it is generally very difficult to invert a matrix by hand. As noted in Eq. (4-10), the invert required finding nine unknown values in the A^{-1} matrix. This is equivalent to nine simultaneous equations or $M \times M$ equations, where M is the size of the matrix. The only practical means of doing this formidable task is to use the computer. Even on the computer, problems arise because of storage within the working part of the computer (core) and roundoff errors from using large numbers and large quantities of numbers. A large matrix size may require that significant digits be carried to a larger number of digits than the computer's capability (usually 16 in double precision). This type of error can be found, however, by either of two methods:

1 $A^{-1}A \neq I$ indicates an error.
2 $X = A^{-1}C$ with back substitution into the original equations to find that $AX \neq C$, indicating that the X values obtained from A^{-1} are in error.

Special techniques are available to correct the X values. One technique is to obtain a value

$$\Delta C = C_{\text{computed}} - C_{\text{given}}$$

Now compute the quantity ΔX as

$$\Delta X = A^{-1}\,\Delta C$$

The more refined value of X is computed as

$$X_c = X - \Delta X$$

In some cases the errors may not require correction, depending on how precise the input data are.

Much effort has been (and is currently being) expended to obtain methods of inverting large matrices. For example assume computer core is:

Core 20,000 words
Program 5,000 words
*Available** 15,000 words

A matrix of $M = 100$ is of computer usage size of M^2

$$M^2 = \begin{cases} 10{,}000 & \text{single precision} \\ 20{,}000 & \text{double precision} \end{cases}$$

One must either use single-precision computations with roundoff errors or some other technique, such as matrix partitioning, product inverse [Orchard-Hays (1968)], iterative, or relaxation [Mantell and Marron (1962)] techniques.

The problems in this text can (with two exceptions) be solved without using special techniques since the matrix to be inverted is generally less than $M = 70$. The included computer programs are reasonably efficient in conserving computer core space, so that maximum core consumption is less than 20,000 words including the use of double-precision computations where required.

* Actually less, since certain operational and bookkeeping requirements will consume part of this; every computer has this problem.

4-4 COMPUTER PROGRAM FOR INVERTING A MATRIX

This computer program uses the Gauss-Jordan method [James et al. (1965)] of elimination. It does not check for a zero on the diagonal; therefore, the user must scan the input data to alter the order of equations to use this program. This program can solve 30 to 35 equations using double precision with very little loss of accuracy. It can solve about 100 equations with little loss of computational accuracy if the matrix to be inverted is sparse (has mostly zeros), as is the case with most structural and foundation engineering problems.

Matrix-Inversion Routine

The following program inverts any matrix with no zero on the diagonal. As given, the size is controlled by the DIMENSION (50,50) to size 50 × 50.

Line	Operation
4	READ TITLE, N (two cards) N = size of matrix, as 4 for 4 × 4, 8 for 8 × 8, etc.
13	READ A(I,J) the matrix one row at a time at up to eight values per card. Start each row with a new data card
15	READ C(I) constants
21–30	Inversion routine
36–39	Computes X values

```
              C         MATRIX INVERSION SUBROUTINE --- IF DESIRED TO INCREASE
              C           SIZE OF MATRIX CHANGE DIMENSION STATEMENT
0001                    DIMENSION A(50,50), C(50), X(50),TITLE(20)
0002                    DOUBLE PRECISION A,C,X
0003                    JJ = 0
0004              5000  READ(1,1000,END=150)TITLE,N
0005              1000  FORMAT(20A4,/,I5)
0006                    JJ = JJ+1
0007                    IF(JJ.GT.1)GO TO 10
0008                    WRITE(3,1001)TITLE
0009              1001  FORMAT('1',T5,20A4,//)
0010                10  IF(JJ.GT.1)WRITE(3,1002)TITLE
0011              1002  FORMAT(///,T5,20A4,//)
0012                    DO 320  I = 1,N
0013               320  READ(1,1004)(A(I,J),J=1,N)
0014              1004  FORMAT(8F10.4)
0015               521  READ(1,1004) (C(I),I = 1,N)
0016                    WRITE(3,1012)
0017              1012  FORMAT(//,T5, 'THE A-MATRIX IS')
0018                    DO 522  I = 1,N
0019               522  WRITE(3,1006) (A(I,J),J=1,N), C(I)
0020              1006  FORMAT(10F10.4,5X,F12.4)
0021                    DO 562  L = 1,N
0022                    DO 563  I = 1,N
0023               563  IF(I.NE.L)A(L,I) = A(L,I)/A(L,L)
0024                    DO 565  J = 1,N
0025                    IF(J.EQ.L)GO TO 565
0026                    DO 565  I = 1,N
0027               565  IF(I.NE.L)A(J,I) = A(J,I)-A(L,I)*A(J,L)
0028                    DO 567  J = 1,N
0029               567  IF(J.NE.L)A(J,L) = -A(J,L)/A(L,L)
0030               562  A(L,L) = 1./A(L,L)
0031                    WRITE(3,1015)
0032              1015  FORMAT(//,T5, 'THE INVERSE A-MATRIX IS',//)
0033                    DO 568  I = 1,N
0034               568  WRITE(3,1007) (A(I,J),J=1,N)
0035              1007  FORMAT(10F10.5)
0036                    DO 572  I = 1,N
0037                    X(I) = 0.
0038                    DO 572  J = 1,N
0039               572  X(I) = X(I) + A(I,J)*C(J)
0040                    WRITE(3,1017)
0041              1017  FORMAT(//,T5, 'THE X-VALUES ARE')
0042                    WRITE(3,1018)((I,X(I)),I=1,N)
0043              1018  FORMAT(T10, 'X(',I2,') =',F10.5)
0044                    GO TO 5000
0045               150  STOP
0046                    END
```

PROBLEMS

4-1 Invert the following matrices by hand

(*a*) $\begin{bmatrix} 2 & 4 \\ 6 & 3 \end{bmatrix}$ (*b*) $\begin{bmatrix} 3 & 5 \\ 1 & 4 \end{bmatrix}$

and check that $A^{-1}A = I$.

4-2 Assume that

(*a*) $2X_1 + 4X_2 = 16$
$6X_1 + 3X_2 = 21$

(*b*) $3X_1 + 5X_2 = 8$
$X_1 + 4X_2 = 10$

Using the A^{-1} matrices of Prob. 4-1, find X_1, X_2. *Ans.* (*a*) $X_1 = 2$

4-3 What is the product of the two matrices of Prob. 4-1?

4-4 What is the difference of the two matrices of Prob. 4-1? The sum?

4-5 Using the invert program of Sec. 4-4, find the inverse and the values of $X(I)$ for the following matrix.

$$\begin{bmatrix} 3 & 5 & 7 & 9 \\ 8 & 6 & 5 & 2 \\ 7 & 4 & 1 & 3 \\ 1 & 2 & 4 & 6 \end{bmatrix} \begin{bmatrix} X_1 \\ X_2 \\ X_3 \\ X_4 \end{bmatrix} = \begin{bmatrix} 18 \\ 32 \\ 19 \\ 5 \end{bmatrix}$$

Partial Ans. $A^{-1}(1,1) = -0.60870$, $X(1) = 0.49068$; $X(4) = -1.14286$

4-6 Using the invert program of Sec. 4-4, find the inverse and the values of $X(I)$ for the following matrix:

$$\begin{bmatrix} 9 & 8 & 7 & 6 & 3 \\ 8 & 5 & 4 & 1 & 2 \\ 7 & 6 & 3 & 9 & 2 \\ 6 & 4 & 3 & 9 & 7 \\ 3 & 6 & 8 & 4 & 12 \end{bmatrix} \begin{bmatrix} X_1 \\ X_2 \\ X_3 \\ X_4 \\ X_5 \end{bmatrix} = \begin{bmatrix} 10 \\ 42 \\ 87 \\ 93 \\ 51 \end{bmatrix}$$

REFERENCES

BOWLES, J. E. (1968): "Foundation Analysis and Design," chap. 5, McGraw-Hill, New York.

JAMES, M. L., ET AL. (1965): "Analog and Digital Computer Methods in Engineering Analysis," chap. 5, International Textbook, Scranton, Pa.

MANTELL, M. I., and J. F. MARRON (1962): "Structural Analysis," chap. 1, Ronald, New York.

ORCHARD-HAYS, WILLIAM (1968): "Advanced Linear-Programming Computing Techniques," chap. 4, McGraw-Hill, New York.

TIMOSHENKO, S., and S. WOINOWSKY-KRIEGER (1959): "Theory of Plates and Shells," 2d ed., pp. 351–363, McGraw-Hill, New York.

U.S. BUREAU OF RECLAMATION (1954): Moments and Reactions for Rectangular Plates Fixed along Three Edges and Free along the Fourth, *Rep.* 30, Denver, Colo.

5

THE BEAM ON AN ELASTIC FOUNDATION: MATRIX SOLUTION

5-1 INTRODUCTION

In recent years treating continuous footings, i.e., footings with two or more columns in a line, as beams on an elastic foundation has been receiving more and more attention. Two major factors justify this trend: (1) the resulting solution provides a more realistic distribution of longitudinal bending moment in the member, and (2) engineers currently are tending to use more mathematical and refined approaches to the solution of engineering problems.

The author provides still another reason for using the beam on an elastic foundation, namely, the modulus-of-subgrade reaction is not a significant factor in the practical range of footing flexural rigidity *EI* and the modulus-of-subgrade reaction k_s. The use of

$$k_s = \text{SF} \times 12q_a \qquad \text{fps units}$$

has been found to give consistently reliable and reasonable values of both deflections and computed soil pressure to compare with the given allowable soil pressure q_a.

Historically, there are three basic approaches to the problem of a beam on an elastic foundation: (1) the so-called *Winkler approach*, proposed by E. Winkler in about 1867, treats the soil mass supporting the foundation as a series of springs on which the structural member is supported; (2) the second, generally credited to Biot (1937) with elaboration by Ohde [see appendix of Vesić and Johnson (1963)] treats the foundation bed as an elastic solid; (3) the third solves the differential equation of the soil-structure interaction problem. Hetenyi (1946) contributed an entire text to the third solution and considered various boundary conditions. None of the solutions just cited, however, considered or accounted for computed soil tension or for foundation deflections into the nonlinear range of soil behavior. The Hetenyi solution is treated in Chap. 6.

Due to the complexity of the problem, especially where beams of finite length are used (as in most practical problems), many engineers have advocated using office aids in the form of tabulated solutions [Dodge (1964), Gazis (1958), Iyengar et al. (1965), Reti (1967), Wölfer (1969)]. Others have proposed approximate solutions [Cheung and Nag (1968), Levinton (1949), Malter (1960), Popov (1951)]. Of the approximate solutions, the finite-difference method [Malter (1960), Bowles (1968)] using the Winkler foundation concept has so far been the easiest to use in the author's opinion, especially since a computer program has been readily available [Bowles (1968), appendix].

The finite-difference solution, however, has several disadvantages, of which the principal ones are as follows:

1 It is troublesome to account for general boundary conditions because of the formulation of the coefficient matrix.
2 It is difficult to correct for negative deflections, i.e., eliminating the Winkler springs when the footing tends to separate from the soil foundation.
3 It is fairly difficult to write a computer program to generate a general coefficient matrix.
4 It is extremely difficult to account for different load conditions.
5 It is difficult to account for nonlinear soil deformation.

The finite-element method proposed by the author is somewhat similar to the finite-difference solution but eliminates the five major difficulties just cited.

5-2 THE MATRIX (OR FINITE-ELEMENT) SOLUTION

At a selected node of any structure (Fig. 5-1), the equation

$$P_i = A_i F_i \tag{a}$$

is valid. This equation simply equates the external force P to the internal force F using a constant of proportionality A.

For a set of nodes and introducing matrix notation (deleting subscripts and omitting braces and brackets sometimes used), this becomes

$$P = AF \tag{5-1}$$

Also relating the internal deformation of the structural members at the node to the external nodal displacement and considering the same set of nodes for Eq. (5-1) above gives

$$e = BX \tag{5-2}$$

It can be shown [Wang (1970), Laursen (1969)] that the B matrix is the transpose of the A matrix ($B = A^T$); thus

$$e = A^T X$$

The internal force in the ith member F_i is related to the internal member displacements e_i as

$$F_i = S_i e_i$$

and for all the members this becomes in matrix notation

$$F = Se \tag{5-3}$$

Equations (5-1) to (5-3) are the fundamental equations of the displacement, or stiffness, method of matrix analysis.

Substituting Eq. (5-2) into (5-3), we obtain

$$F = SA^T X \tag{b}$$

Substitution of Eq. (*b*) into (5-1) gives

$$P = ASA^T X \tag{c}$$

Note the order of terms and the use of $A^T = B$. Equation (*c*) is solved for X by inverting the square matrix ASA^T of size $P \times P$ to obtain

$$X = [ASA^T]^{-1} P \tag{d}$$

By substitution of the X's obtained from Eq. (*d*) into Eq. (*b*) the desired internal member forces at the selected nodes are obtained.

5-3 THE *A* MATRIX

Consider the beam supported on a system of springs with constant K, as shown in Fig. 5-1. In Fig. 5-1*b* the beam of Fig. 5-1*a* has been *coded* for applying *external* joint moments P_1 through P_6 with corresponding joint rotations X_1 through X_6

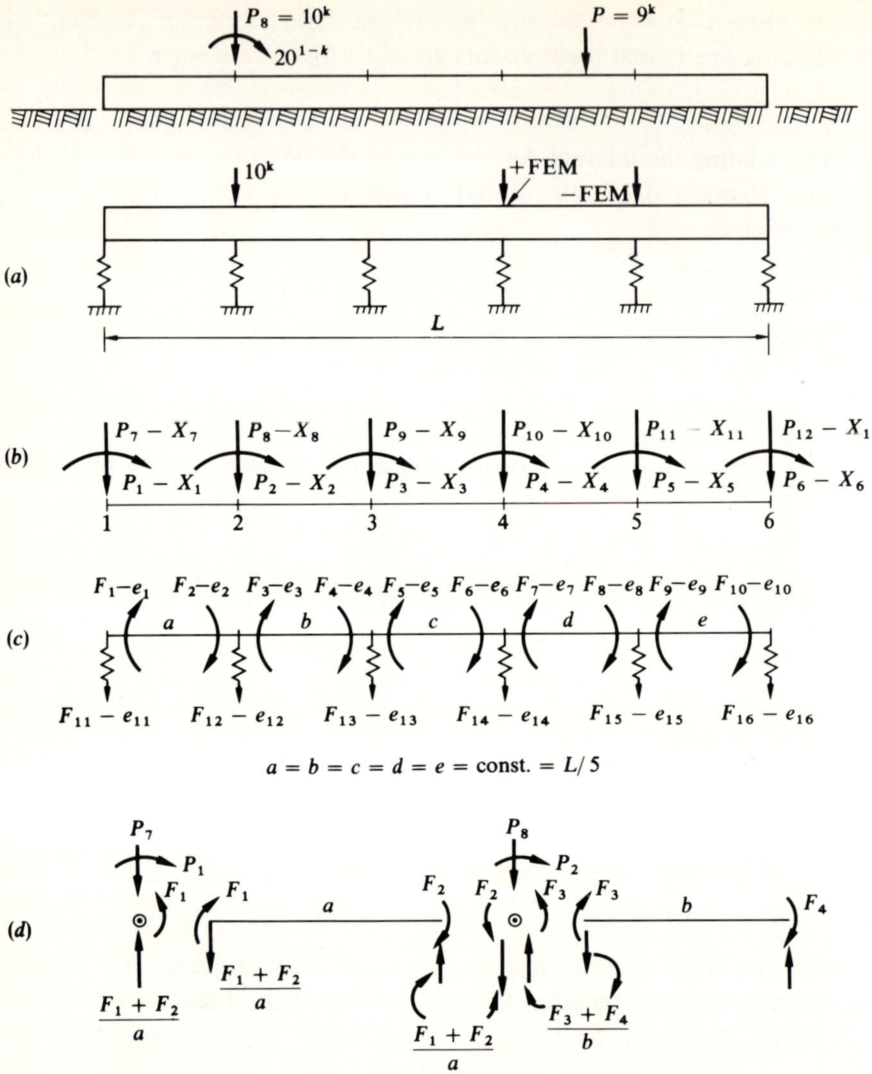

FIGURE 5-1
The beam on an elastic foundation: (*a*) Winkler model with the load between nodes prorated to adjacent nodes and including fixed-end moments; (*b*) the *P-X* coding; (*c*) the internal-force deformation *F-e* coding; (*d*) forming the statics matrix *A* considering free bodies at joints (nodes) 1 and 2.

and external joint forces in the vertical direction P_7 through P_{12} with corresponding joint vertical translations X_7 through X_{12}. This is called the *P-X* diagram.

Note that P may be either an external moment or force and X is either a rotation or translation. The order of numbering the P's is for convenience in building the S matrix of Sec. 5-5.

Next an examination of Fig. 5-1*c* shows that we have applied internal-member forces at each end of each of the six node points, which divides the beam into five finite elements.

Note that F_1 through F_{10} are internal-element end moments; F_{11} through F_{16} are internal "spring" forces. Also e_1 through e_{10} are element end rotations; e_{12} through e_{16} are spring compressions.

It should be evident that there is a relationship between the external and internal nodal forces. From statics we can write this relationship in condensed form as

$$P = AF$$

where A relates the external to internal forces. Expanding this equation by selected examples, we have (Fig. 5-1*d*) at joint 1 (and on the node not on the member; note *soil spring acts on node*)

$$P_1 - F_1 = 0 \tag{a}$$

or

$$P_1 = F_1 \tag{b}$$

likewise

$$P_7 - \frac{F_1}{a} - \frac{F_2}{a} + F_{11} = 0 \tag{c}$$

$$P_7 = \frac{F_1}{a} + \frac{F_2}{a} - F_{11} \tag{d}$$

At joint 2 to satisfy moments

$$P_2 = F_2 + F_3 \tag{e}$$

and to satisfy $\sum F_v = 0$ (note that all segment lengths are equal in derivation and as used in computer program)

$$P_8 = -\frac{F_1}{a} - \frac{F_2}{a} + \frac{F_3}{a} + \frac{F_4}{a} - F_{12} \tag{f}$$

At joint 6, summing moments gives

$$P_6 = F_{10} \tag{g}$$

and summing vertical forces, we have

$$P_{12} = -\frac{F_9}{a} - \frac{F_{10}}{a} - F_{16} \qquad (h)$$

Figure 5-2*a* displays the complete *A* matrix for the beam. Note that zeros (not shown) are used to fill out the locations which are blank, analogous to Eq. (4-5). The particular coding system is used so that a convenient pattern is formed as illustrated. If more divisions are used, the pattern is exactly the same only the resulting matrix is larger. The size of this one is found as follows. Let

$$N = \text{number of elements} = 5$$

$$P = 2N + 2 = 12 = NP$$

$$F = 3N + 1 = 16 = NF$$

or A is of size $NP \times NF$.

5-4 THE *B* MATRIX

If joint 1 rotates $X = 1$ rad, it is evident that since the soil spring cannot resist rotation (small-deflection theory), e_1 rotates as follows:

$$e_1 = X_1 + \frac{X_7}{a} - \frac{X_8}{a}$$

Likewise,

$$e_2 = X_2 + \frac{X_7}{a} - \frac{X_8}{a}$$

$$e_3 = X_2 + \frac{X_8}{a} - \frac{X_9}{a}$$

$$e_4 = X_3 + \frac{X_8}{a} - \frac{X_9}{a}$$

The internal spring deformations e_{11} through e_{16} are

$$e_{11} = -X_7$$

$$e_{12} = -X_8$$

. . . .

$A =$

P \ F	1	2	3	4	5	6	7	8	9	10	11	12	13	14	15	16
1	1															
2			1													
3				1	1											
4						1	1									
5								1	1							
6										1						
7	Z	Z									−1					
8	−Z	−Z	Z	Z								−1				
9			−Z	−Z	Z	Z							−1			
10					−Z	−Z	Z	Z						−1		
11							−Z	−Z	Z	Z					−1	
12									−Z	−Z						−1

(*a*)

$A^T =$

e \ X	1	2	3	4	5	6	7	8	9	10	11	12
1	1						Z	−Z				
2		1					Z	−Z				
3		1						Z	−Z			
4			1					Z	−Z			
5			1						Z	−Z		
6				1					Z	−Z		
7				1						Z	−Z	
8					1					Z	−Z	
9					1						Z	−Z
10						1					Z	−Z
11							−1					
12								−1				
13									−1			
14										−1		
15											−1	
16												−1

(*b*)

$S =$

F \ e	1	2	3	4	5	6	7	8	9	10	11	12	13	14	15	16
1	A	B														
2	B	A														
3			A	B												
4			B	A												
5					A	B										
6					B	A										
7							A	B								
8							B	A								
9									A	B						
10									B	A						
11											K_1					
12												K_2				
13													K_3			
14														K_4		
15															K_5	
16																K_6

(*c*)

F	1	2
1	A	B
2	B	A
3	A	B
4	B	A
5	A	B
6	B	A
7	A	B
8	B	A
9	A	B
10	B	A
11	K_1	0
12	K_2	0
13	K_3	0
14	K_4	0
15	K_5	0
16	K_6	0

(*d*)

Segment length, a = constant

$$Z = \frac{1}{a}$$

$$A = \frac{4EI}{a}$$

$$B = \frac{2EI}{a}$$

FIGURE 5-2
The necessary finite-element matrices (except the P matrix) for Fig. 5-1: (*a*) statics matrix; (*b*) deformation matrix, $B = A^T$; (*c*) stiffness matrix S; (*d*) S matrix in two columns.

The completed B matrix is illustrated in Fig. 5-2*b*. Note that it is precisely A transpose and it generally need not be formed since

$$B = A^T$$

$$A^T(I,J) = A(J,I)$$

and the computer can be programmed to obtain the needed $A^T(I,J)$ values directly from the $A(I,J)$ matrix.

5-5 THE S MATRIX

Consider Fig. 5-3 and recall from conjugate-beam principles that the end slopes e_1 and e_2 are

$$\frac{F_1 L}{3EI} - \frac{F_2 L}{6EI} = e_1 \tag{i}$$

$$-\frac{F_1 L}{6EI} + \frac{F_2 L}{3EI} = e_2 \tag{j}$$

Solving Eqs. (g) and (h) simultaneously for the first segment of Fig. 5-1, where $a = L$, we obtain

$$F_1 = \frac{4EI}{a} e_1 + \frac{2EI}{a} e_2$$

$$F_2 = \frac{2EI}{a} e_1 + \frac{4EI}{a} e_2$$

and similarly

$$F_3 = \frac{4EI}{b} e_3 + \frac{2EI}{b} e_4$$

$$F_4 = \frac{2EI}{b} e_3 + \frac{4EI}{b} e_4$$

. . . .

The force F_{11} is simply

$$F_{11} = K_1 e_{11}$$

The symbol K is used here since the spring-deflection equation is

$$F = K\delta$$

The soil "spring" will have units of FL^{-1} obtained from the modulus-of-subgrade reaction k_s, footing width B, and segment length as

$$K_1 = aBk_s$$

$$K_2 = \frac{a + b}{2} Bk_s$$

. . . .

$$K_6 = eBk_s$$

and if $a = b = c \cdots = e = h$,

$$K_\iota = Bhk_s$$

Note the use of a full contributing end area rather than one-half, as has been used in the past [Bowles (1968)]. The reason is an attempt to allow for increased edge

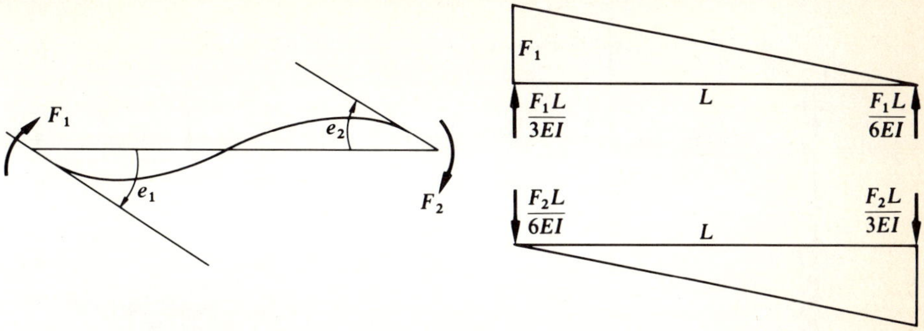

FIGURE 5-3
Relationship between internal forces and deformations using conjugate-beam principles.

pressure (Sec. 3-6) and to provide better computed agreement with measured values (Sec. 5-10).

One could use some other method of building spring constants. If one wished to assume a linear or parabolic variation of k_s along the foundation, the use of some procedure such as in Bowles (1968, chap. 5) could be used. The author has found that, in general, the constant k_s assumed here is adequate. Local variation in k_s as soft spots, holes, etc., can be accounted for by reading particular value(s) into the *S* matrix. The included computer program allows this to be done.

The complete *S* matrix is shown in Fig. 5-2*c* and in two columns in Fig. 5-2*d*. Forming and using the *S* matrix in two columns reduces computer storage from $16^2 = 256$ down to $2 \times 16 = 32$ records.

5-6 THE *P* MATRIX

The coding of Fig. 5-1*b* displays P_i as the external force (moment) acting at a joint depending on the *i* subscript. Expanding Eqs. (*a*) through (*h*), we have

$$P_1 = F_1 + 0F_2 + 0F_3 + \cdots + 0F_{16}$$

$$P_2 = 0F_1 + F_2 + F_3 + \cdots + 0F_{16}$$

$$P_3 = 0F_1 + 0F_2 + 0F_3 + F_4 + F_5 + \cdots + 0F_{16}$$

. . . .

$$P_7 = \frac{F_1}{a} + \frac{F_2}{b} + \cdots - F_{11} + \cdots + 0F_{16}$$

. . . .

$$P_{12} = 0F_1 + 0F_2 \cdots - \frac{F_9}{e} - \frac{F_{10}}{e} + \cdots - F_{16}$$

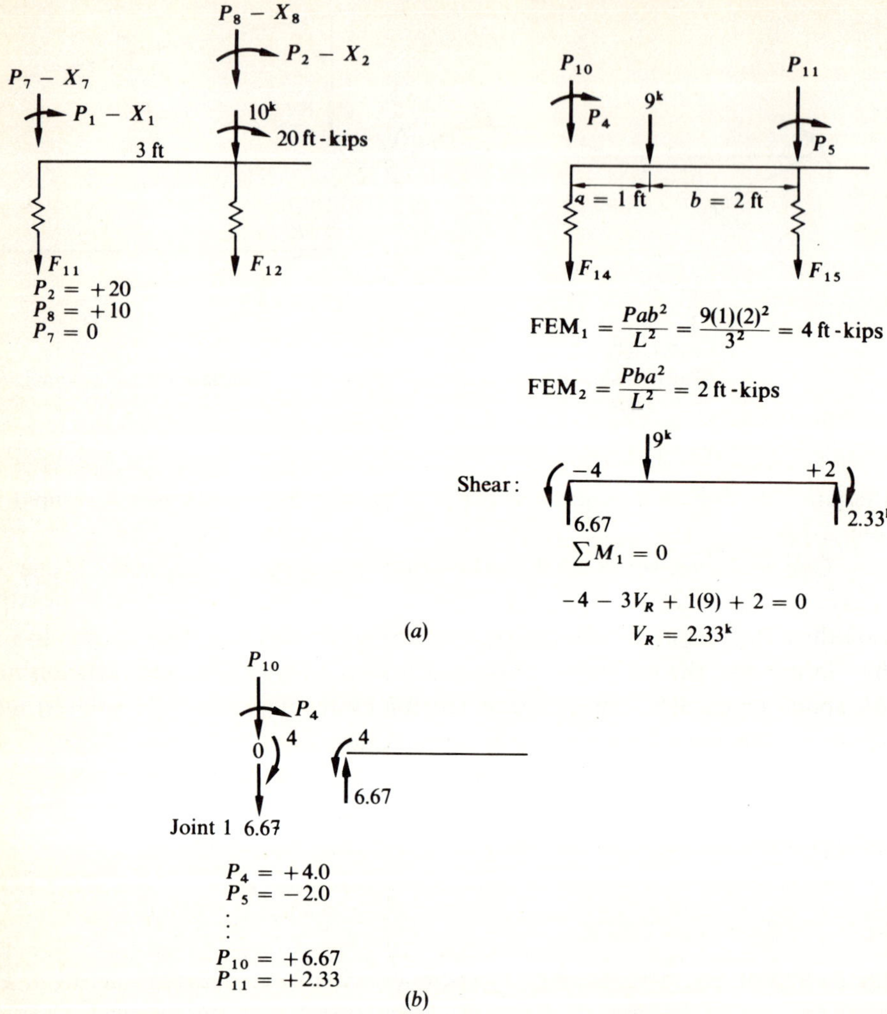

FIGURE 5-4
Relationship between external beam loads and the P matrix for beam of Fig. 5-1.

From this expansion it is evident that the P values are the external forces on the nodes and may be zero if no forces exist. It is also evident that if the external forces are in the direction of the P coding, they have plus signs in the P matrix.

The reader should also be aware that loads located between node points will contribute both *shear* and *fixed-end-moment* P forces at the node. Consider the two examples in Fig. 5-4. In Fig. 5-4*a* the load is $P_2 = 20$ and is a moment by inspection.

P_8 is 10 kips since the 10 is acting on the node (or joint) in the same direction as the coding of P_8. Figure 5-4*b* has the load between nodes. It is prorated as in the figure, using

$$\mathrm{FEM}_1 = \frac{Pab^2}{L^2} \qquad \text{or} \qquad \mathrm{FEM}_2 = \frac{Pba^2}{L^2} \tag{5-4}$$

with a and b identified in Fig. 5-4*b*. The reader may recall that the fixed-end moment for a uniformly distributed load is

$$\mathrm{FEM} = \frac{wL^2}{12} \tag{5-5}$$

Signs are as shown in Fig. 5-4 for computation of shear.

If FEMs are used, the resulting F values of moment at any node will not compute equal when using Eq. (*b*), but will differ by the FEM. The correct moment value is obtained by *adding algebraically* the F value from Eq. (*b*) and the FEM.

5-7 BEAM WEIGHTLESS OR WITH WEIGHT?

When the conventional design procedure of Chap. 3 is used, the beam may properly be considered weightless as the increase in soil pressure due to weight exactly cancels the beam weight. The conventional procedures, however, do not usually consider the columns applying moment to the footing as well as axial load.

Now consider the situation of Fig. 5-5*a*, where a beam is loaded with a couple. If the beam is weightless, the situation of Fig. 5-5*b* is obtained no matter how small the couple. This is not the physical situation, where the condition of Fig. 5-5*b* is obtained only if the couple is larger than the bending moment along the beam due to its self weight. It may be added that the situation of Fig. 5-5*b* is obtained only because with the method outlined herein we are able to remove the "negative" soil springs since no tension is allowed when the footing separates from the soil.

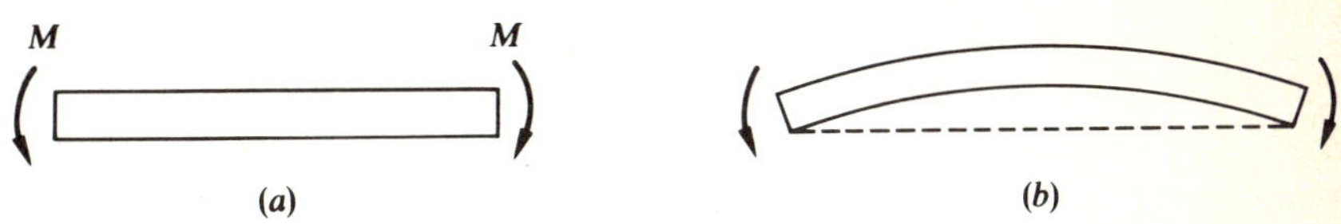

FIGURE 5-5
Weightless beam with equal end moments.

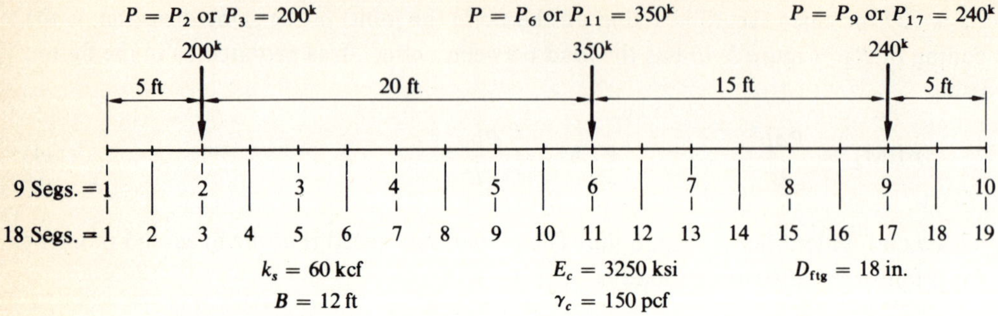

FIGURE 5-6
Problem used to illustrate number of segments on solution.

As the weight per linear foot of a concrete footing may be appreciable, the weight has been included in the computer program to be more realistic if the footing tends to separate from the soil. The beam weight is simply prorated to the node points.

A comparison of several weightless beams versus the same beams with beam weight indicates that the beams with weight compute with about 1 to 3 percent *smaller* bending moments (as would be expected) than the weightless beam. Obviously the soil pressure for the weightless beam is slightly less than for the beam with weight, and the computed deflections are slightly larger as the "springs" must also carry the footing weight.

5-8 FINITE-ELEMENT SOLUTION OF THE BEAM ON AN ELASTIC FOUNDATION

Normally about 10 divisions of a beam should be taken. The number of divisions will depend somewhat on length and loading. Figure 5-6 illustrates a situation where 10 and 19 nodes were used with three columns. The results were as follows:

Segments	M_{max}	Node	M_{min}	Node
9	581.7	6	−274.8	4
18	595.8	11	−280.3	7

The average error when the larger number of divisions is assumed more correct was about −2 percent in this case. Other factors may also influence the result, however, since a comparison of the results of the experiments of Vesić and Johnson (1963) using 10 or 20 divisions gave much larger computed discrepancies using 20 divisions.

It is logical to assume that if the elastic line is essentially in single curvature or the beam length is such that a smooth, slow transition from concave to convex curvature can be made, then fewer divisions are needed to describe the curvature; 10 to 15 divisions generally are adequate. Ten divisions require a matrix inversion of 22 × 22. Twenty divisions require a matrix inversion of 42 × 42, which is within the capability of most digital computers currently available.

5-9 EXAMPLES

The finite-element method will be illustrated with the following examples.

EXAMPLE 5-1 Referring to Fig. E5-1.1 (same number of nodes as Fig. 5-1), which is the beam of Example 3-3, obtain a finite-element solution. Note that the beam has been drawn on the first page of the computer output. This problem is shown in complete detail including statics checks in Figs. E5-1.1 to E5-1.4.

SOLUTION Data cards as follows (also shown on output):

Card	Data
1	TITLE JE Bowles···etc.
	UT1 UT2 UT3 UT4 UT5 UT6 FU1 FU2 FU3 FU4
2	FT IN KIPS FT-KIPS KIPS/SQ FT KIPS/CU FT 12. 1.0 144. 0
	Units card for fps units for format and certain internal program conversions (FU1–FU4
	KL JJS LIST
3	5 0 1 5 divisions, full list with LIST = 1
	XL BX DX XK ELAS XMAX UNITWT L1 L2
4	20. 10. 1.667 48. 468000. 1.50 .150 2 0
	Note use of all foot units except XMAX; since both columns are between nodes L1 = 2 and L2 = 0
	M1 N1 T Y COLLD AMOM (line 109)
5	1 2 2. 2. 310. 0.0 (column 1)
6	5 6 2. 2. 310. 0.0 (column 2)

This is input data; program gives output shown.

```
J E BOWLES EXAMPLE CH 5   EX 5-1, BEAM 5-DIV 10X20X20

    BEAM ON ELASTIC FOUNDATION BY DISPLACEMENT METHOD

         BEAM LENGTH = 20.00 FT
         BEAM WIDTH = 10.00 FT
         BEAM DEPTH = 1.6667 FT
         BEAM MOD OF ELAS =    468000. K/SQ FT
         MOD. OF SUBGRADE REAC. =       48. K/CU FT
         THE MAX VALUE OF LINEAR SOIL DEFL =   1.50 IN
         THE UNIT WEIGTH OF BEAM =    0.150 K/CU FT

         MOMENT OF INERTIA(FT**4)       SEG LENGTH,FT
SEG NO   1        3.858045              4.0000
SEG NO   2        3.858045              4.0000
SEG NO   3        3.858045              4.0000
SEG NO   4        3.858045              4.0000
SEG NO   5        3.858045              4.0000

              THE STATICS MATRIX IS AS FOLLOWS
```

1	1.000	0.0	0.0	0.0	0.0	0.0	0.0	0.0	0.0	0.0	0.0	0.0	0.0	0.0	0.0	0.0
2	0.0	1.000	1.000	0.0	0.0	0.0	0.0	0.0	0.0	0.0	0.0	0.0	0.0	0.0	0.0	0.0
3	0.0	0.0	0.0	1.000	1.000	0.0	0.0	0.0	0.0	0.0	0.0	0.0	0.0	0.0	0.0	0.0
4	0.0	0.0	0.0	0.0	0.0	1.000	1.000	0.0	0.0	0.0	0.0	0.0	0.0	0.0	0.0	0.0
5	0.0	0.0	0.0	0.0	0.0	0.0	0.0	1.000	1.000	0.0	0.0	0.0	0.0	0.0	0.0	0.0
6	0.0	0.0	0.0	0.0	0.0	0.0	0.0	0.0	0.0	1.000	0.0	0.0	0.0	0.0	0.0	0.0
7	0.250	0.250	0.0	0.0	0.0	0.0	0.0	0.0	0.0	0.0	-1.000	0.0	0.0	0.0	0.0	0.0
8	-0.250	-0.250	0.250	0.250	0.0	0.0	0.0	0.0	0.0	0.0	0.0	-1.000	0.0	0.0	0.0	0.0
9	0.0	0.0	-0.250	-0.250	0.250	0.250	0.0	0.0	0.0	0.0	0.0	0.0	-1.000	0.0	0.0	0.0
10	0.0	0.0	0.0	0.0	-0.250	-0.250	0.250	0.250	0.0	0.0	0.0	0.0	0.0	-1.000	0.0	0.0
11	0.0	0.0	0.0	0.0	0.0	0.0	-0.250	-0.250	0.250	0.250	0.0	0.0	0.0	0.0	-1.000	0.0
12	0.0	0.0	0.0	0.0	0.0	0.0	0.0	0.0	-0.250	-0.250	0.0	0.0	0.0	0.0	0.0	-1.000

```
    THE COL POSITIONS AND LOADS:
                   DIST FROM NODE TO COL
COL NO   ADJ NODES   LT NODE       RT NODE    COL LOAD,KIPS    MOMENT,FT-K
   1       1    2     2.000         2.000       310.00             0.0
   2       5    6     2.000         2.000       310.00             0.0
```

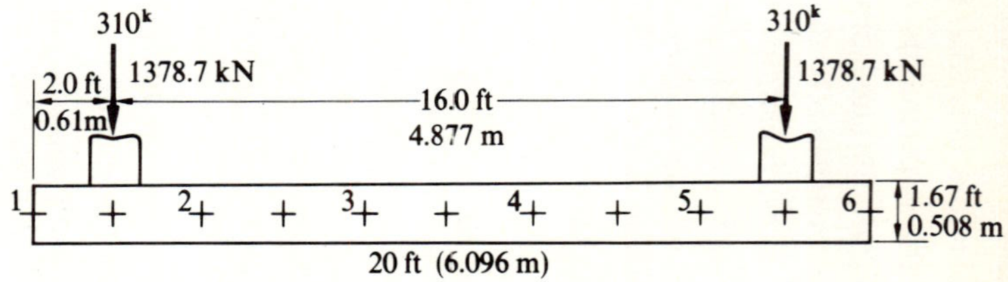

Note alternate coding and metric units

FIGURE E5-1.1

```
                    THE FIXED-END MOMENTS ARE AS FOLLOWS
1     -155.000  2     155.000  3        0.0   4      0.0     5        0.0
6        0.0    7       0.0    8        0.0   9   -155.000  10     155.000 <-- Note signs &
                                                                            F-numbers

      P-MATRIX     BM WT MATRIX     SOIL MATRIX     SUM = PM(I) <-- P matrix used to
  1    155.000         0.0             0.0           155.000        compute X matrix
  2   -155.000         0.0             0.0          -155.000
  3      0.0           0.0             0.0             0.0
  4      0.0           0.0             0.0             0.0
  5    155.000         0.0             0.0           155.000
  6   -155.000         0.0             0.0          -155.000
  7    155.000         5.000           0.0           160.000
  8    155.000        10.000           0.0           165.000
  9      0.0          10.000           0.0            10.000
 10      0.0          10.000           0.0            10.000
 11    155.000        10.000           0.0           165.000
 12    155.000         5.000           0.0           160.000

      THE INITIAL STIFFNESS MATRIX IN 2-COLS IS
       1       1805564.00        902782.25 <------ 2EI/4
       2        902782.25       1805564.00
       3       1805564.00        902782.25
       4        902782.25       1805564.00 <------ 4EI/4
       5       1805564.00        902782.25
       6        902782.25       1805564.00
       7       1805564.00        902782.25
       8        902782.25       1805564.00
       9       1805564.00        902782.25
      10        902782.25       1805564.00
      11          1920.00            0.0
      12          1920.00            0.0
      13          1920.00            0.0
      14          1920.00            0.0
      15          1920.00            0.0
      16          1920.00 <--------- 0.0 <------ 48(10)(4)
```

FIGURE E5-1.2

THE SAT MATRIX X 10**3

1	1805.5640	902.7823	0.0	0.0	0.0	0.0	677.0866	-677.0866	0.0	0.0	0.0	0.0
2	902.7823	1805.5640	0.0	0.0	0.0	0.0	677.0866	-677.0866	0.0	0.0	0.0	0.0
3	0.0	1805.5640	902.7823	0.0	0.0	0.0	0.0	677.0866	-677.0866	0.0	0.0	0.0
4	0.0	902.7823	1805.5640	0.0	0.0	0.0	0.0	677.0866	-677.0866	0.0	0.0	0.0
5	0.0	0.0	1805.5640	902.7823	0.0	0.0	0.0	0.0	677.0866	-677.0866	0.0	0.0
6	0.0	0.0	902.7823	1805.5640	0.0	0.0	0.0	0.0	677.0866	-677.0866	0.0	0.0
7	0.0	0.0	0.0	1805.5640	902.7823	0.0	0.0	0.0	0.0	677.0866	-677.0866	0.0
8	0.0	0.0	0.0	902.7823	1805.5640	0.0	0.0	0.0	0.0	677.0866	-677.0866	0.0
9	0.0	0.0	0.0	0.0	1805.5640	902.7823	0.0	0.0	0.0	0.0	677.0866	-677.0866
10	0.0	0.0	0.0	0.0	902.7823	1805.5640	0.0	0.0	0.0	0.0	677.0866	-677.0866
11	0.0	0.0	0.0	0.0	0.0	0.0	-1.9200	0.0	0.0	0.0	0.0	0.0
12	0.0	0.0	0.0	0.0	0.0	0.0	0.0	-1.9200	0.0	0.0	0.0	0.0
13	0.0	0.0	0.0	0.0	0.0	0.0	0.0	0.0	-1.9200	0.0	0.0	0.0
14	0.0	0.0	0.0	0.0	0.0	0.0	0.0	0.0	0.0	-1.9200	0.0	0.0
15	0.0	0.0	0.0	0.0	0.0	0.0	0.0	0.0	0.0	0.0	-1.9200	0.0
16	0.0	0.0	0.0	0.0	0.0	0.0	0.0	0.0	0.0	0.0	0.0	-1.9200

THE ASAT MATRIX X 10**3

1	1805.5640	902.7823	0.0	0.0	0.0	0.0	677.0866	-677.0866	0.0	0.0	0.0	0.0
2	902.7823	3611.1280	902.7823	0.0	0.0	0.0	677.0866	0.0	-677.0866	0.0	0.0	0.0
3	0.0	902.7823	3611.1280	902.7823	0.0	0.0	0.0	677.0866	0.0	-677.0866	0.0	0.0
4	0.0	0.0	902.7823	3611.1280	902.7823	0.0	0.0	0.0	677.0866	0.0	-677.0866	0.0
5	0.0	0.0	0.0	902.7823	3611.1280	902.7823	0.0	0.0	0.0	677.0866	0.0	-677.0866
6	0.0	0.0	0.0	0.0	902.7823	1805.5640	0.0	0.0	0.0	0.0	677.0866	-677.0866
7	677.0866	677.0866	0.0	0.0	0.0	0.0	340.4633	-338.5433	0.0	0.0	0.0	0.0
8	-677.0866	0.0	677.0866	0.0	0.0	0.0	-338.5433	679.0065	-338.5433	0.0	0.0	0.0
9	0.0	-677.0866	0.0	677.0866	0.0	0.0	0.0	-338.5433	679.0065	-338.5433	0.0	0.0
10	0.0	0.0	-677.0866	0.0	677.0866	0.0	0.0	0.0	-338.5433	679.0065	-338.5433	0.0
11	0.0	0.0	0.0	-677.0866	0.0	677.0866	0.0	0.0	0.0	-338.5433	679.0065	-338.5433
12	0.0	0.0	0.0	0.0	-677.0866	-677.0866	0.0	0.0	0.0	0.0	-338.5433	340.4633

FIGURE E5-1.3

THE LOAD MATRIX (KIPS OR FT-K) IS

LOAD DIR.	1	155.0000
LOAD DIR.	2	-155.0000
LOAD DIR.	3	0.0
LOAD DIR.	4	0.0
LOAD DIR.	5	155.0000
LOAD DIR.	6	-155.0000
LOAD DIR.	7	155.0000
LOAD DIR.	8	155.0000
LOAD DIR.	9	0.0
LOAD DIR.	10	0.0
LOAD DIR.	11	155.0000
LOAD DIR.	12	155.0000

THE JOINT DEFLECTIONS (FT OR RADIANS) ARE

JOINT DIR.	1	-0.00117408	
JOINT DIR.	2	-0.00134427	
JOINT DIR.	3	-0.00058555	
JOINT DIR.	4	0.00058557	
JOINT DIR.	5	0.00134434	
JOINT DIR.	6	0.00117415	— θ
JOINT DIR.	7	0.06297374	← Δ
JOINT DIR.	8	0.05782162	
JOINT DIR.	9	0.05368692	} Sym. Δ ✓
JOINT DIR.	10	0.05368704	} Sym. Δ ✓
JOINT DIR.	11	0.05782177	
JOINT DIR.	12	0.06297415	

THE FORCE MATRIX (KIPS OR FT-K) IS

MOMENT	1	154.9727	
MOMENT	2	1.3242	+ FEM
MOMENT	3	-156.2383	
MOMENT	4	528.7227	
MOMENT	5	-528.6836	
MOMENT	6	528.5781	} Equal ✓
MOMENT	7	-528.6406	} Equal ✓
MOMENT	8	156.3594	
MOMENT	9	-1.3281	+ FEM
MOMENT	10	-154.9648	
FORCE	11	-120.9096	
FORCE	12	-111.0175	
FORCE	13	-103.0789	
FORCE	14	-103.0791	
FORCE	15	-111.0178	
FORCE	16	-120.9104	

SHEAR AT EACH SEGMENT,KIPS		BEND. MOMENT AT EACH ORDINATE FT-K (ORIG. FEMS ADDED)		SOIL REACTION AT EA. ORD., KIPS		SOIL PRESSURE K/SQ FT
1	-39.0904	1	-0.0273	120.9096		3.023
2	-93.0729	2	-156.2383	111.0175		2.775
3	0.0060	3	-528.6836	103.0789		2.577
4	93.0851	4	-528.6406	103.0791		2.577
5	39.1029	5	-156.3281	111.0178		2.775
		6	-0.0352	120.9104		3.023
			SUM OF SOIL REACTIONS =	670.0127	(670.00)	

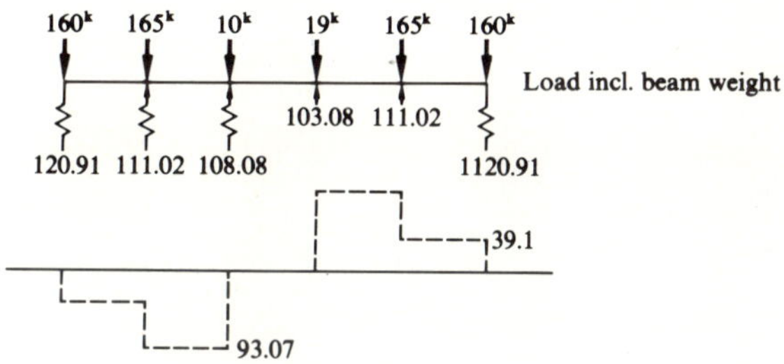

FIGURE E5-1.4

////

EXAMPLE 5-2 Repeat Example 5-1 using 10 divisions.

SOLUTION Only the data cards requiring changes are shown.

Card	Data
3	10 0 1
	LIST = 1 gives full listing (not shown)
4	20. 10. 1.6667 48. 468000. 1.50 .150 0 2
	Note both columns at nodes, L2 = 2, L1 = 0
	If beam weight is not to be included, use 0. instead of 0.150
5	2 310. 0. (Read on line 131)
6	10 3.10 0.

Partial output is shown in Figs. E5-2.1 and E5-2.2.

```
        J E BOWLES CONT FTG FOR TEXT CH 5, EX 5-2   10 DIV 10X20X20

           BEAM ON ELASTIC FOUNDATION BY DISPLACEMENT METHOD

               BEAM LENGTH = 20.00 FT
               BEAM WIDTH = 10.00 FT
               BEAM DEPTH = 1.6667 FT
               BEAM MOD OF ELAS =     468000. K/SQ FT
               MOD. OF SUBGRADE REAC. =         48. K/CU FT
               THE MAX VALUE OF LINEAR SOIL DEFL =   1.50 IN
               THE UNIT WEIGTH OF BEAM =     0.150 K/CU FT

               MOMENT OF INERTIA(FT**4)          SEG LENGTH,FT
 SEG NO    1          3.858026                   2.0000
 SEG NO    2          3.858026                   2.0000
 SEG NO    3          3.858026                   2.0000
 SEG NO    4          3.858026                   2.0000
 SEG NO    5          3.858026                   2.0000
 SEG NO    6          3.858026                   2.0000
 SEG NO    7          3.858026                   2.0000
 SEG NO    8          3.858026                   2.0000
 SEG NO    9          3.858026                   2.0000
 SEG NO   10          3.858026                   2.0000

 COL NODE POINT  =   2      P =   310.00 KIPS    COL MOMENT =      0.0     FT-K

 COL NODE POINT  =  10      P =   310.00 KIPS    COL MOMENT =      0.0     FT-K

     P-MATRIX    BM WT MATRIX     SOIL MATRIX     SUM = PM(I)
  1     0.0          0.0             0.0             0.0
  2     0.0          0.0             0.0             0.0
  3     0.0          0.0             0.0             0.0
  4     0.0          0.0             0.0             0.0
  5     0.0          0.0             0.0             0.0
  6     0.0          0.0             0.0             0.0
  7     0.0          0.0             0.0             0.0
  8     0.0          0.0             0.0             0.0
  9     0.0          0.0             0.0             0.0
 10     0.0          0.0             0.0             0.0
 11     0.0          0.0             0.0             0.0
 12     0.0          2.500           0.0             2.500
 13   310.000        5.000           0.0           315.000
 14     0.0          5.000           0.0             5.000
 15     0.0          5.000           0.0             5.000
 16     0.0          5.000           0.0             5.000
 17     0.0          5.000           0.0             5.000
 18     0.0          5.000           0.0             5.000
 19     0.0          5.000           0.0             5.000
 20     0.0          5.000           0.0             5.000
 21   310.000        5.000           0.0           315.000
 22     0.0          2.500           0.0             2.500
```

FIGURE E5-2.1

```
SHEAR AT EACH     BEND. MOMENT AT EACH ORDINATE     SOIL REACTION AT          SOIL PRESSURE
 SEGMENT,KIPS      FT-K (ORIG. FEMS ADDED)            EA. ORD., KIPS               K/SQ FT

 1    65.3410       1         0.6250 <-- Single precision   67.8410             3.392
 2  -185.2658       2       131.0000                        64.3930             3.220
 3  -129.4212       3      -240.2500                        60.8446             3.042
 4   -76.6552       4      -496.6875 <-- Roundoff -->       57.7659             2.888
 5   -25.9488       5      -648.2500                        55.7065             2.785
 6    24.0419       6      -699.1875                        54.9906             2.750
 7    74.7679       7      -648.5625                        55.7260             2.786
 8   127.5730       8      -496.8750 <--          -->       57.8051             2.890
 9   183.4771       9      -239.6875                        60.9041             3.045
10   -67.0500      10       130.7500                        64.4726   Chk ✓     3.224
                   11         0.0625                        67.9404             3.397
                     SUM OF SOIL REACTIONS =               668.3887 <-- ( --> 670.00)
```

FIGURE E5-2.2

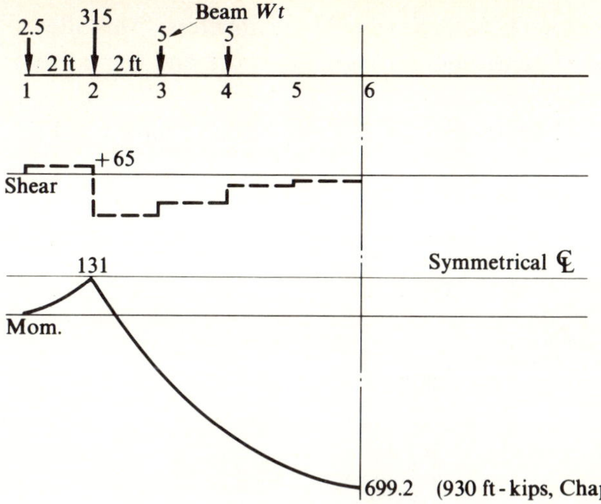

(930 ft-kips, Chapter 3, by conventional method)

$$\Sigma M_2 = ?\curvearrowright +$$
$$2.5(z) - 67.84(z) + M_2 = 0$$
$$M_2 = -5 + 135.68 = 130.68\ (131.00)$$

FIGURE E5-2.2 (*Continued*) ////

EXAMPLE 5-3 Repeat Example 5-2 using metric units and 10 divisions.

SOLUTION Convert dimensions, etc., to meters and kilonewtons, using Table 2-8. Partial conversions are as follows:

$$DX = 1.6667(0.3048) = 0.508 \text{ m}$$

$$BX = 10(0.3048) = 3.048 \text{ m}$$

$$k_s = 48(157.09) = 7{,}540.32 \text{ kN/cu m}$$

$$P = 310(4.4475) = 1{,}378.7 \text{ kN}$$

$$E = 468{,}000(4.7882) = 22{,}408{,}750 \text{ kN/sq m}$$

Data cards are:

Card	Data									
1	TITLE JE Bowles· · ·etc. (see Fig. E5-3.1)									
	UT1	UT2	UT3	UT4	UT5	UT6	FU1	FU2	FU3	FU4
2	M	CM	KN	KN-M	KN/SQ M	KN/CU M	100.	.3	10.	0
	KL	JJS	LIST							
3	10	0	0	A, ASAT not listed (LIST = 0)						
	XL	BX	DX	XK	ELAS	XMAX	UNITWT	L1	L2	
4	6.096	3.048	.508	7540.32	22408750.	3.81	23.6	0	2	
5	2	1378.7	0.	Column node and loading						
6	10	1378.7	0.							

Note use of meters and kilonewtons except XMAX, which is centimeters. This allows the program to compute either fps or metric units by use of card 2 and the correct data-card input units.

Partial output is shown in Figs. E5-3.1 and E5-3.2; the beam is shown in Fig. E5-1.1.

```
J E BOWLES EXAMPLE 5-3--BEAM OF EXAMPLE 5-1 W/10 DIV AND METRIC UNITS

    BEAM ON ELASTIC FOUNDATION BY DISPLACEMENT METHOD

         BEAM LENGTH =   6.10 M
         BEAM WIDTH =   3.05 M
         BEAM DEPTH = 0.5080 M
         BEAM MOD OF ELAS = 22408736.  KN/SQ M
         MOD. OF SUBGRADE REAC. =       7540.  KN/CU M
         THE MAX VALUE OF LINEAR SOIL DEFL =    3.81 CM
         THE UNIT WEIGTH OF BEAM =   23.600 KN/CU M

         MOMENT OF INERTIA(M **4)          SEG LENGTH,M
SEG NO    1         0.033298               0.6096
SEG NO    2         0.033298               0.6096
SEG NO    3         0.033298               0.6096
SEG NO    4         0.033298               0.6096
SEG NO    5         0.033298               0.6096
SEG NO    6         0.033298               0.6096
SEG NO    7         0.033298               0.6096
SEG NO    8         0.033298               0.6096
SEG NO    9         0.033298               0.6096
SEG NO   10         0.033298               0.6096

COL NODE POINT =  2       P = 1378.70 KN      COL MOMENT =      0.0     KN-M

COL NODE POINT = 10       P = 1378.70 KN      COL MOMENT =      0.0     KN-M

    P-MATRIX    BM WT MATRIX    SOIL MATRIX    SUM = PM(I)
 1       0.0          0.0             0.0           0.0
 2       0.0          0.0             0.0           0.0
 3       0.0          0.0             0.0           0.0
 4       0.0          0.0             0.0           0.0
 5       0.0          0.0             0.0           0.0
 6       0.0          0.0             0.0           0.0
 7       0.0          0.0             0.0           0.0
 8       0.0          0.0             0.0           0.0
 9       0.0          0.0             0.0           0.0
10       0.0          0.0             0.0           0.0
11       0.0          0.0             0.0           0.0
12       0.0         11.138           0.0          11.138
13    1378.700       22.276           0.0        1400.976
14       0.0         22.276           0.0          22.276
15       0.0         22.276           0.0          22.276
16       0.0         22.276           0.0          22.276
17       0.0         22.276           0.0          22.276
18       0.0         22.276           0.0          22.276
19       0.0         22.276           0.0          22.276
20       0.0         22.276           0.0          22.276
21    1378.700       22.276           0.0        1400.976
22       0.0         11.138           0.0          11.138
```

FIGURE E5-3.1

THE INITIAL STIFFNESS MATRIX IN 2-COLS IS

1	4896173.00	2448085.00
2	2448085.00	4896173.00
3	4896173.00	2448085.00
4	2448085.00	4896173.00
5	4896173.00	2448085.00
6	2448085.00	4896173.00
7	4896173.00	2448085.00
8	2448085.00	4896173.00
9	4896173.00	2448085.00
10	2448085.00	4896173.00
11	4896173.00	2448085.00
12	2448085.00	4896173.00
13	4896173.00	2448085.00
14	2448085.00	4896173.00
15	4896173.00	2448085.00
16	2448085.00	4896173.00
17	4896173.00	2448085.00
18	2448085.00	4896173.00
19	4896173.00	2448085.00
20	2448085.00	4896173.00
21	14010.36	0.0
22	14010.36	0.0
23	14010.36	0.0
24	14010.36	0.0
25	14010.36	0.0
26	14010.36	0.0
27	14010.36	0.0
28	14010.36	0.0
29	14010.36	0.0
30	14010.36	0.0
31	14010.36	0.0

FIGURE E5-3.1 (*Continued*)

THE LOAD MATRIX (KN OR KN-M) IS

LOAD DIR. 1	0.0
LOAD DIR. 2	0.0
LOAD DIR. 3	0.0
LOAD DIR. 4	0.0
LOAD DIR. 5	0.0
LOAD DIR. 6	0.0
LOAD DIR. 7	0.0
LOAD DIR. 8	0.0
LOAD DIR. 9	0.0
LOAD DIR. 10	0.0
LOAD DIR. 11	0.0
LOAD DIR. 12	0.0
LOAD DIR. 13	1378.7000
LOAD DIR. 14	0.0
LOAD DIR. 15	0.0
LOAD DIR. 16	0.0
LOAD DIR. 17	0.0
LOAD DIR. 18	0.0
LOAD DIR. 19	0.0
LOAD DIR. 20	0.0
LOAD DIR. 21	1378.7000
LOAD DIR. 22	0.0

THE JOINT DEFLECTIONS (M OR RADIANS) ARE

JOINT DIR. 1	-0.00178004
JOINT DIR. 2	-0.00185246
JOINT DIR. 3	-0.00179212
JOINT DIR. 4	-0.00138393
JOINT DIR. 5	-0.00074858
JOINT DIR. 6	-0.00000125
JOINT DIR. 7	0.00074618
JOINT DIR. 8	0.00138182
JOINT DIR. 9	0.00179049
JOINT DIR. 10	0.00185110
JOINT DIR. 11	0.00177876
JOINT DIR. 12	0.02157150
JOINT DIR. 13	0.02047165
JOINT DIR. 14	0.01934008
JOINT DIR. 15	0.01835753
JOINT DIR. 16	0.01769904
JOINT DIR. 17	0.01746760
JOINT DIR. 18	0.01769743
JOINT DIR. 19	0.01835456
JOINT DIR. 20	0.01933599
JOINT DIR. 21	0.02046677
JOINT DIR. 22	0.02156578

THE FORCE MATRIX (KN OR KN-M) IS

MOMENT 1	0.2500	Round-off
MOMENT 2	-177.0625	
MOMENT 3	175.6250	
MOMENT 4	323.3125	✓
MOMENT 5	-325.2500	
MOMENT 6	674.0625	
MOMENT 7	-675.1875	
MOMENT 8	880.1875	
MOMENT 9	-879.9375	
MOMENT 10	949.6250	
MOMENT 11	-948.4375	
MOMENT 12	881.3125	
MOMENT 13	-880.6250	
MOMENT 14	675.5000	
MOMENT 15	-675.0625	
MOMENT 16	325.3750	
MOMENT 17	-325.1250	
MOMENT 18	-176.7500	
MOMENT 19	177.3750	
MOMENT 20	0.2500	
FORCE 21	-302.2244	
FORCE 22	-286.8152	
FORCE 23	-270.9612	
FORCE 24	-257.1956	
FORCE 25	-247.9699	
FORCE 26	-244.7273	
FORCE 27	-247.9474	
FORCE 28	-257.1538	
FORCE 29	-270.9041	
FORCE 30	-286.7466	
FORCE 31	-302.1443	

Checking: @ M_6 = 948.44/1.356 = 699.4 (699.2 E5-2·3)
@ ΣF_v = 2974.8/4.4475 = 668.9^k (668.4^k)

SHEAR AT EACH SEGMENT, KN		BEND. MOMENT AT EACH ORDINATE KN-M (ORIG. FEMS ADDED)		SOIL REACTION AT EA. ORD., KN	SOIL PRESSURE KN/SQ M
1	291.0864	1	0.2500	302.2244	162.656
2	-823.0742	2	175.6250	286.8152	154.363
3	-574.3889	3	-325.2500	270.9612	145.830
4	-339.4692	4	-675.1875	257.1956	138.422
5	-113.7753	5	-879.9375	247.9699	133.456
6	108.6761	6	-948.4375	244.7273	131.711
7	334.3477	7	-880.6250	247.9474	133.444
8	569.2256	8	-675.0625	257.1538	138.399
9	817.8538	9	-325.1250	270.9041	145.799
10	-296.3755	10	177.3750	286.7466	154.326
		11	-0.2500	302.1443	162.613
		SUM OF SOIL REACTIONS =		2974.7891	(2980.15)

FIGURE E5-3.2

////

5-10 LIMITATIONS OF MATRIX SOLUTION FOR BEAM ON ELASTIC FOUNDATION

This partial Winkler foundation solution is as valid as any of the proposals such as Biot (1937) or Ohde [see Vesić and Johnson (1963)]. The solution compares well with the classical (Hetenyi) solution and the finite-difference solution. A solution of a given beam shows:

	ft-kips		
Node	Matrix	Finite-difference	Hetenyi
6	−698.71	−706.90	856.1

The comparison is not strictly valid, however, since the matrix solution includes the beam weight and the other two do not. As stated earlier, this would tend to reduce the moments 2 to 3 percent.

There has been little reported actual testing of continuous beams to establish the validity of these solutions. The most notable is the series of tests reported by Vesić and Johnson (1963). The author has analyzed these, as shown in Tables 5-1 and 5-2. From these data it appears that the finite-element solution proposed is one of the better solutions.

These tabular data indicate that the matrix solution by the author is not sensitive to k_s. The author has solved a large number of problems and has found this statement to be consistently correct; and, as indicated earlier, also illustrates that the Vesic equation

$$k_s' = 0.65 \sqrt[12]{\frac{E_s B^4}{E_b I_b}} \frac{E_s}{1 - \mu^2} \tag{2-25}$$

does not require correction for L/B ratios greater than 1, as originally proposed by Vesić.

5-11 CORRECTING FOR NONLINEAR SOIL BEHAVIOR

The matrix solution is easy to correct for holes, changes in soil properties, excessive deflection, or footing separation as follows:

1 Holes or change in k_s: read in zero or modified K_i value into S matrix.

2 Excessive deflection: apply a force in the P matrix of $K \times \delta$, where δ is the predetermined deflection to enter the nonlinear range of soil deformation.

3 Footing separation: set $K = 0$ in the S matrix and do the problem over.

Table 5-1 COMPARISON OF MATRIX ANALYSIS OF BEAM ON ELASTIC FOUNDATION WITH VALUES REPORTED BY VESIĆ AND JOHNSON (1963)

Beam	I, in^4	Load, kips	Vesić* k_s, kcf	Vesić* Meas	Vesić* Comp	Vesić* Ohde's	Vesić* Δ, in	Test no.	Bowles‡ M_{comp}	Bowles‡ Δ, in	Vary k_s k_s	Vary k_s M_{comp}	Vary k_s Δ, in
WF 8 × 31	109.7	16.5^k (center), 6 ft	98.2	+172.0	+148.5	+160.2	0.59	1	+160.9	0.47	70.0	+161.8	0.66
		8.25, 8.25		−113.2	−132.0	−117.9	0.60*cl*† 0.68*e*	2	−117.0	0.45*cl* 0.48*e*		−117.4	0.64*cl* 0.66*e*
		6.19, 12.38, 6.19		+36.6 −33.7	+12.4 −42.8	+31.7 −30.9	0.08 const	3	+32.9 −28.0	0.69 const		+32.9 −28.0	1.0 const
8 in Std Ship Channel 21.4 lb	6.9	8.25	123.3	+61.4	+67.5	+71.0	0.28*cl*	7	+71.0	0.24*cl*	175.0	+67.8	0.18*cl*
		4.13, 4.13		−48.4	−56.2	−49.4	0.18*cl* 0.31*e*	8	−48.1	0.13*cl* 0.25*e*		−44.5	0.08*cl* 0.19*e*
		6.19, 12.38, 6.19		+40.0 −30.0	+17.0 −40.1	+32.1 −30.4	1.0*cl* 0.87*e*	9	+26.2 −33.2	0.57*cl* 0.60*e*		+27.1 −32.2	0.40*cl* 0.43*e*
1 in. × 8 in.	0.67	8.25	150.0	+37.0	+41.6	−42.0	0.36*cl*	4	+38.4	0.35*cl*	54.0	+52.7	0.72*cl*
		6.19, 6.19, 6.19		+5.0 −27.6	+1.9 −29.9	+0.1 −29.7	0.48*cl* 0.82*e*	6	+6.15 −25.4	0.21*cl* 0.61*e*		−0.82 −34.8	0.70*cl* 1.5*e*
		6.19, 12.38, 6.19		+43.0 −20.4	+32.2 −28.3	+31.1 −29.4	0.9*cl* 1.0*e*	6	+35.9 −26.0	0.47*cl* 0.32*e*		+31.8 −29.4	1.26*cl* 1.54*e*

* Vesić deflections taken from graphs.

† *cl* refers to centerline deflections; *e* is end deflection.

‡ Bowles moment computations use same k_s as Vesić.

When reworking the Vesić test data the author found that considering excessive deflections or footing separations does not improve the computations; in fact the computed values deviated more when considering nonlinear soil properties than when ignoring them. This phenomenon is contrary to what most soil engineers would expect; the author offers no explanation since he has none.

This solution does not consider footing-to-soil friction. A finite-element solution by Cheung and Nag (1968) attempted to account for base friction with results which may be questionable. There is no doubt that friction is a factor and could account for part of the discrepancy between some of the measured values of Vesić and computed ones. A major factor to overcome in applying the friction effect is to determine the direction. Except for the case of a reasonably flexible beam loaded with a single load near the center, the direction is uncertain because the direction depends on the strain difference in the beam and soil at their interface. The situation becomes even more formidable if the beam curvature reverses. A solution which merely satisfies the problem statics may not be correct.

Comparing the Vesić work (Table 5-2), which is all that is available, *and assuming that the measured values are correct*, it appears that the computed moments are unsafe by a factor of about 20 percent maximum. The author, having also loaded members on soil beds, is aware that there is considerable difficulty in loading by using a hydraulic jack against a beam, as opposed to a dead-load type of loading device to keep the load constant as long as 15 min, the time reported by Vesić as that required to take the strain-gage readings. Considering these facts, model problems, uncertainty

Table 5-2 PERCENT ERROR OF MOMENTS COMPUTED BY VARIOUS METHODS AND VARYING THE MODULUS-OF-SUBGRADE REACTION COMPARED TO THE REPORTED VALUES BY VESIĆ AND JOHNSON (1963)*

Test no.	Vesić	Ohde	Bowles	$k_s = 250$	$k_s = 175$	$k_s = 200$
1	−13.66	−6.86	−6.43	−7.49		
2	+16.61	+4.15	+3.35	+1.55		
3	−66.12	−13.39	−10.11	−9.84		
	+27.00	−8.31	−17.03	−17.51		
7	+9.93	+15.64	+15.64	+3.75	+10.42	
8	+16.12	+2.07	−0.62	−16.95	−8.06	
9	−57.50	−19.00	−34.50	−29.00	−32.25	
	+33.67	+1.33	+10.67	+5.67	+7.33	
4	+12.43	+13.51	+3.86			+1.08
6	−62.00	−98.00	+23.00			+86.40
	+8.33	+7.61	−8.01			−15.94
6	−25.12	−27.67	−16.63			−15.12
	+38.73	+44.12	+27.35			+21.57

$$\% \text{ error} = \frac{\text{comp. value} \times 100}{\text{measured value}} - 100$$

* Part of values from Table 5-1.

of friction, etc., it appears that the Winkler model is adequate for computational purposes.

5-12 DESIGNING THE FOOTING AS A BEAM ON AN ELASTIC FOUNDATION

The design of a footing as a beam on an elastic foundation requires the analysis to proceed initially as for the conventional (rigid) design of Chap. 3. This establishes depth for shear and the width to satisfy, initially at least, the allowable bearing capacity.

With the overall depth D including shear depth plus 3 in (7.5 cm) of steel cover as

$$D = d + 3 \text{ in} \qquad \text{or} \qquad D = d + 7.5 \text{ cm}$$

and the column loads converted to P_{ult} the solution is made. Note that one must convert loads to ultimate for ultimate-strength design. One should not convert k_s by a load-ratio factor as this would make the soil too stiff. The resulting computed deflections will be too large by the ultimate load-ratio factor, as expected.

The bending moments for the beam on an elastic foundation are always smaller than the rigid solution, as illustrated in Fig. 5-7; thus, some economy in steel requirements for bending results. This solution probably yields a more realistic description of soil pressure and is therefore to be recommended.

It is further recommended that the design moments be increased at least 10 percent to account for the fact the computed moments tend more often to be under the measured values than over. This increase will require slightly more steel but still considerably less than required in the conventional design.

Transverse steel should be computed using either the method of Chap. 3, of considering a zone centered on and containing the column of width

$$a + 3d$$

or the method of treating the footing as a mat (Chap. 7). If computed moments are too small, it may be necessary to decrease the appropriate footing dimensions and reanalyze.

5-13 THE COMPUTER SOLUTION

The steps in the solution of a beam on an elastic foundation by the finite-element method are as follows:

1 Make a sketch of the footing system and code the structure for *P-X* and *F-e* as in Fig. 5-1. The included computer program is set up for the coding shown in the figure, including the directions shown as positive.

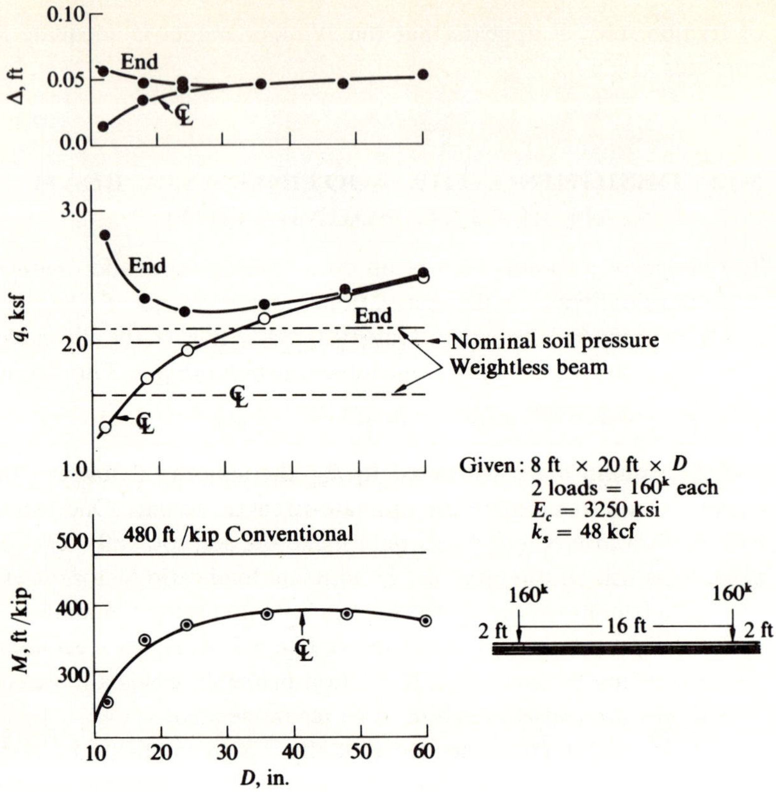

FIGURE 5-7
Influence of beam depth on computed values. Soil pressure increases partly because footing weight is included in analysis and varies from 24 kips at $D = 12$ in to 120 kips at $D = 60$ in. Deflections and moments include beam weight.

2 Generate the A matrix. Note that it is not necessary to generate the A^T matrix since the SA^T matrix can be formed directly from the A matrix by appropriate subscripting. This conserves computer core.

3 Generate the S matrix, preferably in two columns to conserve computer core.

4 Form the SA^T matrix. This matrix must be retained (stored), as it is needed both immediately to form

$$ASA^T = A \times SA^T$$

and after inverting the ASA^T to compute the F matrix as

$$F = SA^T \times X$$

5 Form the ASA^T matrix and invert it. Note that the ASA^T will always be of size

$$P \times P \qquad \text{or} \qquad 2N + 2$$

where N is number of beam segments.

6 Compute the nodal displacement X as

$$X = [ASA^T]^{-1}P$$

Note that X_i may be a rotation (radians) or translation (units of length) determined by inspection of the P-X diagram.

7 Check the translation elements of the X matrix for zero, negative, or values which are in nonlinear deflection range.

a For negative or zero values of translation, zero out the appropriate K values in the S matrix.

b For X values larger than the predetermined value X_{max}, which represents soil deformation in the linear range, multiply the appropriate K value in the S matrix to obtain a force

$$G_i = S_i \times X_{max}$$

Now zero the K value out of the S matrix (requiring storage of all the soil K values under an alternate identity).

8 Apply the G_i force at the ith node as a negative P force in the P matrix so that the resulting P value at that node is

$$P_i + G_i = PM_i$$

9 Recycle with the modified S matrix and reform SA^T, ASA^T, and X.

10 Repeat steps 8 and 9 until:

a Either the same or a smaller number of nodes in the current solution have zero, negative, or maximum deflections than in the immediate preceding solution; i.e., either convergence or oscillation has been obtained.

b Or a predetermined number of cycles has been made.

c It should be noted that if all the K values are made zero, an unstable situation will exist and the program will "blow up" unless this provision is taken into account.

11 Note that the use of the second data card (after TITLE) with UT1 through UT6 and FU1 through FU4 (and used with the same FORMAT in all the included computer programs, which optionally compute in fps or metric

units) allows the printing of I/O consistent with the units used in the program. For convenience to the user the units are defined below:

UT1	UT2	UT3	UT4	UT5	UT6	FU1	FU2	FU3	FU4
FT	IN	KIPS	FT–KIPS	KIPS/SQ FT	KIPS/CU FT	12.	1.	144.	0
M	CM	KN	KN–M	KN/SQ M	KN/CU M	100.	0.3	10.	0

Note that KN is used for kilonewtons and that FU4 is left for the user to fill in with any desired constant of conversion.

The Computer Program

Line	Operation
1–5	Bookkeeping; note storing the ASA^T over the A matrix and double precision of UT5 and UT6 so that eight spaces in A-format code can be used
6	READ TITLE, UNIT card (two cards)
8	READ KL = number of beam divisions; JJS = number of nodes requiring correction with separate data cards in the S matrix; LIST = output control; if LIST > 1, prints A, SA^T, and ASA^T
12–18	Computes control counters
19	READ (7F10.4, 2I5) XL,BX,DC = beam length, width, and depth; XK = soil modulus; EC = modulus of elasticity; XMAX = maximum linear soil deflection (inches or centimeters); UNITWT = unit weight of beam (use 0. for weightless beams); L1 = number of columns between nodes; L2 = number of columns on nodes. Note only XMAX is in units other than feet or meter units; use kips or kilonewtons for all force units
24	Computes moment of inertia (ft^4 or m^4)
35	Converts XMAX to feet or meters using FU1
36–74	Builds and writes A matrix. For nonlinear soil effect loops back to here since ASA^T is stored over the A matrix
75–93	Builds S matrix in two columns. Note JJS > 0 is used on line 88
94–150	Builds P matrix. Note if L1 > 0, program computes fixed-end moments. READ line 108 for L1 > 0; READ line 130 if L2 > 0. Unit weight of beam is concentrated at nodes and added to P matrix, as are nonlinear soil effects (line 144) on recycles
159–178	Builds and writes (LIST > 0)SA^T
179–199	Builds and writes ASA^T, but if written, it is factored by 10^3
200–211	Inverts the ASA^T matrix
212–215	Computes X matrix (deflections, feet or meters)
216–222	Computes F matrix and adjusts if nonlinear soil values are used [G(I–KD)] in earlier cycles
266–283	Tests for nonlinear soil effects

```
      C       J E BOWLES  MATRIX ANALYSIS OF BEAM ON AN ELASTIC FOUNDATION
      C       KL = NUMBER OF BEAM ELEMENTS
      C       JJS = NO OF CORRECTIONS IS S-MATRIX S(I,1) OR S(I,2)
      C       L1 = NO OF COLS BETWEEN NODES;  M1 = LT NODE NO; N1 = RT NODE NO
      C       T = DIST LT NODE TO COL;  Y = DIST RT NODE TO COL, FT
      C       L2 = NO OF COLS AT NODE POINTS
      C       LIST = CONTROL AMOUNT OF OUTPUT LIST IF LIST >0 LIST A,ASAT,SAT
      C       ** PROG WORKS ONLY 1 LOAD CONDITION SINCE ASAT IS STORED OVER A-MA
      C       UNITWT = UNIT WT OF BEAM MATERIAL---PROG CONSIDERS BEAM WT.
0001          DIMENSION X(32),P(32),F(46),F1(32),SS(15),G(22),PM(23)
0002          DIMENSION E(32,32),A(32,46),S(46,2),C(46,32),SOIR(15),V(15),LS(15)
0003          DIMENSION EE(32),Q(22),PW(32),TITLE(20)
0004          EQUIVALENCE (E(1,1), A(1,1))
```

```
           DOUBLE PRECISION UT5,UT6
      6000 READ (1,1000,END=150)TITLE,UT1,UT2,UT3,UT4,UT5,UT6,FU1,FU2,FU3,FU4
      1000 FORMAT (20A4/4(A4,6X),A8,2X,A8,2X,4F5.1)
           READ (1,1002) KL,JJS,LIST
      1002 FORMAT(3I5)
           WRITE (3,1001) TITLE
      1001 FORMAT('1',T5, 20A4)
           M = 3*KL + 1
           N = 2*KL + 2
           KD = M-N
           KP = KL + 1
           KK = 2*KL
           NM1 = N-1
           KLP2 = KL+2
           READ(1,1003)XL,BX,DX,XK,ELAS,XMAX,UNITWT,L1,L2
      1003 FORMAT(7F10.4, 2I5)
           WRITE(3,1004)XL,UT1,BX,UT1,DX,UT1,ELAS,UT5, XK,UT6,XMAX,UT2,
          1UNITWT,UT6
      1004 FORMAT(//T10,'BEAM ON ELASTIC FOUNDATION BY DISPLACEMENT METHOD'//
          1T15,'BEAM LENGTH = ',F5.2,1X,A2/T15,'BEAM WIDTH = ',F5.2,1X,A2 /
          2T15,'BEAM DEPTH =',F7.4,1X,A2/T15,'BEAM MOD OF ELAS =',F10.0,1X,A7
          3             /T15,'MOD. OF SUBGRADE REAC. = ',F8.0,1X,A7/T15,'THE MA
          4X VALUE OF LINEAR SOIL DEFL = ',F5.2,1X,A2/T15,'THE UNIT WEIGTH OF
          5 BEAM = ',F7.3,1X,A7)
           IF(DX.LE.0.0001)READ(1,1003)XI
           IF(DX.GT.0.0001)XI=BX*DX**3/12.
           AKL = KL
           H = XL/AKL
           WRITE (3,1007) UT1,UT1
      1007 FORMAT(//T15,'MOMENT OF INERTIA(',A2,'**4)',5X,'SEG LENGTH,'A2)
           DO 512  I = 1,KL
       512 WRITE(3,1008)I, XI, H
      1008 FORMAT(T5,'SEG NO',2X,I2,4X,F12.6,12X,F7.4)
           JK = 2
           LS(1) = 0
           LS(JK) = 0
           XMA = AMAX/FU1
C     A    MATRIX FORMULATION
       498 DO 501  I = 1,N
           DO 501  J = 1,M
       501 A(I,J) = 0.
           A(1,1)=1.0
           A(KP,KK) = 1.
           NN = 2
           DO 503 J=2,KL
           DO 503 I=2,3
           A(J,NN)=1.0
       503 NN=NN+1
           KM = N - 1
           DO 504  J = KLP2,N
           A(J,KM) = -1.0
       504 KM = KM+1
           NN=1
           KM = N-1
           DO 506 J=KLP2,KM
           DO 506 I=2,3
           A(J,NN)=1./H
       506 NN = NN+1
           NN=1
           KLP3 = KL+3
           DO 508 J=KLP3,N
           DO 508 I=2,3
           A(J,NN)=-1./H
       508 NN = NN + 1
C          ******* BYPASS TO #3 ON NON-LINEAR SOIL LOOP
           IF(JK.GT.2)GO TO 3
           IF(LIST.LE.0)GO TO 7010
           WRITE(3,8011)
      8011 FORMAT(    /, T20, 'THE STATICS MATRIX IS AS FOLLOWS',//)
           M2 = 0
       815 M1 = M2+1
           M2 = MIN0(M1+15,M)
           WRITE(3,8015)(I,(A(I,J),J=M1,M2),I=1,N)
      8015 FORMAT(T3,I2,3X,16F7.3)
           IF(M2.GE.M)GO TO 7010
           WRITE(3,8016)
      8016 FORMAT('1',/, T10, 'ADDITIONAL PART OF STATICS MATRIX',//)
           GO TO 815
C     S-MATRIX FORMULATION
      7010 DO 511  I = 1,KK
           S(I,1) = 4.*ELAS*XI/H
           S(I,2) = 2.*ELAS*XI/H
           IF(I/2*2.NE.I)GO TO 511
           SAVE = S(I,2)
           S(I,2) = S(I,1)
           S(I,1) = SAVE
       511 CONTINUE
           DO 514  I = NM1,M
           S(I,1) = H*BX*XK
           S(I,2) = 0.
       514 CONTINUE
           SUML = 0.
C          MODIFICATION OF S-MATRIX FOR POOR SOIL CONDITIONS OR HOLES
           IF(JJS)116,116,122
```

```
  122 DO 115  J = 1,JJS
  115 READ(1,1042) I, S(I,1)
 1042 FORMAT(I5,F10.4)
  116 DO 515  I = 1,KP
  515 SS(I) = S(I+KK,1)
C      END OF S-MATRIX FORMULATION
C      CALCULATION OF P MATRIX WITH EXTERNALLY APPLIED MOM. AND COL. LOAD
      DO 516  I = 1,N
      PW(I) = 0.
      G(I) = 0.
      F1(I) =.0.
  516 P(I) = 0.
      WFTB = DX*BX*UNITWT
      IF(DX.LE.0.001)WFTB = UNITWT
      PW(1+KP) = WFTB*H/2
      PW(N) = PW(1+KP)
      DO 15  I = 2,KL
   15 PW(I+KP) = WFTB*H
      IF(L1.EQ.0) GO TO 5
      WRITE (3,1010) UT3,UT4
 1010 FORMAT(//,T10,'THE COL POSITIONS AND LOADS:',/,T23, 'DIST FROM NOD
     1E TO COL',/,T5,'COL NO', 2X,'ADJ NODES',2X,'LT NODE',7X,'RT NODE',
     22X,'COL LOAD,',A4,2X,'MOMENT,',A4)
C      **COMPUTE FIXED END MOMENTS AND SHEAR FOR COLS BETWEEN NODES
      DO 518  I = 1,L1
      READ(1,1011)M1,N1,T,Y,COLLD,AMOM
      SUML = SUML + COLLD
 1011 FORMAT(2I5,4F10.5)
      WRITE(3,1012)I,M1,N1,T,Y,COLLD,AMCM
 1012 FORMAT(T7,I2,T13,I2,T18,I2,5X,F6.3,8X,F6.3,3X,F8.2,6X,F9.2)
      RMOMA=AMOM*(2.*T*T*Y+T*Y*Y-Y**3)/((T+Y)**3)
      RMOMB=AMOM*(2.*T*Y*Y+T*T*Y-T**3)/((T+Y)**3)
      P(M1)=(COLLD*T*Y*Y)/((T+Y)**2)+(-RMOMA)+P(M1)
      P(N1)=-(COLLD*T*T*Y)/((T+Y)**2)-RMCMB+P(N1)
      MX2=2*M1
      MX1=MX2-1
      F1(MX1)=-P(M1)
      F1(MX2)=-P(N1)
      M1 = M1 + KP
      N1 = N1 + KP
      P(M1)=COLLD*Y*Y*(3.*T+Y)/((T+Y)**3)-(RMOMA+RMOMB+AMOM)/(T+Y)+P(M1)
  518 P(N1)=COLLD*T*T*(3.*Y+T)/((T+Y)**3)+(RMOMA+RMOMB+AMOM)/(T+Y)+P(N1)
      WRITE(3,1014)
 1014 FORMAT('1',//,T20,'THE FIXED-END MOMENTS ARE AS FOLLOWS')
      WRITE(3,1015)(I,F1(I),I = 1,KK)
 1015 FORMAT(5(I3,2X,F10.3))
    5 IF(L2.EQ.0) GO TO 520
      DO 520  I = 1,L2
      READ(1,1016)MRD,COLLD,AMOM
 1016 FORMAT(I5,2F10.5)
      WRITE (3,1032) MRD,COLLD,UT3,AMOM,UT4
 1032 FORMAT(//T5,'COL NODE POINT =',I3,5X,'P =',F8.2,1X,A4,3X,'COL MOME
     1NT =',F9.3,1X,A4)
      P(MRD)=AMOM
      M1 = MRD + KP
      P(M1)=COLLD
      SUML = SUML + COLLD
  520 CONTINUE
C****** BUILD P-MATRIX---LOOP RE-ENTERED HERE FOR NON-LINEAR COMPUTATION
    3 SUMWT = 0.
      DO 6  I = 1,N
      SUMWT = SUMWT+PW(I)
    6 PM(I) = P(I)+PW(I)+G(I)
      SUM1 = SUML+SUMWT
      WRITE(3,1030)
 1030 FORMAT(//,T8,'P-MATRIX',3X,'BM WT MATRIX',3X,'SOIL MATRIX',3X,'SUM
     1 = PM(I)')
      WRITE(3,1031)(I,P(I),PW(I),G(I),PM(I), I = 1,N)
 1031 FORMAT(T5,I2,F11.3,3X,F10.3,4X,F10.3,4X,F10.3)
      IF(JK.LE.2)GO TO 7
      JKM2 = JK-2
      WRITE(3,1017)JKM2
 1017 FORMAT(//,T10, 'THE MODIFIED STIFFNESS MATRIX FOR CYCLE NUMBER',I3
     1,' IS')
      GO TO 8
    7 WRITE(3,1018)
 1018 FORMAT(     //,T10,'THE INITIAL STIFFNESS MATRIX IN 2-COLS IS')
    8 WRITE(3,8018)(I,(S(I,J),J=1,2),I=1,M)
 8018 FORMAT(T10,I3,5X,F12.2,5X,F12.2)
C      FORM SAT MATRIX*********************         *********************
      DO 529  I = 1,KK
      KA = I
      IF(I/2*2.EQ.I)KA = I-1
      DO 529  J = 1,N
  529 C(I,J) = S(I,1)*A(J,KA) + S(I,2)*A(J,KA+1)
      DO 530  I = KM,M
      DO 530  J = 1,N
  530 C(I,J) = S(I,1)*A(J,I)
      IF(LIST.LE.0)GO TO 6221
      WRITE(3,6215)
 6215 FORMAT('1',//,T5,'THE SAT MATRIX X 10**3')
      M2 = 0
 6298 M1 = M2+1
      M2 = M1+11
```

```
          IF(M2.GT.N)M2=N
          DO 6216  I = 1,M
     6216 WRITE(3,6241)I,  (C(I,J),J=1,N)
          IF(M2.GE.N)GO TO 6221
          WRITE(3,6270)
          GO TO 6298
    C     FORM ASAT MATRIX AND STORE IN STATICS MATRIX CORE AREA
     6221 DO 5532  I = 1,N
          DO 531  J = 1,N
          EE(J) = 0.
          DO 531  K = 1,M
      531 EE(J) = EE(J) + A(I,K)*C(K,J)
          DO 5532  J = 1,N
     5532 A(I,J) = EE(J)
          IF(LIST.LE.0)GO TO 6219
          WRITE(3,6217)
     6217 FORMAT(//,T10,'THE ASAT MATRIX X 10**3')
          M2 = 0
     6300 M1 = M2+1
          M2 = M1+11
          IF(M2.GT.N)M2=N
          DO 6218  I = 1,N
     6218 WRITE(3,6241)I,  (A(I,J),J=1,N)
     6241 FORMAT(T2,I2,-3P12F10.4)
          IF(M2.GE.N)GO TO 6219
          WRITE(3,6270)
     6270 FORMAT(//)
          GO TO 6300
    C     INVERT MATRIX USING GAUSS-JORDAN METHOD
     6219 DO 25 K=1,N
          DO 20 J=1,N
       20 IF(J.NE.K)E(K,J)=E(K,J)/E(K,K)
          DO 21 I=1,N
          IF(I.EQ.K) GO TO 21
          DO 21 J=1,N
          IF(J.EQ.K) GO TO 21
          E(I,J)=E(I,J)-E(K,J)*E(I,K)
       21 CONTINUE
          DO 22 I=1,N
       22 IF(I.NE.K)E(I,K)=-E(I,K)/E(K,K)
       25 E(K,K)=1./E(K,K)
          DO 26  I = 1,N
          X(I)=0.
          DO 26  K = 1,N
       26 X(I)=X(I)+E(I,K)*PM(K)
    C     ****COMPUTE FORCE MATRIX****
          DO 27  I = 1,M
          F(I)=0.
          DO 27  K = 1,N
       27 F(I) = F(I) + C(I,K)*X(K)
          IF(JK.LE.2)GO TO 29
          DO 28  I = NM1,M
       28 F(I) = F(I) + G(I-KD)
       29 WRITE (3,1020) UT3,UT4,UT1,UT3,UT4
     1020 FORMAT('1',T9,'THE LOAD MATRIX',18X,'THE JOINT DEFLECTIONS',17X,'T
         1HE FORCE MATRIX'/T8,'(',A4,' OR ',A4,') IS',16X,'(',A2,' OR RADIAN
         2S) ARE',20X,'(',A4,' OR ',A4,') IS'/)
          DO 30  I = 1,KK
       30 WRITE(3,1021) I, P(I), I, X(I), I, F(I)
     1021 FORMAT(T7,'LOAD DIR.',I3,F10.4,T39,'JOINT DIR.',I3,F13.8,T79,'MOME
         1NT',I3,F10.4)
          DO 35  I = NM1,N
          NUM=I+1
       35 WRITE(3,1022) I, P(I), I, X(I), I, F(I)
     1022 FORMAT(T7,'LOAD DIR.',I3,F10.4,T39,'JOINT DIR.',I3,F13.8,T79,'FORC
         1E ',I3,F10.4)
          IF(NUM.EQ.(M+1))GO TO 43
          WRITE(3,1023)(I,F(I),I=NUM,M)
     1023 FORMAT(T79,'FORCE ',I3,F10.4)
    C     CALCULATION OF SHEARS AND FINAL FIXED-END MOMENTS
       43 IF(KL.GE.10)WRITE(3,1025)
     1025 FORMAT('1')
          WRITE (3,1024) UT3,UT4,UT3,UT5
     1024 FORMAT(///T5,'SHEAR AT EACH',T22,'BEND. MOMENT AT EACH ORDINATE',T
         155,'SOIL REACTION AT',T80,'SOIL PRESSURE'/T6,'SEGMENT,',A4,T23,A4,
         2' (ORIG. FEMS ADDED)',T58,'EA. ORD., ',A4,T85,A7//)
          DO 45  I = 1,KK
       45 F(I)=F(I)+F1(I)
          VI = 0.
          K = KP
          J = KK
          DO 50  I = 1,KL
          K=K+1
          J=J+1
          V(I) = VI-PM(K)-F(J)
       50 VI = V(I)
          SUM = 0.
          DO 52  I = 1,KP
          SOIR(I) = -F(I+KK)
          Q(I) = XK*X(I+KP)
          IF(Q(I).LE.0.0)Q(I)=0.0
       52 SUM = SUM + SOIR(I)
          L=-1
          DO 54  I = 1,KL
```

```
          L=L+2
       54 WRITE(3,1027)I,V(I),I,F(L),SOIR(I),Q(I)
     1027 FORMAT(T5, I2, F10.4, T24, I2, F15.4, T55, F15.4,T83,F8.3)
          FX = -F(KK)
          WRITE(3,1028)KP,FX,SOIR(KP),Q(KP)
     1028 FORMAT(T24,I2,F15.4,T55, F15.4, T83, F8.3)
          WRITE(3,1029)SUM,SUM1
     1029 FORMAT(T32,'SUM OF SOIL REACTIONS =', T55, F15.4, T72,'(', F10.3,
         1T82,')',//)
C         JK = NON-LINEAR LOOP COUNTER FOR UP TO 5 CYCLES (JK.EQ.7) BELOW
          LC = 0
          DO 59  I = KLP2,N
          IF(X(I).LT.0.000001)S(I+KD,1) = C.0
          IF(X(I).GT.0.000001)GO TO 61
          GO TO 58
       61 IF(X(I)-XMA)59,60,60
       60 G(I) = -SS(I-KP)*XMA
          S(I+KD,1) = 0.0
       58 LS(JK) = LC + 1
          LC = LS(JK)
       59 CONTINUE
          JKM1 = JK-1
          IF(LS(JK).LE.LS(JKM1).OR.LS(JK).EQ.KP.OR.JK.EQ.7)GO TO 6000
          WRITE(3,5636)LS(JK),LS(JKM1),JK
     5636 FORMAT(//,T5,'LS(JK) =',I3,' LS(JKM1) =',I3,' JK =',I3)
          WRITE(3,8005)JK
     8005 FORMAT('1',T10,'COMPUTATIONS FOR CYCLE',I3,' FOLLOW',//)
          JK = JK + 1
          GO TO 498
      150 STOP
          END
```

5-14 RING FOUNDATIONS

Little work appears to have been done on ring foundations, yet there are many structures, e.g., antennas or water towers in which a ring foundation provides a satisfactory solution. The theoretical work of Volterra (1952), Volterra and Chung (1955), and Egorov (1965) appear to be the major published efforts, at least in English.

The matrix method as used for beams can also be used for ring foundations in the following manner (refer to Fig. 5-8). The structure is first coded as if the ring were a linear beam for *P-X* and *F-e*. Since the mean radius does not define the center of area, it is proposed that the computations be based on a radius which does define the center of area as follows. Let

$$A_{\text{total}} = (D_o^2 - D_i^2)\frac{\pi}{4}$$

and the outside diameter to split the total area in half is $2R_m$. Therefore, equating half areas gives

$$[(2R_m)^2 - D_i^2]\frac{\pi}{4} = \tfrac{1}{2}(D_o^2 - D_i^2)\frac{\pi}{4}$$

and, solving for R_m,

$$R_m = \sqrt{(D_o^2 - D_i^2)\tfrac{1}{8}}$$

It is this radius on which the column loads should be applied to reduce twisting to a minimum since it is this radius which defines the resultant soil-pressure line when a uniform soil-pressure distribution is assumed across the footing width.

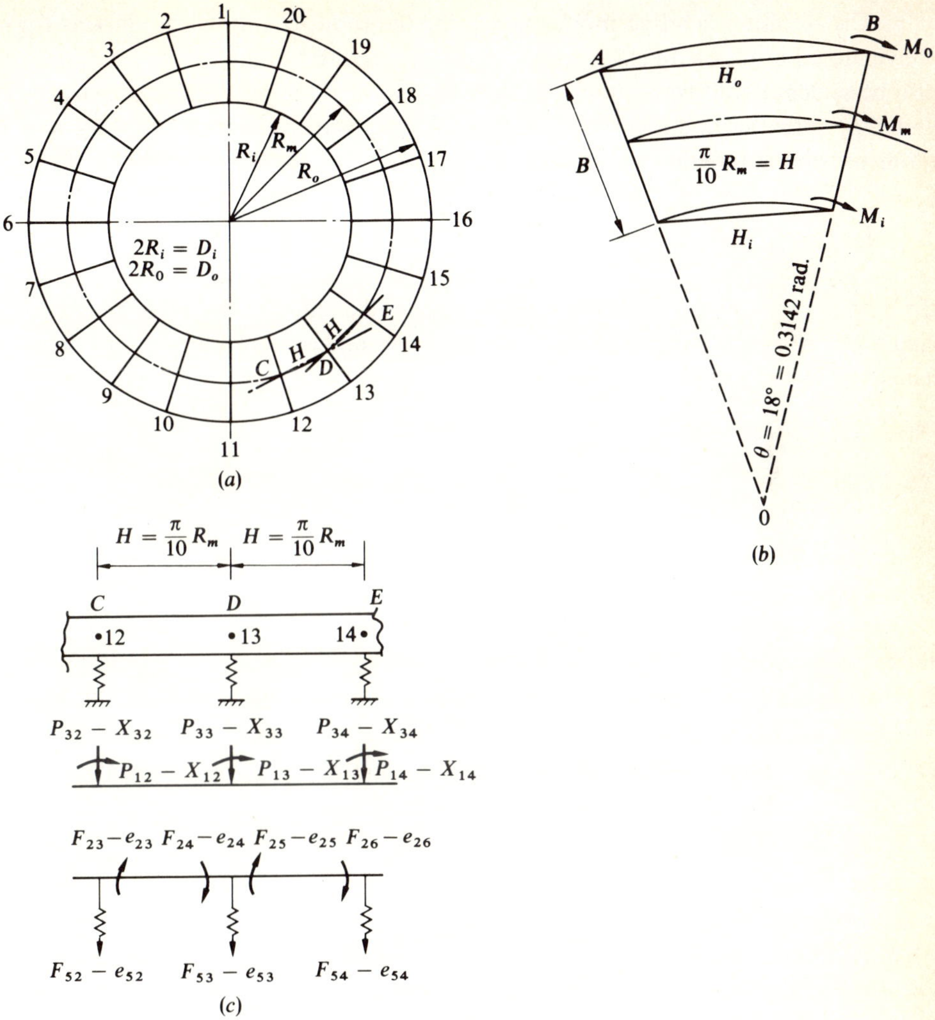

FIGURE 5-8
Matrix solution of ring foundation. Note that R_m is not the average radius using R_i and R_o. (*a*) Ring foundation with 20 segments. (*b*) Tangential moments on one side of the segment. Radial moments perpendicular to the tangential moments are not shown. (*c*) Coding for matrix solution, considering segments *CDE* of part (*a*).

The resulting bending moments are for the total footing width, just as for the conventional beam on an elastic foundation. Referring to Figs. 5-8 and 5-3, it should also be evident that if we assume a constant rotation along any radial line such as OA and OB, with the nodal rotation across the ring width assumed constant and the moment being a function of

$$F_1 = \frac{4EI\theta_1}{L} + \frac{2EI\theta_2}{L}$$

the moment must be larger at the interior edge and smaller at the exterior edge of the ring at approximately the ratio of the inner, outer, and center of area chord distances. Thus,

$$M_i = \frac{M_m}{B}\frac{H_o}{H_i} \qquad \text{inner edge}$$

$$M_m = \frac{M}{B} \qquad \text{center-of-area radius}$$

$$M_o = \frac{M_m H}{BH_o} \qquad \text{outer edge}$$

It is proposed that radial moments be computed analogous to the method for conventional solutions of using a radial zone of width $a + 3d$ centered on the column at the center of area. Alternatively, one may compute the radial moment as

$$M_r = M_m \sin\frac{\phi}{2}$$

where ϕ is the central angle of any segment, as shown in Fig. 5-8.

A comparison of Volterra's problem [Volterra (1952)] is shown in Fig. 5-9. Since the author's method places the column loads on the center of area, no comparison can be made of tangential twisting (or torsional) moments. Volterra did not compute radial moments because his solutions were for beams of very narrow width B.

To compute shear at the nodes use the central-finite-difference expressions for y''' given in Table 4-1.

EXAMPLE 5-4 Compute the tangential bending moments and shears for the 20 node points of Fig. 5-8 for four equally spaced loads of 150 kips each [Volterra (1952)]. Other data: E_c = 468,000 ksf; k_s = 86.4 kcf; ID = 47.5 ft; OD = 52.5 ft; DC =

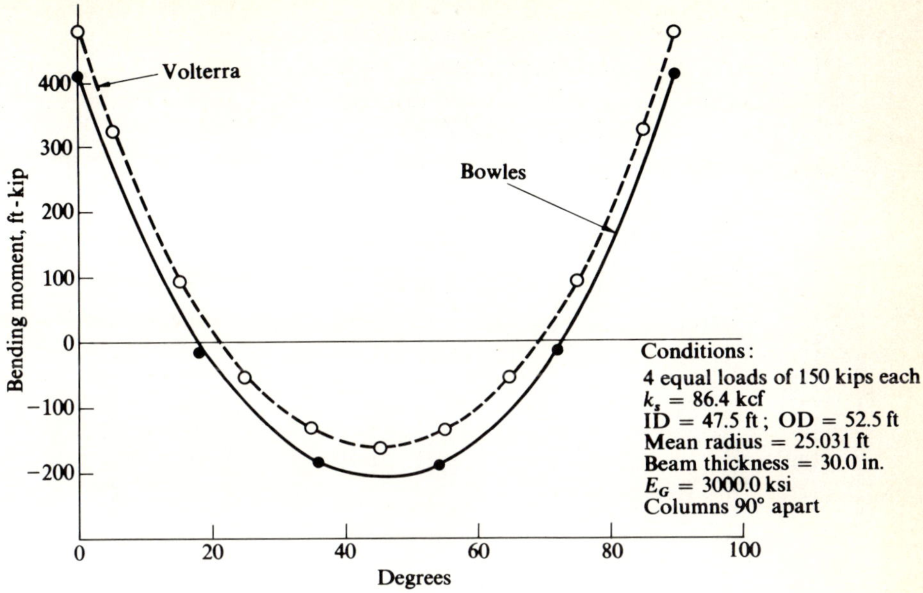

FIGURE 5-9
Comparison of the matrix solution and that of Volterra (1952). Twenty divisions are used in the matrix solution. The difference is not over 15 percent and may be less if more divisions are used in the matrix solution.

2.5 ft; columns spaced at 90°. The load matrix is P(1) = P(6) = P(11) = P(16) = 150 kips.

The data cards for the computer program are as follows:

Card	Data						
1	TITLE (see output of Fig. E5-4.1)						
2	UNITS (UT1–UT6 and FU1–FU4)						
3	86.4	468000.	47.5	52.5	2.5	(5F10.4)	Footing data
4	150.	0.	0.	0.	0.	150.	0.
5	0.	0.	0.	150.	0.	0.	0.
6	0.	150.	0.	0.	0.	0.	0. (7F10.5)

This represents the total input data.

SOLUTION The solution is presented in Fig. E5-4.1 as computer output. Note that the solution has also been plotted in Fig. 5-9 to make a comparison with the published solution. Output is for total footing width, not per ft of width.

```
      J E BOWLES  EXAMPLE 5-4  RING FOUNDATION CHECK VOLTERRA (1952)

    THE INPUT DATA/COMPUTATIONS AND COMPUTATION CONSTANTS:
    SKS =       86.40 K/CU FT   MOD OF ELAS =    468000. K/SQ FT    ID = 47.50 FT
    OD = 52.50 FT    DC = 2.500 FT
    MOM OF INERTIA = 3.25521        FT**4    MEAN RADIUS = 25.031 FT    H =  7.863807
    AREA =     392.698 SQ FT    E =  0.127165    B =     1696.46
    C =  774910.813    D =   387455.375
          THE P-MATRIX IS AS FOLLOWS:
       150.00     0.0       0.0        0.0       0.0   150.00     0.0       0.0       0.0       0.0
       150.00     0.0       0.0        0.0       0.0   150.00     0.0       0.0       0.0       0.0

          *** THE OUTPUT IS AS FOLLOWS ***
   PT    DEFL     MI        MCTR       MEXT       VI        VCT       VEXT      Q      SOIL R
   NOS    IN     FT-K       FT-K       FT-K      KIPS      KIPS       KIPS   K/SQ    KIPS
   1 -.2951    453.81    408.54    371.49     0.00     0.00     0.00 -2.125 -41.72
   2 -.2302    -19.10    -17.19    -15.63   -63.33   -54.10   -46.91 -1.657 -32.54
   3 -.1528   -207.82   -187.09   -170.12   -27.43   -23.43   -20.31 -1.100 -21.60
   4 -.1528   -207.82   -187.09   -170.12    27.43    23.43    20.31 -1.1C0 -21.60
   5 -.2302    -19.11    -17.20    -15.64    63.33    54.10    46.91 -1.657 -32.54
   6 -.2951    453.83    408.56    371.50     0.00     0.00     0.00 -2.125 -41.72
   7 -.2302    -19.10    -17.19    -15.63   -63.33   -54.10   -46.91 -1.657 -32.54
   8 -.1528   -207.81   -187.08   -170.11   -27.43   -23.43   -20.31 -1.100 -21.61
   9 -.1528   -207.82   -187.09   -170.12    27.43    23.43    20.31 -1.100 -21.61
  10 -.2302    -19.11    -17.20    -15.64    63.33    54.10    46.91 -1.657 -32.54
  11 -.2951    453.83    408.56    371.50     0.00     0.00     0.00 -2.125 -41.72
  12 -.2302    -19.09    -17.19    -15.63   -63.33   -54.10   -46.91 -1.657 -32.54
  13 -.1528   -207.81   -187.08   -170.11   -27.43   -23.43   -20.31 -1.100 -21.61
  14 -.1528   -207.82   -187.09   -170.12    27.43    23.43    20.31 -1.100 -21.61
  15 -.2302    -19.10    -17.19    -15.64    63.33    54.10    46.91 -1.657 -32.54
  16 -.2951    453.83    408.56    371.51    -0.00    -0.00    -0.00 -2.125 -41.72
  17 -.2302    -19.09    -17.19    -15.63   -63.33   -54.10   -46.91 -1.657 -32.54
  18 -.1528   -207.81   -187.08   -170.11   -27.43   -23.43   -20.31 -1.100 -21.61
  19 -.1528   -207.82   -187.09   -170.12    27.43    23.43    20.31 -1.100 -21.61
  20 -.2302    -19.09    -17.18    -15.63    63.33    54.10    46.91 -1.657 -32.54
     SUM P =    6C0.000 KIPS    SUM V =   0.00 KIPS    SUM R =      -600.025 KIPS
```

FIGURE E5-4.1
Ring-foundation output. ////

5-15 COMPUTER PROGRAM FOR RING FOUNDATIONS

This computer program will solve any ring foundation with 20 divisions (segments) in either fps or metric units using UT1 through UT6 as the second data card. Read all dimensions in feet or meters, forces in kips or kilonewtons. Any number of column loads may be used, but if columns are not on nodes, prorate the load using simple beam theory, as with (normally) short segments the fixed-end moments will not be significant in the computations. The user must prorate the column loads and can include the beam weight as part of the data read into the P matrix.

Line	Operation
1–6	Bookkeeping, note storage of the ASA^T (Z) over the A matrix
7	READ TITLE, UNITS (use two cards)
11	READ SKS = soil modulus; EC = modulus of elasticity; ID,OD,DC = inside and outside diameters and footing depth (feet or meters)
14	Computes moment of inertia of cross section
15	Computes mean radius of ring (center of area)
16–24	Computation constants
25–44	Builds A matrix
45–57	Builds S matrix in two columns
59–66	Builds SA^T matrix
67–74	Builds ASA^T and stores over A matrix
75	READ P matrix (always three cards at 7F10.5)
85–102	Inverts ASA^T matrix
103–111	Computes X matrix and F matrix
119–124	Computes segment shears using finite-difference equations
141	Converts deflections to inches or centimeters using FU1

```
C   J E BOWLES  FINITE ELEMENT METHOD FOR RING FOUNDATION -- 20 DIVISIONS
C         USE ALL UNITS FT (M) OR KIPS (KILC-NEWTONS) AREAS IN FT OR M
C         IF COL LOAD NOT ON NODE, PRORATE TO ADJACENT NODES
C            USING SIMPLE BEAM THEORY
      DIMENSION A(40,60),S(60,2),SAT(60,40),EE(60),P(40),X(40),F(60)
      DIMENSION Q(20),EMI(20),EMC(20),EMEX(20),VI(20),VC(40)
      DIMENSION VEX(20),R(20),Y(20),Z(40,40),TITLE(20)
      DOUBLE PRECISION UT5,UT6
      REAL*4 ID,INER
      EQUIVALENCE (Z(1,1),A(1,1))
 6000 READ (1,1000,END=150)TITLE,UT1,UT2,UT3,UT4,UT5,UT6,FU1,FU2,FU3,FU4
 1000 FORMAT (20A4/4(A4,6X),A8,2X,A8,2X,4F5.1)
      WRITE(3,1001)TITLE
 1001 FORMAT('1',//,T5,20A4)
      READ(1,4)SKS,EC,ID,OD,DC
    4 FORMAT(5F10.4)
      BBB=(OD-ID)*.5
      INER = BBB*DC**3/12.
      RM =SQRT((OD**2+ID**2)/8.)
      H=RM*0.31416
      HA=2.*RM/ID
      HB=2.*RM/OD
      AREA=(OD**2-ID**2)*3.14159/4.
      E = 1./H
      B=SKS*AREA/20.
      C = (4.*EC*INER)/H
      D=C/2.
      W = (EC*INER)/H**3
      DO 8 I=1,40
      DO 8 J =1,60
      A(I,J) = 0.
    8 CONTINUE
      DO 14 I=1,19
      N=2*I
      A(I+1,N)=1.
      A(I+1,N+1)=1.
      A(I+20,N)=E
      A(I+20,N-1)=E
      A(I+21,N)=-E
      A(I+20,I+40)=-1.
   14 A(I+21,N-1)=-E
      A(1,1)=1.
      A(1,40)=1.
      A(40,60)=-1.
      A(40,39)=E
      A(40,40)=E
      A(21,39)=-E
      A(21,40)=-E
      DO 20 I=1,40
      S(I,1) =C
      S(I,2) =D
      KA =I
      IF(KA/2*2.EQ.I)GO TO 9
      GO TO 20
    9 SAVE=S(I,2)
      S(I,2) =S(I,1)
      S(I,1) =SAVE
   20 CONTINUE
      DO 25 I =41,60
      S(I,1)=B
      S(I,2)= 0.0
   25 CONTINUE
C     FORM SAT MATRIX IN 2-STEPS
      DO 26  I = 1,40
      DO 26 J =1,40
      KA =I
      IF(I/2*2.EQ.I)KA =I-1
   26 SAT(I,J)=S(I,1)*A(J,KA)+S(I,2)*A(J,KA+1)
      DO 73 I =41,60
      DO 73 J=1,40
   73 SAT(I,J) = S(I,1)*A(J,I)
C       FORM ASAT MATRIX AND STORE IN A-MATRIX LOCATION
      DO 75 I=1,40
      DO 27  J = 1,40
      EE(J) = 0.
      DO 27  K = 1,60
   27 EE(J) = EE(J)+A(I,K)*SAT(K,J)
      DO 74 J=1,40
   74 Z(I,J) =EE(J)
   75 CONTINUE
      DO 30 I=1,20
   30 P(I)=0.0
   80 READ(1,31)(P(I),I=21,40)
   31 FORMAT(7F10.5)
      WRITE(3,845)SKS,UT6,EC,UT5,ID,UT1,OD,UT1,DC,UT1,INER,UT1,RM,UT1,H,
     1AREA,UT1,E,B,C,D
  845 FORMAT(/,T5, 'THE INPUT DATA/COMPUTATIONS AND COMPUTATION CONSTANT
     1S:',/,T5, 'SKS =',F10.2,1X,A7,3X,'MOD OF ELAS =' ,F10.0,1X,A7,3X,'
     2ID =',F6.2,1X,A2,/,T5,'OD =',F6.2,1X,A2,3X, 'DC =',F6.3,1X,A2,/,T5
     3, 'MOM OF INERTIA =',G12.6,1X,A2,'**4',3X, 'MEAN RADIUS =',F7.3,1X
     4,A2,3X,'H =',F10.6,/,T5,'AREA =',F10.3,1X,'SQ ',A2,3X,'E =',F10.6,
     53X,'B =',F10.2,/,T5,'C =',F12.3,3X,'D =',F12.3)
      WRITE(3,52)
   52 FORMAT(T10,'THE P-MATRIX IS AS FOLLOWS:')
      WRITE(3,53)(P(I),I=21,30),(P(I),I=31,40)
```

```
   53 FORMAT(T5,10F8.2,/,T5,10F8.2)
C     *** INVERT ASAT MATRIX
      N = 40
      DO 562  K = 1,N
      DO 563  J = 1,N
      IF(J.EQ.K)GO TO 563
      Z(K,J) = Z(K,J)/Z(K,K)
  563 CONTINUE
      DO 565  I = 1,N
      IF(I.EQ.K)GO TO 565
      DO 565  J = 1,N
      IF(J.EQ.K)GO TO 565
      Z(I,J) = Z(I,J)-Z(K,J)*Z(I,K)
  565 CONTINUE
      DO 567  I = 1,N
      IF(I.EQ.K)GO TO 567
      Z(I,K) = -Z(I,K)/Z(K,K)
  567 CONTINUE
      Z(K,K) = 1./Z(K,K)
  562 CONTINUE
C     *******END OF ASAT MATRIX INVERSION
      DO 35 I=1,40
      X(I) = 0.
      DO 35 J=1,40
   35 X(I)=X(I)+Z(I,J)*P(J)
      DO 38 I=1,60
      F(I) = 0.
      DO 38 J=1,40
      F(I) = F(I)+SAT(I,J)*X(J)
   38 CONTINUE
      SUMP = 0.
      DO 40 I=1,20
      N=I+40
      Y(I)=F(N)/B
      M=2*I-1
      SUMP = SUMP+P(N-20)
   40 EMC(I)=F(M)
C     COMPUTE SHEAR USING FINITE DIFFERENCE EXPRESSIONS
      VC(1)=(Y(3)-2.*Y(2)+2.*Y(20)-Y(19))*W
      VC(2)=(Y(4)-2.*Y(3)+2.*Y(1)-Y(20))*W
      VC(19)=(Y(1)-2.*Y(20)+2.*Y(18)-Y(17))*W
      VC(20)=(Y(2)-2.*Y(1)+2.*Y(19)-Y(18))*W
      DO 41 I=3,18
   41 VC(I)=(Y(I+2)-2.*Y(I+1)+2.*Y(I-1)-Y(I-2))*W
      DO 42 I=1,20
      VI(I)=VC(I)*(HA**3)
      VEX(I)=VC(I)*(HB**3)
      EMI(I)=EMC(I)*(HA**2)
   42 EMEX(I)=EMC(I)*(HB**2)
      WRITE(3,61)
   61 FORMAT(// ,T10,'*** THE OUTPUT IS AS FOLLOWS ***')
      WRITE(3,62)UT2,UT4,UT4,UT4,UT3,UT3,UT3,UT5,UT3
   62 FORMAT(T4,'PT',3X,'DEFL',3X,'MI',6X,'MCTR',4X,'MEXT',4X,'VI',5X,'V
     1CT',4X,'VEXT',4X,'Q',4X,'SOIL R',/,T4,'NOS',2X,A2,4X,A4,5X,A4,5X,A
     24,3X,A4,3X,A4,1X,A7,2X,A4,2X,A4)
      SUN=0.
      SUM=0.0
      DO 46 I=1,20
      Q(I)=Y(I)*SKS
      SUN=SUN+VC(I)
      N=I+40
      SUM=SUM+F(N)
      Y(I)=Y(I)*FU1
   46 WRITE(3,65)I,Y(I),EMI(I),EMC(I),EMEX(I),VI(I),VC(I),VEX(I),Q(I),F(
     1N)
   65 FORMAT(T2,I2,1X,F6.4,1X,3F8.2,3F7.2,F7.3,F7.2)
      WRITE(3,66)SUMP,UT3,SUN,UT3,SUM,UT3
   66 FORMAT(T4,'SUM P =',F10.3,1X,A4,3X, 'SUM V =', F6.2,1X,A4,3X, 'SUM
     1 R =',F12.3,1X,A4)
      GO TO 6000
  150 STOP
      END
```

PROBLEMS

5-1 Repeat the assigned part of Prob. 3-5 as a beam on an elastic foundation. Do the problem for the following conditions (and 10 divisions):
(*a*) Three values of k_s including $k_s = 36q_a$.
(*b*) Three values of D (total concrete depth) including the value used in Chap. 3.
(*c*) Three values of f'_c (varies E_c) including the assigned values of Chap. 3.
Comment on the effects of these variables on a solution.

5-2 Repeat the assigned problem of Chap. 3 if *each* column has an applied ultimate moment of

$$M_u = 0.2P_{\text{ult}}$$

5-3 Repeat the assigned problem of Chap. 3 using a number of divisions other than 10 as assigned by the instructor.

5-4 Solve a ring foundation using the included computer program if

$$k_s = 100 \text{ kcf} \qquad B = 6 \text{ ft} \qquad f'_c = 3{,}000 \text{ psi}$$
$$DC = 30 \text{ in} \qquad \text{ID} = 24 \text{ ft}$$

for:
(*a*) Four equally spaced working dead loads of 200 kips.
(*b*) Six equally spaced working dead loads of 200 kips.
(*c*) Three loads at 30° (nonsymmetrical) of 300 kips.
Note: with DL, the load factor = 1.4.

5-5 What are the loads of parts (*a*) and (*b*) of Prob. 5-4 to obtain a soil pressure of not over 2 ksf?

5-6 What value of B with the load of the assigned part (*a*), (*b*), or (*c*) of Prob. 5-4 will reduce the soil pressure to not over 2 ksf?

5-7 What is a reasonable explanation for the fact that using five divisions in Example 5-1 provides such erroneous results?

5-8 Modify the ring foundation to solve a ring with *any* number of divisions. If your computer will invert a matrix of 80 × 80, recompute Example 5-2 using 40 divisions and compare the results with Fig. 5-9.

REFERENCES

BIOT, M. A. (1937): Bending of an Infinite Beam on an Elastic Foundation, *J. Appl. Mech. ASME*, vol. 59, pp. A1–A7.

BOWLES, J. E. (1968): "Foundation Analysis and Design," chap. 5 and p. 632, McGraw-Hill, New York.

CHEUNG, Y. K., and D. K. NAG (1968): Plates and Beams on Elastic Foundations, *Geotech. (Lond.)*, vol. 23, no. 2, June, pp. 250–260.

DODGE, ALEXANDER (1964): Influence Functions for Beams on Elastic Foundations, *J. Struct. Div., ASCE*, vol. 90, ST4, August, pp. 63–101.

EGOROV, K. E. (1965): Calculation of Bed for Foundation with Ring Footing, *Proc. 6th Int. Conf. Soil Mech. Found. Eng., Montreal*, vol. 2, pp. 41–45.

GAZIS, D. C. (1958): Analysis of Finite Beams on Elastic Foundations, *J. Struct. Div., ASCE*, vol. 84, ST4, pt. 1, July, Proc. no. 1722.

HETENYI, M. (1946): "Beams on Elastic Foundation," University of Michigan Press, Ann Arbor, 255 pp.

IYENGAR, K. T., and S. ANANTHARAMU (1965): Influence Lines for Beams on Elastic Foundations, *J. Struct. Div., ASCE*, vol. 91, ST3, June, pp. 45–56.

LAURSEN, HAROLD I. (1969): "Structural Analysis," chap. 14, McGraw-Hill, New York.

LEVINTON, Z. (1949): Elastic Foundations Analyzed by the Method of Redundant Reactions, *Trans. ASCE*, vol. 114, pp. 40–78.

MALTER, H. (1960): Numerical Solutions for Beams on Elastic Foundations, *Trans. ASCE*, vol. 125, pp. 757–791.

POPOV, E. P. (1951): Successive Approximations for Beams on Elastic Foundations, *Trans. ASCE*, vol. 116, pp. 1083–1108.

RETI, A. A. (1967): Suggested Design Procedures for Combined Footings and Mats, *J. Am. Concr. Inst.*, vol. 64, no. 6, pt. 2, pp. 1537–1544.

VESIĆ, A. S., and W. H. JOHNSON (1963): Model Studies of Beams Resting on a Silt Subgrade, *J. Soil Mech. Found. Div., ASCE*, vol. 89, SM1, February, pp. 1–31.

VOLTERRA, ENRICO (1952): Bending of a Circular Beam Resting on an Elastic Foundation, *J. Appl. Mech., ASME*, vol. 19, no. 1, March, pp. 1–4.

———, and RANDALL CHUNG (1955): Constrained Circular Beams on Elastic Foundations, *Trans. ASCE*, vol. 120, pp. 301–310.

WANG, C. K. (1970): "Matrix Methods of Structural Analysis," chap. 3, 2d ed., International Textbook, Scranton, Pa.

WOLFER, K. H. (1969): "Elastically Supported Beams: Tables of Coefficients for Bending Moment, Shear Force and Bearing Pressure," 2d ed., trans. by C. van Amerongen, Bauverlag, Wiesbaden, 576 pp.

6

FINITE-DIFFERENCE AND HETENYI SOLUTION OF BEAM ON ELASTIC FOUNDATION

6-1 FINITE-DIFFERENCE SOLUTION

The finite-difference solution of a simply supported beam was presented in Sec. 4-2. A method presented by Malter (1960) extended the idea of using this technique to obtain bending moments in a beam on a Winkler foundation, i.e., replacing the soil mass with a series of soil springs. Teng (1962) included the method, and Bowles (1968) presented the method along with a modest computer program for a solution.

To have a reasonably valid solution one must use more than the six beam segments illustrated in Sec. 4-2. Generally 10 to 20 segments should be employed to define the elastic curve of the beam adequately. Reverse curvature obviously requires a larger number of beam segments than single curvature.

The solution presented here has one major advantage over the matrix solution of Chap. 5, namely, the size of the matrix to be inverted is small, always equal to $N + 1$, where N is the number of beam segments used. The disadvantages were cited in Chap. 5, the major one being the difficulty of building the P matrix.

In Fig. 6-1 a typical beam with two loads is shown (this is the one for which the solution is illustrated later in Example 6-1). Figure 6-1*b* shows the beam divided into 10 segments with soil reactions or springs at each node point. Note that the end K

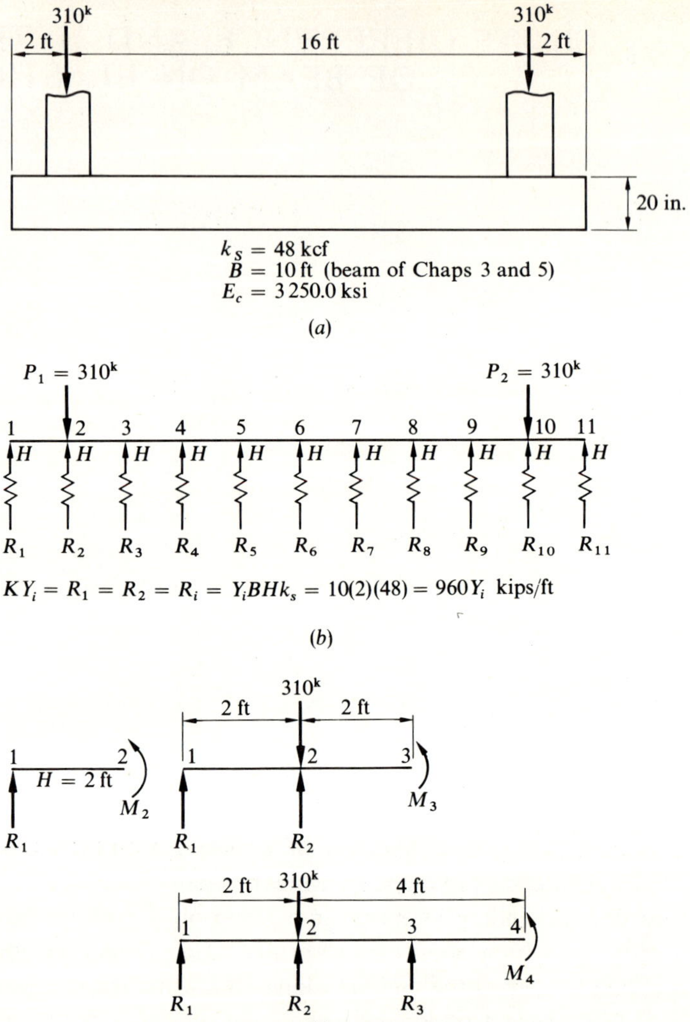

FIGURE 6-1
Beam on an elastic foundation by the finite-difference method: (*a*) concrete combined footing of Chap. 3 with footing total thickness and loads (ultimate) shown; (*b*) beam using 10 segments supported by 11 "springs" as a Winkler foundation; (*c*) free-body diagram to sum moments at nodes 2, 3, and 4.

values are the same as the interior values used in Chap. 5. From Fig. 4-1 and the example of Sec. 4-2, the bending moment at any station is

$$EI\frac{d^2y}{dx^2} = -M$$

or in finite-difference notation (y'' of Table 4-1)

$$\frac{EI}{h^2}(y_{i-1} - 2y_i + y_{i+1}) = -M_i \qquad (6\text{-}1)$$

Now back to the beam and data of Fig. 6-1*b*:

$$E = 3{,}250(144) = 468{,}000 \text{ ksf}$$

$$I = \frac{10(\frac{20}{12})^3}{12} = 3.858 \text{ ft}^4$$

$$h = \tfrac{20}{10} = 2.0 \text{ ft}$$

$$\frac{EI}{h^2} = \frac{3{,}250(144)(3.858)}{2^2} = 451{,}388.0$$

$$R_i = k_s HBy_i = 960.0y_i$$

At point 2 (we could use point 1 with a *forward*-difference expression, but, as we will see, we need only 11 independent equations[1])

$$\sum M_{\text{left}} = 0$$

$$M_2 - 2R_i = 0$$

but

$$M_2 = \frac{EI}{h^2}(y_1 - 2y_2 + y_3)$$

[1] Intermixing central- and forward- (or backward-) difference expressions is mathematically valid but should be done only as a last resort.

Substituting [and recalling $EIy'' = -M$ from Eq. (6-1)], we have

$$\frac{EI}{h^2}(y_1 - 2y_2 + y_3) + 2(R_1) = 0$$

$$451{,}388y_1 - 902{,}776y_2 + 451{,}388y_3 + 2(960y_1) = 0$$

Introducing the remaining y's with zero coefficients to build a legitimate matrix, we obtain

$$453{,}308y_1 - 902{,}776y_2 + 451{,}388y_3 + 0y_4 + \cdots + 0y_{11} = 0 \qquad (1)$$

At point 3, $\sum M_{\text{left}} = 0$ and (note that the column load is in this area of interest)

$$\frac{EI}{h^2}(y_2 - 2y_3 + y_4) + 4R_1 + 2R_2 - 2(P_1) = 0$$

Substituting gives

$$451{,}388y_2 - 902{,}776y_3 + 451{,}388y_4 + 3{,}840y_1 + 1{,}920y_2 = 620$$

and rearranging and including the remaining y's, we have

$$3{,}840y_1 + 453{,}308y_2 - 902{,}776y_3 + 451{,}388y_4 + 0y_5 + \cdots + 0y_{11} = 620 \qquad (2)$$

In a similar manner we can write equations at each node point through node 10 using *central-difference* expressions for moment. This yields nine equations.

The tenth equation will be obtained to satisfy the total problem statics of

$$\sum F_v = 0$$

or

$$\sum_1^n R_i - \sum_1^n P_i = 0$$

or in this case (with 960 factored for writing ease),

$$960(y_1 + y_2 + y_3 + y_4 + y_5 + \cdots + y_{11}) = 620 \qquad (10)$$

To obtain the eleventh equation, let us sum moments about *either* end. The computer program sums about the left end; therefore,

$$960h(0y_1 + 1y_2 + 2y_3 + 3y_4 + 4y_4 + \cdots + 10y_{11}) = 2(310) + 18(310)$$

or

$$0 + 1{,}920y_2 + 3{,}840y_3 + \cdots 192{,}00y_{11} = 6{,}200 \tag{11}$$

We have now formed the necessary number of equations in the general matrix form of

$$A_{11\times 11}Y_{11\times 1} = P_{11\times 1}$$

The A matrix is square, and we can solve for Y by inverting to obtain

$$Y = A^{-1}P$$

EXAMPLE 6-1 Obtain the solution of the example partially formulated and illustrated in Fig. 6-1. Use the included computer program. Note that the computer matrix is slightly different due to a roundoff and smaller by a factor of 10 for output convenience.

SOLUTION The data cards to solve this problem are as follows:

Card	Data
1	TITLE
2	FT IN KIPS FT–KIPS KIPS/SQ FT KIPS/CU FT
	This is the standard UNIT card containing the entries UT1 through UT6
3	48. 468000. 10. 1.6667 2.0 2.0 16.
4	310. 310. 0. 0.

The output is shown on Fig. E6-1.1. Note the checking of M_3 and one-half of shear and moment diagrams. Only one-half of each diagram is shown because of symmetry. ////

```
J E BOWLES   EXAMPLE 6-1--FTG OF CHAP 3 BY FINITE DIFFERENCE

SOIL MODULUS =   48.00 K/CU FT      FTG WIDTH = 10.00 FT        FTG DEPTH =   1.67 FT
MOD OF ELAS =   468CCO. K/SQ FT        H = 2.00 FT      DIST 1ST COL = 2.00 FT
DIST 2ND COL = 16.00 FT        MOM OF INERTIA = 3.8582 FT**4       FTG LENGTH = 20.00 FT

COL LOADS & MOMENTS:        P1 =310.00 KIPS        MOMENT M1 =   0.0   FT-K
                            P2 =310.00 KIPS        MOMENT M2 =   0.0   FT-K

          THE COEFFICIENTS OF THE 11 SIMULTANEOUS EQ ARE (X 10)

45334.  -90283.   45142.       0.       0.       0.       0.       0.       0.       0.       0.      0.
  384.   45334.  -90283.   45142.       0.       0.       0.       0.       0.       0.       0.     62.
  576.     384.   45334.  -90283.   45142.       0.       0.       0.       0.       0.       0.    124.
  768.     576.     384.   45334.  -90283.   45142.       0.       0.       0.       0.       0.    186.
  960.     768.     576.     384.   45334.  -90283.   45142.       0.       0.       0.       0.    248.
 1152.     960.     768.     576.     384.   45334.  -90283.   45142.       0.       0.       0.    310.
 1344.    1152.     960.     768.     576.     384.   45334.  -90283.   45142.       0.       0.    372.
 1536.    1344.    1152.     960.     768.     576.     384.   45334.  -90283.   45142.       0.    434.
 1728.    1536.    1344.    1152.     960.     768.     576.     384.   45334.  -90283.   45142.    496.
   96.      96.      96.      96.      96.      96.      96.      96.      96.      96.      96.     62.
    0.     192.     384.     576.     768.     960.    1152.    1344.    1536.    1728.    1920.    620.

                         THE OUTPUT IS AS FOLLOWS

POINT NOS       1        2        3        4        5        6        7        8        9       10       11
DEFL, FT      0.066    0.063    0.059    0.055    0.053    C.C52    0.053    0.055    0.059    0.063    0.066

SEG SHEAR,KIPS     63.70 -186.07 -129.58  -76.31  -25.19   25.19   76.31  129.58  186.07  -63.70

SHEAR AT FIRST COL, LEFT =    63.70 KIPS      TO RIGHT =  -186.07 KIPS

SHEAR AT SECOND COL, LEFT =   186.07 KIPS      TO RIGHT =    -63.70 KIPS

MOM,FT-K        0.0  -127.43  244.75  503.90  656.56  706.90  656.52  503.91  244.76 -127.40    0.0

Q K/SQ FT       3.18    3.01    2.82    2.66    2.56    2.52    2.56    2.66    2.82    3.01    3.18

SOIL R KIPS     63.7    60.2    56.5    53.3    51.1    50.4    51.1    53.3    56.5    60.2    63.7
```

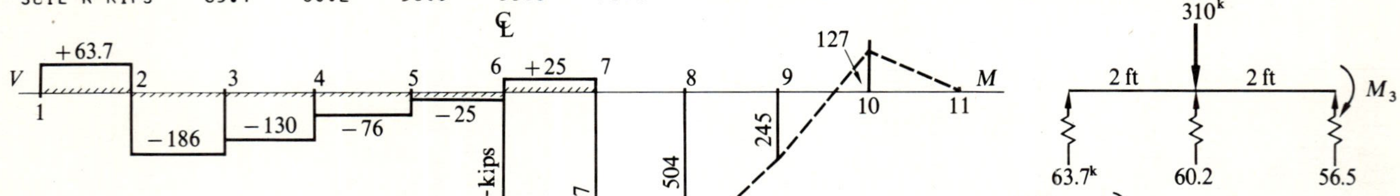

$$\Sigma M_3 = 0 \curvearrowright +$$
$$M_3 + 4(63.7) - 2(310 - 60.2) = 0$$
$$M_3 = 499.6 - 254.8 = \underline{\underline{244.8}}$$
$$\Sigma F_v = 0$$

FIGURE E6-1.1
I/O for finite-difference solution of beam on elastic foundation. Partial shear and moment diagram due to symmetry. Note check of M_3; sum of vertical forces on footing equals zero.

6-2 GENERAL CONSIDERATIONS OF THE FINITE-DIFFERENCE SOLUTION OF THE ELASTIC SUPPORTED BEAM

The computer program will solve a beam with two concentrated loads with moments. The spring constants are computed using the simplest soil condition, that of constant k_s. The necessary method of incorporating the soil spring into the A matrix makes it difficult to modify the solution for holes or changes in K_i or to account for negative or zero deflections through more than one cycle. In the matrix solution, however, these adjustments are rather simple.

It is very difficult to achieve general loading conditions on the footing since the P matrix contains a position factor ($nH \times P_i$) unless it is a moment; thus, the node at which a load is located must be identified early. It is not necessary to prorate intermediate column loads to adjacent nodes because of this fact. Load position and number of loads were no problem in the finite-element solution of Chap. 5.

The outstanding advantage of the finite-difference method, however, is the small size of the matrix to be inverted compared to that of the matrix solution. For special solutions, say, two-column loads or three-column loads with or without column moments (the included program uses two-column loads with moments) and a small computer, the finite-difference program provides a satisfactory means of obtaining a solution.

If a larger computer is available, the matrix solution is greatly to be preferred because of its superior versatility.

6-3 COMPUTER PROGRAM FOR FINITE-DIFFERENCE SOLUTION OF ELASTIC BEAM

The computer-program listing shown will compute the A matrix for a beam loaded with two concentrated column loads and including moments. The two columns may be anywhere along the beam.

This program checks for negative deflections and recycles.

Metric or fps units may be used by using the appropriate UNIT card.

Lines	Operation
1–4	Bookkeeping
5	READ TITLE, UNITS (two cards)
7	READ (7F10.4) SK = soil modulus; EC = modulus of elasticity; B = footing width; DC = thickness; H = length of segment of footing; X = distance from left end to first column; X1 = distance between columns. Use units of feet, meters, kips, and kilonewtons (no inch or centimeter units)
13	READ (7F10.4) Column loads and moments P1, P2, MN1, MN2. Use units of kips, kilonewtons ft-kips, and kN-m
22–53	Computes deflection matrix Z(I,J) and stores as ZZ(I,J) for the test for negative deflections
62	Calls DMINV (standard double-precision inversion subroutine (IBM))
65–67	Computes deflections Y(J)
68–86	Tests and recomputes Z(I,J) if negative deflections are found
87–127	Computes shear at nodes and at column faces
128–131	Computes bending moments
134–141	Computes soil nodal reactions and soil pressure

```
             C      J E BOWLES PROG TO SOLVE CONT FTG BY FINITE DIFFERENCES
             C      PROG USES 10 DIV. AND TWO COL LOADS AT ANY LOC.
0001                DIMENSION Y(11),R(12),V(11),M(11),Q(11),ZZ(11,12),SR(11)
0002                DIMENSION Z(11,12), ZA(11),ZB(11),TITLE(20)
0003                DOUBLE PRECISION Z,ZZ,D,ZA,ZB,UT5,UT6
0004                REAL M,MN1,MN2,INER
0005           2000 READ(1,1000,END=150)TITLE,UT1,UT2,UT3,UT4,UT5,UT6
0006           1000 FORMAT(20A4/4(A4,6X),A8,2X,A8)
0007                READ(1,200)SK,EC,B,DC,H,X,X1
0008            200 FORMAT(7F10.4)
0009                WRITE(3,1001)TITLE
0010           1001 FORMAT('1',T5,20A4,//)
0011                AL = 10.*H
0012                INER = B*(DC**3)/12.
0013                READ(1,200)P1,P2,MN1,MN2
0014                WRITE(3,98)SK,UT6,B,UT1,DC,UT1,EC,UT5,H,UT1,X,UT1,X1,UT1,INER,UT1,
                   1AL,UT1
0015             98 FORMAT(T5,'SOIL MODULUS =',F7.2,1X,A7,5X,'FTG WIDTH =',F6.2,1X,A2,
                   15X,'FTG DEPTH =',F6.2,1X,A2 /T5,'MOD OF ELAS =',F10.0,1X,A7,5X,'H
                   2=',F5.2,1X,A2, 5X,'DIST 1ST COL =',F5.2,1X,A2/T5,'DIST 2ND COL =',
                   3F6.2,1X,A2, 5X,'MOM OF INERTIA =',F7.4,1X,A2,'**4', 5X,'FTG LENGTH
                   4 =',F6.2,1X,A2//)
0016                WRITE(3,99)P1,UT3,MN1,UT4,P2,UT3,MN2,UT4
0017             99 FORMAT(T5,'COL LOADS & MOMENTS:',T30,'P1 =',F6.2,1X,A4, 5X,'MOMENT
                   1 M1 =',F7.2,1X,A4/T30,'P2 =',F6.2,1X,A4, 5X,'MOMENT M2 =',F7.2,1X,
                   2A4//)
0018                SK = SK*B
0019                AA = EC*INER/H**2
0020                BB=(H**2.)*SK
0021                CC=H*SK
0022                DO 3 I=1,11
0023                DO 3 J=1,12
0024              3 Z(I,J)=0.0
0025                DO 8 I=1,9
0026                IF(I*H-X)4,4,5
0027              4 Z(I,12)=0.0
0028                GO TO 21
0029              5 IF(I*H-X-X1)6,6,7
0030              6 Z(I,12)=P1*(I*H-X)+MN1
0031                GO TO 21
0032              7 Z(I,12)=P1*(I*H-X)+P2*(I*H-X-X1)+MN1+MN2
0033             21 Z(I,I)=AA+BB
0034                Z(I,I+1)=-2.0*AA
0035                Z(I,I+2)=AA
0036                Z(10,I+1)=CC
0037                Z(11,I+1)=I*BB
0038              8 Z(I,1) = I*BB
0039                DO 9 I=2,8
0040                C = 1.0
0041                DO 9 J=I,8
0042                C=C+1.0
0043              9 Z(J+1,I)=C*BB
0044                Z(1,1)=Z(1,1)+AA
0045                Z(10,1) = CC
```

```
          Z(10,11) = CC
          Z(11,11) = 10.*BB
          Z(10,12)=P1+P2
          Z(11,12)=P1*X+P2*(X+X1)-MN1-MN2
          NN = 0.
          DO 10  I = 1,11
          DO 10 J=1,12
     10   ZZ(I,J)=Z(I,J)
          INDEX = 0
     22   WRITE(3,100)
    100   FORMAT(T15,'THE COEFFICIENTS OF THE 11 SIMULTANEOUS EQ ARE (X 10)'
         1,/)
          WRITE(3,101)((Z(I,J),J=1,12),I=1,11)
    101   FORMAT(1X,-1P12F9.0)
          WRITE(3,102)
    102   FORMAT('0',T40,'THE OUTPUT IS AS FOLLOWS',/)
          WRITE(3,103)(JJ,JJ=1,11)
    103   FORMAT(T6,'POINT NOS.',T15,11I8)
          CALL DMINV(Z,11,D,ZA,ZB)
          DO 11 J=1,11
          Y(J)=0.0
C         CHECK FOR ZERO OR NEG. VALUES OF DEFLECTION
          DO 11 I=1,11
     11   Y(J)=Y(J)+Z(J,I)*Z(I,12)
          DO 40 J=1,11
          IF(Y(J))12,12,40
     12   NN = NN+1
          DO 41  I = 1,11
     41   ZZ(I,J)=0.0
          ZZ(J,J)=AA
          ZZ(J-1,J)=-2.0*AA
          ZZ(J-2,J)=AA
          ZZ(10,J)=0.0
          ZZ(11,J)=0.0
     40   CONTINUE
          DO 23 I=1,11
          DO 23 K=1,12
     23   Z(I,K)=ZZ(I,K)
          WRITE(3,104)UT1,Y
    104   FORMAT(T6,'DEFL, ',A2,T17,11F8.3/)
          IF(NN.GE.1)INDEX = INDEX+1
          IF(INDEX.GT.3)GO TO 30
          IF(NN.GE.1)GO TO 22
C         COMPUTE SHEAR IN EACH SEGMENT
     30   VC = 0.
          INDEX = 0
          DO 17  I = 1,10
          A = I
          XV = A*H
          IF(XV.GT.X)GO TO 14
          V(I) = VC+H*SK*Y(I)
          GO TO 17
     14   IF(XV.GT.(X1+X))GO TO 15
          IF(INDEX.EQ.1)GO TO 32
          V(I) = VC+H*SK*Y(I) - P1
          INDEX = 1
          GO TO 17
     15   IF(INDEX.EQ.2)GO TO 32
          V(I) = VC+H*SK*Y(I)-P2
          INDEX = 2
          GO TO 17
     32   V(I) = VC+H*SK*Y(I)
     17   VC = V(I)
          WRITE(3,105)UT3,(V(I),I=1,10)
    105   FORMAT(T6,'SEG SHEAR,',A4,T22,10F8.2/)
C         COMPUTE SHEAR TO LEFT & RT OF COLUMN LOAD POINTS
          COL = X/H
          ICOL = COL
          CF = ICOL
          DIFF = COL-CF
          IF(DIFF.LE.0..AND.(ICOL.EQ.1.OR.ICOL.EQ.0))ICOL = 2
          VLEFT1 = V(ICOL-1)
          VRT1 = VLEFT1-P1
          IF(DIFF.EQ.0.)VRT1 = VRT1 + H*SK*Y(ICOL)
          COL2 = (X1+X)/H
          ICOL2 = COL2
          CF2 = ICOL2
          IF(ICOL2.EQ.11)Y(ICOL2+1)=0.
          DIFF2 = COL2-CF2
          VLEFT2 = V(ICOL2)
          VRT2 = VLEFT2-P2
          IF(DIFF2.EQ.0.)VRT2 = VRT2+H*SK*Y(ICOL2+1)
          WRITE(3,108)VLEFT1,UT3,VRT1,UT3
    108   FORMAT(T5,'SHEAR AT FIRST COL, LEFT =',F9.2,1X,A4,5X,'TO RIGHT =',
         1F9.2,1X,A4/)
          WRITE(3,109)VLEFT2,UT3,VRT2,UT3
    109   FORMAT(T5,'SHEAR AT SECOND COL, LEFT =',F9.2,1X,A4,5X,'TO RIGHT ='
         1,F9.2,1X,A4/)
          M(1)=0.0
          M(11)=0.0
          DO 19 I=2,10
```

```
   19 M(I)=AA*(Y(I-1)-2.*Y(I)+Y(I+1))
      WRITE(3,106)UT4,M
  106 FORMAT(T6,'MOM,',A4,T17,11F8.2/)
      DO 20 J=1,11
   20 Q(J)=Y(J)*SK/B
      WRITE(3,107)UT5,(Q(J),J=1,11)
  107 FORMAT (T6,'Q ',A7,T17,11F8.2/)
      DO 26  I = 1,11
   26 SR(I)=H*SK*Y(I)
      WRITE(3,110)UT3,(SR(I),I=1,11)
  110 FORMAT(T6,'SOIL R ',A4,T17,11F8.1/)
      GO TO 2000
  150 STOP
      END
```

6-4 RING FOUNDATIONS BY FINITE DIFFERENCES

The ring foundation of Chap. 5 can be solved by the finite-difference method. Answers using 20 divisions are almost identical to that of the matrix solution; e.g., by finite differences the ring of Example 5-2 is:

Location, deg	Matrix, ft-kips	Finite difference, ft-kips (at column node)
0	403.3	398.9
18	−18.1	−17.8
36	−183.6	−181.7

The matrix values are those plotted in Fig. 5-9.

Referring to Fig. 6-2, the solution would be obtained as follows:

1 Compute the mean radius defining the center of area and locate columns on this mean radius.
2 Compute the chord distance of each segment to use as h.
3 Along any diameter such as AB or CD (Fig. 6-2) sum moments.

Since there are 20 diameters available for 20 segments, this yields 20 equations. Two typical equations obtained are (see Fig. 6-2)

$$\sum M_{AB} = 0$$

$$R_m k_s A(y_2 \sin\alpha + y_3 \sin 2\alpha + y_4 \sin 3\alpha + y_5 \sin 4\alpha + y_6 \sin 5\alpha + y_7 \sin 4\alpha + y_8 \sin 3\alpha + y_9 \sin 2\alpha + y_{10} \sin\alpha) - P_2 R_m \sin 5\alpha = \sum M_{AB} \qquad (a)$$

But

$$\sum M_{AB} = \frac{EI}{h^2}(y_{20} - 2y_1 + y_2) + \frac{EI}{h^2}(y_{10} - 2y_{11} + y_{12}) \qquad (a\text{-}1)$$

It is evident for P_2 at 90° from AB that $\sin 5\alpha = 90°$.

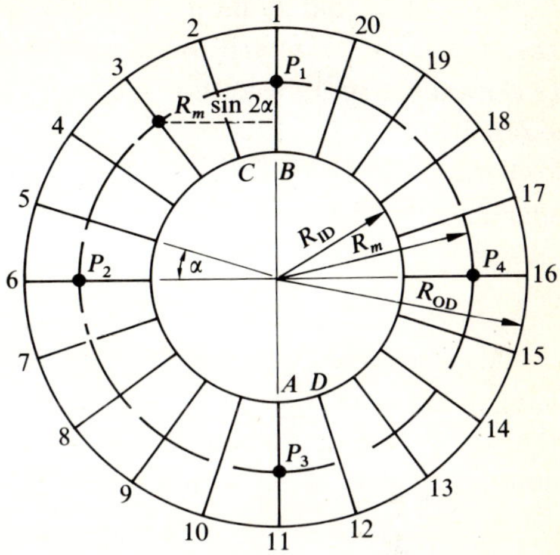

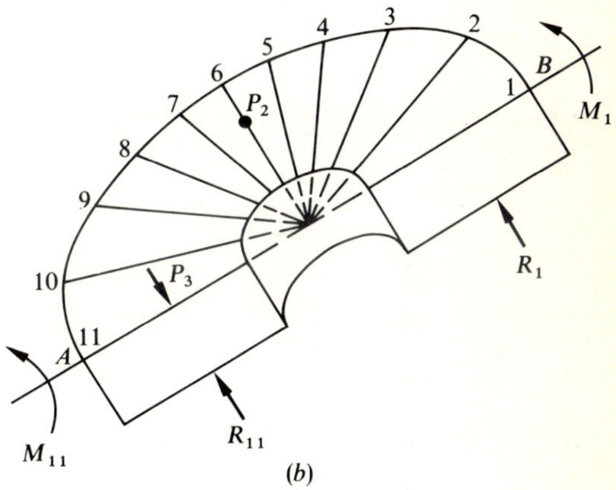

FIGURE 6-2
Ring foundation by finite differences: (*a*) ring divided into 20 segments for segment central angles of 18°; (*b*) semidiameter for summing moments; note that with AB rotated counterclockwise $d\alpha$, the reaction R_1 shown and P_1 (not shown) are not considered in either the moment or shear computation.

We next rotate the diameter to cut the ring at nodes 2 and 12. Summing moments gives

$$R_m k_s A(y_3 \sin\alpha + y_4 \sin 2\alpha + y_5 \sin 3\alpha + y_6 \sin 4\alpha + y_7 \sin 5\alpha + y_8 \sin 4\alpha + y_9 \sin 3\alpha + y_{10} \sin 2\alpha + y_{11} \sin\alpha) - P_2 R_m \sin 4\alpha - P_3 R_m \sin\alpha = \sum M_{CD} \tag{b}$$

But

$$\sum M_{CD} = \frac{EI}{h^2}(y_1 - 2y_2 + y_3) + \frac{EI}{h^2}(y_{11} - 2y_{12} + y_{13}) \tag{b-1}$$

It is evident that 20 moment events such as this can be made to occur and yield enough equations to match unknowns. Unfortunately some of these equations are not independent, and the author was never able to deduce which ones are the culprits. One rapidly discovers this unpleasant event by trying to solve the set of equations.

It may also be evident that one can utilize equations for shear in a somewhat similar manner at each section; thus, for section *AB*, diameter rotated $d\alpha$ counter-clockwise to include R_{11} but exclude R_1, summing the shear of one-half the ring as

$$\sum F_V = 0$$

gives in expanded form

$$k_s A_s(y_2 + y_3 + y_4 + y_5 + y_6 + y_8 + y_9 + y_{10} + y_{11}) - P_2 - P_3 = \sum V \tag{c}$$

where A_s is the area of any segment. But $\sum V = \sum EIy'''$, or in finite differences (Table 4-1)

$$\frac{EI}{2h^2}(y_3 - 2y_2 + y_{20} - y_{19}) + \frac{EI}{2h^3}(y_9 - 2y_{10} + 2y_{12} - y_{13}) \tag{c-1}$$

The author tried various combinations of the shear and moment equations but never found a reliable combination of 20 independent equations.

Incidentally one of the 20 independent equations should be $\sum F_V = 0$ or

$$k_s A_s \sum_1^{20} Y_i - \sum_1^{n} P_i = 0 \tag{d}$$

Combining all possible shear equations (20) plus all possible moment equations (20) and the $\sum F_V = 0$ yields 41 equations, which in matrix notation is

$$A_{41 \times 20} Y_{20 \times 1} = P_{41 \times 1} \tag{e}$$

One may try to solve this for Y as

$$Y = A^{-1}P \tag{f}$$

but A is of size 41 × 20, which is nonsquare and cannot be inverted.

Recalling from Chap. 4 that we can premultiply both sides of any equation (including matrix equations) by a common factor, let us multiply by A^T (we know what that is, and we do have to multiply by something of a matrix size to allow us to cancel interior dimensions, as illustrated in Chap. 4, to end up with a square matrix). Multiplying, we obtain

$$A^T_{20\times 41}A_{41\times 20}Y_{20\times 1} = A^T_{20\times 41}P_{41\times 1} \tag{g}$$

or

$$[A^T A]_{20\times 20}Y_{20\times 1} = A^T P_{20\times 1}$$

which is now square, so that it can be inverted. It also matches the matrix dimensions of the right-hand side of the equation. Remember that the A matrix above was 41×20.

Now solving for Y

$$Y_{20\times 1} = [A^T A]^{-1}_{20\times 20}A^T P_{20\times 1} \tag{h}$$

and canceling interior terms, we see that an equality of matrix size (20×1) is obtained. This also yields the solution of the ring.

A computer program is not furnished for this solution because the computer program in Chap. 5 is more efficient (matrix solution) in solving this problem.

6-5 THE HETENYI SOLUTION OF A BEAM ON AN ELASTIC FOUNDATION

The theoretical solution of a beam on an elastic foundation was treated in considerable detail by Hetenyi (1947). The solution (refer to Fig. 6-3) is based on the differential equation of a beam loaded with an intensity of pressure q. For a beam on soil, q is the soil-reaction pressure using the concept of the modulus-of-subgrade reaction as

$$q = k_s y \qquad (FL^{-2})$$

The differential equation is

$$EI\frac{d^4y}{dx^4} = -q = -k_s y \tag{6-2}$$

This is a linear fourth-order differential equation. The general solution is standard, with four arbitrary constants of the form

$$y = e^{\lambda x}(A\cos\lambda x + B\sin\lambda x) + e^{-\lambda x}(C\cos\lambda x + D\sin\lambda x) \tag{6-3}$$

$$\lambda = \sqrt[4]{\frac{k_s B}{4EI}} \tag{6-4}$$

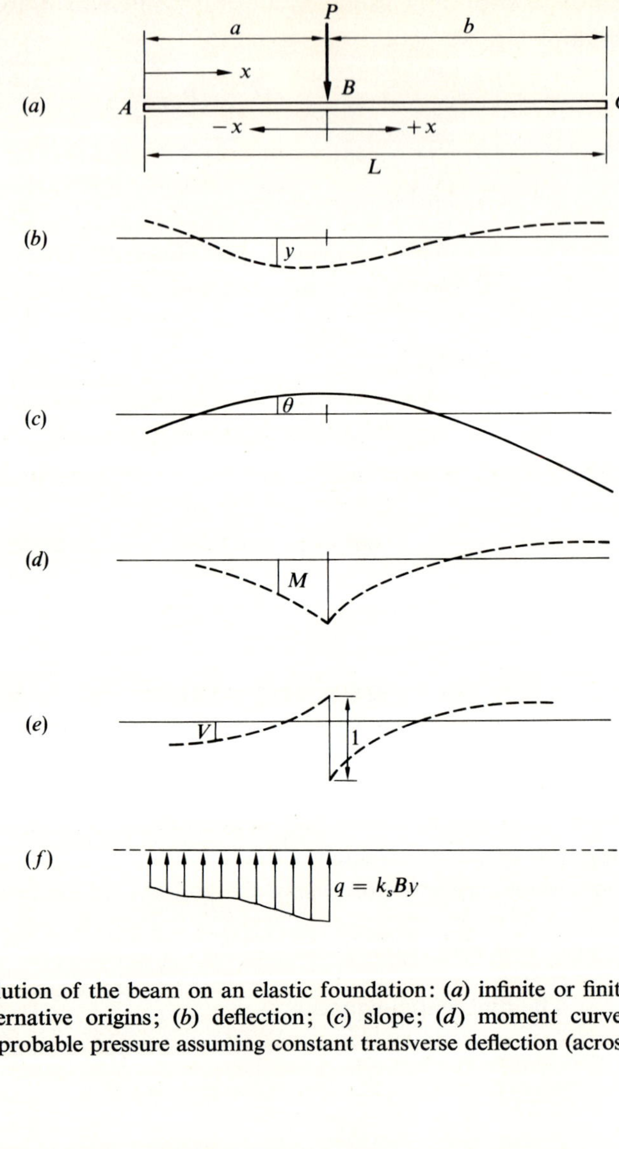

FIGURE 6-3
Theoretical solution of the beam on an elastic foundation: (*a*) infinite or finite beam and alternative origins; (*b*) deflection; (*c*) slope; (*d*) moment curve; (*e*) shear; (*f*) probable pressure assuming constant transverse deflection (across width B).

where B, E, I = footing width, material, and section properties

k_s = subgrade modulus

With consistent units used under the radical, λ will have units of L^{-1}.

We can differentiate Eq. (6-3) to obtain

$$y' = \text{slope} \qquad y'' = \text{moment} \qquad y''' = \text{shear} \qquad y'''' = \text{load}$$

Taking the alternate origin of $x = 0$ at A and inserting general boundary conditions, we find:

At x	Shear	Moment	Slope	Deflection
0	0	0	Unknown	Unknown
a (load)	To left +			
	To right −	$M_L = -M_R$	$\theta_L = \theta_R$	Unknown
L	0	0	Unknown	Unknown

A special but limited case exists where $a = b = \infty$ (and origin at B)

$$y' = 0 \qquad \text{at } x = \pm\infty \qquad y'' = 0 \qquad \text{at } x = \pm\infty$$

$$y''' = \pm\frac{P}{2} \qquad \text{at } x = 0 \qquad y'''' = 0 \qquad \text{at } x = \pm\infty$$

An example of the form of solution obtained with this latter condition of an infinitely long beam with a concentrated load at the center is

$$M = \frac{P}{4\lambda} e^{-\lambda x} (\cos \lambda x - \sin \lambda x) \tag{6-5}$$

$$V = \frac{-P}{2} e^{-\lambda x} \cos \lambda x \tag{6-5a}$$

The author [Bowles (1968)] has provided tables of the infinite-beam solution including shear and slope, together with the remainder of the equations. This reference may be consulted should the need arise.

The solution of major interest is the finite-length beam with a load at some distance a from the left end. Using the previously stated boundary conditions, we obtain

$$\begin{aligned} y = {} & \frac{P\lambda}{k_s B} \frac{1}{\sinh^2 \lambda L - \sin^2 \lambda L} \\ & \times \{2 \cosh \lambda x \cos \lambda x (\sinh \lambda L \cos \lambda a \cosh \lambda b - \sin \lambda L \cosh \lambda a \cos \lambda b) \\ & \quad + (\cosh \lambda x \sin \lambda x + \sinh \lambda x \cos \lambda x) \\ & \quad \times [\sinh \lambda L (\sin \lambda a \cosh \lambda b - \cos \lambda a \sinh \lambda b) \\ & \qquad + \sin \lambda L (\sinh \lambda a \cos \lambda b - \cosh \lambda a \sin \lambda b)]\} \end{aligned} \tag{6-6}$$

$$\begin{aligned} V = {} & P \frac{1}{\sinh^2 \lambda L - \sin^2 \lambda L} \\ & \times \{(\cosh \lambda x \sin \lambda x + \sinh \lambda x \cos \lambda x) \\ & \quad \times (\sinh \lambda L \cos \lambda a \cosh \lambda b - \sin \lambda L \cosh \lambda a \cos \lambda b) \\ & \quad + \sinh \lambda x \sin \lambda x \\ & \quad \times [\sinh \lambda L (\sin \lambda a \cosh \lambda b - \cos \lambda a \sinh \lambda b) + \sin \lambda L \\ & \qquad \times (\sinh \lambda a \cos \lambda b - \cosh \lambda a \sin \lambda b)]\} \end{aligned} \tag{6-7}$$

$$M = \frac{P}{2\lambda} \frac{1}{\sinh^2 \lambda L - \sin^2 \lambda L}$$
$$\times \{2 \sinh \lambda x \sin \lambda x (\sinh \lambda L \cos \lambda a \cosh \lambda b - \sin \lambda L \cosh \lambda a \cos \lambda b)$$
$$+ (\cosh \lambda x \sin \lambda x - \sinh \lambda x \cos \lambda x)$$
$$\times [\sinh \lambda L (\sin \lambda a \cosh \lambda b - \cos \lambda a \sinh \lambda b)$$
$$+ \sin \lambda L (\sinh \lambda a \cos \lambda b - \cosh \lambda a \sin \lambda b)]\} \quad (6\text{-}8)$$

These equations can be programmed for

$$\lambda L = \text{const}$$

Load at any point $a = nL$

to obtain coefficients for a solution aid. The computer program of Sec. 6-7 extends this set of equations to solve directly for shear and bending moments at the 0.1 points along a beam for a single point load and for a maximum of two point loads will sum the two sets of computations for a composite solution.

EXAMPLE 6-2 Solve Example 6-1 (also Example 5-1) using the Hetenyi solution. Compare the solution to the finite-element (Example 5-1) and finite-difference (Example 6-1) solutions. Use metric units.

SOLUTION Convert the appropriate dimensions to metric units (use Table 2-8):

SK = 48 kcf = 7,540.5 kN/cu m

EC = 468,000 ksf = 2,2408,736 kN/sq m

B = 10 ft = 3.048 m L = 20 ft = 6.096 m

DC = 20 in = 0.508 m X1 = 2 ft = 0.6096 m

X1 = 18 ft = 5.486 m second column

P(M) = 310 kips = 1,378.7 kN

Data cards for input are as follows:

Card	Data
1	TITLE
2	UT1–UT6 (columns 1, 11, 21, . . . , 51) M CM KN KN–M KN/SQ M KN/CU M
3	1. 0. ALIST > 1 lists extra data
4	1. R
5	7540.5 .6096 6.096 3.048 .508 22408736. 1378.7
6	1. R
7	7540.5 5.486 6.096 3.048 .508 22408736. 1378.7
8	−1. R

These eight data cards represent the input. The partial output is shown on Fig. E6-2.1.

```
J E BOWLES   EXAMPLE 6-2   HETENYI SOLUTION OF EX 6-1 USING METRIC UNITS
    RUN = 1.00


BETA =   1.806     LAMBDA = 0.29623     COL DIST =    0.6096 M
FTG LENGTH =     6.096 M      FTG WIDTH =    3.0480 M      FTG DEPTH = 0.5080 M
SOIL MODULUS =    7540.50 KN/SQ M     MOD ELAST =22408736.0 KN/SQ M
MOM OF INERTIA =0.332985E-01 M **4     COL LOAD =    1378.72 KN

       LOAD AT 0.1 L
 DIST          DEF          MOM          SHEAR


  0.0       1.9793      0.0           0.0
  0.1       1.6693      0.0612        0.3295
  0.1       1.6693      0.0612       -0.6705
  0.2       1.3586     -0.1299       -0.3972
  0.3       1.0559     -0.2324       -0.1794
  0.4       0.7680     -0.2659       -0.0149
  0.5       0.4971     -0.2493        0.0990
  0.6       0.2424     -0.2001        0.1656
  0.7       0.0006     -0.1351        0.1873
  0.8      -0.2324     -0.0700        0.1663
  0.9      -0.4607     -0.0200        0.1037
  1.0      -0.6876     -0.0000        0.0000

BETA =   1.806     LAMBDA = 0.29623     COL DIST =    5.4860 M
FTG LENGTH =     6.096 M      FTG WIDTH =    3.0480 M      FTG DEPTH = 0.5080 M
SOIL MODULUS =    7540.50 KN/SQ M     MOD ELAST =22408736.0 KN/SQ M
MOM OF INERTIA =0.332985E-01 M **4     COL LOAD =    1378.72 KN

       LOAD AT 0.9 L
 DIST          DEF          MOM          SHEAR


  0.0      -0.6876      0.0           0.0
  0.1      -0.4607     -0.0200       -0.1037
  0.2      -0.2324     -0.0700       -0.1663
  0.3       0.0006     -0.1351       -0.1873
  0.4       0.2424     -0.2001       -0.1656
  0.5       0.4971     -0.2493       -0.0990
  0.6       0.7680     -0.2659        0.0149
  0.7       1.0559     -0.2324        0.1794
  0.8       1.3586     -0.1299        0.3972
  0.9       1.6693      0.0612        0.6705
  0.9       1.6693      0.0612       -0.3295
  1.0       1.9793     -0.0000       -0.0000

TOTAL DEFL, MOM AND SHEAR AT 0.1 PTS ARE:
   DIST,M              DEF,M            MOM,KN-M        SHEAR,KN

   0.0                 0.023              0.0             0.0
   0.6096              0.021             95.924         311.264
   0.6096              0.021             95.922       -1067.451
   1.2192              0.020           -465.200        -776.927
   1.8288              0.019           -855.230        -505.608
   2.4384              0.018          -1084.614        -248.887
   3.0480              0.018          -1160.274          -0.000
   3.6576              0.018          -1084.617         248.887
   4.2672              0.019           -855.232         505.607
   4.8768              0.020           -465.199         776.927
   5.4864              0.021             95.921        1067.452
   5.4864              0.021             95.926        -311.264
   6.0960              0.023             -0.001           0.001
```

FIGURE E6-2.1
Partial output for Hetenyi solution for Example 6-1 in metric units. Note that symmetry in deflection, moments, and shear is a check on this particular problem output.

A comparison of the maximum bending moment is as follows:

Method	Maximum bending moment, ft-kips	% difference*	λ_L
Finite-element	−693.7	0	
Finite-difference	−706.9	+2	
Hetenyi	−855.7†	+23	1.81
Conventional	−930.0	+34	

* Finite-element solution taken as correct.
† M = 1,160.27/1.356 = 855.7 (converting metric to fps). ////

6-6 GENERAL COMMENTS ON HETENYI SOLUTION

The theoretical solution using the method proposed by Hetenyi provides reasonable computed values. Inspection of Table 6-1 indicates, however, that the values are in error (assuming the measured values are correct) from about the same amount as the finite-element method of Chap. 5 to somewhat higher orders of magnitude. When the end springs are increased, however, the finite-element solution is considerably more accurate.

The theoretical solution is much more difficult to use for general loading conditions than the finite-element method. If columns are not located at 0.1 points (to take advantage of tables or the included computer program), a theoretical solution is even more of a problem. Column moments will require the user to go back to the basic differential equation to develop a solution. It is not possible to include footing-weight effects or footing separation from the soil in this solution, and it should be noted that when the footing tends to separate from the soil, its self weight will tend to reduce the separation effect.

The theoretical solution appears to be more sensitive to the modulus-of-subgrade reaction than the finite-element solution of Chap. 5.

Table 6-1 COMPARISON OF HETENYI METHOD TO VALUES REPORTED BY VESIĆ

Test no.	k_s	Moment, in-kips: Measured	Moment, in-kips: Hetenyi	λL
1	98.2	+172.0	+147.7	0.982
2	98.2	−113.2	−147.3	0.982
7	123.3	+61.4	+67.58	2.07
8	123.3	−48.4	−63.34	2.07
4	150.0	+37.0	+40.24	3.908

* See Table 5-1 for loading data.

The theoretical solution for more than one load is based on superposition of effects. Many users feel this is invalid; however, it should be evident that the end result is simply a number and since the computations are merely numbers, the answer is *numerically* correct. Whether it is *physically* correct or not will require an examination, just as superposition in any structural design requires the designer's inspection to see if the computed value is in the range where superposition is valid.

It has been proposed [Vesić (1961)] that one may observe the dimensionless parameter λL, defined as

$$\lambda L = L \sqrt[4]{\frac{k_s B}{4EI}}$$

to determine whether the conventional solution of Chap. 3 or the beam-on-an-elastic-foundation solution is the better method for a particular solution.

The proposal is

$\lambda L < 0.8$	conventional solution
$0.8 < \lambda L < \pi$	finite beam on elastic foundation
$\lambda L > \pi$	infinite beam on elastic foundation

Table 6-1, with two types of loads, does not indicate that this criterion is significant, nor does a comparison of the tabulated data earlier presented (see Example 6-2) indicate an easily discernible trend. The original proposal was based on the observation that with small λL values there is little change in computed moments up to some value (which has been found by the author to depend both on the load system and λL). Beyond this point larger changes in moment occur more rapidly with change in λL; beyond some larger limiting value of λL the moment reaches essentially a lower limiting value. Because the λL zone limits depend on load, footing flexibility, and soil modulus, the author makes no recommendations, especially with all the computational techniques presented herein to obtain a solution.

6-7 COMPUTER PROGRAM FOR HETENYI SOLUTION

This program will compute deflections, shears, and moments at the 0.1 points for two loads located anywhere along the beam but the column loads must be placed at the nearest 0.1 points. The beam is assumed weightless. No provision is (or can be) made for excessive deflections or for footing separation. The designer should use ultimate loads if ultimate shears and moments are desired for strength design by ACI 318-71. This program will solve either fps or metric problems.

Line	Operation
1–4	Bookkeeping
5	READ TITLE, UNITS (two cards)
7	READ AA = computation cycle; ALIST increases amount of output if > 0
18	READ R = recycle for second column. If R = 0 or −1, sums shear and bending moments at 0.1 points for one (or both) column loads
21	READ SK = soil modulus; X1 = distance from left end to column; EL = length; BF = width; DC = depth; EC = modulus of elasticity of footing; P(M) = column load. Use foot and kip units (meters and kilonewtons); do not use inches (or centimeters)
26	READ XI = moment of inertia, ft^4 or m^4, if DC is read as zero. If DC is read >0.0, program computes I
31–104	Computes nondimensional coefficients for 0.1 points along beam for deflection, shear, and moment for each load position
119–144	Uses nondimensional coefficients to compute deflections, shear, and bending moment for the actual beam loads

```
      C         J E BOWLES HETENYI  SOLUTION FINITE BEAM ON ELAS. FOUND
      C         ORDER OF DATA CARDS IS 1=TITLE;   2=UNITS;   3=AA;   4=R;   5=FIRST DA
      C          6 = R ; 7 2ND DATA--USE R = 0 FOR 1-COL;   USE R = 0 AT END OF 2ND
      C         READ ALL UNITS AS FT OR METERS, KIPS OR KILO-NEWTONS
0001            DIMENSION A1(5,15),B1(5,15),C1(5,15), F(15), V(15), H(15), P(10)
0002            DIMENSION DIST(15),TITLE(20)
0003            DOUBLE PRECISION UT5,UT6
0004            REAL LAMBDA
0005       1004 READ(1,1000,END=150)TITLE,UT1,UT2,UT3,UT4,UT5,UT6
0006       1000 FORMAT (20A4/4(A4,6X),A8,2X,A8)
0007            READ(1,10)AA,ALIST
0008            WRITE (3,1001)TITLE,AA
0009       1001 FORMAT('1',//,T5,20A4,/,T10,'RUN =',F5.2,/)
0010            N = 0
0011            M = 0
0012            DO 109  I = 1,4
0013            DO 109  J = 1,12
0014            A1(I,J) = 0.
0015            B1(I,J) = 0.
0016            C1(I,J) = 0.
0017        109 CONTINUE
0018       1002 READ(1,10)R
0019            M = M + 1
0020            IF(R)111,111,1
0021          1 READ (1,10)SK,X1,EL,BF, DC, EC, P(M)
0022         10 FORMAT(7F10.4)
0023            XI = BF*DC**3/12.
0024            SKS = SK*BF
0025            IF(DC)13,13,14
0026         13 READ(1,10)XI
0027         14 BETA = EL*((SK*BF)/(4.*EC*XI))**.25
0028            LAMBDA = BETA/EL
0029            WRITE(3,12)BETA,LAMBDA,X1,UT1,EL,UT1,BF,UT1,DC,UT1,SK,UT5,EC,UT5,
               1XI,UT1,P(M),UT3
0030         12 FORMAT(//,T5,'BETA =',F7.3,3X, 'LAMBDA =',F8.5,3X, 'COL DIST =',F8
               1.4,1X,A2,/,T5, 'FTG LENGTH =',F8.3,1X,A2,3X,'FTG WIDTH =',F8.4,1X,
               2A2,3X, 'FTG DEPTH =',F7.4,1X,A2,/,T5, 'SOIL MODULUS =',F9.2,1X,A7,
               33X,'MOD ELAST =',F10.1,1X,A7,/,T5, 'MOM OF INERTIA =',G12.6,1X,A2,
               4'**4',3X, 'COL LOAD =',F9.2,1X,A4)
0031            N = N + 1
0032            G = 1.
0033            W2 = .5*(EXP (BETA*G) + 1./(EXP (BETA*G)))
0034            X2 = .5*(EXP (BETA*G)-1./(EXP (BETA*G)))
0035            Y2 = SIN (BETA*G)
0036            Z2 = COS (BETA*G)
0037            Z = BETA/EL
0038            D = EL/10.
0039            D1 = X1/D
0040            J = D1
0041            D2 = J
0042            DIFF = D1 - D2
0043            IF (DIFF - 0.5)8,8,9
0044          8 A = D2/10.
0045            GO TO 16
0046          9 A = (D2+1.)/10.
0047         16 WRITE (3,17)A
0048         17 FORMAT (/T10, 'LOAD AT ',F3.1,' L',/,T6,'DIST', T16,'DEF',T25, 'MO
               2M',T34, 'SHEAR',//)
0049            DO 104  L = 1,12
0050            I = L
0051            K=1
0052            IF(N.EQ.2)K=3
```

```
      X = (I - 1.)/10.
      DIS = X
      IF(DIS - 1.)19,19,1002
   19 IF (X - (A+0.002))20,20,50
   20 B = 1. - A
      WO = .5*(EXP (BETA*A)+1./(EXP (BETA*A)))
      XO = .5*(EXP (BETA*A)-1./(EXP (BETA*A)))
      YO = SIN (BETA*A)
      ZO = COS (BETA*A)
      W1 = .5*(EXP (BETA*B)+1./(EXP (BETA*B)))
      X1 = .5*(EXP (BETA*B)-1./(EXP (BETA*B)))
      Y1 = SIN (BETA*B)
      Z1 = COS (BETA*B)
      W3 = .5*(EXP (BETA*X)+1./(EXP (BETA*X)))
      X3 = .5*(EXP (BETA*X)-1./(EXP (BETA*X)))
      Y3 = SIN (BETA*X)
      Z3 = COS (BETA*X)
      Q = (X2*(YO*W1-ZO*X1) + Y2*(XO*Z1 - WO*Y1))
      S = (1./(X2**2 - Y2**2))
      A1(K,I) = S*(2.*W3*Z3*(X2*ZO*W1 - Y2*WO*Z1) + (W3*Y3 + X3*Z3)*Q)
      B1(K,I) = S*(2.*X3*Y3*(X2*ZO*W1-Y2*WO*Z1)+(W3*Y3-X3*Z3)*Q)
      C1(K,I) = S*((W3*Y3+X3*Z3)*(X2*ZO*W1-Y2*WO*Z1)+X3*Y3*Q)
      GO TO 104
   50 K=2
      IF(N.EQ.2)K=4
      A = 1. - A
      B = 1. - A
      X = A
   55 IF(X)1002,61,61
   61 WO = .5*(EXP (BETA*A)+1./(EXP (BETA*A)))
      XO = .5*(EXP (BETA*A)-1./(EXP (BETA*A)))
      YO = SIN (BETA*A)
      ZO = COS (BETA*A)
      W1 = .5*(EXP (BETA*B)+1./(EXP (BETA*B)))
      X1 = .5*(EXP (BETA*B)-1./(EXP (BETA*B)))
      Y1 = SIN (BETA*B)
      Z1 = COS (BETA*B)
      W3 = .5*(EXP (BETA*X)+1./(EXP (BETA*X)))
      X3 = .5*(EXP (BETA*X)-1./(EXP (BETA*X)))
      Y3 = SIN (BETA*X)
      Z3 = COS (BETA*X)
      Q = (X2*(YO*W1-ZO*X1) + Y2*(XO*Z1 - WO*Y1))
      S = (1./(X2**2 - Y2**2))
      A1(K,I) = S*(2.*W3*Z3*(X2*ZO*W1 - Y2*WO*Z1) + (W3*Y3 + X3*Z3)*Q)
      B1(K,I) = S*(2.*X3*Y3*(X2*ZO*W1-Y2*WO*Z1)+(W3*Y3-X3*Z3)*Q)
      C1(K,I) = S*((W3*Y3+X3*Z3)*(X2*ZO*W1-Y2*WO*Z1)+X3*Y3*Q)*(-1.)
      DIS = 1. - X
      X = X - .1
      WRITE(3,101)DIS,A1(K,I),B1(K,I),C1(K,I)
      I = I + 1
      GO TO 55
  104 WRITE (3, 101)DIS, A1(K,I), B1(K,I), C1(K,I)
  101 FORMAT(5X, F4.1, 3F10.4)
  111 IF(ALIST.LE.0.)GO TO 140
      WRITE (3,124)
  124 FORMAT('1', T10, 'A1(1,J)', T20, 'A1(2,J)', T30, 'A1(3,J)', T40, '
     1A1(4,J)', T60, 'B1(1,J),',T70, 'B1(2,J)', T80, 'B1(3,J)',T90,'B1(4
     2,J)'//)
      WRITE (3,127) ((A1(K,I), B1(K,I), K = 1,4), I = 1,12)
  127 FORMAT (T9, 4F10.5, T59, 4F10.5)
      WRITE (3,126)
  126 FORMAT(//,T10,'C1(1,J)',3X,'C1(2,J)',3X,'C1(3,J)',3X,'C1(4,J)'//)
      WRITE (3,125) ((C1(K,I),  K = 1,4), I = 1,12)
  125 FORMAT (T9, 4F10.5)
  140 WRITE (3,21)
   21 FORMAT(//,T5,'TOTAL DEFL, MOM AND SHEAR AT 0.1 PTS ARE:')
      WRITE (3,22) UT1,UT1,UT4,UT3
   22 FORMAT(T8,'DIST,',A2,T20,'DEF,',A2,T33,'MOM,',A4,T45,'SHEAR,'A4/)
C     COMPUTE DEFL F(J);  BENDING MOMENT V(J);  SHEAR H(J)
      DO 108  J = 1,12
  108 DIST(J) = 0.
      DO 29 J = 1,13
      IF (J-13)32,25,25
   32 IF (A1(1,J))24,25,24
   24 F(J) = (A1(1,J)*P(1) + A1(3,J)*P(2))*Z/(SKS)
      V(J)=((B1(1,J)*P(1)+B1(3,J)*P(2))/(2.*Z))
      H(J)= (C1(1,J)*P(1)+C1(3,J)*P(2))
      A = J - 1
      DIST (J) = DIST (J) + A*EL/10.
      GO TO 29
   25 IF (A1(3,J-1))26,27,26
   26 F(J)=(A1(2,J)*P(1) +A1(3,J-1)*P(2))*Z/(SKS)
      V(J) =((B1(2,J)*P(1)+B1(3,J-1)*P(2))/(2.*Z))
      H(J)=(C1(2,J)*P(1)+C1(3,J-1)*P(2))
      A = J - 2
      DIST(J) = A*EL/10.
      GO TO 29
   27 F(J) = (A1(2,J-1)*P(1) + A1(4,J-1)*P(2))*Z/(SKS)
      V(J)=((B1(2,J-1)*P(1)+B1(4,J-1)*P(2))/(2.*Z))
      H(J)=(C1(2,J-1)*P(1)+C1(4,J-1)*P(2))
      A = J - 3
      DIST (J) = A*EL/10.
   29 WRITE (3,23)DIST(J), F(J), V(J), H(J)
   23 FORMAT (3X, F10.4, 4X, F10.3, 3X, F10.3, 3X, F10.3)
      GO TO 1004
  150 STOP
      END
```

PROBLEMS

6-1 Repeat the assignment of Prob. 3-5 using the finite-difference method.

6-2 Repeat the assignment of Prob. 3-5 using the Hetenyi method.

6-3 Check the tabulated values in Table 6-1 using the Hetenyi solution.

REFERENCES

BOWLES, J. E. (1968): "Foundation Analysis and Design," chap. 5, McGraw-Hill, New York.

HETENYI, M. (1946): "Beams on Elastic Foundation," University of Michigan Press, Ann Arbor, 255 pp.

MALTER, H. (1960): Numerical Solutions for Beams on Elastic Foundations, *Trans. ASCE*, vol. 125, pp. 757–791.

TENG, W. C. (1962): "Foundation Design," chap. 7, Prentice-Hall, Englewood Cliffs, N.J.

VESIĆ, A. S. (1961): Bending of Beams Resting on Isotropic Elastic Solid, *J. Eng. Mech. Div., ASCE*, vol. 87, EM2, pp. 35–53.

7

ECCENTRICALLY LOADED FOOTINGS, NOTCHED FOOTINGS, AND MATS

7-1 INTRODUCTION

This chapter considers both conventional analysis procedures of eccentrically loaded footings, footings with notches (or holes), and mats as well as computer solutions. Actually for the computer solutions considered here these are all the same problem, namely, a plate on an elastic foundation. The conventional analysis, however, properly considers these three cases separately. The computer solutions utilize both the finite-difference solution considered in Chap. 4 and a finite-element solution, the details of which are introduced here.

7-2 ECCENTRICALLY LOADED FOOTINGS

Design limitations may require that a footing be loaded not at the center of area, or the column may transmit both axial force and moment (Fig. 7-1). In either case from statics $\sum F_v = 0$ results in

$$R = P$$

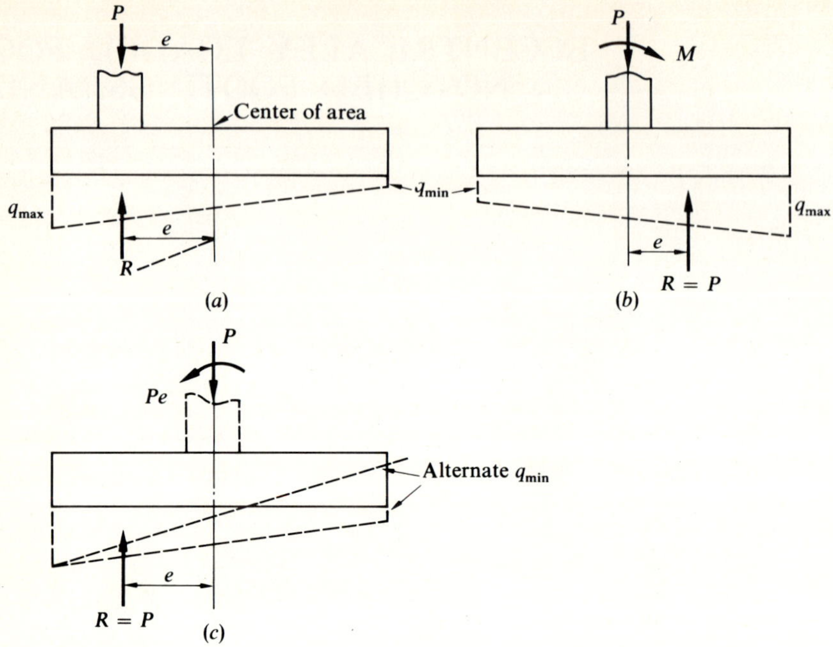

FIGURE 7-1
Footings with eccentricity (or moment). Eccentricity with respect to one axis is shown, but it may be about both the X and Y axes. (*a*) Eccentrically placed column. (*b*) Column with moment. (*c*) Equivalent of part (*a*).

and $\sum M = 0$ with respect to a centroidal axis gives

$$e = \frac{M}{P}$$

Obviously the volume of the soil-pressure diagram is

$$R = \int_0^A q \, dA$$

Conventional design treats this problem as if the footing were a rigid body (Fig. 7-2) with superposition applicable; thus, the pressure intensity at footing corners is

$$q = \frac{P}{A} \pm \frac{M_x}{S_x} \pm \frac{M_y}{S_y}$$

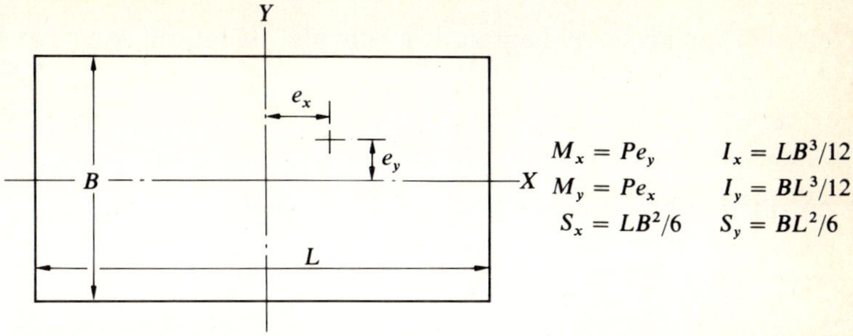

FIGURE 7-2
Eccentricity with respect to either or both axes and the footing as a rigid body.

which simplifies for a rectangle to

$$q = \frac{P}{BL}\left(1 \pm \frac{6e_y}{B} \pm \frac{6e_x}{L}\right) \tag{7-1}$$

as obtained from Fig. 7-2. As long as

$$e_x \leq \frac{L}{6} \qquad \text{and} \qquad e_y \leq \frac{B}{6}$$

the entire footing is considered effective. When the eccentricity is on only one axis, say the X axis as in Fig. 7-3, the maximum soil pressure when $e > L/6$ can be computed as follows. Let L' = effective footing length; then

$$\frac{qL'}{2} = R = P$$

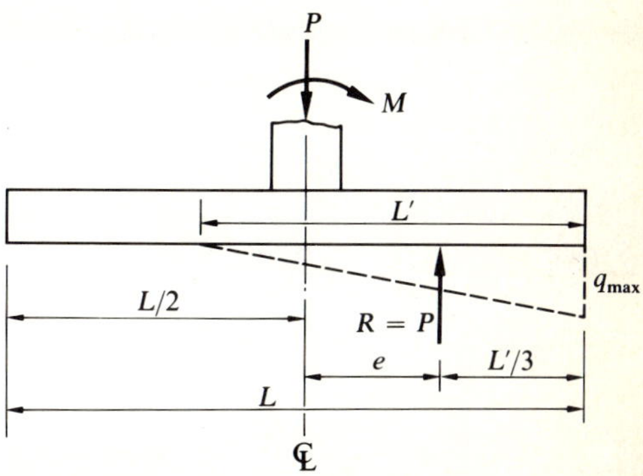

FIGURE 7-3
Conditions for eccentricity out of middle third of base and rigid-footing assumption.

Note that the pressure diagram is a triangle; therefore, R acts at $L'/3$ from q_{max}, and

$$e = \frac{L}{2} - \frac{L'}{3}$$

by inspection of the figures. Solving for L' and q, we obtain

$$q = \frac{2P}{3(L/2 - e)} \tag{7-2}$$

with terms identified in Fig. 7-3.

When the eccentricity is with respect to both axes and is larger than $L/6$ and $B/6$, the method of Example 7-4 can be used. An alternative solution given by Bowles (1968) is readily available in the cited reference.

7-3 MATS: CONVENTIONAL ANALYSIS

The conventional design of mat foundations uses the same basic approach as the spread footing. The principal difference is that the mat is larger and carries more column loads.

Mats are used where the soil is low in bearing capacity or the loads are such that the footings use 50 percent or more of the total site area.[1] They may also be used in combination with basement walls to "float" the structure, i.e., on an excavation which removes the approximate building weight to reduce settlements. By virtue of their bridging action, mats tend toward much less differential settlement, and thus the designer may use higher allowable soil pressures. This bridging action is an asset also where the site contains both firmer material and pockets, or lenses, of soft material.

Mats may be ribbed where the column spacing is irregular and/or for economy in using a relatively thin plate over most of the site. Alternatively, mats may be thickened at the column locations for economy and to provide a depth sufficient to resist shear. Figure 7-4 illustrates several mat foundations.

The conventional design of mat foundations assumes rigidity of the mat [Teng (1962), ACI (1966), Bowles (1968)], just like the spread footing. Therefore, taking $\sum$ (column loads) $= P_t$, we have

$$q = \frac{P_t}{BL}\left(1 \pm \frac{6e_y}{B} \pm \frac{6e_x}{L}\right)$$

[1] This is a rule of thumb, and what must be considered is the cost of footing formwork versus the cost of additional reinforcing steel required to maintain continuity.

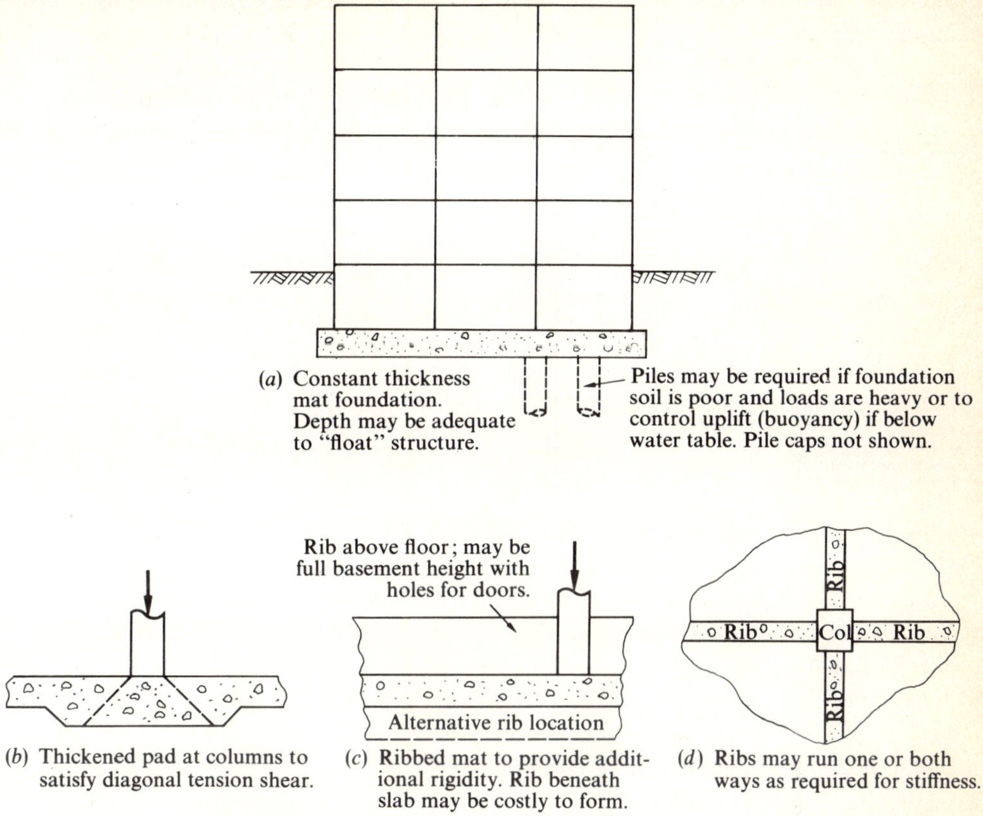

FIGURE 7-4
Mat foundations.

as in Eq. (7-1). The eccentricities of P_t with respect to the center of area are obtained using methods of statics.

With the soil pressure obtained it is usual practice to divide the mat up into column strips each way and, using approximate moment factors of $wL^2/10$ or $wL^2/12$, to compute the bending moments in the strips. Depth for diagonal-tension shear is obtained by the method of Chap. 3 (although with low soil pressures it is conservative to neglect the soil pressure in the punching shear zone). Alternatively one may use moment distribution [Goodman and Karol (1968)] to refine the computations somewhat.

EXAMPLE 7-1 For the given mat-foundation layout, design the thickness and obtain the steel-reinforcing-bar requirements using the conventional design procedures, ACI 318-71 and ultimate-strength design (see Fig. E7-1.1). Other data:

$f'_c = 3{,}000$ psi $\quad f_y = 60{,}000$ psi $\quad$ all columns = 15 × 15 in

$q_a = 1.2$ ksf $\quad$ use average load factor = 1.55 for given loads

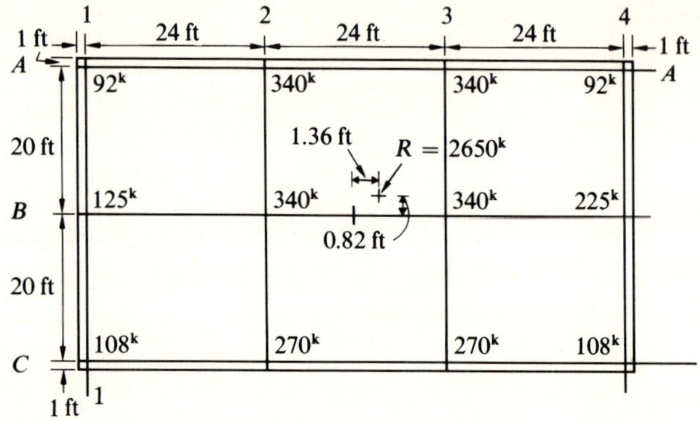

FIGURE E7-1.1
Mat-foundation layout.

SOLUTION Mat foundation:

$\sum P = 2{,}650$ kips for all 12 columns

Locate resultant:

First $\sum M$ line 1-1

$$2{,}650\bar{X} = 24(340 + 340 + 270) + 48(340 + 340 + 270) + 72(92 + 225 + 108)$$
$$= 24(950) + 48(950) + 72(425)$$
$$= 950(72) + 425(72) = 1{,}375(72)$$
$$\bar{X} = 37.36 \text{ ft} \qquad \text{or } 1.36 \text{ ft from cga}$$

Second $\sum M$ line A-A

$$2{,}650\bar{Y} = 20(125 + 340 + 340 + 225) + 40(108 + 270 + 270 + 108)$$
$$= 20(1{,}030) + 40(756) = 20{,}600 + 30{,}240$$
$$\bar{Y} = \frac{50{,}840}{2{,}650} = 19.18 \text{ ft} \qquad \text{or } 0.82 \text{ ft from cga}$$

Compute soil pressure at critical locations:

$$I_x = \frac{74(42^3)}{12} \qquad\qquad I_y = \frac{42(74^3)}{12}$$

$$= \frac{74(74{,}088)}{12} = 456{,}876\ \text{ft}^4 \qquad\qquad = \frac{42(405{,}224)}{12} = 1{,}418{,}284\ \text{ft}^4$$

$$A = 74(42) = 3{,}108\ \text{ft}^2$$

$$\frac{P}{A} = \frac{2{,}650}{3{,}108} = 0.853\ \text{ksf}$$

$$M_x = Py_0 = 2{,}650(0.82) = 2{,}173\ \text{ft-kips}$$

$$M_y = Px_0 = 2{,}650(1.36) = 3{,}604\ \text{ft-kips}$$

At corner A-1

$$\sigma = 0.853 - \frac{3{,}604(37)}{1{,}418{,}284} + \frac{2{,}173(21)}{456{,}876} = 0.853 - 0.094 + 0.0999 = 0.859\ \text{ksf}$$

At corner A-4

$$\sigma = 0.853 + 0.094 + 0.099 = 1.046\ \text{ksf} < 1.2 \qquad \text{O.K.}$$

At corner C-1

$$\sigma = 0.853 - 0.094 - 0.099 = 0.660\ \text{ksf}$$

At corner C-4

$$\sigma = 0.853 + 0.094 - 0.099 = 0.848\ \text{ksf}$$

Due to such small differences in the computed soil pressure assume

$$\sigma = 0.9\ \text{ksf}$$

In long direction (and take L as center to center of column distance)

$$+M = -M = \frac{wL^2}{10}$$

For any strip in long direction

$$M_{\text{design}} = \frac{0.9(24)^2}{10} = 51.8\ \text{ft-kips/ft of width}$$

For any strip in short direction take $M = wL^2/8$ since there is only a two-span equivalent beam compared to a three-span beam in the long direction:

$$M_{\text{design}} = \frac{0.9(20)^2}{8} = 45\ \text{ft-kips/ft of width}$$

Compute required mat thickness d ($D \approx d + 3\frac{1}{2}$ in); since diagonal tension will control,

$$v_c = 4\phi\sqrt{f'_c}$$

$$v_c = 186.2 \text{ psi} = 26.81 \text{ ksf} \qquad \text{(Table 3-2)}$$

For a corner load (neglect upward soil pressure of 0.9 ksf on area in diagonal tension as conservative) see Fig. E7-1.2:

$$\text{Perimeter} = 2(d/2 + 1.625) = d + 3.25$$

and

$$(d + 3.25)26.81 = 108(1.55)$$

$$d + 3.25 = 6.24$$

$$d = 6.24 - 3.25 = 2.99 \text{ ft}$$

Check side load and include upward soil pressure (interior loads not critical)

$$\text{Perimeter} = 2(d/2 + 1.625) + d + 1.25$$

$$= 2d + 4.50$$

$$\text{and Area} = A = (d + 1.25)(d/2 + 1.625)$$

$$(2d + 4.50)26.81 = 340(1.55)$$

$$-.9(1.55)(d^2/2 + 2.25d + 2.03)$$

$$.697d^2 + 56.76d = 403.52$$

$$d^2 + 81.43d = 578.94$$

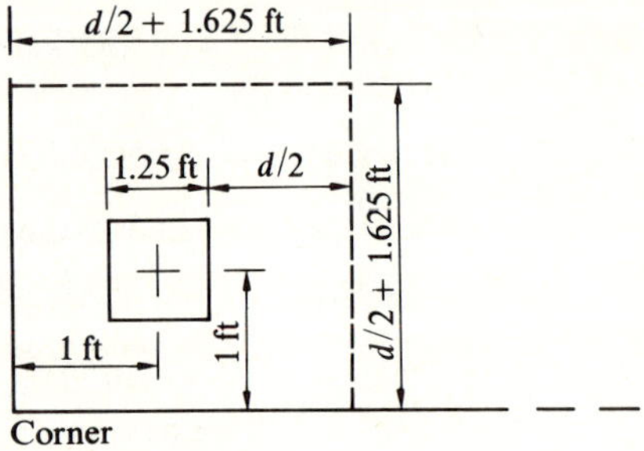

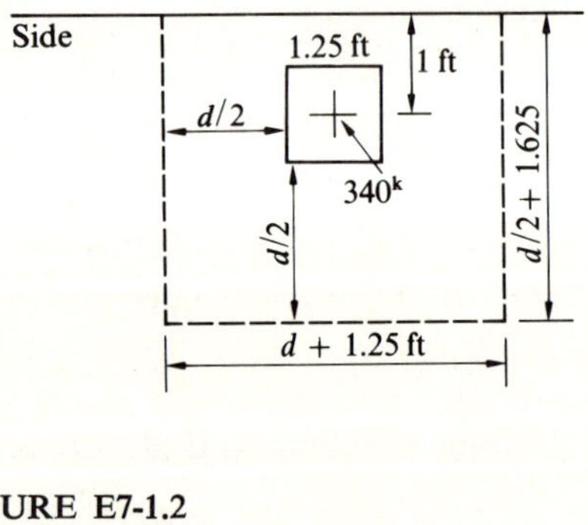

FIGURE E7-1.2

and completing the square

$$d = 47.29 - 40.72 = 6.57 \text{ ft} \qquad \text{say 6 ft 7 in}$$

Select steel long direction (for $f_y = 60$, $f'_c = 3$ ksi, $a = 1.96A_s$)

$$M_u = \phi A_s f_y \left(d - \frac{a}{2}\right) \qquad \frac{M_u}{0.9f_y} = \frac{51.8(1.55)}{0.9(60)} = 1.487$$

$$A_s[6.57(12) - 0.98A_s] = 1.487(12)$$

$$0.98A_s^2 - 78.84A_s = -17.844$$

$$A_s = -40.00 + 40.22 = 0.22 \text{ sq in/ft}$$

Check minimum amount (use ACI, art. 10-5.1, as conservative):

$$P = \frac{200}{f_y} = 0.00333$$

$$A_s = 0.0033(42)(12) = 1.68 \text{ sq in/ft} \qquad \text{controls both directions}$$

Use two no. 8 bars each way top and bottom at 1 ft center to center

$$A_s = 2(0.79) = 1.58 < 1.68 \qquad \text{but O.K.}$$

////

7-4 MATS: FINITE-DIFFERENCE SOLUTION

The finite-difference solution holds considerable promise for mat analysis. This approach has been used [Deryck and Severn (1960, 1961), Severn (1966), Bowles (1968)] to analyze large flat slabs on an elastic medium. The finite-difference solution utilizes thin-plate theory, but when the plan dimensions are reasonable compared to the thickness, the error [Frederick (1957) "Discussion"] is neglecting the plate thickness effect is very small.

A major problem of the finite-difference method (and a much more serious problem using the author's finite-element method of Sec. 7-7) is the large matrix to be solved. Included in the author's computer program is the product-inverse method, which can solve perhaps a 1,000 × 1,000 matrix if the user can afford paying for that much computer time. This method took the author about $5\frac{1}{2}$ hr of IBM 360 computation time to solve the 315 × 315 matrix used in the included example. This method is slow, both because of the very large number of computations and because the matrix is stored on disk (or tape) and only one column at a time is operated on. Relatively slow solutions are a fact of life on small to medium computers with limited core space. Although alternative special equation-solving techniques such as relaxation have been used, the product-inverse method appears to be as rapid as any (see also Sec. 7-7).

The grid values given in Fig. 4-3 can be used to solve any rectangular (or square) grid problem if they are suitably modified. For example, if a corner is as shown in Fig. 7-5, as would occur with a notch, one can obtain the finite-difference equation for the point using the center-point grid system and the moment relationship

$$\frac{\partial^2 w}{\partial x^2} + \mu \frac{\partial^2 w}{\partial y^2} = 0 \tag{7-3}$$

Also one may use the shear relationship of

$$\frac{\partial^3 w}{\partial x^3} + (2 - \mu) \frac{\partial^3 w}{\partial x \, \partial y^2} = 0 \tag{7-4}$$

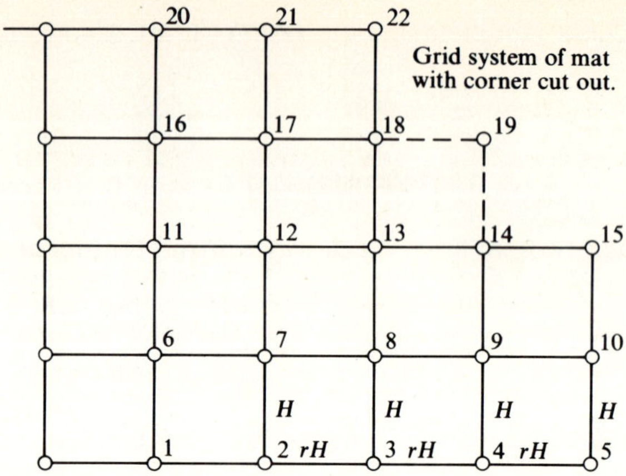

FIGURE 7-5
An arbitrary grid and numbering scheme for node points to illustrate the modification of the central-point finite-difference equation of Fig. 4-3*i* for point 13 above. Point 19 is off the plate.

i.e., moments and shears perpendicular to free edges are zero, but this is not needed in the example to be illustrated.

If point 19 were on the plate, no problem would exist, but since it is not, it will be taken as a fictitious point and the relationship of Eq. (7-3) will be used to obtain its value. Expanding Eq. (7-3) with central finite differences at node 18 (the moment perpendicular to the edge equals 0), we obtain

$$\frac{1}{(rH)^2}(w_{17} - 2w_{18} + w_{19}) + \frac{\mu}{H^2}(w_{22} - 2w_{18} + w_{13}) = 0$$

Canceling H^2 and combining terms, we have

$$w_{17} - 2w_{18} + w_{19} + \mu r^2 w_{22} - 2\mu r^2 w_{18} + \mu r^2 w_{13} = 0$$

Solving for w_{19} gives

$$w_{19} = -\mu r^2 w_{13} - w_{17} + (2 + 2\mu r^2) w_{18} - \mu r^2 w_{22}$$

Substitution of this value of w_{19} into the difference expression at node 13 and applying $2r^2/r^4$, we obtain the finite-difference equation for node 13 (using the notation of Table 4-2 where possible and $*$ as product)

$$\begin{aligned} &X10 * w_3 + X18 * w_7 + X15 * w_8 + X18 * w_9 + X27 * w_{11} + X16 * w_{12} \\ &\quad + (X22 - 2\mu) * w_{13} + X16 * w_{14} + X27 * w_{15} + 0 * w_{17} \\ &\quad + (-4 + 4\mu) * w_{18} + (1 - 2\mu) * w^{22} \\ &\qquad = \frac{1}{Dr^5}[PH^2 - k_s(rH)^2 * w_{13}] \qquad (7\text{-}5) \end{aligned}$$

If necessary, the user can obtain the finite-difference equations when the notch is one node point from the end or other configuration not shown in Fig. 4-3. With this preliminary discussion, let us proceed to a simple example to illustrate the computer program. Due to space limitations, the computer output listing of the equations, etc., will not be shown.

From Fig. 4-3 and Table 4-2 and referring to the sketch in Example 7-4, at point (1,1), we write (using the *computer notation* and omitting all zero terms)

$$\begin{aligned} &\mathrm{X3} * \mathrm{W(1,1)} + \mathrm{X2} * \mathrm{W(1,2)} + \mathrm{X1} * \mathrm{W(1,3)} + \mathrm{X5} * \mathrm{W(2,1)} + \mathrm{X4} * \mathrm{W(2,2)} \\ &\qquad + \mathrm{X6} * \mathrm{W(3,1)} = \mathrm{P(1,1)} - \frac{k_s(rH^2)}{D} * \mathrm{W(1,1)} \end{aligned}$$

Simplifying, we see that the soil-reaction term is additive to W(1,1), and so we obtain

$$\left[\mathrm{X3} + \frac{k_s(rH^2)}{D}\right] * \mathrm{W(1,1)} + \mathrm{X2} * \mathrm{W(1,2)} + \cdots = \mathrm{P(1,1)}$$

as the first of 49 equations required to solve the problem. The value of P(1,1), either zero or nonzero, is entered in the constant matrix.

At point (1,2) we have

$$\begin{aligned} &\mathrm{X2} * \mathrm{W(1,1)} + \mathrm{X8} * \mathrm{W(1,2)} + \mathrm{X7} * \mathrm{W(1,3)} + \mathrm{X1} * \mathrm{W(1,4)} + \mathrm{X9} * \mathrm{W(2,1)} \\ &\qquad + \mathrm{X12} * \mathrm{W(2,2)} + \mathrm{X9} * \mathrm{W(2,3)} + \mathrm{X11} * \mathrm{W(3,2)} \\ &\qquad\qquad = \mathrm{P(1,2)} - \frac{k_s(rH^2)}{D} * \mathrm{W(1,2)} \end{aligned}$$

which is the second of 49 equations. The remaining equations are formed in a like manner and solved using the enclosed computer program.

The computer program can include the weight of the footing if desired. If nodal deflections are negative (upward) or zero, the soil spring effect is removed from the computations, i.e., by not adding the term

$$\frac{k_s(rH^2)}{D}$$

to that W(I,J) node. If negative deflections are encountered, it will be more correct to include the footing weight.

EXAMPLE 7-2 Repeat Example 7-1 as a mat by finite differences. Figure E7-2.2 illustrates the grid using 3.7 × 3 ft or $r = 1.233$. The soil modulus is 36 kcf, $E_c =$ 468,000 ksf, and $D = 3.833$ ft; $\mu = 0.150$, and there will be 40 P-matrix entries due to prorating column loads to the four adjacent nodes using simple beam theory (Fig. E7-2.1). Note that D is not that from Example 7-1.

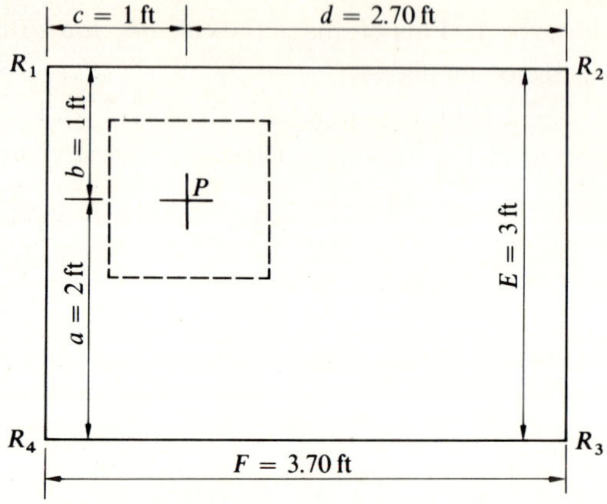

$$R_1 = \left(\frac{a}{E}\right)\frac{d}{F}P = 0.486P$$

$$R_2 = \left(\frac{a}{E}\right)\frac{c}{F}P = 0.18P$$

FIGURE E7-2.1
Prorating column loads to adjacent nodes using simple beam analysis.

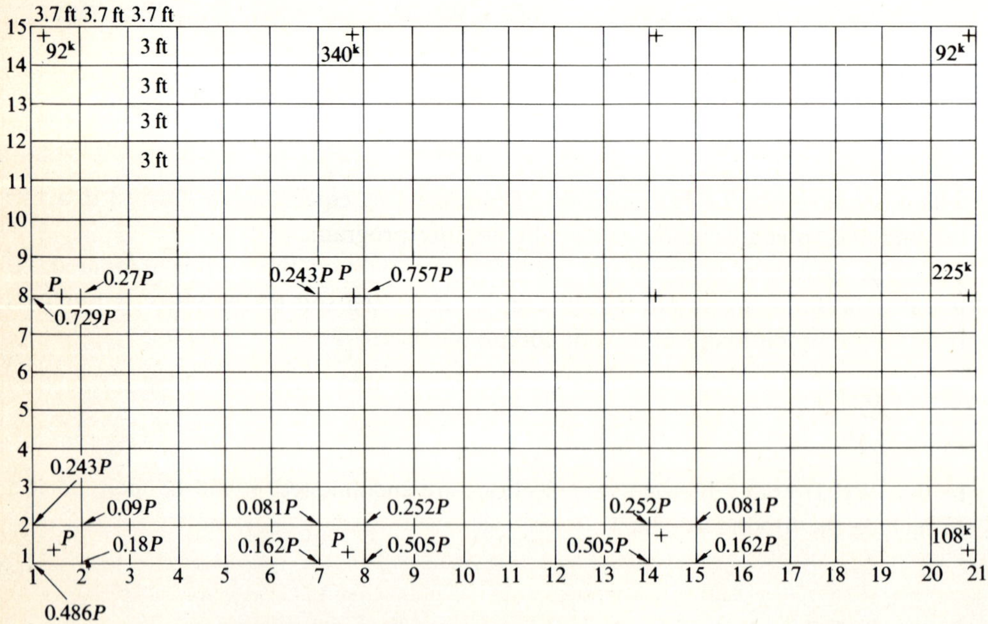

FIGURE E7-2.2
Grid for finite-difference solution with 315 nodes for 74 × 42 ft mat with columns as shown.

SOLUTION The data cards for the finite-difference solution are as follows:

Card	Data
1	TITLE
2	UNITS UT1–UT6 standard FU1 = 12. leave FU2–FU4 = 0
3	15 21 40 1
	H RH T E SM UNITWT XMU
4	3.00 3.70 3.8333 468000. 36. .150 .150
	Sample *P*-matrix entries are [I,J, P(I,J), kips]:
5	1 1 52.490
	1 2 19.440
	1 7 43.740
	. . . .
	15 21 44.760

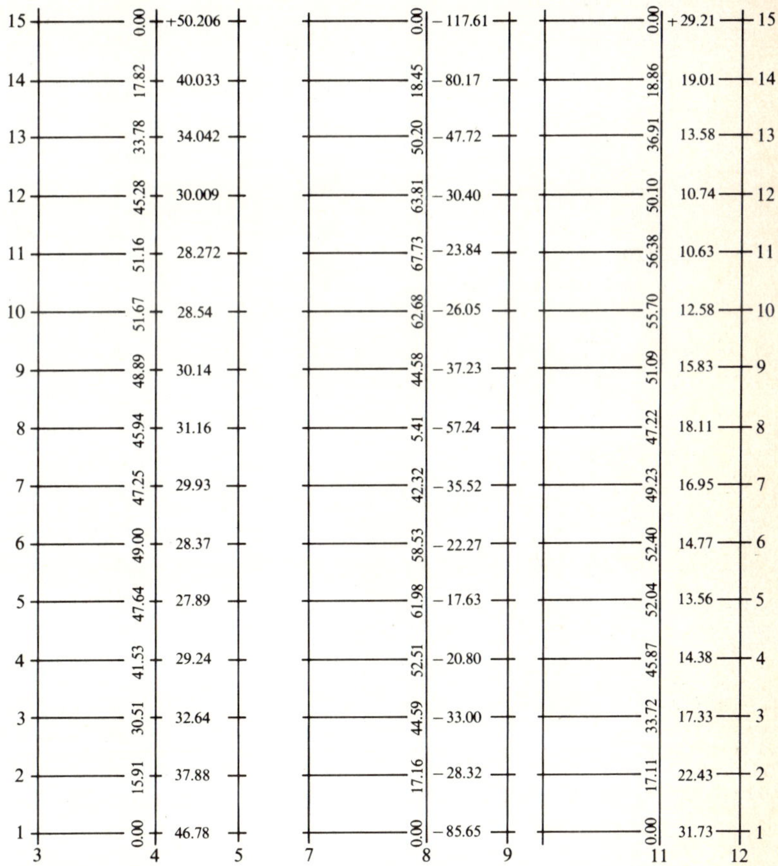

FIGURE E7-2.3
Partial finite difference output for selected nodes for comparison with Example 7-1.

The resulting Y matrix is 315 × 315. Output for columns 3 to 5 and 7 to 12 (all rows shown) is given in Fig. E7-2.3 (p. 221). Moments shown are in foot-kips. This solution summed forces to give

$$\text{Sum soil reactions} = 4{,}459.2 \text{ kips}$$

$$\text{Actual vertical load} = 4{,}436.8 \text{ kips}$$

indicating that the computation error was negligible. ////

EXAMPLE 7-3 Compute the bending moments of the rectangular spread footing (metric) shown in Fig. E7-3.1. Use a grid of 0.3 × 0.3 m. Take E_c = 22,408,730

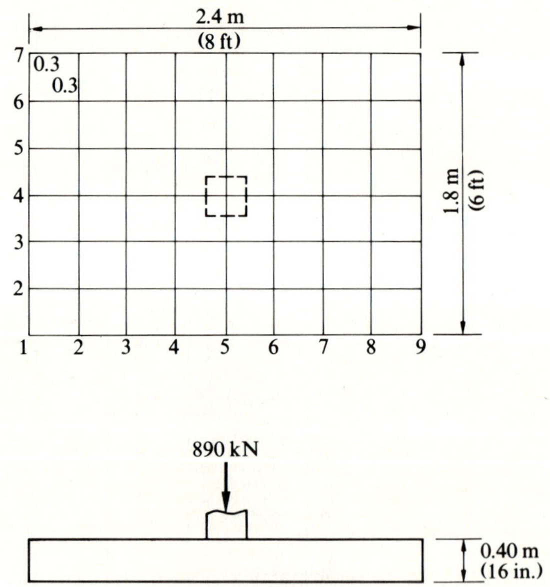

FIGURE E7-3.1

kN/sq m (3,250 ksi). The allowable soil pressure is 2 kg/sq cm, from which one can compute $k_s \approx$ 23,536 kN/cu m. Concrete at 150 pcf gives 23.56 kN/cu m. Poisson's ratio is taken as 0.15.

SOLUTION Data cards consist of:

Card	Data
1	TITLE (Fig. E7-3.2)
2	UNITS (M CM··· KN/CU M 100.) UT1–UT6 and FU1 = 100
3	.30 0.3 .40 22408730. 23536. 23.56 .150
4	4 5 890. load, KN)

Figure E7-3.2 illustrates *I/O* and the deflection matrix including counters for zero deflections. Figure E7-3.3 illustrates computer output of bending moments (kilo-newton-meters), as well as nodal soil reactions and pressures. Computational checks consists in checking values for symmetry and $\sum F_v = 0$ (figures on pages 224–227). ////

EXAMPLE 7-4 Solve the footing shown in Fig. E7-4.1*a* by the finite-difference method. The footing is 6 × 6 ft × 1.50 ft thick. The footing is concentrically loaded with 60 kips and bending moments about both axes of 120 ft-kips. Compare the results to the hand solution of Bowles (1968, chap. 5). Take k_s = 60 kcf and E_c = 468,000 ksf. Note that the dashed line in footing sketch is the Bowles' line of zero pressure, which is "exact" using calculus. The soil pressure of 22.5 ksf at node (7,7) is obtained from the hand solution (figures on pages 228 and 229).

SOLUTION Solve as a mat on an elastic foundation by finite difference both *including the footing weight* (which the hand solution does not) and as a weightless footing. Converting the concentric column load with moments to an eccentric column load without moments results in the equivalent column location of node (6,6) with a load of 60 kips.

The data cards are:

Card	Data
1	TITLE
2	UNITS (FT, IN-KIPS/CU FT 12.) UT1–UT6 and FU1 = 12.
3	1. 1. 1.50 468000. 60. .150* .150
4	6 6 60.

* For the weightless case UNITWT = 0.0.

```
 J E BOWLES   EXAMPLE 7-3 BEND. MOMENTS IN RECT. SPREAD FTG--METRIC  W/WEIGHT


          MAT FOUNDATION INPUT DATA
NO OF ROWS, N=  7      NO. OF COLS, M=  9      NO OF NON-ZERO Q-VALUES =  1
POISSONS RATIO =.150    MAT THICKNESS =   0.40 M     MOD ELAS = 22408730.0 KN/SQ M
GRID DIMENSION =  0.30 M     SOIL MODULUS = 23536.00 KN/CU M
UNIT WT OF FTG = 23.560 KN/CU M      R =1.00000
D =   122264.        DD = 0.736111D-06

MAT DIMENSIONS ARE:  X=  2.400 M       Y =  1.800 M

THE FOUNDATION LOADS AND COORDS ARE

4    5   890.000
INVERSION COUNT (EVERY 10) =     10 OF    63
INVERSION COUNT (EVERY 10) =     20 OF    63
INVERSION COUNT (EVERY 10) =     30 OF    63
INVERSION COUNT (EVERY 10) =     40 OF    63
INVERSION COUNT (EVERY 10) =     50 OF    63
INVERSION COUNT (EVERY 10) =     60 OF    63

          THE LOAD MATRIX & FTG WEIGHT
     INCL DDIV          FOOTING WT         LOAD MATRIX        FOOTING WT
       1   0.000000     0.212040       33   0.000001       0.848159
       2   0.000000     0.424079       34   0.000001       0.848159
       3   0.000000     0.424079       35   0.000001       0.848159
       4   0.000000     0.424079       36   0.000000       0.424079
       5   0.000000     0.424079       37   0.000000       0.424079
       6   0.000000     0.424079       38   0.000001       0.848159
       7   0.000000     0.424079       39   0.000001       0.848159
       8   0.000000     0.424079       40   0.000001       0.848159
       9   0.000000     0.212040       41   0.000001       0.848159
      10   0.000000     0.424079       42   0.000001       0.848159
      11   0.000001     0.848159       43   0.000001       0.848159
      12   0.000001     0.848159       44   0.000001       0.848159
      13   0.000001     0.848159       45   0.000000       0.424079
      14   0.000001     0.848159       46   0.000000       0.424079
      15   0.000001     0.848159       47   0.000001       0.848159
      16   0.000001     0.848159       48   0.000001       0.848159
      17   0.000001     0.848159       49   0.000001       0.848159
      18   0.000000     0.424079       50   0.000001       0.848159
      19   0.000000     0.424079       51   0.000001       0.848159
      20   0.000001     0.848159       52   0.000001       0.848159
      21   0.000001     0.848159       53   0.000001       0.848159
      22   0.000001     0.848159       54   0.000000       0.424079
      23   0.000001     0.848159       55   0.000000       0.212040
      24   0.000001     0.848159       56   0.000000       0.424079
      25   0.000001     0.848159       57   0.000000       0.424079
      26   0.000001     0.848159       58   0.000000       0.424079
      27   0.000000     0.424079       59   0.000000       0.424079
      28   0.000000     0.424079       60   0.000000       0.424079
      29   0.000001     0.848159       61   0.000000       0.424079
      30   0.000001     0.848159       62   0.000000       0.424079
      31   0.000001     0.848159       63   0.000000       0.212040
      32   0.000656     0.848159
```

THE DEFLECTION MATRIX IS (M)

	1	2	3	4	5	6	7	8	9
1	C.00886	0.00900	0.00912	0.00921	0.00925	0.00921	0.00912	0.00900	0.00886
2	C.00890	0.00904	0.00917	0.00927	0.00931	0.00927	0.00917	0.00904	0.00890
3	0.00893	0.00908	0.00922	0.00933	0.C0939	0.00933	0.00922	0.00908	0.00893
4	0.C0895	0.00909	0.00924	0.00937	0.00945	0.00937	0.00924	0.00909	0.00895
5	0.C0893	0.00908	0.00922	0.00933	0.00939	0.00933	0.00922	0.00908	0.00893
6	0.00890	0.00904	0.00917	0.00927	0.C0931	0.00927	0.00917	0.00904	0.00890
7	0.00886	0.00900	0.00912	0.00921	0.00925	0.00921	0.00912	0.00900	0.00886

THE NON-LINEAR DEFLECTION MATRIX IS (CM)

	1	2	3	4	5	6	7	8	9
1	0.88643	0.89973	0.91195	0.92109	0.92465	0.92109	0.91195	0.89973	0.88643
2	0.89037	0.90404	0.91698	0.92707	0.93127	0.92707	0.91698	0.90404	0.89037
3	0.89344	0.90762	0.92160	0.93337	0.93906	0.93337	0.92160	0.90762	0.89344
4	0.89465	0.90912	0.92382	0.93711	C.94529	0.93711	0.92382	0.90912	0.89465
5	C.89344	0.90762	0.92160	0.93337	0.93906	0.93337	0.92160	0.90762	0.89344
6	0.89037	0.90404	0.91698	0.92707	0.93127	0.92707	0.91698	0.90404	0.89037
7	0.88643	0.89973	0.91195	0.921C9	C.92465	0.92109	0.91195	0.89973	0.88643

NN = 3   LCOUN(NN) = 0   LCOUN(NN-1) = 0

FIGURE E7-3.2
Input data and load matrix for Example 7-3. Also shown are output deflection matrices.

THE BENDING MOMENTS IN SLAB ARE AS FOLLOWS

COORDS		X-AXIS	Y-AXIS
1	1	0.0	0.0
1	2	-16.2486	0.0
1	3	-42.5486	0.0
1	4	-75.1063	0.0
1	5	-94.5094	0.0
1	6	-75.1063	0.0
1	7	-42.5486	0.0
1	8	-16.2486	0.0
1	9	0.0	0.0
2	1	0.0	-13.2687
2	2	-11.4208	-11.3422
2	3	-39.3576	-11.2186
2	4	-79.6091	-7.9276
2	5	-111.5566	-1.0933
2	6	-79.6091	-7.9276
2	7	-39.3576	-11.2186
2	8	-11.4208	-11.3422
2	9	0.0	-13.2687
3	1	0.0	-25.6848
3	2	-6.8705	-28.6743
3	3	-35.1486	-37.3836
3	4	-87.5601	-46.9084
3	5	-157.9818	-44.4842
3	6	-87.5601	-46.9084
3	7	-35.1486	-37.3836
3	8	-6.8705	-28.6743
3	9	0.0	-25.6848
4	1	0.0	-32.3637
4	2	-3.0026	-40.2532
4	3	-28.0423	-62.9307
4	4	-84.8139	-112.2446
4	5	-247.5834	-202.5515
4	6	-84.8139	-112.2446
4	7	-28.0423	-62.9307
4	8	-3.0026	-40.2532
4	9	0.0	-32.3637

CCORDS		X-AXIS	Y-AXIS
5	1	0.0	-25.6848
5	2	-6.8705	-28.6743
5	3	-35.1486	-37.3836
5	4	-87.5601	-46.9084
5	5	-157.9818	-44.4842
5	6	-87.5601	-46.9084
5	7	-35.1486	-37.3836
5	8	-6.8705	-28.6743
5	9	0.0	-25.6848
6	1	0.0	-13.2687
6	2	-11.4208	-11.3422
6	3	-39.3576	-11.2186
6	4	-79.6091	-7.9276
6	5	-111.5566	-1.0933
6	6	-79.6091	-7.9276
6	7	-39.3576	-11.2186
6	8	-11.4208	-11.3422
6	9	0.0	-13.2687
7	1	0.0	0.0
7	2	-16.2486	0.0
7	3	-42.5486	0.0
7	4	-75.1063	0.0
7	5	-94.5094	0.0
7	6	-75.1063	0.0
7	7	-42.5486	0.0
7	8	-16.2486	0.0
7	9	0.0	0.0

THE NODAL REACTIONS (KN) ARE AS FCLLCWS

	1	2	3	4	5	6	7	8	9
1	4.69419	9.52924	9.65859	9.75540	9.79318	9.75540	9.65859	9.52924	4.69419
2	9.43007	19.14969	19.42372	19.63765	19.72650	19.63765	19.42372	19.14969	9.43007
3	9.46259	19.22553	19.52178	19.77091	19.89159	19.77091	19.52178	19.22553	9.46259
4	9.47539	19.25727	19.56861	19.85028	20.02352	19.85028	19.56861	19.25727	9.47539
5	9.46259	19.22553	19.52178	19.77091	19.89159	19.77091	19.52178	19.22553	9.46259
6	9.43007	19.14969	19.42372	19.63765	19.72650	19.63765	19.42372	19.14969	9.43007
7	4.69419	9.52924	9.65859	9.75540	9.79318	9.75540	9.65859	9.52924	4.69419

THE NCDAL SOIL PRESSURE, KN/SQ M, IS

	1	2	3	4	5	6	7	8	9
1	208.63083	211.76114	214.63555	216.78692	217.62641	216.78692	214.63555	211.76114	208.63083
2	209.55741	212.77449	215.81930	218.19636	219.18352	218.19636	215.81930	212.77449	209.55741
3	210.27996	213.61717	216.90885	219.67694	221.01783	219.67694	216.90885	213.61717	210.27996
4	210.56435	213.96990	217.42926	220.55890	222.48372	220.55890	217.42926	213.96990	210.56435
5	210.27996	213.61717	216.90885	219.67694	221.01783	219.67694	216.90885	213.61717	210.27996
6	209.55741	212.77449	215.81930	218.19636	219.18352	218.19636	215.81930	212.77449	209.55741
7	208.63083	211.76114	214.63555	216.78692	217.62641	216.78692	214.63555	211.76114	208.63083

```
TOTAL SUM OF FOOTING EXTERNAL LOADS =    890.000 KN
          FOOTING WEIGHT =      40.711 KN
          SOIL REACTIONS =     932.181 KN
```

FIGURE E7-3.3
Bending moments, nodal reactions, and soil pressures for Example 7-3. Note symmetry of values (as expected) in this problem. Sum of vertical forces are also satisfied (930.7 versus 932.2 kN).

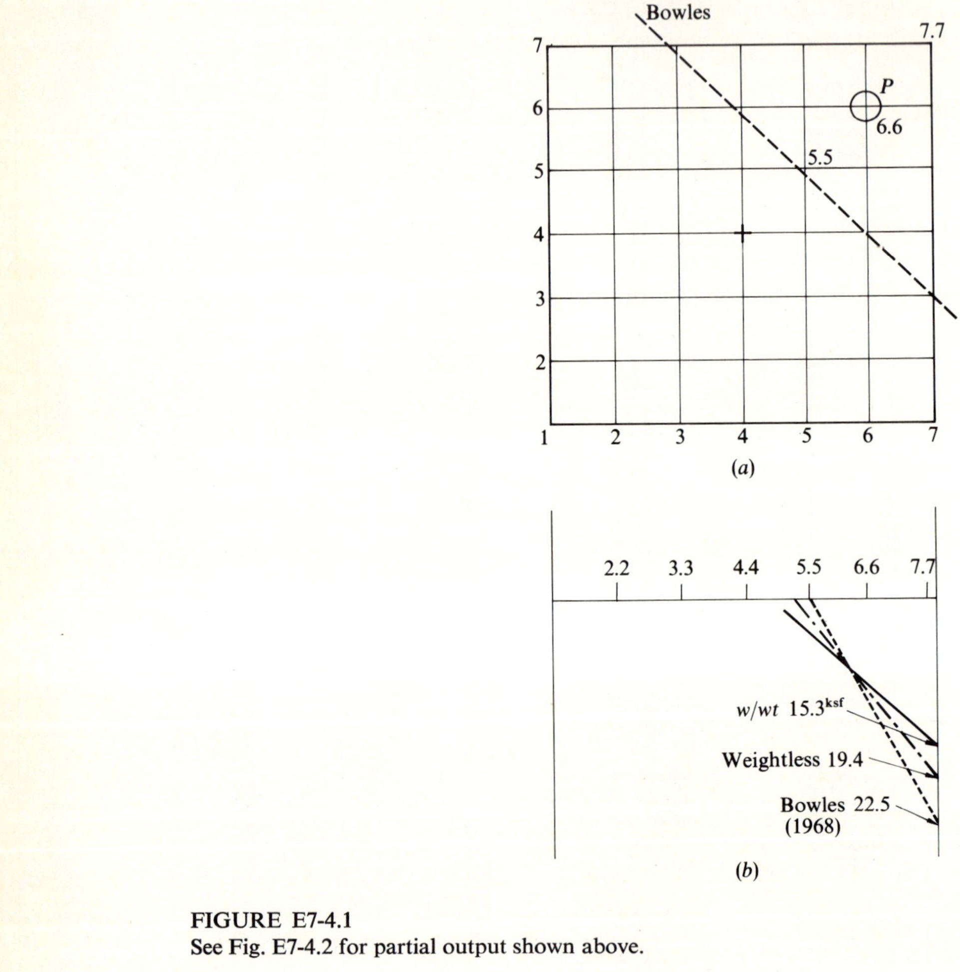

FIGURE E7-4.1
See Fig. E7-4.2 for partial output shown above.

The solid line of Fig. E7-4.1*a* represents the output. As one would expect, the inclusion of footing weight tends to reduce the maximum pressure. With a weightless footing the pressures are nearly equal. Note also that the line of zero pressure is about the same in all cases. Figure E7-4.2 illustrates the partial computer output for given input data. Note the approximate line of zero pressure and soil pressure of 15.3 ksf at node (7, 7) are shown on the computer printout.

THE BENDING MOMENTS IN SLAB ARE AS FOLLOWS

COORDS		X-AXIS	Y-AXIS	COORDS		X-AXIS	Y-AXIS
1	1	0.0	0.0	5	1	0.0	-0.3542
1	2	-0.3806	0.0	5	2	0.2740	-0.4976
1	3	-0.4935	0.0	5	3	0.6349	-0.6326
1	4	-0.4840	0.0	5	4	0.7413	-0.6912
1	5	-0.3542	0.0	5	5	-0.2821	-0.2821
1	6	-0.1021	0.0	5	6	-3.6203	0.7221
1	7	0.0	0.0	5	7	0.0	-0.1319
2	1	0.0	-0.3806	6	1	0.0	-0.1021
2	2	-0.2481	-0.2481	6	2	0.4968	-0.3310
2	3	-0.3880	-0.1412	6	3	1.3520	-0.6174
2	4	-0.4783	0.0301	6	4	2.2240	-1.2462
2	5	-0.4976	0.2740	6	5	0.7221	-3.6203
2	6	-0.3310	0.4968	6	6	-9.6672	-9.6672
2	7	0.0	0.5039	6	7	0.0	-7.8710
3	1	0.0	-0.4935	7	1	0.0	0.0
3	2	-0.1412	-0.3880	7	2	0.5039	0.0
3	3	-0.2433	-0.2433	7	3	1.7302	0.0
3	4	-0.4050	0.0687	7	4	2.4049	0.0
3	5	-0.6326	0.6349	7	5	-0.1319	0.0
3	6	-0.6174	1.3520	7	6	-7.8710	0.0
3	7	0.0	1.7302	7	7	0.0	0.0
4	1	0.0	-0.4840				
4	2	0.0301	-0.4783				
4	3	0.0687	-0.4050				
4	4	-0.0985	-0.0985				
4	5	-0.6912	0.7413				
4	6	-1.2462	2.2240				
4	7	0.0	2.4049				

THE NODAL REACTIONS (KIPS) ARE AS FOLLOWS

	1	2	3	4	5	6	7
1	0.0	0.0	0.0	0.0	0.0	0.0	0.0
2	0.0	0.0	0.0	0.0	0.0	0.0	0.08349
3	0.0	0.0	0.0	0.0	0.0	0.17019	1.60055
4	0.0	0.0	0.0	0.0	0.17287	3.20539	3.11833
5	0.0	0.0	0.0	0.17287	3.20777	6.24248	4.63712
6	0.0	0.0	0.17019	3.20539	6.24248	9.28056	6.15609
7	0.0	0.08349	1.60055	3.11833	4.63712	6.15609	3.83625

THE NODAL SOIL PRESSURE, K/SQ FT, IS

	1	2	3	4	5	6	7
1	0.0	0.0	0.0	0.0	0.0	0.0	0.0
2	0.0	0.0	0.0	0.0	0.0	0.0	0.16698
3	0.0	0.0	0.0	0.0	0.0	0.17019	3.20110
4	0.0	0.0	0.0	0.0	0.17287	3.20539	6.23666
5	0.0	0.0	0.0	0.17287	3.20777	6.24248	9.27425
6	0.0	0.0	0.17019	3.20539	6.24248	9.28056	12.31218
7	0.0	0.16698	3.20110	6.23666	9.27425	12.31218	15.34502

```
TOTAL SUM OF FOOTING EXTERNAL LOADS =     60.000 KIPS
          FOOTING WEIGHT =       6.750 KIPS
          SOIL REACTIONS =      67.097 KIPS
```

FIGURE E7-4.2
Partial output for Example 7-4, including footing weight.

////

7-5 GENERAL COMMENTS ON MAT SOLUTIONS BY FINITE DIFFERENCES

From solving a large number of problems the author has found that:

1 The linear solution for soil pressure currently used as Eq. (7-1) is valid for footings; i.e., the soil pressures obtained by finite difference are linear and only slightly affected by footing thickness or soil modulus k_s.

2 The situation of large eccentricity as shown in Example 7-4 also displays a linear pressure distribution. Again the conventional solution is valid.

3 Conclusions 1 and 2 were obtained for an extremely wide range of footing thickness. A typical case of a 12×12 footing was varied from 12 to 60 in in thickness with little change in line of zero pressure.

4 Including the footing weight does not move the zero-pressure line much where footings are subjected to large eccentricities.

5 The computations appear to become unstable if an unstable load condition is imposed; i.e., if a load is put only on an edge or corner, the $\sum F_v = 0$ may not be satisfied. Contrary to what one would expect, the footing weight does not seem to contribute to footing stability.

6 Column loads not at node points are prorated to the adjacent two or four nodes, using any reasonable method.

7-6 COMPUTER PROGRAM FOR MAT FOUNDATION BY FINITE DIFFERENCES

This program will solve any square or rectangular mat using a square ($rh = h$) or rectangular ($rh \neq h$) grid. A notched (or reentrant) corner can be solved by modifying the program to zero out the unwanted nodes and to allow reading in the nonzero coefficients of the affected nodes.

A subroutine is used (and included) for the inversion of the coefficient matrix. This subroutine is based on the product-inverse method of inverting a matrix. The product inverse operates on one column at a time of the coefficient matrix. The matrix is developed one row at a time, but it can be shown that the matrix generated in this computer program is of the form

$$AX = C$$

and that

$$A = A^T$$

which is most helpful in computer bookkeeping since the program develops a row of the A matrix at a time, whereas a column is operated upon in the inversion subroutine.[1]

This program will solve either fps or metric problems through use of appropriate entries and the correct unit data card.

[1] Only if notches or holes are not present is the matrix always symmetrical. Using this product-inverse method when holes or notches are present may not always be possible.

Line	Operation
1–6	Bookkeeping
6	COMMON statement and variables must be exactly in order shown
8	READ TITLE, UNITS (two cards)
10	READ (4I5) N = number of rows (horizontal lines); M = number of columns (vertical lines); NQ = number of nonzero entries in *P* matrix; LIST = switch to write *A* matrix and certain other data
14	READ (7F10.4) H = vertical grid; RH = horizontal grid length; T = total mat thickness, all in feet or meters; E = modulus of elasticity; SM = subgrade modulus; UNITWT = unit weight of mat and XMU = Poisson's ratio
16–21	Computation constants
26–51	Builds *P* matrix
44	READ I,J, P(I,J) coordinates of load and load
54–84	Computation constants of Table 4-2
85–208	Builds *Y* matrix one line or row at a time. Note PRINT (W,I,J) subroutine both sets up rows and also writes values if LIST > 0. MATZER initializes the row to zero
209	Calls inversion subroutine PROINV; this gives the values of deflection [X(I)]
225–254	Sets deflections W(I,J) = X(I) and builds W1(I,J) matrix. W1 is used for soil reactions and pressures; W is used for moment computations. W1 will have zeros for zero or negative values of W. If any negative or zero W values are found, they are counted and the program is recycled to statement 23 until previous count and current count are the same. CALL OUTPUT (W,5), CALL OUTPUT (W,7) is subroutine to write the deflection matrix from disk work areas 5 and 7
255–285	Computes bending moments in X and Y directions in slab
301–325	Computes and writes nodal soil reactions and soil pressures

```
       C      J E BOWLES -- MAT FOUNDATION BY FINITE DIFFERENCE -- SQ OR RECT
       C      UNITS = FPS OR METRIC; USE KIPS, KN; FT OR M; KCF OR KN/CU M, ETC
0001          DIMENSION W(15,21),W1(15,21),XMX(15,21),XMY(15,21),SOILP(15,21),S
             1OILR(15,21),Q(320),WT(320),LCCUN(50),Y(320),X(320),TITLE(20)
0002          DOUBLE PRECISION E,D,DD,W,Y,X,Q,QIN,W1,SOILP,SOILR,WT
0003          DOUBLE PRECISION KCON,C,CC,XMX,XMY,UT5,UT6
C004          COMMON M,N,NN,K,X,LIST
C005          EQUIVALENCE(WT(1),XMX(1,1),SCILP(1,1)), (Q(1),XMY(1,1),SOILR(1,1))
0006     5500 READ(1,1000,END=150)TITLE,UT1,UT2,UT3,UT4,UT5,UT6,FU1,FU2,FU3,FU4
0007     1C00 FORMAT(20A4/4(A4,6X),A8,2X,A8,2X,4F5.0)
CC08          READ (1,1002)N,M,NQ,LIST
0009     1002 FORMAT(4I5)
0010          WRITE(3,1001)TITLE
0011     1001 FORMAT('1',//, T5,20A4)
0012          READ(1,1004)H,RH,T,E,SM,UNITWT,XMU
0013     1004 FORMAT(7F10.4)
0014          A = (M-1)*RH
0015          B = (N-1)*H
0016          R = RH/H
0017          NSAVE = N
0018          MSAVE = M
0019          D=E*T**3/(12.0*(1.0-XMU**2))
0020          DD = H**2/(D*R)
0021          WRITE(3,8001)N,M,NQ,XMU,T,UT1,E,UT5,H,UT1,SM,UT6,UNITWT,UT6,R,D,DD
0022     8001 FORMAT(///T15,'MAT FOUNDATION INPUT DATA'/T5,'NO OF ROWS, N=',I3,
             1 5X,'NO. OF COLS, M=',I3, 5X,'NO OF NON-ZERO Q-VALUES =',I3/T5,
             2'POISSONS RATIO =',F4.3, 3X,'MAT THICKNESS =',F6.2,1X,A2,3X,'MOD E
             3LAS =',F11.1,1X,A7,/,T5, 'GRID DIMENSION =',F6.2,1X,A2,3X,'SOIL MO
             4DULUS =',F9.2,1X,A7,/,T5, 'UNIT WT OF FTG =',F7.3,1X,A7,5X,'R =',F
             57.5,/,T5, 'D =',G13.6,3X,'DD =',G13.6,//)
0023          WRITE(3,8003)A,UT1,B,UT1
0024     8003 FORMAT(T5,'MAT DIMENSIONS ARE:  X=',F7.3,1X,A2,5X,'Y =',F7.3,1X,
             1A2)
0025          SUMFTG = 0.
0026          DO 687  I = 1,N
0027          DO 687  J = 1,M
0028          W1(I,J) = 1.
0029          LL = (I-1)*M+J
0030          WEIGHT = T*UNITWT*(H**2)*R
0031          Q(LL) = WEIGHT
0032          IF((I.EQ.1.OR.I.EQ.N).AND.(J.EQ.1.OR.J.EQ.M))Q(LL) = .25*WEIGHT
0033          IF((I.EQ.1.OR.I.EQ.N).AND.(J.NE.1.AND.J.NE.M))Q(LL)=.5*WEIGHT
0034          IF((I.NE.1.AND.I.NE.N).AND.(J.EQ.1.OR.J.EQ.M))Q(LL)=.5*WEIGHT
0035          SUMFTG = SUMFTG+Q(LL)
0036          WT(LL) = Q(LL)
0037      687 Q(LL) = Q(LL)*DD
0038          SUMQ = 0.
```

```
      IF(NQ.EQ.0)GO TO 3
      WRITE(3,6240)
 6240 FORMAT(//,T5, 'THE FOUNDATION LOADS AND COORDS ARE',//)
      DO 2  II = 1,NQ
      READ(1,688)I,J,QIJ
  688 FORMAT(2I5,F10.4)
      WRITE(3,6250)I,J,QIJ
 6250 FORMAT(2I5,F10.3)
      SUMQ = SUMQ + QIJ
      LL = (I-1)*M+J
    2 Q(LL) = Q(LL) + QIJ*DD
    3 LCOUN(2) = 0
      NN = 3
      LCOUN(NN) = 0
      XMU1 = 1-XMU
      XMU2 = 1-XMU**2
      R2 = R**2
      R4 = R**4
      X1 =(.5*XMU2)/R4
      X2 =(-1*XMU2-2*XMU1*R2)/R4
      X3 = (.5*XMU2+2*XMU1*R2+.5*XMU2*R4)/R4
      X4 =(2*XMU1*R2)/R4
      X5 =(-2*XMU1*R2-XMU2*R4)/R4
      X6 =(.5*XMU2*R4)/R4
      X7 =(-2*XMU2-2*XMU1*R2)/R4
      X8 =(2.5*XMU2+4*XMU1*R2+R4)/R4
      X9 =((2-XMU)*R2)/R4
      X10 = R4/R4
      X11 -(3*XMU2+4*XMU1*R2+R4)/R4
      X12 =(-2*(2-XMU)*R2-2*R4)/R4
      X13 =(-2-2*(2-XMU)*R2)/R4
      X14 =(1+4*(XMU1)*R2+2.5*(XMU2)*R4)/R4
      X15 =(-4*R2-4*R4)/R4
      X16 =(-4-4*R2)/R4
      X17 =(6+8*R2+5*R4)/R4
      X18 =(2*R2)/R4
      X19 =(-2*XMU1*R2-2*XMU2*R4)/R4
      X20 =(5+8*R2+6*R4)/R4
      X21 =(5 + 8*R2+5*R4)/R4
      X22 =(6 +8*R2+6*R4)/R4
      X23 =(1 + 4*XMU1*R2 + 3*XMU2*R4)/R4
      X24 = H**4*(SM/D)
      X25 = X24/4.
      X26 = X24/2.
   23 K = NSAVE*MSAVE
      REWIND 4
      LL = 1
C     STAT NOS 63,64,65,64,63 TAKE BOTTOM & TOP ROWS
C     STAT NOS 68,69,70,69,68 TAKES CARE OF 2ND AND NEXT-TO LAST ROWS
C     STAT NOS 73,74,75,74,73 TAKES CARE OF ALL INTERMED ROWS
      DO 99 I = 1,N
      DO 99 J = 1,M
      CALL MATZER(W,N,M)
      KI = 1.
      KJ = 1.
      IF(LL.GE.16.AND.LL.LE.25) KI= -1.0
      IF(LL.EQ.4.OR.LL.EQ.5.OR.LL.EQ.9.OR.LL.EQ.10.OR.LL.EQ.14) KJ = -1.
      IF(LL.EQ.15.OR.LL.EQ.19.OR.LL.EQ.20.OR.LL.EQ.24.OR.LL.EQ.25)KJ=-1.
      GO TO (63,64,65,64,63,68,69,70,69,68,73,74,75,74,73,68,69,70,69,68
     1,63,64,65,64,63),LL
   63 W(I,J) = X3+X25
      IF(W1(I,J).LE.0.) W(I,J) = X3
      W(I,J+KJ) = X2
      W(I,J+2*KJ) = X1
      W(I+KI,J) = X5
      W(I+KI,J+KJ) = X4
      W(I+2*KI,J) = X6
      GO TO 98
   64 W(I,J) = X8 + X26
      IF(W1(I,J).LE.0.) W(I,J) = X8
      W(I,J+2*KJ) = X1
      W(I,J+KJ) = X7
      W(I,J-KJ) = X2
      W(I+KI,J) = X12
      W(I+KI,J+1) = X9
      W(I+KI,J-1) = X9
      W(I+2*KI,J) = X10
      GO TO 98
   65 W(I,J) = X11 + X26
      IF(W1(I,J).LE.0.) W(I,J) =  X11
      W(I,J-1) = X7
      W(I,J+1) = X7
      W(I,J-2) = X1
      W(I,J+2) = X1
      W(I+KI,J) = X12
      W(I+KI,J-1) = X9
      W(I+KI,J+1) = X9
      W(I+2*KI,J) = X10
      IF(J.LT.M-2) LL = LL-1
      GO TO 98
   68 W(I,J) = X14 + X26
      IF(W1(I,J).LE.0.) W(I,J) =  X14
      W(I-KI,J) = X5
      W(I+KI,J) = X19
      W(I+2*KI,J) = X6
      W(I,J+KJ) = X13
```

```
      W(I-1,J+KJ) = X9
      W(I+1,J+KJ) = X9
      W(I,J+2*KJ) = 1./R4
      IF(J.EQ.M.AND.N.EQ.4.AND.LL.NE.20) LL = 15
      GO TO 98
   69 W(I,J) = X21 + X24
      IF(W1(I,J).LE.0.) W(I,J) =  X21
      W(I,J+KJ) = X16
      W(I,J+2*KJ) = 1./R4
      W(I,J-KJ) = X13
      W(I-KI,J) = X12
      W(I-KI,J+KJ) = X9
      W(I-KI,J-KJ) = X4
      W(I+KI,J) = X15
      W(I+KI,J+KJ) = X18
      W(I+KI,J-KJ) = X9
      W(I+2*KI,J) = X10
      GO TO 98
   70 W(I,J) = X17 + X24
      IF(W1(I,J).LE.0.) W(I,J) =  X17
      W(I,J-1) = X16
      W(I,J+1) = X16
      W(I,J-2) = 1./R4
      W(I,J+2) = 1./R4
      W(I-KI,J) = X12
      W(I-KI,J+1) = X9
      W(I-KI,J-1) = X9
      W(I+KI,J+1) = X18
      W(I+KI,J-1) = X18
      W(I+2*KI,J) = X10
      W(I+KI,J) = X15
      IF(J.LT.M-2) LL = LL-1
      GO TO 98
   73 W(I,J) = X23+ X26
      IF(W1(I,J).LE.0.) W(I,J) =  X23
      W(I,J+KJ) = X13
      W(I,J+2*KJ) = 1./R4
      W(I-1,J) = X19
      W(I+1,J) = X19
      W(I-1,J+KJ) = X9
      W(I+1,J+KJ) = X9
      W(I-2,J) = X6
      W(I+2,J) = X6
      IF(J.EQ.M.AND.I.NE.N-2) LL = 10
      GO TO 98
   74 W(I,J) = X20+ X24
      IF(W1(I,J).LE.0.) W(I,J) =  X20
      W(I,J+KJ) = X16
      W(I,J+2*KJ) = 1./R4
      W(I,J-KJ) = X13
      W(I-1,J+KJ) = X18
      W(I+1,J+KJ) = X18
      W(I+1,J) = X15
      W(I-1,J) = X15
      W(I-2,J) = X10
      W(I+2,J) = X10
      W(I-1,J-KJ) = X9
      W(I+1,J-KJ) = X9
      GO TO 98
   75 W(I,J) = X22+ X24
      IF(W1(I,J).LE.0.) W(I,J) =  X22
      W(I,J-1) = X16
      W(I,J+1) = X16
      W(I,J+2) = 1./R4
      W(I,J-2) = 1./R4
      W(I+1,J-1) = X18
      W(I-1,J+1) = X18
      W(I+1,J+1) = X18
      W(I-1,J-1) = X18
      W(I-2,J) = X10
      W(I+2,J) = X10
      W(I-1,J) = X15
      W(I+1,J) = X15
      IF(J.LT.M-2) LL = LL-1
   98 CALL PRINT(W,I,J)
   99 LL = LL+1
      WRITE (4) (Q(I),I=1,K)
      REWIND 4
      CALL PROINV(Y)
C************************
   83 IF(NN.GT.3)GO TO 185
      WRITE(3,1057)
 1057 FORMAT(//,T15, 'THE LOAD MATRIX & FTG WEIGHT',/,T10,'INCL DDIV',
     15X,'FOOTING WT',5X,  'LOAD MATRIX',5X,'FOOTING WT')
      KO2 = K/2
      IF(KO2*2.NE.K)KO2 = KO2+1
      DO 180  J = 1,KO2
      JP = KO2+J
      IF(JP.GT.K)GO TO 179
      WRITE(3,1058)J,Q(J),WT(J),JP,Q(JP),WT(JP)
      GO TO 180
  179 WRITE(3,1058)J,Q(J),WT(J)
 1058 FORMAT(T5,I5,1X,F10.6,2X,F10.6,3X,I5,1X,F10.6,2X,F10.6)
  180 CONTINUE
  185 N = NSAVE
      DO 152  I = 1,N
      DO 152  J = 1,M
```

```
          K2 = (I-1)*M + J
          W(I,J) = X(K2)
      152 W1(I,J) = W(I,J)*FU1
      354 ICOUN=0
          DO 127 I=1,N
          DO 127  J = 1,M
          IF(W1(I,J).LE.0.)W1(I,J) = 0.
          IF(W1(I,J).EQ.0.)LCOUN(NN) = ICOUN+1
      127 ICOUN = LCOUN(NN)
          WRITE(3,1056)UT1
     1056 FORMAT(///T5,'THE DEFLECTION MATRIX IS (',A2,')'//)
          CALL OUTPUT(W)
          WRITE(3,1069)UT2
     1069 FORMAT(//T5,'THE NON-LINEAR DEFLECTION MATRIX IS (',A2,')'//)
          CALL OUTPUT(W1)
          WRITE(3,1073)NN,LCOUN(NN),LCOUN(NN-1)
     1073 FORMAT(//,T5,'NN =',I3,T15,'LCOUN(NN) =',I3,T35,'LCOUN(NN-1) =',I3
         A,//)
          IF(LCOUN(NN).LE.LCOUN(NN-1))GO TO 355
          NN = NN+1
          IF(NN.GT.8)GO TO 355
          GO TO 23
C         COMPUTE BENDING MOMENTS IN SLAB
      355 WRITE(3,1084)
     1084 FORMAT(//,T5, 'THE BENDING MOMENTS IN SLAB ARE AS FOLLOWS',/, T3,
         1 'COORDS',4X,'X-AXIS',6X,'Y-AXIS',7X,'COORDS',4X,'X-AXIS',6X,'Y-AX
         2IS',/)
          C = D/(RH**2)
          CC = XMU*D/(H**2)
          F = D/H**2
          FF = XMU*C
          DO 200  J = 1,N
          DO 200  K = 1,M
          IF(J.EQ.1.OR.J.EQ.N) GO TO 204
          IF(J.GT.1)GO TO 212
      204 IF(K.EQ.1.OR.K.EQ.M) GO TO 208
          IF(J.EQ.N)GO TO 206
          XMX(J,K) = C*(W(J,K-1) - 2.*W(J,K) + W(J,K+1)) + CC*(W(J+2,K) - 2.
         A*W(J+1,K) + W(J,K))
          XMY(J,K) = 0.
          GO TO 200
      206  XMX(J,K) = C*(W(J,K-1)  - 2.*W(J,K) + W(J,K+1)) + CC*(W(J-2,K) -
         A2.*W(J-1,K) + W(J,K))
          XMY(J,K) = 0.
          GO TO 200
      208 XMX(J,K) = 0.
          XMY(J,K) = 0.
          GO TO 200
      212 IF(K.EQ.1)GO TO 214
          IF(K.EQ.M)GO TO 216
          XMX(J,K) = C*(W(J,K-1) -2.*W(J,K) + W(J,K+1)) + CC*(W(J-1,K) -2.*W
         A(J,K) + W(J+1,K))
          XMY(J,K) = F*(W(J-1,K) -2.*W(J,K) + W(J+1,K)) + FF*(W(J,K-1)-2.*W(
         1J,K)+W(J,K+1))
          GO TO 200
      214 XMX(J,K) = 0.
          XMY(J,K) = F*(W(J-1,K)- 2.*W(J,K) + W(J+1,K)) +FF*(W(J,K+2) -2.*W(
         AJ,K+1) + W(J,K))
          GO TO 200
      216 XMX(J,K) = 0.
          XMY(J,K) = F*(W(J-1,K) - 2.*W(J,K) + W(J+1,K)) + FF*(W(J,K-2) - 2.
         A*W(J,K-1) + W(J,K))
      200 CONTINUE
          NN=N/2
          IF(NN*2.NE.N)NN = NN+1
          DO 401 J=1,NN
          JJ=J+NN
          DO 401 K=1,M
          IF(JJ.GT.N)GO TO 400
          WRITE(3,402)J,K,XMX(J,K),XMY(J,K),JJ,K,XMX(JJ,K),XMY(JJ,K)
          GO TO 401
      400 WRITE(3,402)J,K,XMX(J,K),XMY(J,K)
      402 FORMAT(2I4,2(2X,F10.4),3X,2I4,2(2X,F10.4))
      401 CONTINUE
      403 WRITE(3,1080)UT3
     1080 FORMAT(//T20,'THE NODAL REACTIONS (',A4,') ARE AS FOLLOWS')
C         COMPUTE SOIL REACTIONS AND SOIL PRESSURE
          SUMR = 0.
          DO 300 J = 1,N
          DO 300  K = 1,M
          KCON = 1.0
          SOILP(J,K) = W1(J,K)*SM/FU1
          IF((J.EQ.1).OR.(J.EQ.N)) KCON = .5*KCON
          IF((K.EQ.1).OR.(K.EQ.M)) KCON = .5*KCON
          SOILR(J,K) = KCON*SOILP(J,K)*R*H**2
      298 SUMR = SUMR + SOILR(J,K)
      300 CONTINUE
          CALL OUTPUT(SOILR)
          WRITE (3,1082)UT5
     1082 FORMAT(//T20,'THE NODAL SOIL PRESSURE, ',A7,', IS')
          CALL OUTPUT(SOILP)
          WRITE(3,1083)SUMQ,UT3,SUMFTG,UT3,SUMR,UT3
     1083 FORMAT(//T10,'TOTAL SUM OF FOOTING EXTERNAL LOADS =',F10.3,1X,A4/
         1T20,'FOOTING WEIGHT =',F10.3,1X,A4/T20,'SOIL REACTIONS =',F10.3,
         21X,A4)
          GO TO 5500
      150 STOP
          END
```

```
      SUBROUTINE PRINT(W,I,J)
      DIMENSION W(15,21),Y(320),X(320)
      DOUBLE PRECISION W,Y,X
      COMMON M,N,NN,K,X,LIST
      IF(LIST.EQ.0)GO TO 31
      IF(M.GT.10)GO TO 31
      IF(NN.GT.6) GO TO 31
      WRITE(3, 20)I,J,((W(N1,M1),M1=1,M),N1=1,N)
   20 FORMAT(1H0,2I6/(12F10.5))
   31 KK = (I-1)*M+J
      KKK = 0
      DO 30  I2 = 1,N
      DO 30  I3 = 1,M
      KKK = KKK+1
   30 Y(KKK) = W(I2,I3)
      WRITE (4)  (Y(LL),LL= 1,K)
      RETURN
      END
```

```
      SUBROUTINE PROINV(ASAT)
C     PRODUCT FORM--SOLUTION OF SYSTEM OF EQUATIONS.
C     THE SYSTEM AND INVERSE ARE STORED ON DISK.
C     THE COLUMNWISE LARGEST SCALED PIVOT IS USED AT EACH STEP.
      DIMENSION IXR(320),IXK(320),H(320),B(320),ASAT(320),X(320)
      DOUBLE PRECISION ASAT,X,H,B
      COMMON M,N,NN,K,X,LIST
      TEST = 0.00001
      N = K
      N1 = N+1
      SMLPIV = 999.9
      REWIND 5
      DO 2 I=1,N
      IXR(I) = 0
      IXK(I) = 0
    2 X(I) = 0.0
      DO 4 J=1,N
      READ (4)  (ASAT(I),I=1,N)
      DO 4 I=1,N
    4 X(I) = X(I) + ASAT(I)**2
      REWIND 4
      DO 5 I=1,N
    5 X(I) = DSQRT(X(I))
C     CALL IN FIRST COLUMN OF THE ASAT MATRIX
      READ (4)  (ASAT(I),I=1,N)
      DO 20 K=1,N
      IF(K/10*10.EQ.K)WRITE(3,950)K,N
  950 FORMAT(T5,'INVERSION COUNT (EVERY 10) =',I6, ' OF',I5)
      BIGPIV = 0.0
      K1 = K+1
      DO 9 I=1,N
      IF(IXR(I).NE.0) GO TO 9
      TPIV = DABS(ASAT(I)/X(I))
      IF(TPIV.LE.BIGPIV) GO TO 9
      BIGPIV=TPIV
      IK=I
    9 CONTINUE
      IXR(IK) = K
      IXK(K) = IK
      IF(SMLPIV.LE.BIGPIV) GO TO 13
   10 SMLPIV = BIGPIV
      ISML = IK
      KSML = K
      IF(SMLPIV.GE.TEST) GO TO 13
      WRITE(3,12)
   12 FORMAT(T8,'THE SYSTEM IS TOO ILL-CONDITIONED TO SOLVE',//)
      GO TO 33
   13 DO 15  I = 1,N
   15 IF(I.NE.IK)ASAT(I)=-ASAT(I)/ASAT(IK)
      ASAT(IK) = 1.0/ASAT(IK)
      DO 16 I=1,N
   16 B(I)=ASAT(I)
      READ (4)  (ASAT(I),I=1,N)
      WRITE(5)  (B(I),I=1,N)
C     FROM HERE TO 20 UPDATES THE NEXT COLUMN OF A.
      REWIND 5
  101 DO 20 J=1,K
      READ (5)(H(L),L=1,N)
C     NC IS THE INDEX OF THE NON UNIT VECTOR IN H(I).
      NC = IXK(J)
      DO 19 I=1,N
   19 IF(I.NE.NC)ASAT(I)=ASAT(I)+ASAT(NC)*H(I)
   20 ASAT(NC) = ASAT(NC)*H(NC)
C     THE SCRAMBLED SOLUTION IS IN ASAT(I)
      DO 23 I=1,N
      DO 22 J=1,N
      IF(I.NE.IXR(J)) GO TO 22
      X(I)=ASAT(J)
      GO TO 23
   22 CONTINUE
   23 CONTINUE
C     THE SOLUTION IS IN X(I)
   33 RETURN
      END
```

```
      SUBROUTINE OUTPUT(B)
C     SUBROUTINE TO WRITE MAT OUTPUT IN FORM EASY TO INTERPRET
      DIMENSION B(15,21)
      DOUBLE PRECISION B
      COMMON M,N
   15 M2 = 0
  818 M1 = M2+1
      M2 = MIN0(M1+9,M)
      WRITE(3,8017) (JJ,JJ=M1,M2)
 8017 FORMAT(//,T7, 10I12)
      DO 819  KK = 1,N
  819 WRITE(3,8018)KK,(B(KK,JJ),JJ=M1,M2)
 8018 FORMAT(T5,I3,2X,10F12.5)
      IF(M2.GE.M)GO TO 821
      WRITE(3,8019)
 8019 FORMAT(////,T10, 'ADDITIONAL PART OF OUTPUT MATRIX',//)
      GO TO 818
  821 RETURN
      END
```

```
      SUBROUTINE MATZER(W,N,M)
      DIMENSION W(15,21)
      DOUBLE PRECISION W
      DO 5 II=1,N
      DO 5 JJ = 1,M
    5 W(II,JJ) = 0.0
      RETURN
      END
```

7-7 MAT FOUNDATIONS BY FINITE ELEMENT

This section is based heavily on the direct element method of grid frameworks of Wang (1970). Considering Fig. 7-6, we shall subdivide the mat into a series of grid members with a torsional resistance as well as bending. Properties of the grid members will be determined by their dimensions, which in turn are determined by the member location within the grid. For example members 1 to 3 have widths one-half that of members 4, 8, 10, and 11.

Refer to Fig. 7-7, which illustrates a typical element *P-X*, *F-e* diagram. The *A* and *S* matrices (general case for diagonal as well as rectangular members) are

$$A = \begin{array}{c} P\backslash F \\ \\ \\ \\ \\ \\ \\ \end{array}\begin{array}{c} \end{array}\left[\begin{array}{cccccc} & 1 & 2 & 3 & 4 & 5 \\ 1 & -\sin\alpha & 0 & -\cos\alpha & 0 & 0 \\ 2 & +\cos\alpha & 0 & -\sin\alpha & 0 & 0 \\ 3 & +\dfrac{1}{L} & +\dfrac{1}{L} & 0 & -1.0 & 0 \\ 4 & 0 & -\sin\alpha & +\cos\alpha & 0 & 0 \\ 5 & 0 & +\cos\alpha & +\sin\alpha & 0 & 0 \\ 6 & -\dfrac{1}{L} & -\dfrac{1}{L} & 0 & 0 & -1.0 \end{array}\right] \tag{7-6}$$

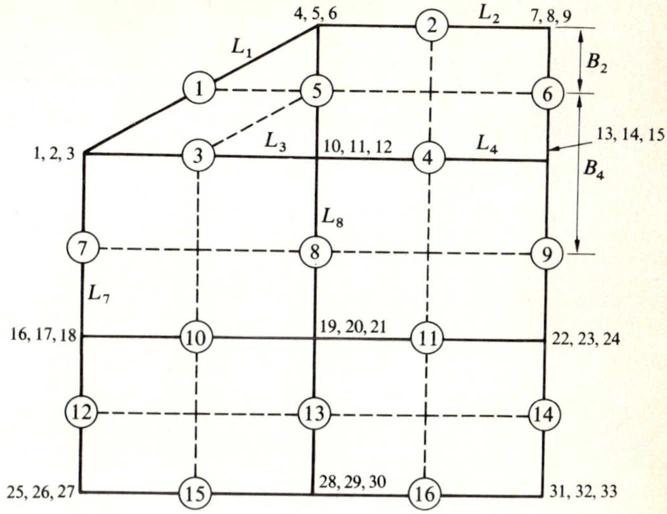

FIGURE 7-6
Plate or mat divided into a grid framework of dimensions shown. Use average length L for diagonal-member properties. All members are subjected both to bending and torsion, and each node is supported by a soil "spring." Nodes show the node coordinates (as 1, 2, 3) or P-X identification for ASA^T matrix.

and

$$S = \begin{array}{c|ccccc} F\backslash e & 1 & 2 & 3 & 4 & 5 \\ \hline 1 & \frac{4E1}{L} & \frac{2E1}{L} & & & \\ 2 & \frac{2E1}{L} & \frac{4E1}{L} & & & \\ 3 & & & \frac{GJ}{L} & & \\ 4 & & & & K & \\ 5 & & & & & K \end{array} \qquad (7\text{-}7)$$

The soil spring is computed as $k_s LB/4$ (the 4 is used because of the summation process involved at the joints in building the ASA^T matrix). Now the element SA^T and ASA^T matrices can be built. Next, by employing the proper counting devices via node coding similar to that shown in Fig. 7-6 one can superimpose the contributing element ASA^T values at any node to build the total ASA^T of the mat.

This method has two major disadvantages: (1) the matrix is of size $NP \times NP$,

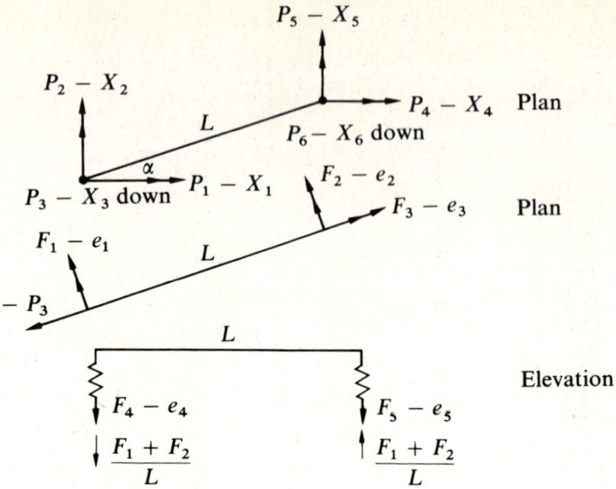

FIGURE 7-7
A typical element from the grid of Fig. 7-6. With the general element orientation shown, one can easily allow for bevels, corners, slots, etc.

which is three times as large as the finite-difference matrix for the same grid layout,[1] and (2) for practical reasons it is best to read the member data of each member on separate data cards. Thus, a large number of data cards are required just for members. Two major advantages, however, are (1) that *any* foundation configuration may be used, including notches, holes, slots, bevels, etc., and (2) that both column axial loads and moments can be applied. Loads not on node points are prorated as in the finite-difference method.

Ribbed footings may be used in this method by reading appropriate values into the member data. (This feature is not in the included program.)

The included computer program computes the torsion rigidity J of the grid elements using Seely and Smith's (1952) values of β as a best fit over three regions. The largest of element thickness or width dimension is used as element width B to obtain

$$\frac{B}{T} \leq 2 \qquad \beta = \frac{0.087B}{T} + 0.054$$

[1] The included computer program does not have the feature, however, the recently developed method of reducing a banded matrix (as the ASA^T of most problems of the type in this text) can reduce storage and inversion time by a very large factor. Using the banded technique the author inverted a 192 × 192 matrix in less than 10 min where the product-inverse method required $1\frac{3}{4}$ hr. This method may be used for finite-difference problems also if the matrix is symmetrical.

$$2 < \frac{B}{T} < 4.5 \qquad \beta = \frac{0.0288B}{T} + 0.174$$

$$\frac{B}{T} > 4.5 \qquad \beta = \frac{0.00218B}{T} + 0.2902$$

for $J = \beta B T^3$. For ribbed footings, the finite element segments may be tee-shaped, and an additional modification of J would be necessary.

EXAMPLE 7-5 Compare the solution of a footing with a corner notch cut out as shown in Fig. E7-5.1 with the finite-element solution. Show partial I/O for the finite-element solution. Given:

$$P = 500 \text{ kips} \qquad \text{Footing} = 10 \times 10 \qquad D = 2 \text{ ft}$$

$$E_c = 468{,}000 \text{ ksf} \qquad k_s = 96 \text{ kcf}$$

Required: soil pressure after notch is cut.

SOLUTION 1 By rigid method [Bowles (1968)] before the notch cut

$$q = \frac{500}{12^2} = 5 \text{ ksf}$$

Compute new properties of footing with notch (I_x,I_y,I_{xy}):

$$A' = 10^2 - 2(4) = 92 \text{ sq ft}$$

The center of the new area is

$$\bar{X} = \frac{-4(-8)}{92} = 0.348 \text{ ft}$$

$$\bar{Y} = \frac{3(-8)}{92} = -0.261 \text{ ft}$$

M_y (about Y' axis)
$= 500(0.261) = +130.5$ ft-kips

M_x (about X' axis)
$= 500(0.348) = +174.0$ ft-kips

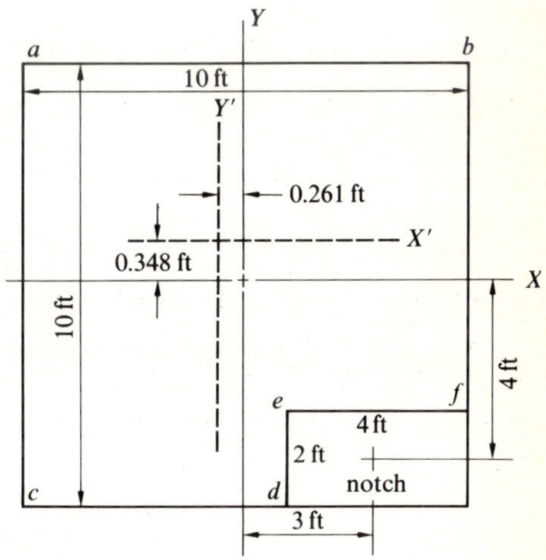

FIGURE E7-5.1
Notched footing.

Make a table to compute properties with respect to new axis through cga:

Part	A	X	Y	AX^2	AY^2	I_{0X}	I_{0Y}
Uncut	100.0	0.261	−0.348	6.81	12.11	833.33	833.33
Notch	−8.0	−4.348	3.261	−151.24	−85.072	−2.67	−10.67
	92.0			−144.43	−72.96		

New axis:

$$I_X = I_{0X} + I_{0X,\,\text{notch}} + \sum AY^2 = 833.33 - 2.67 - 72.96 = 757.70 \text{ ft}^4$$

New axis:

$$I_Y = I_{0Y} + I_{0Y,\,\text{notch}} + \sum AX^2 = 833.33 - 10.67 - 144.43 = 678.23 \text{ ft}^4$$

Compute product of inertia with respect to new axis:

$$I_{XY} = I^*_{0XY} + \sum A\bar{X}\bar{Y}$$

$$= 0.00 + (-8)(4.348)(3.261) + 100(-0.261)(-0.348)$$

$$= 0.00 - 113.432 + 9.10 = -104.33 \text{ ft}^4$$

$$I_Y - \frac{{I_{XY}}^2}{I_X} = 678.23 - \frac{(104.33\ ^2}{757.70} = 678.23 - 14.37 = 663.86$$

$$I_X - \frac{{I_{XY}}^2}{I_Y} = 757.70 - \frac{(104.33)^2}{678.23} = 757.70 - 16.05 = 741.65$$

$$M_Y - M_X \frac{I_{XY}}{I_X} = 130.5 - \frac{174(-104.33)}{757.70} = 130.5 + 23.96 = 154.46$$

$$M_X - M_Y \frac{I_{XY}}{I_Y} = 174.0 - \frac{130.5(-104.33)}{678.23} = 174.00 + 20.07 = 194.07$$

Solving for soil pressure at selected points

$$q = \frac{500}{92} + \frac{154.46}{663.86} X + \frac{194.07}{741.65} Y$$

$$= 5.43 + 0.233X + 0.262Y$$

Point	X'	Y'	P/A	$0.233X'$	$0.262Y'$	q	$q_{\text{w/ftg wt}}$
a	−4.739	−4.652	5.43	−1.10	−1.22	3.11	3.30
b	+5.261	−4.652	5.43	+1.23	−1.22	5.44	5.77
c	−4.739	+5.348	5.43	−1.10	+1.40	5.73	6.07
d	+1.261	+5.348	5.43	+0.29	+1.40	7.12	7.55
e	+1.261	+3.348	5.43	+0.29	+0.88	6.60	7.00
f	+5.261	+3.348	5.43	+1.23	+0.88	7.54	7.99

* Note I_{0XY} is with respect to the original XY axis before the notch was cut out.

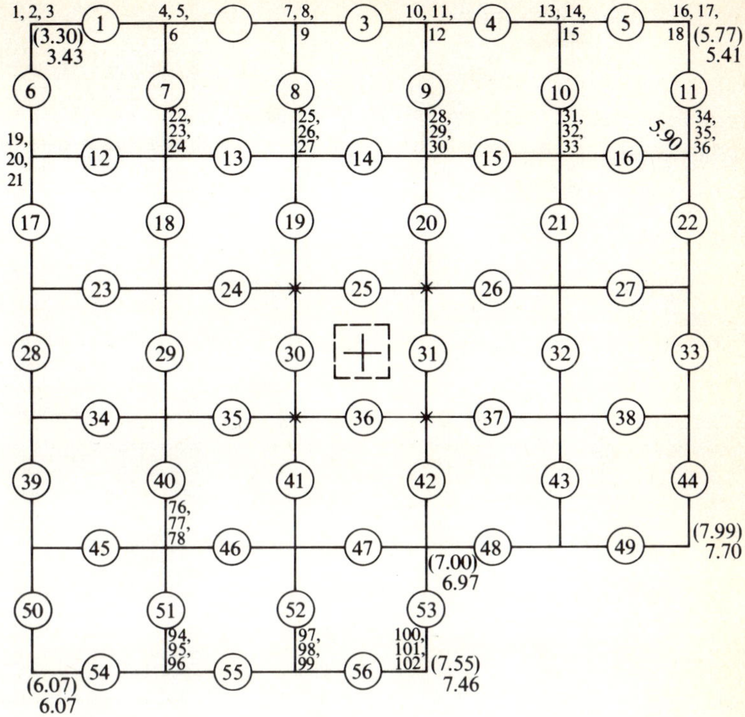

FIGURE E7-5.2
Partial coding of notched footing worked by hand in Example 7-5. Element numbers are shown in circles; soil pressures from the hand solution shown in parentheses, for example, (7.99); node coordinates as 1, 2, 3. Note that there are 56 members with a 102 × 102 matrix.

SOLUTION 2 By finite-element solution (refer to Fig. E7-5.2 for coding) the partial data cards are:

Card	Data
1	TITLE
2	UNITS (UT1, UT3, UT5, UT6)
3	102 56 4 0
4	468000. 219600. 100. 2.00 0.
5	1 1 2 3 4 5 6 2.00 0.00 1.00
↓	
60	56 97 98 99 100 101 102 2.0 0. 1.00
61	45 132.5
62	48 132.5
63	63 132.5
64	66 132.5

The partial output is shown in Figs. E7-5.3, E7-5.4, and E7-5.5. ////

J E BOWLES EXAM 7-5 BY FIN. ELEMENT 10 X 10 X 2 FT W/CORNER NOTCH 2 X 4 FT

E = 468000.0 K/SQ FT G = 219600.0 K/SQ FT SOIL MODULUS = 100.00 K/CU FT
UNIT WT = 0.0 K/CU FT

MEMNO	NP1	NP2	NP3	NP4	NP5	NP6	H	V	LEN	B	T	SM	INERTIA	POLAR I
1	1	2	3	4	5	6	2.000	0.0	2.000	1.000	2.000	50.000	0.6666666	1.8239985
2	4	5	6	7	8	9	2.000	0.0	2.000	1.000	2.000	50.000	0.6666666	1.8239985
3	7	8	9	10	11	12	2.000	0.0	2.000	1.000	2.000	50.000	0.6666666	1.8239985
4	10	11	12	13	14	15	2.000	0.0	2.000	1.000	2.000	50.000	0.6666666	1.8239985
5	13	14	15	16	17	18	2.000	0.0	2.000	1.000	2.000	50.000	0.6666666	1.8239985
6	1	2	3	19	20	21	0.0	-2.000	2.000	1.000	2.000	50.000	0.6666666	1.8239985
7	4	5	6	22	23	24	0.0	-2.000	2.000	2.000	2.000	100.000	1.3333330	2.2559986
8	7	8	9	25	26	27	0.0	-2.000	2.000	2.000	2.000	100.000	1.3333330	2.2559986
9	10	11	12	28	29	30	0.0	-2.000	2.000	2.000	2.000	100.000	1.3333330	2.2559986
10	13	14	15	31	32	33	0.0	-2.000	2.000	2.000	2.000	100.000	1.3333330	2.2559986
11	16	17	18	34	35	36	0.0	-2.000	2.000	1.000	2.000	50.000	0.6666666	1.8239985
12	19	20	21	22	23	24	2.000	0.0	2.000	2.000	2.000	100.000	1.3333330	2.2559986
13	22	23	24	25	26	27	2.000	0.0	2.000	2.000	2.000	100.000	1.3333330	2.2559986
14	25	26	27	28	29	30	2.000	0.0	2.000	2.000	2.000	100.000	1.3333330	2.2559986
15	28	29	30	31	32	33	2.000	0.0	2.000	2.000	2.000	100.000	1.3333330	2.2559986
16	31	32	33	34	35	36	2.000	0.0	2.000	2.000	2.000	100.000	1.3333330	2.2559986
17	19	20	21	37	38	39	0.0	-2.000	2.000	1.000	2.000	50.000	0.6666666	1.8239985
18	22	23	24	40	41	42	0.0	-2.000	2.000	2.000	2.000	100.000	1.3333330	2.2559986
19	25	26	27	43	44	45	0.0	-2.000	2.000	2.000	2.000	100.000	1.3333330	2.2559986
20	28	29	30	46	47	48	0.0	-2.000	2.000	2.000	2.000	100.000	1.3333330	2.2559986
21	31	32	33	49	50	51	0.0	-2.000	2.000	2.000	2.000	100.000	1.3333330	2.2559986
22	34	35	36	52	53	54	0.0	-2.000	2.000	1.000	2.000	50.000	0.6666666	1.8239985
23	37	38	39	40	41	42	2.000	0.0	2.000	2.000	2.000	100.000	1.3333330	2.2559986
24	40	41	42	43	44	45	2.000	0.0	2.000	2.000	2.000	100.000	1.3333330	2.2559986
25	43	44	45	46	47	48	2.000	0.0	2.000	2.000	2.000	100.000	1.3333330	2.2559986
26	46	47	48	49	50	51	2.000	0.0	2.000	2.000	2.000	100.000	1.3333330	2.2559986
27	49	50	51	52	53	54	2.000	0.0	2.000	2.000	2.000	100.000	1.3333330	2.2559986
28	37	38	39	55	56	57	0.0	-2.000	2.000	1.000	2.000	50.000	0.6666666	1.8239985
29	40	41	42	58	59	60	0.0	-2.000	2.000	2.000	2.000	100.000	1.3333330	2.2559986

30	43	44	45	61	62	63	0.0	-2.000	2.000	2.000	2.000	100.000	1.3333330	2.2559986
31	46	47	48	64	65	66	0.0	-2.000	2.000	2.000	2.000	100.000	1.3333330	2.2559986
32	49	50	51	67	68	69	0.0	-2.000	2.000	2.000	2.000	100.000	1.3333330	2.2559986
33	52	53	54	70	71	72	0.0	-2.000	2.000	1.000	2.000	50.000	0.6666666	1.8239985
34	55	56	57	58	59	60	2.000	0.0	2.000	2.000	2.000	100.000	1.3333330	2.2559986
35	58	59	60	61	62	63	2.000	0.0	2.000	2.000	2.000	100.000	1.3333330	2.2559986
36	61	62	63	64	65	66	2.000	0.0	2.000	2.000	2.000	100.000	1.3333330	2.2559986
37	64	65	66	67	68	69	2.000	0.0	2.000	2.000	2.000	100.000	1.3333330	2.2559986
38	67	68	69	70	71	72	2.000	0.0	2.000	2.000	2.000	100.000	1.3333330	2.2559986
39	55	56	57	73	74	75	0.0	-2.000	2.000	1.000	2.000	50.000	0.6666666	1.8239985
40	58	59	60	76	77	78	0.0	-2.000	2.000	2.000	2.000	100.000	1.3333330	2.2559986
41	61	62	63	79	80	81	0.0	-2.000	2.000	2.000	2.000	100.000	1.3333330	2.2559986
42	64	65	66	82	83	84	0.0	-2.000	2.000	2.000	2.000	100.000	1.3333330	2.2559986
43	67	68	69	85	86	87	0.0	-2.000	2.000	2.000	2.000	100.000	1.3333330	2.2559986
44	70	71	72	88	89	90	0.0	-2.000	2.000	1.000	2.000	50.000	0.6666666	1.8239985
45	73	74	75	76	77	78	2.000	0.0	2.000	2.000	2.000	100.000	1.3333330	2.2559986
46	76	77	78	79	80	81	2.000	0.0	2.000	2.000	2.000	100.000	1.3333330	2.2559986
47	79	80	81	82	83	84	2.000	0.0	2.000	2.000	2.000	100.000	1.3333330	2.2559986
48	82	83	84	85	86	87	2.000	0.0	2.000	1.000	2.000	50.000	0.6666666	1.8239985
49	85	86	87	88	89	90	2.000	0.0	2.000	1.000	2.000	50.000	0.6666666	1.8239985
50	73	74	75	91	92	93	0.0	-2.000	2.000	1.000	2.000	50.000	0.6666666	1.8239985
51	76	77	78	94	95	96	0.0	-2.000	2.000	2.000	2.000	100.000	1.3333330	2.2559986
52	79	80	81	97	98	99	0.0	-2.000	2.000	2.000	2.000	100.000	1.3333330	2.2559986
53	82	83	84	100	101	102	0.0	-2.000	2.000	1.000	2.000	50.000	0.6666666	1.8239985
54	91	92	93	94	95	96	2.000	0.0	2.000	1.000	2.000	50.000	0.6666666	1.8239985
55	94	95	96	97	98	99	2.000	0.0	2.000	1.000	2.000	50.000	0.6666666	1.8239985
56	97	98	99	100	101	102	2.000	0.0	2.000	1.000	2.000	50.000	0.6666666	1.8239985

45 P = 132.5000

48 P = 132.5000

63 P = 132.5000

66 P = 132.5000

FIGURE E7-5.3
Input for Example 7-5.

MEMNO	BENDING MOMENTS		TORSION MOMENT	ELEMENT SOIL REACTIONS	
1	8.640	-12.312	8.503	-1.715	-1.943
2	27.616	-23.501	10.731	-1.943	-2.159
3	35.579	-28.011	4.325	-2.159	-2.356
4	26.603	-18.574	-3.231	-2.356	-2.536
5	11.168	-4.264	-4.075	-2.536	-2.706
6	8.503	-11.693	-8.640	-1.715	-2.007
7	2.228	-25.558	-15.304	-3.886	-4.481
8	-6.406	-14.318	-12.078	-4.317	-4.926
9	-7.557	-11.750	1.408	-4.712	-5.325
10	-0.843	-18.317	7.406	-5.071	-5.679
11	4.075	-7.993	4.264	-2.706	-3.007
12	2.804	-27.468	15.522	-4.014	-4.481
13	33.950	-76.409	21.162	-4.481	-4.926
14	83.163	-77.990	6.826	-4.926	-5.325
15	77.655	-27.806	-9.269	-5.325	-5.679
16	25.486	-0.008	-7.685	-5.679	-6.014
17	27.214	-21.795	-11.444	-2.007	-2.287
18	31.198	-72.580	-21.786	-4.481	-5.056
19	-0.017	-107.747	-18.832	-4.926	-5.521
20	-4.345	-102.240	1.743	-5.325	-5.927
21	19.902	-60.127	9.726	-5.679	-6.273
22	15.678	-18.174	4.272	-3.007	-3.301
23	-5.772	-16.821	13.236	-4.574	-5.056
24	3.964	-113.837	19.593	-5.056	-5.521
25	107.901	-105.083	10.731	-5.521	-5.927
26	120.801	-4.769	-2.177	-5.927	-6.273
27	20.382	6.077	-2.793	-6.273	-6.601
28	35.031	-25.317	-5.672	-2.287	-2.550
29	78.937	-73.483	-8.930	-5.056	-5.587
30	98.885	-98.507	-12.897	-5.521	-6.061
31	89.332	-91.547	-13.975	-5.927	-6.479
32	59.511	-60.348	-5.887	-6.273	-6.832
33	20.966	-23.408	-1.805	-3.301	-3.582
34	-7.359	-13.585	0.642	-5.099	-5.587
35	-1.707	-109.515	1.741	-5.587	-6.061
36	100.923	-109.345	17.285	-6.061	-6.479
37	131.645	-14.299	22.047	-6.479	-6.832
38	28.083	4.571	9.557	-6.832	-7.164
39	25.959	-15.697	1.687	-2.550	-2.796
40	74.582	-23.544	6.361	-5.587	-6.075
41	114.051	0.039	-4.305	-6.061	-6.545
42	96.309	-11.121	-36.276	-6.479	-6.979
43	47.857	-18.662	-19.671	-6.832	-7.360
44	13.851	-12.294	-6.376	-3.582	-3.852
45	-1.273	-18.268	-5.238	-5.591	-6.075
46	19.286	-61.653	-4.565	-6.075	-6.545
47	59.039	-70.692	19.898	-6.545	-6.979
48	53.991	-39.895	30.956	-3.490	-3.680
49	20.224	-6.376	12.294	-3.680	-3.852
50	10.459	-3.021	2.961	-2.796	-3.033
51	24.217	1.043	5.343	-6.075	-6.546
52	24.425	6.591	-1.690	-6.545	-7.012
53	22.178	-4.614	-19.575	-3.490	-3.728
54	2.961	-7.654	-3.021	-3.033	-3.273
55	12.997	-18.614	-1.977	-3.273	-3.506
56	16.924	-19.575	4.614	-3.506	-3.728

THE SUM OF VERTICAL COL LOADS = 530.0000 KIPS

FIGURE E7-5.4
Bending moments for Example 7-5.

THE DEFORMATION MATRIX, FT OR RAD--EVERY 3RD = DEFL

1	0.00295	2	0.00231	3	0.03430	4	0.00299	5	0.00224	6	0.03886	7	0.00304	8	0.00208
9	0.04317	10	0.00307	11	0.00187	12	0.04712	13	0.00305	14	0.00173	15	0.05071	16	0.00303
17	0.00168	18	0.05411	19	0.00288	20	0.00235	21	0.04014	22	0.00295	23	0.00230	24	0.04481
25	0.00303	26	0.00213	27	0.04926	28	0.00306	29	0.00187	30	0.05325	31	0.00302	32	0.00170
33	0.05679	34	0.00299	35	0.00166	36	0.06014	37	0.00273	38	0.00241	39	0.04574	40	0.00278
41	0.00239	42	0.05056	43	0.00286	44	0.00220	45	0.05521	46	0.00290	47	0.00186	48	0.05927
49	0.00289	50	0.00166	51	0.06273	52	0.00288	53	0.00164	54	0.06601	55	0.00253	56	0.00244
57	0.05099	58	0.00254	59	0.00243	60	0.05587	61	0.00254	62	0.00225	63	0.06061	64	0.00261
65	0.00192	66	0.06479	67	0.00270	68	0.00168	69	0.06832	70	0.00274	71	0.00165	72	0.07164
73	0.00240	74	0.00243	75	0.05591	76	0.00238	77	0.00240	78	0.06075	79	0.00236	80	0.00227
81	0.06545	82	0.00244	83	0.00206	84	0.06979	85	0.00259	86	0.00176	87	0.07360	88	0.00266
89	0.00168	90	0.07703	91	0.00236	92	0.00241	93	0.06066	94	0.00234	95	0.00238	96	0.06546
97	0.00233	98	0.00228	99	0.07012	100	0.00235	101	0.00216	102	0.07457				

THE NODAL SOIL PRESSURE (KIPS)

1	3.43048	2	3.88592	3	4.31743	4	4.71181	5	5.07127	6	5.41143	7	4.01394	8	4.48080
9	4.92607	10	5.32529	11	5.67939	12	6.01381	13	4.57430	14	5.05556	15	5.52083	16	5.92707
17	6.27301	18	6.60131	19	5.09915	20	5.58676	21	6.06091	22	6.47860	23	6.83249	24	7.16374
25	5.59125	26	6.07537	27	6.54502	28	6.97926	29	7.36050	30	7.70314	31	6.06598	32	6.54592
33	7.01243	34	7.45679												

FIGURE E7-5.5
Deformation matrix and nodal pressure.

7-8 COMPUTER PROGRAM FOR FINITE-ELEMENT METHOD

This program requires coding the mat as shown for the square-footing example used to compare the finite-element and finite-difference solutions (Fig. 7-9) and as the partial coding shown for the corner notched footing of Example 7-5 (Fig. E7-5.2). Since the ASA^T is built directly, a minimum of computer core is used; however, inversion using double precision will still limit the size of the matrix. The program included here writes the ASA^T on disk and calls an inversion subroutine (product inverse). Since the product-inverse subroutine is similar (except for a few variable identifications) to the finite-difference program subroutine, it is not included in the following listing. This program will solve metric problems by simply reading the input data in metric rather than fps units.

Line	Operation
1–4	Bookkeeping
5	READ TITLE, UNITS (note only UT1, UT3, UT5, and UT6 used)
7	READ NP = number of P's in P matrix, also the size of ASA^T to be inverted; NM = number of members; NZP = number of nonzero P values to be read into the P matrix; LIST > 0 lists additional data
11	READ E = modulus of elasticity, ksf (kN/sq m); G = shear modulus, ksf (kN/sq m); SK = soil modulus; T = mat thickness; UNITWT = unit weight of mat
15–19	Zero ASA^T matrix and put on disk for use in statements 81–92
27	READ member data one card per member. Card contains member number, NP numbers at each end of member [NPE(I), I = 1,6], the horizontal (H) and vertical (V) coordinates of the member and its width B. H, V, and B are in feet or meters. (Refer to the partial output of Example 7-5 which illustrates member input)
29	Puts member data on disk
32–46	Computes member properties, inertia, torsion inertia, length, and slope angle and writes data back for check
48–64	Zeros and builds element A matrix [EA(I,J)]
65–74	Zeros and builds element S matrix [ES(I,J)]
75–79	Zeros and builds element SA^T matrix [ESAT(I,J)]
80	Tests counter; if II = 1, control is transferred to statement 605 to compute element forces. If II = 0, we continue to statement 203 to build structure ASA^T
81–92	Builds structure ASA^T based on each member's contribution and puts on disk
93	GO TO 94 to read next member data, make computations, etc.
94	Test IF: II = 0, continue; II > 0, GO TO 605
110–120	Zeros and builds P matrix, makes it column $NP + 1$, and puts on disk for inversion subroutine
112	READ nonzero P values
120	Calls inversion subroutine (CALL PROINV)
124	Sets II = 1 (remember it is a test above)
125	Rewinds disk to first member data
127	Tests if member data are all read (statement 605)
128–133	Computes and writes member forces using the element SA^T, which is recomputed, and the X matrix from the inversion subroutine
143–145	Computes nodal soil pressure
146–152	Punches soil-deformation matrix onto cards

```
C       J E BOWLES  DIRECT ELEMENT METHOD OF MAT ON ELASTIC FOUNDATION
C       B = MEM WIDTH, T = MEM THICK;  H,V = HOR AND VERT DIST FOR LENGTH
C       PROGRAM COMPUTES I AND J (PCLAR I);  E = MOD OF ELAST; G = SHEAR M
C       NM = NO OF MEMBERS, NP = SIZE CF MATRIX; NZP = NO OF NON ZERO P-VA
        DIMENSION P(216),NPE(6),EA(6,5),X(216),ASAT(216),ERR(216),ES(5,5),
       1ESAT(5,6),EASAT(6,6),INDEX(75),F(5),TITLE(20),SM(150),SOILP(150),
       2XX(75)
        COMMON X,ASAT,ERR,NP,IR,IS
        DOUBLE PRECISION ES,ESAT,EASAT,ASAT,X,ERR,F,P,UT5,UT6
        DEFINE FILE 10(200,432,U,IR),11(2C0,432,U,IS)
  6000  READ(1,1000,END=150) TITLE,UT1,UT3,UT5,UT6
  1000  FORMAT(20A4/A4,6X,A4,6X,A8,2X,A8)
        READ(1,1C05)NP,NM,NZP,LIST
  1005  FORMAT(4I5)
   101  WRITE(3,1002) TITLE
  1C02  FORMAT('1',//,T5,20A4)
        READ(1,1007)E,G,SK,T,UNITWT
  1007  FORMAT(5F10.4)
        WRITE(3,1008)E,UT5,G,UT5,SK,UT6,UNITWT,UT6
  1C08  FORMAT(//T5,'E =',F10.1,1X,A7,5X,'G =',F10.1,1X,A7,5X,'SOIL MODULU
       1S =',F9.2,1X,A7/ T5,'UNIT WT =',F7.3,1X,A7//)
        IR = 1
        DO 75  MM = 1,NP
    75  ASAT(MM) = 0.
        DO 103  I = 1,NP
   103  WRITE(10'IR)(ASAT(MM),MM=1,NP)
        WRITE(3,104)
   104  FORMAT(//,T5,'MEMNO',2X,'NP1',2X,'NP2',2X,'NP3',2X,'NP4',2X,'NP5',
       +2X, 'NP6', 7X,'H',8X,'V',8X,'LEN',8X,'B',8X,'T',6X,'SM', 9X,'INERT
       +IA',9X,'POLAR I',//)
        II = 0
        IR = 1
        SUMR = 0.
        SUMP = 0.
   106  IF(II.GT.0)GO TO 11
        READ(1,107)MEMNO,(NPE(I),I=1,6),H,V,B
   107  FORMAT(7I5,3F10.4)
        WRITE(5) MEMNO,(NPE(I),I=1,6),H,V,B
    11  IF(II.GT.0)READ(5)MEMNO,(NPE(I),I=1,6),H,V,B
        IF(MEMNO)301,301,108
   108  C = B
        T1 = B
        IF(B.LT.T)C=T
        IF(B.GT.T)T1=T
        IF(B/T.LE.2.)BE = .087*C/T1 + .C54
        IF(B/T.GT.2..AND.B/T.LT.4.5)BE = 0.0288*C/T1 + .174
        IF(B/T.GT.4.5)BE = 0.00218*C/T1 + 0.2902
        XJ = BE*C*T1**3
        IF(C.GT.T1)XJ=BE*T1*C**3
        XL = SQRT(H**2+V**2)
        XI = B*T**3/12.
        COSA = H/XL
        SINA = V/XL
        SM(MEMNO)=SK*B*XL/4.0
        IF(II.LE.0)WRITE(3,110)MEMNC,(NPE(I),I=1,6),H,V,XL,B,T,SM(MEMNO),
       1XI,XJ
   110  FORMAT(T5,7I5,3X,6F9.3,5X,2F12.7)
        DO 80  I = 1,6
        DO 80  J = 1,5
    80  EA(I,J) = 0.
        EA(1,1) = -SINA
        EA(1,3) =-COSA
        EA(2,1) = COSA
        EA(2,3) = -SINA
        EA(3,1) = 1./XL
        EA(3,2) = 1./XL
        EA(3,4) = -1.
        EA(4,2) = -SINA
        EA(4,3) = COSA
        EA(5,2) = COSA
        EA(5,3) = SINA
        EA(6,1) = -1./XL
        EA(6,2) = -1./XL
        EA(6,5) = -1.
        DO 81 I=1,5
        DO 81 J =1,5
    81  ES(I,J) =0.
        ES(1,1) = 4.*E*XI/XL
        ES(1,2) = 2.*E*XI/XL
        ES(2,1) =ES(1,2)
        ES(2,2)=ES(1,1)
        ES(3,3) = G*XJ/XL
        ES(4,4) = SM(MEMNO)
        ES(5,5) = SM(MEMNO)
        DO 202 I =1,5
        DO 202 J =1,6
        ESAT(I,J)=0.0
        DO 202  K = 1,5
   202  ESAT(I,J) =ESAT(I,J) +ES(I,K)*EA(J,K)
        IF(II)203,203,605
   203  DO 204 I=1,6
        DO  204 J =1,6
        EASAT(I,J) =0.
        DO 204 K= 1,5
   204  EASAT(I,J) =EASAT(I,J) +EA(I,K)*ESAT(K,J)
        DO 205  I = 1,6
        NS1 = NPE(I)
        READ (10'NS1)(ASAT(MM),MM=1,NP)
```

```
      DO 78  J = 1,6
      NS2 = NPE(J)
   78 ASAT(NS2) = ASAT(NS2) + EASAT(I,J)
  205 WRITE(10'NS1) (ASAT(MM),MM=1,NP)
      GO TO 106
C END OF ASAT MATRIX FORMATION
  301 IF(II)302,302,605
  302 CONTINUE
      IF(LIST.LE.0)GO TO 401
      WRITE(3,8823)
 8823 FORMAT('1',//,10X,'THE ASAT MATRIX')
      IZ=1
 8824 IP=IZ+10
      IF(IP.GT.NP)IP=NP
      DO 8825 JJ=1,NP
      READ(10'JJ)(ASAT(MM),MM=1,NP)
 8825 WRITE(3,8828)JJ,(ASAT(MM),MM=IZ,IP)
 8828 FORMAT(T3,I3,2X,-2P(11F10.2))
      IF(IP.LT.NP)WRITE(3,8895)
 8895 FORMAT('1',//,T5,'ASAT MATRIX CONTINUED')
      IZ=IP+1
      IF(IP.LT.NP)GO TO 8824
C ZERO & BUILT P-MATRIX
  401 DO 402 I=1,NP
  402 P(I) =0.
      DO 404 JJ=1,NZP
      READ(1,405)I,P(I)
  405 FORMAT(I5,F10.4)
      IF(I.EQ.I/3*3)SUMP  =  SUMP+P(I)
  404 WRITE(3,3333)I,P(I)
 3333 FORMAT(//,T5, I3, 2X,'P =',F10.4)
      IR = NP+1
      WRITE(10'IR)(P(I),I=1,NP)
      NPP1 = NP+1
      CALL PROINV
C      *****END OF ASAT INVERSION
 4500 WRITE(3,604)
  604 FORMAT(//,T5,'MEMNO',9X,'BENDING MOMENTS',T41,'TORSION MOMENT'T57,
     + 'ELEMENT SOIL REACTIONS',//)
      II = 1
      REWIND 5
      GO TO 106
C      COMPUTE MEMBER FORCES
  605 IF(MEMNO)957,957,606
  606 DO 607  I = 1,5
      F(I) =0.
      DO 607 K =1,6
      N = NPE(K)
  607 F(I) =F(I) +ESAT(I,K)*X(N)
      WRITE(3,608)MEMNO,(F(I),I=1,5)
  608 FORMAT(T5,I5,5F14.3)
      GO TO 106
  957 WRITE(3,610)SUMP,UT3
  610 FORMAT(//T10,'THE SUM OF VERTICAL COL LOADS =',F10.4,1X,A4///)
      WRITE(3,423)    UT1
  423 FORMAT(T8,'THE DEFORMATION MATRIX,',1X,A2,' OR RAD--EVERY 3RD = DE
     1FL'//)
      WRITE(3,426)(I,X(I),I = 1,NP)
  426 FORMAT((T4,8(I4,2X,F9.5),//))
      J = 0
      DO 428  I = 1,NP
      IF(I/3*3.EQ.I)J=J+1
  428 IF(I/3*3.EQ.I)SOILP(J) = X(I)*SK
      JJ1 = 0
      DO 9372  I = 1,NP
      IF(I/3*3.NE.I)GO TO 9372
      JJ1 = JJ1+1
 9372 XX(JJ1) = X(I)
 9975 FORMAT(7F10.7)
      WRITE(2,9975)(XX(I),I=1,JJ1)
      WRITE(3,431)UT3
  431 FORMAT(//,T8,'THE NODAL SOIL PRESSURE (',A7,')',//)
      WRITE(3,426)(I,SOILP(I),I=1,J)
      GO TO 6000
  150 STOP
      END
```

7-9 COMPARISON OF FINITE-DIFFERENCE AND FINITE-ELEMENT SOLUTIONS

There are no published measured values to compare with the analytical results. Teng (1949) reported on some deflections, but no dimensions or loads were given and all deflections were presented graphically. No reported mat failures have occurred, probably because mats are generally very conservatively designed (by the rigid method). At this time not many people are familiar with the techniques cited herein, and because of the programming difficulty probably have not been enthusiastic about taking this task on when the rigid solution works.

The author makes a few comparisons in the following paragraphs to determine at least if the solutions are reasonable. First, Fig. 7-8 illustrates a comparison between the beam on an elastic foundation (Chap. 5) as solved by the matrix solution and by treating the "beams" as mats using the finite-difference solution. Also shown are the short direction moments in the vicinity of the columns so that the reader can estimate what an effective short-side width for principal bending should be.

The finite-element solution was not used, as the matrices are three times as big, and alternative means are used to decide on its merits. Figure 7-8 illustrates about 22 percent difference between the finite-difference solution and beam on an elastic foundation. Most of this can be accounted for by the fact the beam solution uses full "springs" at the ends, whereas the finite-difference solution uses half "springs." The author found in the Vesić test data that increasing the end "springs" on the beam lowers the bending moments about 17 percent (see Table 5-1 for channel with two end loads). It should be noted, therefore, that in line with this reasoning the mat solution should yield moments *larger* than the beam on an elastic foundation with full-value end "springs," which Fig. 7-8 displays.

Next consider the finite-difference and finite-element solutions. Figure 7-9 is a $10 \times 10 \times 2$ ft mat with a 500-kip center load. Finite-element coding is shown. The grid size is the same for both methods (2×2 ft). In both methods the load (including 30 kips for footing weight) is prorated to the four adjacent nodes at 132.5 kips to each node as shown. Figure 7-10 shows a typical node (node 8) with positive-torsion (right-hand rule) directions for any element. From a computer output (similar to Fig. E7-5.3 and not included in the text) the moments shown on the figure were obtained.

If torsion is opposite to the positive direction shown, it is written with a minus sign. One can now compute the design moments in each direction at the nodes, as shown in Fig. 7-10. To make a valid comparison, as in Fig. 7-11, the moments are *divided by the element width.* Figure 7-11 shows a comparison of both moments and soil pressures. From symmetry only one quadrant is used. Large discrepancies exist in moments ranging from 10 to nearly 100 percent. Note, however, that very little

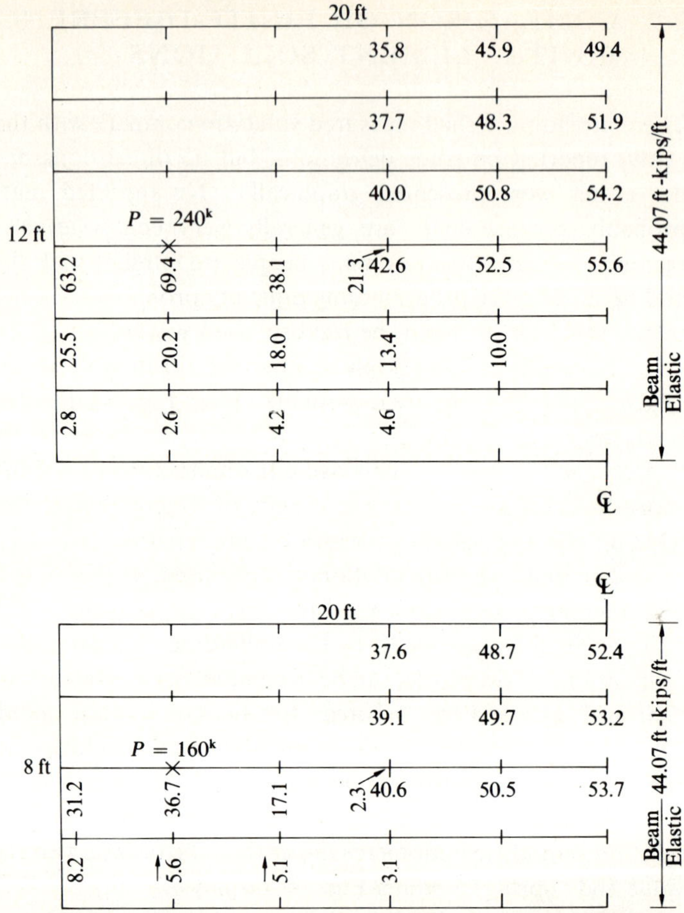

FIGURE 7-8
Comparison of the beam on an elastic foundation and the same beam as a mat. The grid used for the finite-difference solution is 2 × 2 ft. In both cases the mat solution is about 22 percent greater than the beam solution. The beam solution is the same even though both beams are not the same width because the column loads increase in direct proportion to the beam size. Short-direction moments in the vicinity of the column are shown and longitudinal moments for nodes 4 to 6. All moments are per foot of width or length.

discrepancy exists in soil pressure (and hence in computed deflections). These discrepancies are of the same order of magnitude as found by the author in comparing the finite-difference solution herein to a slightly different finite-element solution proposed by Lucas (1970), for which the program of Sec. 7-8 can be adapted with only slight modification.

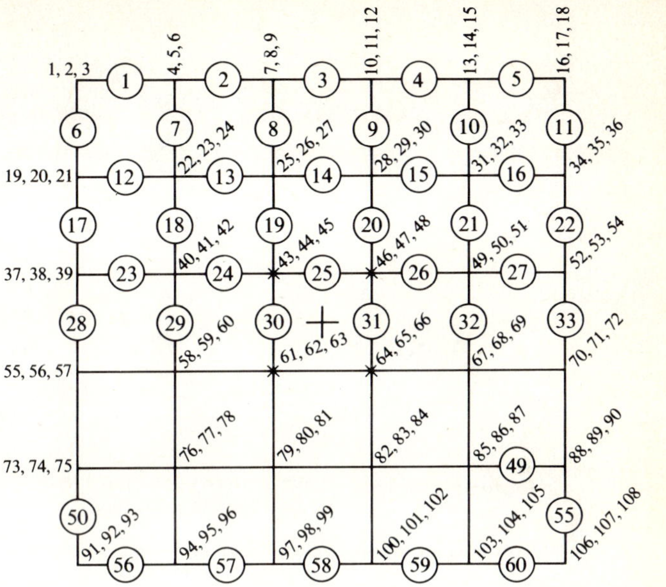

$T = 2$ ft, $E = 468000.0$ ksf; $G = 219600.0$ ksf; $k_s = 100$ kcf

FIGURE 7-9
Grid and coding used in the finite-element program to compare the finite-element solution with the finite-difference solution. Computer input used P values for NP = 45, 48, 63, and 66 with coding given above. Column load = 530 kips, including 30 kips for footing weight.

If we analyze the bending moment using the ACI 318-71 method which allows computation of moment at the column face and assume that a 500-kip column load requires a 2 × 2 ft column (referring to Fig. 7-11 again),

$$q = \tfrac{500}{10} = 5.0 \text{ ksf}$$

$$M \text{ per foot} = \frac{5.0}{2}(4)^2 = 40.0 \text{ ft-kips/ft} \ll \begin{cases} 51.5 \text{ finite difference (max value)} \\ 55.0 \text{ finite element (max value)} \end{cases}$$

Now if we find the average moment at the column face across the 10-ft width of footing, the finite-difference solution gives $4(51.49 + 34.14) + 2(28.56) = 399.64 \approx 40.0$ ft-kips/ft. The finite-element solution gives 45.9 ft-kips/ft.

It appears that the finite-difference solution is somewhat more correct (or at least reasonable) than the finite-element solution for bending. This may be because the torsion effects are forced into making an artificially large contribution to the solution due to the method of forming the mathematical model. It also indicates

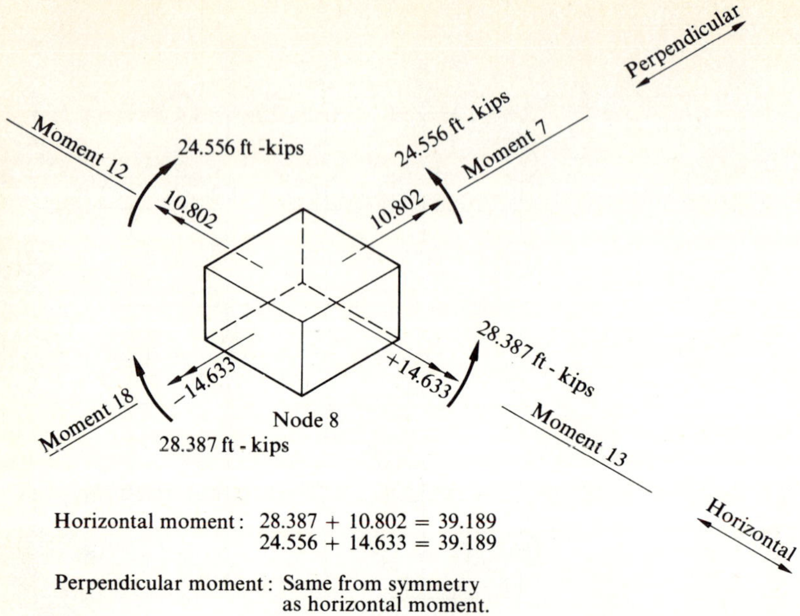

FIGURE 7-10
Computing the nodal bending moments for design using the computer output.

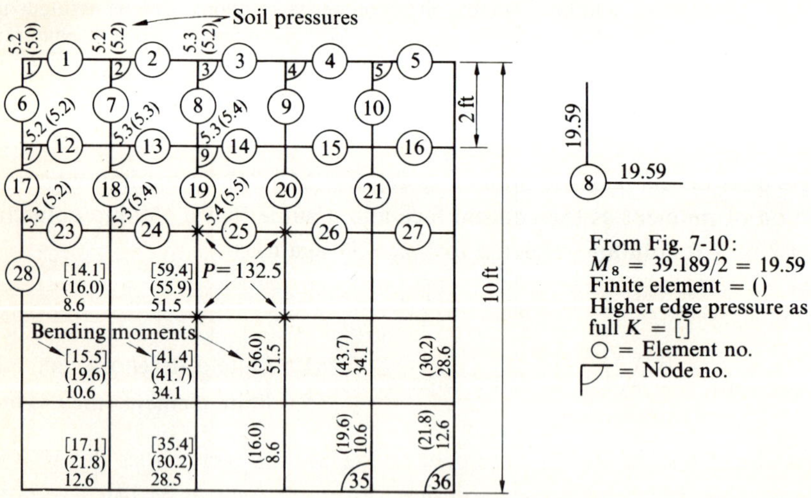

FIGURE 7-11
Comparison of the finite-element and finite-difference solutions. Footing is 10 × 10 ft square with 500-kip center load. P value of 132.5 kips is for finite-element solution to include footing weight. Note that the moments are symmetrical, as are the soil pressures. Moment values are per foot of width, not necessarily the computer-output value.

that the ACI 318-71 method is on the verge of being unsafe, especially using strength design since the maximum is considerably larger than the average moment.

Either solution provides good (and reasonable) soil pressures; hence, the finite-element method can be used for footings with holes and notches to obtain the soil pressures, since this method is easier to input to the computer.

PROBLEMS

7-1 Verify the finite-difference expression [Eq. (7-5)] for the inside corner.

7-2 Derive the finite-difference expression for an inside corner one node from another corner.

7-3 Repeat Example 7-4 for the load concentrically located and study the effect of grid spacing on the computed bending moments.

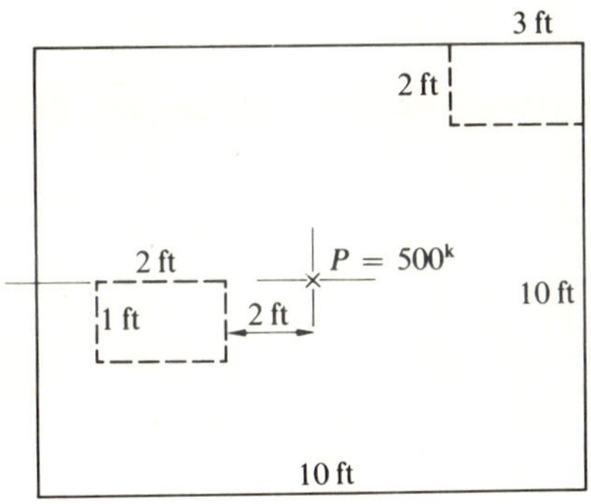

7-4 Refer to the figure above.

(*a*) Find the soil pressure of the footing uncut.

(*b*) Find the soil pressure if a corner is cut as shown.

(*c*) Find the soil pressure if a 1 × 2 ft hole is sawed out as shown.

What conclusions can you draw?

7-5 Apply a moment of $M_x = 100$ ft-kips to the figure in addition to the axial load.

(*a*) Find the soil pressure and plot the soil-pressure diagram.

(*b*) Find the soil pressure if a 1 × 2 ft hole is sawed out as in part (*c*) of Prob. 7-4.

7-6 Referring to Fig. 7-8, is the zone $a + 3d$ proposed by the author in Chap. 3 reasonable? If not, what do you recommend?

7-7 Repeat Prob. 7-4 using metric units.

7-8 Repeat Example 7-3 if the load is at node (4,6); at (5,5).

REFERENCES

ACI COMMITTEE 436 (1966): Suggested Design Procedures for Combined Footings and Mats, *J. Am. Concr. Inst.*, vol. 63, no. 10, October, pp. 1041–1057.

BOWLES, J. E. (1968): "Foundation Analysis and Design," chap. 5, McGraw-Hill, New York.

DERYCK, N., and R. T. SEVERN (1960, 1961): Stresses in Foundation Rafts, *Proc. Inst. Civ. Eng.* (*Lond.*), pt. I, vol. 15, January, pp. 35–48; pt. II, vol. 20, October, pp. 293–304.

FREDERICK, DANIEL (1957): Thick Rectangular Plates on an Elastic Foundation, *Trans. ASCE*, vol. 122, pp. 1069–1087.

GOODMAN, L. J., and R. H. KAROL (1968): "Theory and Practice of Foundation Engineering," chap. 5, Macmillan, New York.

LUCAS, WILLIAM M., JR. (1970): Grid Analysis of Plates Resting on Elastic Foundations, *Univ. Kans. Eng. Bull.* 62, 70 pp.

SEELY, F. B., and J. O. SMITH (1952): "Advanced Mechanics of Materials," p. 271, Wiley, New York.

SEVERN, R. T. (1966): The Solution of Foundation Mat Problems by Finite Element Methods, *Struct. Eng.* (*Lond.*), vol. 44, no. 6, June, pp. 223–238.

TENG, W. C. (1949): A Study of Contact Pressure against a Large Raft Foundation, *Geotech.* (*Lond.*), vol. 1, no. 4, December, p. 222.

——— (1962): "Foundation Design," chap. 7, Prentice-Hall, Englewood Cliffs, N.J.

WANG, C. K. (1970): "Matrix Methods of Structural Analysis," 2d ed., chap. 15, International Textbook, Scranton, Pa.

8
RETAINING WALLS

8-1 INTRODUCTION

A retaining wall is used to stabilize any material where conditions are such that the mass cannot be allowed to form a natural slope. Commonly retaining walls are used to hold earth slopes to a vertical or near vertical face, but they may also be used to contain ore, grain, coal, or even water.

Typical retaining walls are shown in Fig. 8-1. The *gravity* wall (Fig. 8-1*a*) depends upon self weight for stability. *Cantilever* walls (Fig. 8-1*b*, *c*, and *d*) achieve most of their stability by utilizing the weight of soil on the heel portion of the base slab. The *bridge abutment* of Fig. 8-2 is a special type of retaining wall, not only containing the approach fill but serving as a support for part of the bridge superstructure.

Retaining walls are commonly supported by the soil (or rock) underlying the base slab but are also supported on piles; this is especially true of bridge abutments. Pile supports are also used where water may erode or undercut the base soil, typically in waterfront structures.

This chapter will be primarily concerned with the simple cantilever retaining wall of Fig. 8-1*b*. The design of a counterfort wall, which is somewhat more complicated (Fig. 8-1*c*), will be considered in Sec. 8-9. Terms used to describe parts (or location) of the cantilever retaining wall are shown in Fig. 8-1*b*.

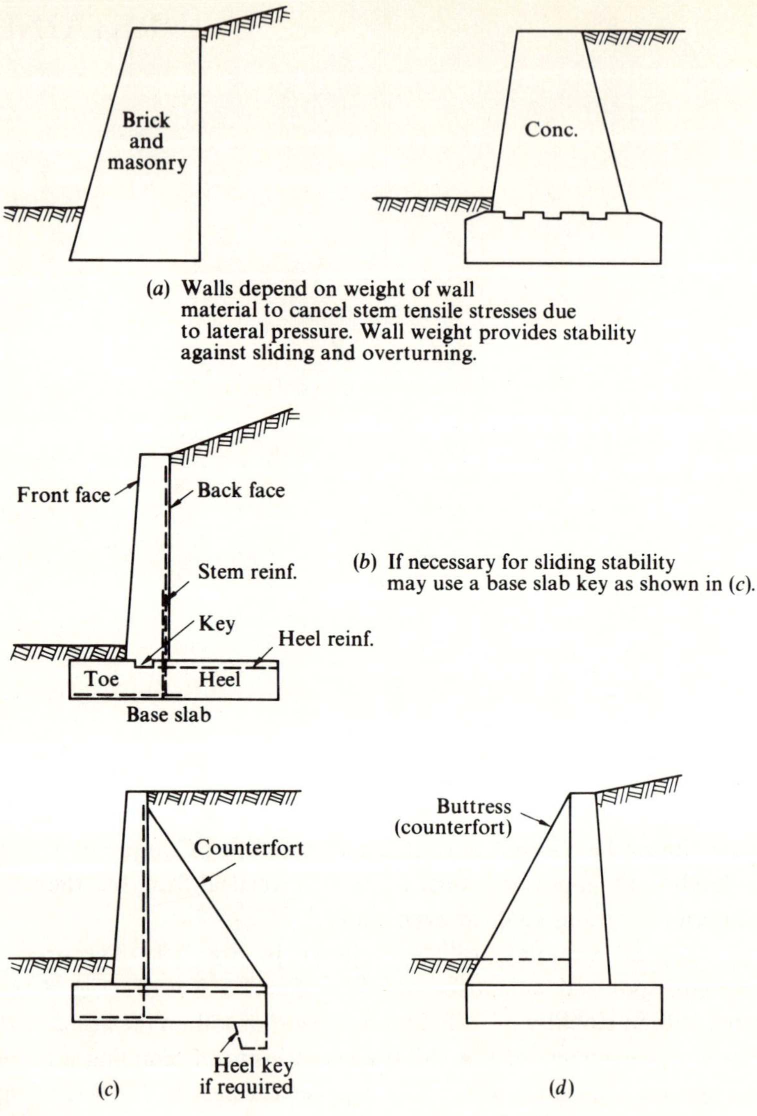

FIGURE 8-1
Types of retaining wall: (*a*) gravity; (*b*) cantilever; (*c*) counterfort; (*d*) buttressed.

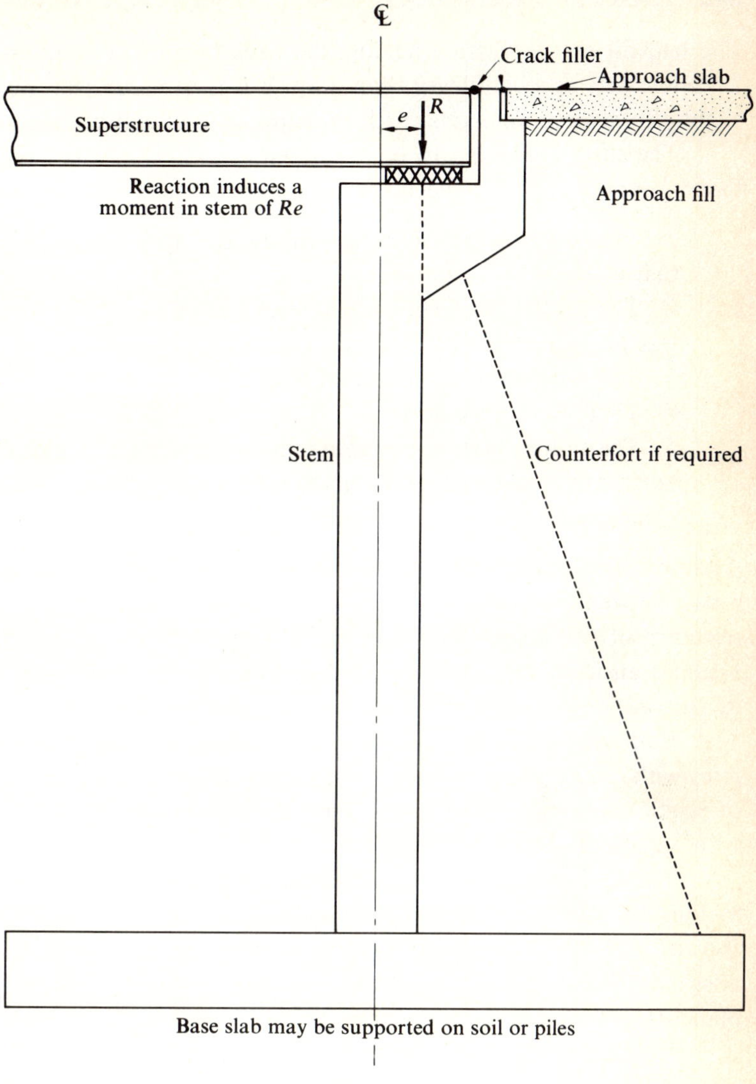

FIGURE 8-2
Bridge abutment as retaining wall.

8-2 EARTH PRESSURE ON RETAINING WALLS

The backfill exerts pressure against the back face of the wall, as shown in Fig. 8-3. The stem acts as a cantilever beam, and if it is of modest proportions, it will deflect slightly to achieve an *active* earth-pressure state of soil pressure against the wall.

The earth pressure can be evaluated (Fig. 8-4) by the Coulomb earth-pressure theory[1] (about 1773) or the Rankine method. The Coulomb method assumes that:

1 The soil is isotropic, homogenous, and with internal friction and cohesion, that is, $s = c + \sigma_n \tan \phi$.
2 The rupture surface is a plane surface (although Coulomb suspected that it was probably curved).
3 The failure surface is a rigid body.
4 There is wall friction.
5 The failure surface for active pressure is at an angle $\rho = 45 + \phi/2$ to the horizontal for horizontal backfills, the angle being independent of the soil cohesion.

Additionally Coulomb stressed the importance of drainage of the backfill. He correctly stated the critical height of a vertical bank of earth and emphasized the importance of soil tests. The Rankine[2] theory is a later theory (about 1857), which assumes an ideal soil with no cohesion (because in time weathering often destroys or markedly reduces the cohesion) and no wall friction.

Both the Coulomb and Rankine theories are invalid if the wall in any way interferes with the slip planes which must form if the retained soil mass is to reach a state of plastic equilibrium (Fig. 8-4*a*). This situation is often ignored in practice, however, with apparently little adverse effect.

The case of incipient slip, a cohesionless soil, and active earth pressure is shown in Fig. 8-5. At incipient slip the idealized failure wedge *ABC* has the body forces shown. P_a is the active-earth-pressure resultant of the wall to just hold the wedge in position. The weight vector W can be computed; the R vector on the idealized failure plane AC is the resultant of the normal force N and the friction resistance on the plane.

Combining P_a, R, and W we obtain the force polygon of Fig. 8-5*b*. Now it is evident that we could try many failure surfaces at various ρ angles and take the maximum value of P_a; or for the conditions of Fig. 8-5*a* (planar backfill and no cohesion) we may take the derivative

$$\frac{dP_a}{d\rho} = 0$$

[1] H. Golder provides a discussion of the Coulomb earth-pressure theory in *Geotech.* (Lond.), vol. 1, pp. 66–71, 1948.

[2] G. Cook discusses Rankine in *ibid.*, vol. 2, no. 4, 1950–1951.

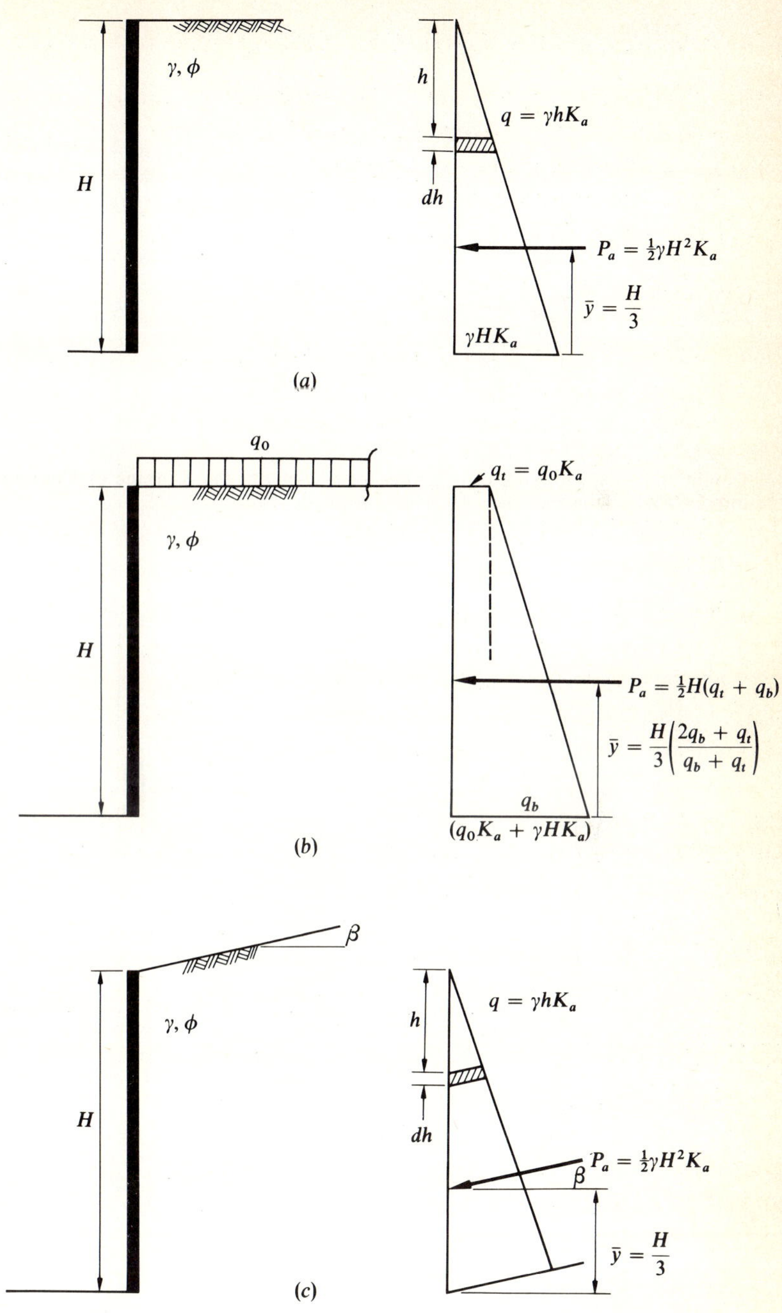

FIGURE 8-3
Wall pressures using Rankine theory: (*a*) cohesionless soil, and horizontal backfill; (*b*) when surcharge is present; (*c*) when backfill slopes.

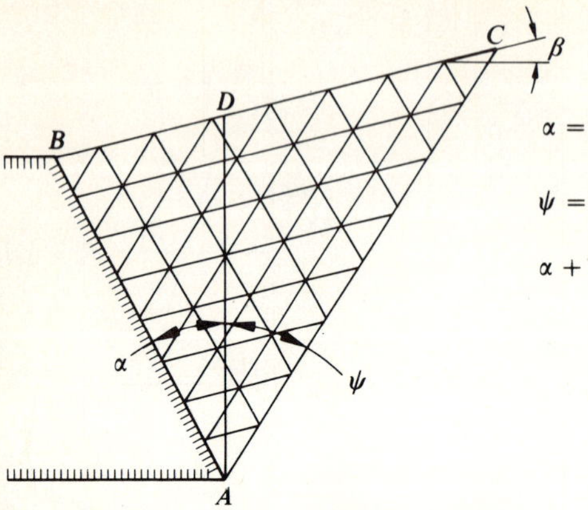

$$\alpha = 45 - \phi/2 - \tfrac{1}{2}\left(\sin^{-1}\frac{\sin\beta}{\sin\phi} - \beta\right)$$

$$\psi = 45 - \phi/2 + \tfrac{1}{2}\left(\sin^{-1}\frac{\sin\beta}{\sin\phi} - \beta\right)$$

$$\alpha + \psi = 90^\circ - \phi$$

(*a*) Conditions in soil in back of retaining wall to obtain the Rankine or Coulomb pressure. If wall interferes with failure surface *AB*, theory is invalidated.

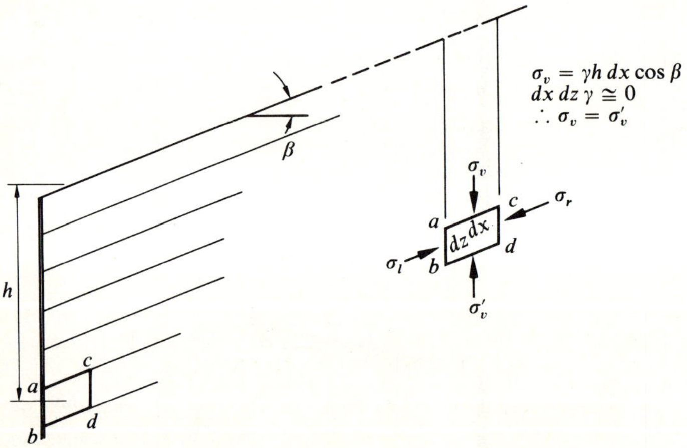

(*b*) Stress conditions to determine direction of stresses and resultant force against retaining wall when backfill slopes and wall friction is 0.

FIGURE 8-4
Stresses in retained soil mass.

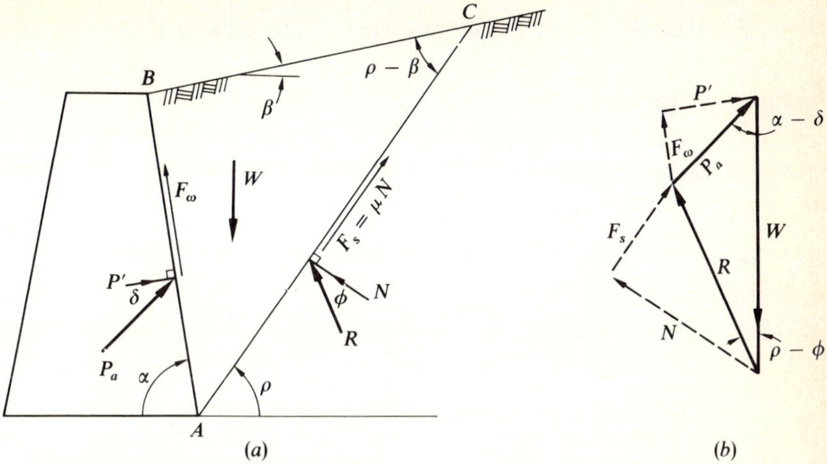

FIGURE 8-5
(*a*) Idealized failure wedge behind a retaining wall and (*b*) the resultant force polygon.

to find the maximum value of P_a to be

$$P_a = \frac{\gamma H^2}{2} \frac{\sin^2(\alpha + \phi)}{\sin^2\alpha \sin(\alpha - \delta)\left[1 + \sqrt{\dfrac{\sin(\phi + \delta)\sin(\phi - \beta)}{\sin(\alpha - \delta)\sin(\alpha + \beta)}}\right]} \tag{8-1}$$

The Rankine solution assumes a vertical wall and no wall friction. Removing these terms from Eq. (8-1) and simplifying, we obtain

$$P_a = \frac{\gamma H^2}{2}\left(\cos\beta \frac{\cos\beta - \sqrt{\cos^2\beta - \cos^2\phi}}{\cos\beta + \sqrt{\cos^2\beta - \cos^2\phi}}\right) \tag{8-2}$$

We may also write for either Eq. (8-1) or (8-2)[1]

$$P_a = \frac{\gamma H^2}{2} K_a$$

where K_a is the trigonometric remainder of either equation, depending on the theory being used. Tables 8-1 and 8-2 give typical values of K_a for both theories. More complete tables are available [Bowles (1968)], or the user may program the equations on the computer. Values are shown as zero for $\beta > \phi$ since the theory is based on equations which were derived for a cohesionless soil. Natural slopes formed with cohesionless soils of $\beta > \phi$ are already in a nearly unstable condition.

[1] One obtains the passive pressure by using $[1 - \sqrt{\cdots}]$ in Eq. (8-1) and reversing signs top and bottom in Eq. (8-2).

For cohesive soils,[1] Coulomb worked the case of horizontal backfill (see Fig. 8-6) and no wall friction to obtain an equation which can be transformed for the *active*-pressure soil state to

$$P_a = \tfrac{1}{2}\gamma H^2 \tan^2\left(45 - \frac{\phi}{2}\right) - 2cH \tan\left(45 - \frac{\phi}{2}\right) \tag{8-3}$$

By extrapolation one can obtain the *passive*-pressure soil state as

$$P_p = \tfrac{1}{2}\gamma H^2 \tan^2\left(45 + \frac{\phi}{2}\right) + 2cH \tan\left(45 + \frac{\phi}{2}\right) \tag{8-4}$$

It is known, of course, that earth slopes will stand at some angle, the approximate

[1] One may also obtain Eqs. (8-3) and (8-4) from a Mohr's circle construction using $\sigma_3 = q_h$ of Eq. (8-6) and integrating for the total pressure against the wall, as Eq. (8-7).

Table 8-1 COULOMB ACTIVE-EARTH-PRESSURE COEFFICIENTS FOR SELECTED VALUES OF β AND ϕ, AND $\alpha = 90°$

$\delta \setminus \phi =$	28°	30°	32°	34°	36°	38°	40°
				$\beta = 0°$			
0°	0.361	0.333	0.307	0.283	0.260	0.238	0.217
5°	0.345	0.319	0.294	0.271	0.250	0.229	0.210
10°	0.333	0.308	0.285	0.263	0.243	0.223	0.204
15°	0.325	0.301	0.279	0.258	0.238	0.219	0.201
20°	0.320	0.297	0.276	0.255	0.235	0.217	0.199
25°	0.319	0.296	0.274	0.254	0.235	0.217	0.199
				$\beta = 10°$			
0°	0.407	0.374	0.343	0.314	0.286	0.261	0.238
5°	0.391	0.359	0.330	0.302	0.277	0.252	0.230
10°	0.380	0.350	0.321	0.294	0.270	0.246	0.225
15°	0.373	0.343	0.315	0.289	0.265	0.243	0.221
20°	0.370	0.340	0.313	0.287	0.263	0.241	0.220
25°	0.369	0.340	0.313	0.287	0.264	0.241	0.221
				$\beta = 20°$			
0°	0.488	0.441	0.399	0.361	0.327	0.295	0.267
5°	0.474	0.428	0.387	0.350	0.317	0.287	0.259
10°	0.465	0.420	0.379	0.343	0.311	0.281	0.254
15°	0.461	0.415	0.375	0.339	0.307	0.278	0.251
20°	0.460	0.414	0.374	0.338	0.306	0.277	0.250
25°	0.464	0.417	0.376	0.340	0.308	0.278	0.252

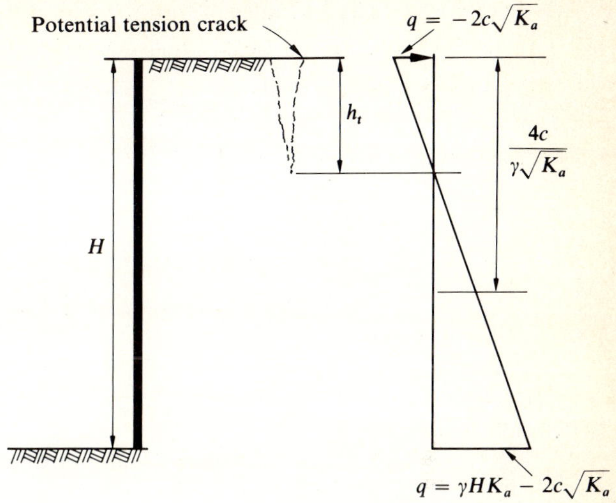

FIGURE 8-6
Retaining wall with cohesive soil.

limiting angle being the angle of internal friction for cohesionless soils and 90° for cohesive slopes of low heights.

Since neither Coulomb or Rankine considered the case of sloping backfill with cohesive soil, the value of earth-pressure coefficient to use when *both* cohesion and angle of internal friction are present is uncertain. One may use trial wedges [Bowles (1968)] as one solution. One may also use the earth-pressure coefficients including the slope angle and be approximately correct. The user should be aware, however, of this limitation. This problem is not expected to arise often since good practice calls for backfilling a zone like Fig. 8-7 with granular material whenever possible to provide drainage and control the lateral earth pressure.

Table 8-2 RANKINE ACTIVE-EARTH-PRESSURE COEFFICIENTS

$\beta \setminus \phi =$	28°	30°	32°	34°	36°	38°	40°
0°	0.361	0.333	0.307	0.283	0.260	0.238	0.217
5°	0.366	0.337	0.311	0.286	0.262	0.240	0.217
10°	0.380	0.350	0.321	0.294	0.270	0.246	0.225
15°	0.409	0.373	0.341	0.311	0.283	0.258	0.235
20°	0.460	0.414	0.374	0.338	0.306	0.277	0.250
25°	0.573	0.494	0.434	0.385	0.343	0.307	0.275
30°	0.000	0.866	0.574	0.478	0.411	0.358	0.315
35°	0.000	0.000	0.000	0.000	0.597	0.468	1.391

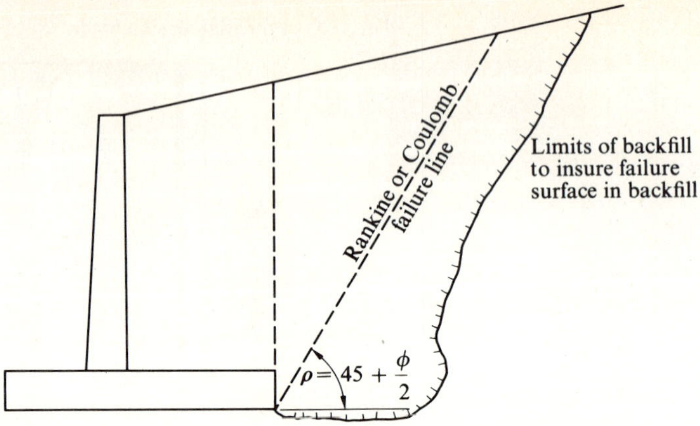

FIGURE 8-7
Backfill zone to achieve Rankine or Coulomb pressure conditions. Decrease ρ by at least β when $\beta > 0$.

8-3 TENSION CRACK, FORCE ON A WALL, AND DIRECTION OF WALL FORCE

The height of stability of an unsupported earth cut would be obtained approximately (SF = 1) by setting $P_a = 0$ in Eq. (8-3), giving

$$H = \frac{4c}{\gamma \tan (45 - \phi/2)} \tag{8-5}$$

Further inspection of Eq. (8-3) indicates that (refer to Fig. 8-6) if the lateral pressure is written as

$$q_h = \gamma h K_a - 2c\sqrt{K_a} \tag{8-6}$$

the wall pressure is

$$dP_a = q\, dh$$

and integrating over H

$$P_a = \int_0^H (\gamma h K_a - 2c\sqrt{K_a})\, dh \tag{8-7}$$

we obtain

$$P_a = \tfrac{1}{2}\gamma H^2 K_a - 2cH\sqrt{K_a} \tag{8-7a}$$

as before.

This computation shows that the wall pressure is triangular and further that a *negative* pressure is exerted on the wall, varying from a value of $-2c\sqrt{K_a}$ at $h = 0$ to a value of zero at a height of

$$h_t = \frac{2c}{\gamma\sqrt{K_a}} \tag{8-8}$$

This height represents a "tension" zone in the soil mass, and as soil may not carry tension stresses, a crack may form, visible at the ground surface, for this depth.

Since the basic derivation was for cohesionless soils and we are extrapolating, the value of $\sqrt{K_a}$ in Eq. (8-8) should be taken as

$$\sqrt{K_a} = \tan\left(45 - \frac{\phi}{2}\right)$$

to obtain a conservative tension-crack depth.

If a surcharge exists, as in Fig. 8-4*b*, for a cohesionless soil

$$q_h = (q + \gamma h)K_a$$

This gives an intensity of pressure at $h = 0$ of

$$q_h = qK_a$$

with a resulting trapezoidal pressure diagram.

A triangular pressure diagram will locate the resultant P_a at the one-third height of wall above the base. The resultant pressure is not located at $H/3$ in either Fig. 8-3*b* or 8-6, but it can be determined by statics. For a trapezoidal pressure diagram (cohesionless soil with surcharge) the resultant is at

$$\bar{y} = \frac{H}{3}\,\frac{2q_b + q_t}{q_b + q_t} \tag{8-9}$$

Figure 8-4*b* illustrates the concept, which inclines the resultant P_a at the angle β for the Rankine solution. Referring to the figure and recalling from the Rankine theory that the wall friction is zero, we see that the soil element *abcd* has no friction on face $\overline{ab}$. The element is assumed 1 unit perpendicular to the plane of the paper. The vertical stress on plane $\overline{ac}$ and $\overline{bd}$ (if we assume $\gamma\, dz\, dx \approx 0$) is

$$\sigma_v = \gamma h\, dx \cos\beta$$

Since the vertical stresses are in equilibrium, it follows that the stresses on the vertical plane $\overline{ab}$ and $\overline{cd}$ must be equal *and* collinear; this condition is satisfied if they are parallel to the ground surface (and to stress planes $\overline{ac}$, $\overline{bd}$).

8-4 DESIGN OF CANTILEVER RETAINING WALLS

Cantilever retaining walls require consideration of two earth-pressure values, as shown in Fig. 8-8. The stem may shear, as in Fig. 8-8*a*, or the system may slide, as in Fig. 8-8*b*.

Figure 8-9 illustrates typical proportions of cantilever retaining walls. The values shown may be used as tentative proportions with final dimensions to satisfy structural stability.

Figure 8-10 illustrates the critical sections for design of the various parts of a cantilever retaining wall. The three parts (stem, toe, and heel) are designed to satisfy shear and moment at the sections shown.

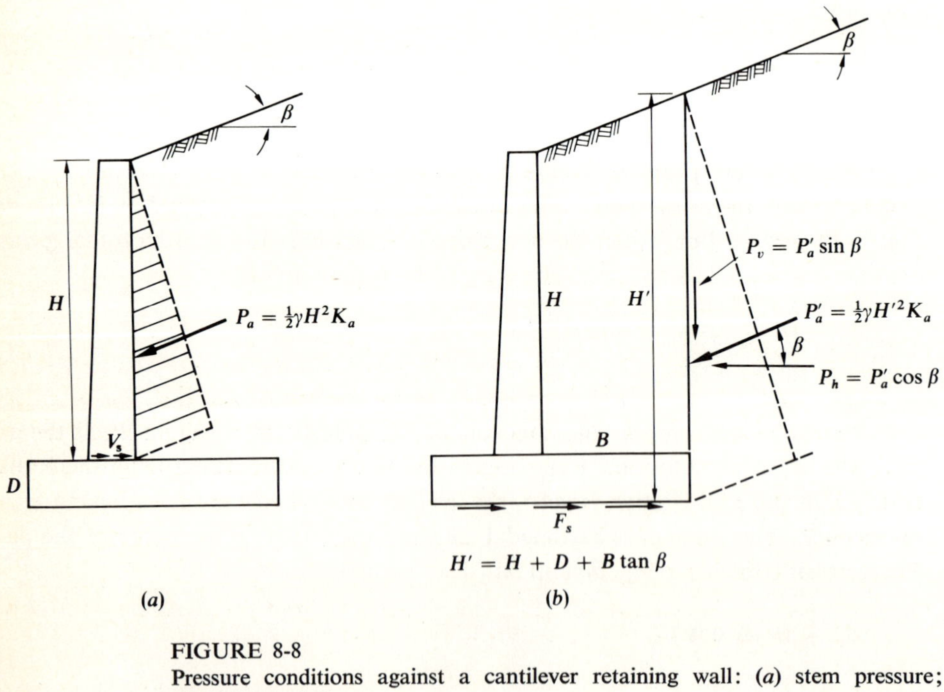

FIGURE 8-8
Pressure conditions against a cantilever retaining wall: (*a*) stem pressure; (*b*) sliding stability.

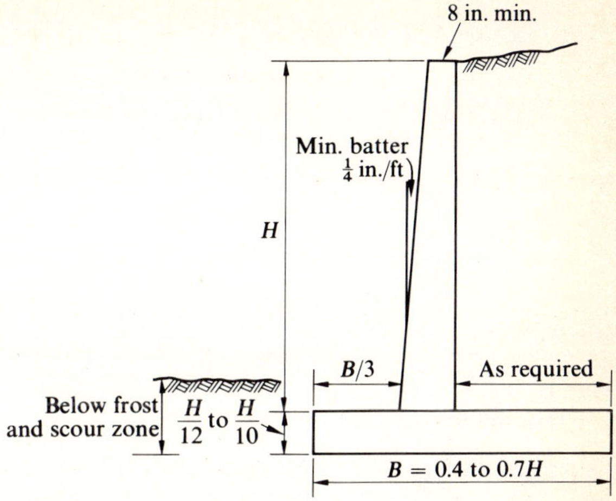

FIGURE 8-9
Trial wall proportions. Final proportions must satisfy structural and wall stability.

Stem Design

Considering the parts in detail and the stem first, it is evident that shear at the top is zero (or some small value with a surcharge[1] present) and increasing to a maximum at the juncture with the base slab. The proposal that the shear section be taken at d from the base slab is, in the author's opinion, unrealistic.[2] Taking the shear as shown is conservative. Another point to consider is that the stem is invariably poured separately from the base slab; thus, designing the stem for a thickness (as usual in most textbooks) to satisfy wide-beam shear (see Chap. 3) of

$$v_c = 2\phi\sqrt{f'_c}$$

is also somewhat unrealistic when the shear resistance at the joint will actually be developed based on the *friction* resistance of the weight of the stem on sections $\overline{ab}$ and $\overline{cd}$ plus the concrete *shear* resistance along $\overline{bc}$ plus the shear resistance of the stem tension (and shrinkage) steel (Fig. 8-10*a*). Why does the current design procedure work? If we take the coefficient of friction of concrete as $\mu = 0.5$ (rough surface), the

[1] Two points should be made here concerning the American Association of State Highway Officials (AASHO) (1969, sec. 1.2.19) design requirements: (1) the product of $\gamma K_a \leq 30$ pcf; (2) if highway traffic is present at the top of the wall within a distance of $H/2$, a surcharge of 2γ shall be used in computing the lateral earth pressure.

[2] See also Commentary on Building Code Requirements for Reinforced Concrete, ACI 318-71, art. 11.2.2.

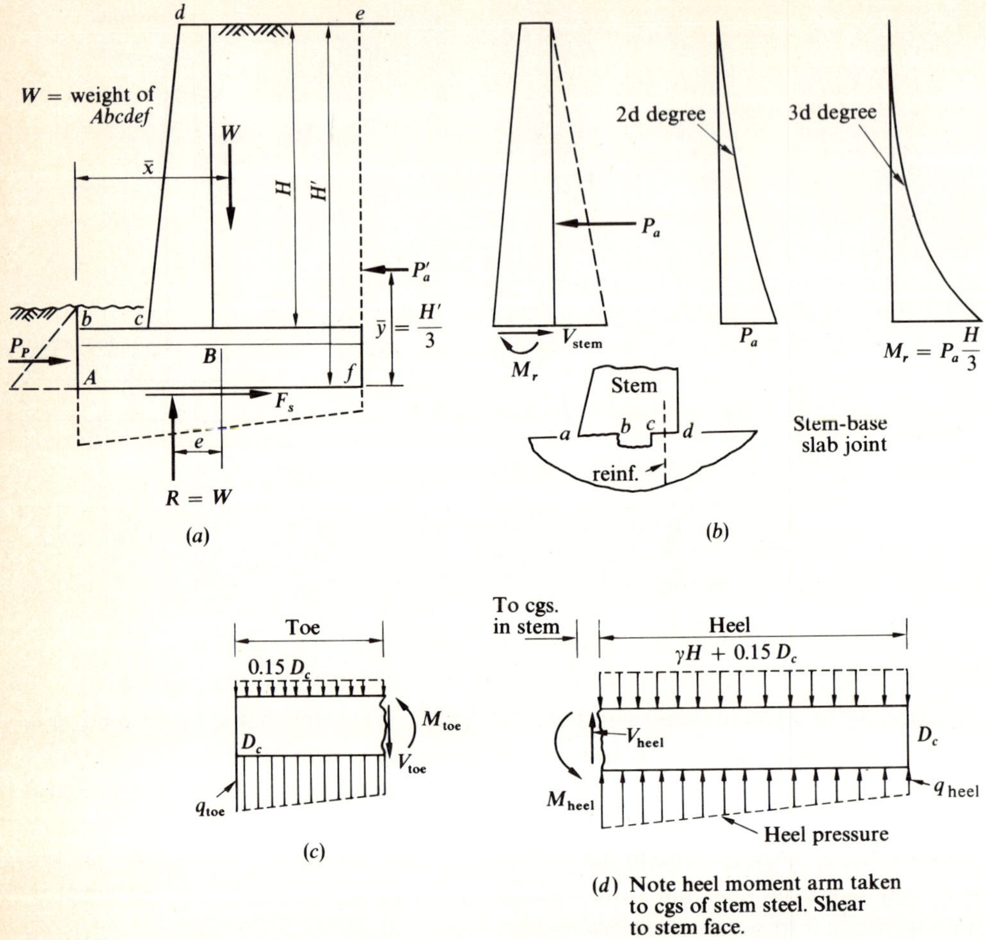

FIGURE 8-10
Structural stability of various parts of a cantilever retaining wall: (*a*) overall stability; (*b*) steam stability; (*c*) toe design; (*d*) heel design.

friction resistance per square inch of stem at the interface of the separate concrete pours is

$$f_s = \mu N = (\text{LF}) \frac{H\gamma_c}{144}(0.5) = 0.75H \qquad \text{psi}$$

when we take the load factor as 1.5; the unit weight of concrete as 144 pcf. For $f'_c =$ 3,000 psi the allowable design shear stress is

$$v_c = 2\phi\sqrt{f'_c} = 93 \text{ psi}$$

When we equate these two values (considering the weight of the stem as a compressive force), the height of wall necessary to provide the same friction resistance as a non-jointed wall base slab is

$$H = \frac{93}{0.75} = 124 \text{ ft} \qquad (\text{if } \mu = 1.0, H = 62 \text{ ft})$$

This computation indicates that the stem shearing resistance depends heavily on the reinforcing steel in the footing. The amount of steel in the stem across the shear zone is based on bending; thus, this steel is actually in a combined stress condition.

Considering that the stem key will carry approximately 20 percent of P_a and friction about 30 percent, shear steel should be provided to carry the remaining 50 percent of the horizontal shear. The allowable shear stress[1] in steel is $0.4F_y$, but increasing by load factor of 1.5, it becomes

$$f_v \leq 0.6F_y$$

Thus, for $F_y = 60$ ksi the steel shear stress would be 36 ksi. For a driving force $P_a = 10$ kips (including the load factor)

$$0.20 \times 10 = 2 \text{ kips shear}$$

$$0.30 \times 10 = 3 \text{ kips friction}$$

There then remains $10 - 5 = 5$ kips to be carried by shear steel requiring an area for $F_y = 60$ ksi of $\frac{5}{36} = 0.138$ sq in/ft.

The bending moment varies as a third-degree curve; therefore, it is evident that cutoff points should be used so that the same amount of steel is not run the full wall height.

Wall Stability

The next step in retaining-wall design is to compute the overturning and sliding stability. The recommended safety factors are:

Soil	Overturning	Sliding
Cohesive	2.0	2.0
Cohesionless	1.5	1.5

A study by Peck et al. (1948) indicates that the majority of recorded retaining-wall failures were founded on and/or retained cohesive soil. This would seem to justify a higher safety factor with cohesive soils.

[1] Using American Institute of Steel Construction, "Manual of Steel Construction," 7th ed., New York, 1971.

In Fig. 8-10*a* (note $\beta = 0$ in the figure), the SF against overturning with respect to the toe at point A is

$$\text{SF} = \frac{W\bar{x}}{P'_a \bar{y} \cos \beta}$$

The SF against sliding is

$$\text{SF} = \frac{F_s + P_p}{P'_a \cos \beta}$$

where P_p is the passive pressure, which should be used only if the designer is certain of no soil erosion or scour at the toe. The friction resistance F_s is computed as

$$F_s = cB(1) + \mu R$$

At this point differences of opinion again arise, since some authorities use only the cohesion ($\phi = 0$) for cohesive soil and $\mu = \tan \phi$ for cohesionless soil. Others use a combination of 0.67ϕ and $0.67c$ as design values. Others, such as the American Railway Engineering Association (AREA), specify the friction coefficient as:

Soil	μ	Corresponding ϕ,* deg
Clean granular	0.55	28.8
Granular with silt	0.45	24.2
Silt	0.35	19.3
Rock	0.60	30.9

* Author's computation.

The author recommends using

$$\phi' = 0.67\phi \text{ to } \phi$$

$$c' = 0.67c$$

or

$$F_s = c'B(1) + W \tan \phi'$$

as reasonably realistic. In granular-base soils, there is strong justification for using $\mu = \tan \phi$ since concrete poured directly on these soils will be very rough at the interface.

To avoid excessively high toe pressures it is desirable to keep the resultant (Fig. 8-10) within the middle third of the base slab. It has been found that if the middle-third criterion is satisfied the overturning SF is usually satisfied. A resultant at the middle-third point results in a soil pressure at the toe of twice the average soil pressure, which may cause an undesirable amount of wall tilt.

The location of the resultant on the base slab can be obtained either by summing moments about the center of base slab area (this requires recomputing the moments) or by finding the location of the resultant with respect to the toe as

$$x = \frac{\sum [\text{moments (net) with respect to toe}]}{R}$$

and the eccentricity with respect to the centroid of the base slab is

$$e = \frac{B}{2} - x$$

Base-Slab Bearing Capacity

The base slab, in addition to being proportioned to satisfy wall stability, must be wide enough to ensure that the allowable soil pressure q_a is not exceeded.

The allowable soil capacity is computed from the ultimate soil pressure[1]

$$q_{\text{ult}} = cN_c s_c d_c i_c + qN_q s_q d_q i_q + \tfrac{1}{2}\gamma B N_\gamma s_\gamma d_\gamma i_\gamma \tag{2-27}$$

Since we will consider a strip of base slab 1 ft $\times$ B, the shape factors are all taken as

$$s_c = s_q = s_\gamma = 1$$

$B' = B - 2e$ is used for all B values. The depth and inclination factors are given in Table 2-7. It should be noted that in the computer program the given values of angle of internal friction ϕ and cohesion c of the base soil are used in the computations for depth and inclination factors.

The allowable soil pressure is

$$q_a = \begin{cases} \dfrac{q_{\text{ult}}}{3} & \text{cohesive soil} \\[2ex] \dfrac{q_{\text{ult}}}{2} & \text{cohesionless soil} \end{cases}$$

The actual soil pressure is computed as for a rigid footing with overturning

$$q = \frac{P}{B}\left(1 \pm \frac{6e}{B}\right)$$

[1] Using Eq. (2-27) with ground and base inclination terms taken as 1.0.

Base-Slab Design

The base slab constitutes two cantilever parts with the stem as the fixed end (Fig. 8-10). The toe should be checked for adequate shear resistance without shear reinforcement at the stem face (not a distance of d out) as a conservative design. Steel is provided for moment resistance. The weight of the base slab should be included in this computation since it is included in the soil pressure. The soil overlying the toe should be neglected.

The heel is analyzed similarly for shear at the stem face using

$$V = \int_0^{\text{heel}} q_{\text{net}}\, dx$$

and for moment at the center of the tension steel in the stem [heel distance plus approximately $3\frac{1}{2}$ in (9 cm)]. For shear resistance the depth is to satisfy

$$v_c = 2\phi\sqrt{f'_c}$$

The AREA specifications require checking for loss of heel pressure. The AASHO specifications (1969, sec. 1.4-8B) state: "The rear projection or heel of base slabs shall be designed to support the entire weight of the superimposed materials, unless a more exact method is used."

The author has provided some interpretation of this latter mode of failure (in computer program) by:

1 Applying $0.67R/B$ as a rectangular toe pressure and simultaneously evaluating the full loss of heel pressure without using a load factor.

2 Rechecking both toe and heel for these two new shear conditions. The base slab is increased in thickness if the computed concrete shear stress is

$$v_{\text{computed}} > 3.1\phi\sqrt{f'_c}$$

The stress is obtained by considering that if 2ϕ corresponds to the situation of using a LF = 1.55 (average with live and dead load), then without a load factor the stress may be increased as $2 \times 1.55 = 3.1$. This increased allowable stress will be about the same as applying ACI eq. (11-4).

3 Rechecking both toe and heel moments and if the new moments (without load factors) exceed the originally computed moments at either location using load factors, the steel area is selected based on the loss of heel-pressure design conditions (and without the load factor).

These alternative conditions are included, but it should be realized that this mode of failure is highly unlikely. It is difficult to conceive of a situation where loss of heel pressure could occur without the stem or toe breaking away. For these reasons the author has used what amounts to very low safety factors for this failure mode.

8-5 COMPUTER-DESIGNED CANTILEVER RETAINING WALL

EXAMPLE 8-1 The computer printout includes sketches so that the reader can follow the design steps. The computer design is exactly as outlined in the preceding section.

SOLUTION Data-card input is as follows:

Card	Data
1	TITLE (see Fig. E8-1.1)
2	FT IN KIPS FT–K K/SQ FT K/CU FT LB/SQ IN SQ IN
	Note two additional entries on this card (start entry in column 1, 11, 21, . . . , 61)
3	12. 144. 030. 2.00 3.5 1000. 4000. 87000.
4	1000. .001 200. .150 .50 .144 3.0
	Cards 3 and 4 represent FU1 through FU15 (8F10.4) in order. These cards include metric and ACI Code values to work program in either fps or metric units by changing entries on cards 2, 3, and 4
5	H G1 G2 β ϕ1 ϕ2 KPP
	20. .112 .120 5.0 34. 32. 1.
	Note with KPP > 0. passive (Rankine) earth pressure is computed
	F1C FY TOE HEEL DC D COH
6	3500. 60000. 3.25 7.50 1.917 4.00 .800
	FAC SURCHG HLOSS TOP
7	1.50 .500 0. 10.
	Using a load factor of 1.5 and specifying a top thickness of 10 in. No loss of heel pressure is considered

These seven data cards represent the input. The output is shown in Fig. E8-1.1 and the final design sketch in Fig. E8–1.2.

To complete the stem design (Fig. E8-1.1) in the first 5 ft above the base slab (4 + 1 ft extension, approx. ACI, art. 12.1.4) use as follows:

two no. 8 bars and one no. 9 bar ($A_s = 2.58 > 2.52$)

Terminate no. 9 bar 5 ft from base. Anchor stem steel by running d into the base slab and 90° bend. In the next 5 ft of wall use

two no. 8 bars and terminate one no. 8 bar

For remainder of wall use one no. 8 bar per foot ($A_s = 0.79 > 0.26$). Shrinkage and temperature requirements perpendicular to plane of paper are as follows. Base slab:

$$0.002(1.9)(12) = 0.456 \text{ sq in}$$

Use two no. 5 bars/ft. Wall (same throughout):

$$0.002(1.25)(12) = 0.30$$

```
  J E BOWLES   EXAMPLE 8-1  RET. WALL--NO LOSS OF HEEL PRESSURE
RETAINING WALL DESIGN BY USD ACI 318-71

     GENERAL INPUT DATA

SOIL DATA:         UNIT WT BACKFILL =    0.112 K/CU FT
                 UNIT WT. BASE SOIL =    0.120 K/CU FT
                     SLOPE BACKFILL =   5.000 DEG
                 INT FRICT BACKFILL = 34.000 DEG
                INT FRICT BASE SOIL = 32.000 DEG
                COHES OF BASE SOIL =    0.80 K/SQ FT

  WALL HEIGHT ABOVE FTG = 20.000 FT
    28-DAY CONCRETE STR = 3500.0 LB/SQ IN
YIELD STENGTH OF STEEL = 60000.0 LB/SQ IN
            LOAD FACTOR = 1.50
              SURCHARGE =   0.500 K/SQ FT

TRIAL FOOTING DIMENSIONS:       TOE = 3.250 FT
                               HEEL = 7.500 FT
                                 DC = 1.917 FT
                              DEPTH = 4.000 FT
********  WALL COMPUTATIONS FOLLOW:

THE RANKINE EARTH PRESSURE COEFF,KA =.28552

THE HOR SHEAR FORCE BASE OF WALL =   9.216    KIPS
REVISED UNIT WEIGHT OF BACKFILL =   0.112 K/CU FT

THE ACTUAL CONCRETE SHEAR STRESS = 100.17 LB/SQ IN
  THE ALLOWABLE SHEAR STRESS = 100.57 LB/SQ IN
CALC STEM THICK:         TOP = 10.00 IN
                      BOTTOM = 15.00 IN
```

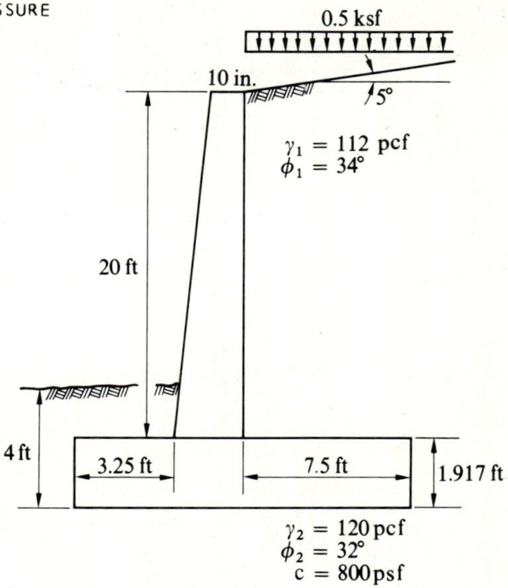

ULTIMATE MOMENTS IN STEM AT 0.1 POINTS AND STEEL REQUIREMENTS PER FT OF WIDTH

POINT FROM TOP	MOMENT, FT-K	AS SQ IN	PERCENT STEEL	WALL THICK. IN	MAX. CODE %
0	0.0	0.260	.003	10.00	.019
1	0.490	0.280	.003	10.50	.019
2	2.216	0.300	.003	11.00	.019
3	5.560	0.320	.003	11.50	.019
4	10.904	0.340	.003	12.00	.019
5	18.631	0.482	.004	12.50	.019
6	29.122	0.728	.006	13.00	.019
7	42.760	1.041	.009	13.50	.019
8	59.928	1.433	.011	14.00	.019
9	81.007	1.917	.015	14.50	.019
10	106.380	2.519	.018	15.00	.019

```
WALL STABILITY

PART      WT OF PART, KIPS    ARM FT      MOMENT FT-K

   1            20.826          8.250         171.811
   2             2.500          4.083          10.208
   3             0.625          3.528           2.205
   4             3.451          6.000          20.704
  PAV            0.991         12.000          11.891
 SUM OF WTS =   28.392    SUM RESIST MOM =    216.819 FT-K

YBAR =   8.583 FT
OVERTURN MOMENT =   97.212 FT-K      SF OVERTURN = 2.230

NET RESIST MOM =  119.607 FT-K      X-BAR = 4.213 FT
 ECCENTRICITY = 1.787 FT       MAX ECCENTRICITY (L/6) = 2.000 FT

FRIC RESIST = 11.09 KIPS      COHES RESIST =  6.40 KIPS
PASSIVE RESIS = 14.43 KIPS      TOTAL SLIDING RESIST =  31.92 KIPS

DRIVING FORCE =  11.33 KIPS      *** SF SLIDING = 2.818

THE BEARING CAPACITY AND OTHER FACTORS ARE:
 NC =  35.490       NQ =  23.177       NGAM =  24.943
 IC =   0.434       IQ =   0.458       IGAM =   0.323
 FDC=   1.190       DQ =   1.122
ULT. SOIL PRESS =  24.443    ALLOW SOIL PRESS =   8.148 K/SQ FT

ACTUAL SOIL PRESSURES:  TOE =    4.48 K/SQ FT
                       HEEL =    0.25 K/SQ FT

SHEAR AT STEM FACES:
     TOE SHEAR =   11.77 KIPS
    HEEL SHEAR =   11.18 KIPS

THE ACTUAL SHEAR STRESS IN FTG. BASED ON  A SHEAR OF  11.766 KIPS IS  75.405 LB/SQ IN
     THE ALLOWABLE SHEAR STRESS IS 100.573 LB/SQ IN

CONVENTIONAL TOE & HEEL MOMENTS
     TOE MOMENT =    20.13 FT-K
    HEEL MOMENT =    57.60 FT-K

THE FINAL SLAB DIMENSIONS ARE:     TOE = 3.250 FT        TOTAL BASE WIDTH =12.000 FT
                                  HEEL = 7.500 FT        THICK BASE SLAB = 1.917 FT
                                        EFF DEPTH BASE SLAB = 1.625 FT
 TOE MOMENT =      20.13      FT-K  STEEL REQD =  0.780 SQ IN        % STEEL = 0.0033

HEEL MOMENT =      57.60 FT-K       STEEL REQD =  1.030 SQ IN        % STEEL = 0.0044
```

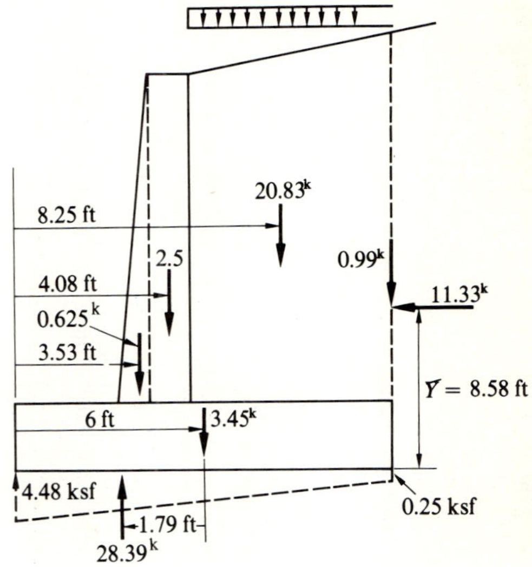

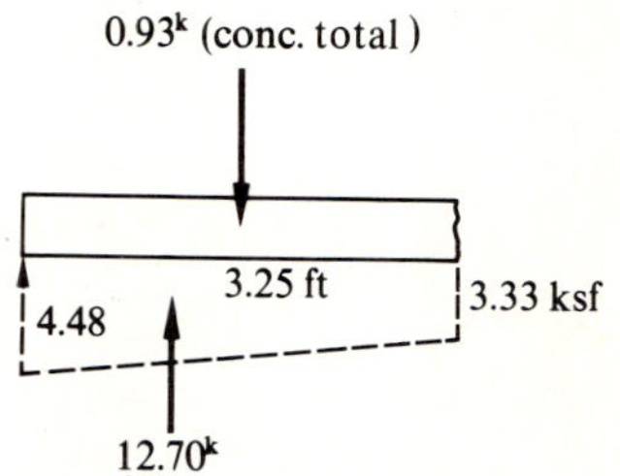

VTOE = 12.70 − 0.93 = 11.77k

FIGURE E8-1.1
Computer I/O.

Use two no. 4 bars/ft. Shear dowels (front face):

$$0.50P_a = 0.5(9.2) = 4.6 \text{ kips}$$

Since this is the actual and not ultimate shearing force, use $f_v = 0.4F_y$.

$$A_s = \frac{4.60}{0.4(60)} = 0.192 \text{ sq in/ft}$$

Use two no. 4 bars at 6 in center to center

$$A_s = 0.40 \text{ sq in} > 0.192$$

Cut one bar at 5 ft; run one bar full height to position alternate shrinkage steel bars inside front face. From computer printout

Heel steel = 1.03 sq in/ft use two no. 9 bars ($A_s = 2.00$)

Toe steel = 0.780 sq in/ft use two no. 6 bars ($A_s = 0.88$) ////

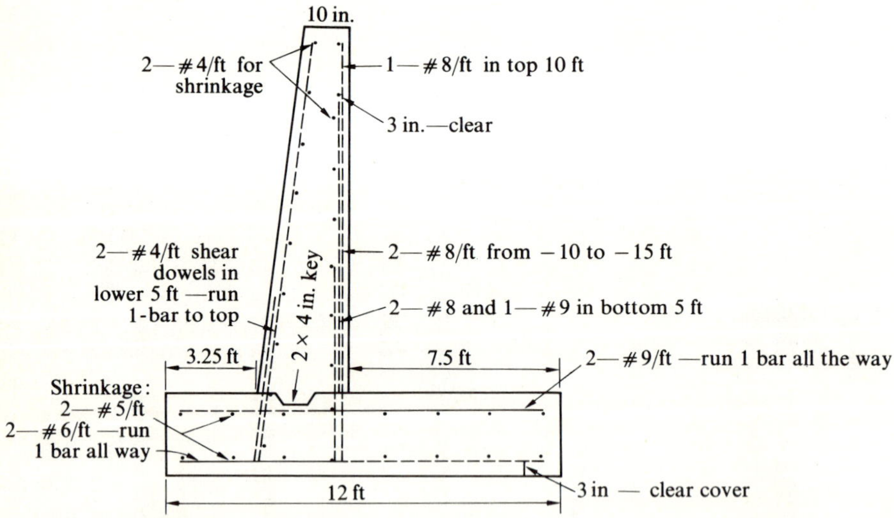

FIGURE E8-1.2
Final-design sketch (not to scale).

EXAMPLE 8-2 Establish dimensions and steel requirements for the retaining wall shown on Fig. E8-2.1. Use metric units.

SOLUTION Data-card input is as follows:

Card	Data
1	TITLE
2	M CM KN KN-M KN/SQ M KN/CU M KG/SQ CM SQ CM
	Note units on "units" card
3	100. 1000. 4.713 .530 9.0 70.3 281. 6117
4	101.968 .009807 14.06 23.564 .150 98.07 7.5
	These two cards are the metric equivalents of FU1–FU15 used in Example 8-1
	H G1 G2 β ϕ1 ϕ2 KPP
5	6.7 17.6 18.9 10. 32. 28. 0.
	No passive resistance considered in sliding SF
	F1C FY TOE HEEL DC D COH
6	211. 4219. .9 2.65 .61 1.20 38.30
	Note f_c' and f_y in kg/sq cm (3000 and 60000 psi)
	FAC SURCHG HLOSS TOP
7	1.5 24.0 0.0 20.
	Note load factor and top width of 20 cm (8 in)

These seven cards represent the input. The output is shown in Figs. E8-2.1 and E8-2.2 with a final sketch. In this example the overturning SF was not adequate for the trial dimensions given, and the program incremented the heel 0.15 m (0.5 ft) to satisfy overturning.

Comments:

1 Final overturning SF = 2.287 (not rounded).

2 Final sliding SF = 1.328 and may (*a*) require additional consideration of a heel key, (*b*) tacitly assume some passive resistance, or (*c*) increase the base width.

3 All forces, moments, and steel requirements are for a 1-m (unit) width.

4 Note that the percentage steel required in most of the stem and toe is for $14.06/f_y$ (or $200/f_y$), which is conservative. The 14.06 is FU11. ////

8-6 COMPUTER PROGRAM FOR RETAINING-WALL DESIGN

This program uses the ACI 318-71 concrete code and strength design. The designer must specify the load factor (FAC) to use. The author recommends 1.5; however, the designer may wish to use 1.2, 1.4, or even 2.0, depending on the amount of uncertainty of loads. The program will consider passive wall pressure if the designer specifies (KPP $>$ 0) after consideration of probability of loss of toe soil. The program considers the AREA (and AASHO) loss of heel pressure using the method outlined by the author in Sec. 8-4 if the designer specifies this alternative check (HLOSS $>$ 0.).

```
   J E BOWLES EXAMPLE 8-2  RETAINING WALL USING METRIC UNITS
RETAINING WALL DESIGN BY USD ACI 318-71

      GENERAL INPUT DATA

SOIL DATA:        UNIT WT BACKFILL =  17.600 KN/CU M
              UNIT WT. BASE SOIL =  18.900 KN/CU M
                  SLCPE BACKFILL = 10.000 DEG
             INT FRICT BACKFILL = 32.000 DEG
            INT FRICT BASE SOIL = 28.000 DEG
            COHES OF BASE SOIL =  38.30 KN/SQ M

   WALL HEIGHT ABOVE FTG =   6.700 M
     28-DAY CONCRETE STR =   211.0 KG/SQ CM
YIELD STENGTH OF STEEL =  4219.0 KG/SQ CM
              LOAD FACTOR = 1.50
                 SURCHARGE = 24.000 KN/SQ M

TRIAL FOOTING DIMENSIONS:        TOE = 0.900 M
                                HEEL = 2.650 M
                                  DC = 0.610 M
                               DEPTH = 1.200 M
********  WALL COMPUTATIONS FOLLOW:

THE RANKINE EARTH PRESSURE COEFF,KA =.32097

THE HOR SHEAR FORCE BASE OF WALL =175.696     KN
REVISED UNIT WEIGHT OF BACKFILL =   17.600 KN/CU M

THE ACTUAL CONCRETE SHEAR STRESS =    6.40 KG/SQ CM
  THE ALLOWABLE SHEAR STRESS =   6.54 KG/SQ CM
CALC STEM THICK:        TOP = 37.00 CM
                     BCTTOM = 51.00 CM
```

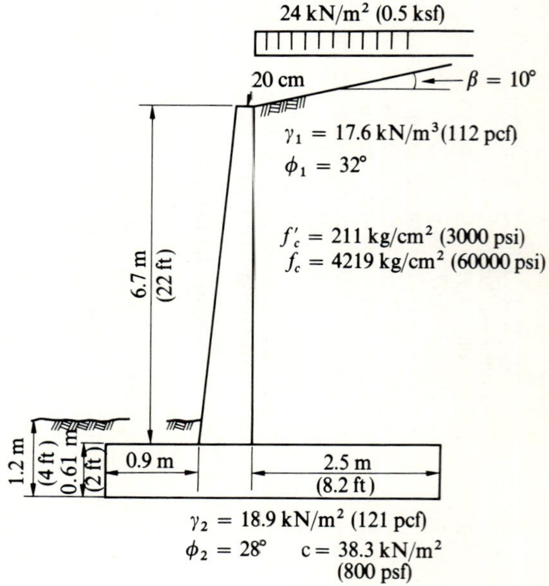

```
ULTIMATE MOMENTS IN STEM AT 0.1 POINTS AND STEEL REQUIREMENTS PER M  OF WIDTH

  POINT FRCM        MOMENT,        AS        PERCENT    WALL THICK.    MAX. CODE %
     TOP              KN-M        SQ CM       STEEL         CM

      0               0.0          9.331      .003        37.00          .016
      1               2.972        9.798      .003        38.40          .016
      2              13.563       10.264      .003        39.80          .016
      3              34.281       10.731      .003        41.20          .016
      4              67.638       11.197      .003        42.60          .016
      5             116.141       11.664      .003        44.00          .016
      6             182.303       14.091      .004        45.40          .016
      7             268.631       20.376      .005        46.80          .016
      8             377.637       28.268      .007        48.20          .016
      9             511.830       38.048      .009        49.60          .016
     10             673.719       50.108      .012        51.00          .016

     WALL STABILITY

  PART      WT OF PART, KN      ARM M      MOMENT KN-M

   1            386.984          2.735       1058.401
   2             58.415          1.225         71.558
   3             11.052          0.993         10.978
   4             58.359          2.030        118.468
  PAV            40.070          4.060        162.685
  SUM OF WTS =  554.879   SUM RESIST MOM =   1422.090 KN-M

YBAR =  2.919 M
OVERTURN MOMENT = 663.283 KN-M     SF OVERTURN = 2.144

NET RESIST MOM =  758.807 KN-M     X-BAR = 1.368 M
 ECCENTRICITY = 0.662 M       MAX ECCENTRICITY (L/6) = C.677 M

FRIC RESIST =187.56 KN        COHES RESIST =103.72 KN
PASSIVE RESIS =  0.0  KN         TOTAL SLIDING RESIST = 291.27 KN

DRIVING FORCE = 227.25 KN        *** SF SLIDING = 1.282

THE BEARING CAPACITY AND OTHER FACTORS ARE:
 NC =  25.803     NQ =  14.720     NGAM =  13.131
 IC =   0.400     IQ =   0.441     IGAM =   0.305
 FDC=   1.175     DQ =   1.124
ULT. SOIL PRESS = 733.484   ALLOW SOIL PRESS = 244.495 KN/SQ M

ACTUAL SOIL PRESSURES:  TOE = 270.47 KN/SQ M
                        HEEL =   2.87 KN/SQ M
```

FIGURE E8-2.1
Input and partial output for Example 8-2.

```
WALL UNSTABLE--HEEL INCREASED 0.15 M  ***

      WALL STABILITY
   PART      WT OF PART, KN      ARM M       MOMENT KN-M

    1             409.541          2.810       1150.809
    2              58.415          1.225         71.558
    3              11.052          0.993         10.978
    4              60.515          2.105        127.383
   PAV             40.308          4.210        169.696
   SUM OF WTS =   579.829     SUM RESIST MOM =  1530.424 KN-M

YBAR =  2.928 M
OVERTURN MOMENT = 669.295 KN-M      SF OVERTURN = 2.287

NET RESIST MOM =   861.128 KN-M       X-BAR = 1.485 M
 ECCENTRICITY = 0.620 M         MAX ECCENTRICITY (L/6) = 0.702 M

FRIC RESIST =195.99 KN          COHES RESIST =107.55 KN
PASSIVE RESIS =   0.0  KN          TOTAL SLIDING RESIST = 303.54 KN

DRIVING FORCE = 228.60 KN          *** SF SLIDING = 1.328

THE BEARING CAPACITY AND OTHER FACTORS ARE:
 NC =   25.803        NQ =   14.720       NGAM =   13.131
 IC =    0.420        IQ =    0.460       IGAM =    0.324
 FDC=    1.162        DQ =    1.115
ULT. SOIL PRESS = 773.054    ALLOW SOIL PRESS = 257.685 KN/SQ M

ACTUAL SOIL PRESSURES:   TOE = 259.40 KN/SQ M
                        HEEL =  16.06 KN/SQ M


WALL UNSTABLE--HEEL INCREASED 0.15 M  ***

      WALL STABILITY
   PART      WT OF PART, KN      ARM M       MOMENT KN-M

    1             432.167          2.885       1246.801
    2              58.415          1.225         71.558
    3              11.052          0.993         10.978
    4              62.671          2.180        136.622
   PAV             40.546          4.360        176.780
   SUM OF WTS =   604.850     SUM RESIST MOM =  1642.740 KN-M

YBAR =  2.937 M
OVERTURN MOMENT = 675.344 KN-M      SF OVERTURN = 2.432

NET RESIST MOM =   967.396 KN-M       X-BAR = 1.599 M
 ECCENTRICITY = 0.581 M         MAX ECCENTRICITY (L/6) = C.727 M

FRIC RESIST =204.45 KN          COHES RESIST =111.38 KN
PASSIVE RESIS =   0.0  KN          TOTAL SLIDING RESIST = 315.83 KN

DRIVING FORCE = 229.95 KN          *** SF SLIDING = 1.373

THE BEARING CAPACITY AND OTHER FACTORS ARE:
 NC =   25.803        NQ =   14.720       NGAM =   13.131
 IC =    0.439        IQ =    0.477       IGAM =    0.343
 FDC=    1.150        DQ =    1.107
ULT. SOIL PRESS = 811.091    ALLOW SOIL PRESS = 270.364 KN/SQ M

ACTUAL SOIL PRESSURES:   TOE = 249.57 KN/SQ M
                        HEEL =  27.89 KN/SQ M

SHEAR AT STEM FACES:
      TOE SHEAR = 191.08 KN
     HEEL SHEAR = 171.07 KN
THE ACTUAL SHEAR STRESS IN FTG. BASED ON  A SHEAR OF 191.083 KN    IS    5.620 KG/SQ CM
     THE ALLOWABLE SHEAR STRESS IS    6.544 KG/SQ CM

CONVENTIONAL TOE & HEEL MOMENTS
      TOE MOMENT =    89.08 KN-M
     HEEL MOMENT =   376.43 KN-M

THE FINAL SLAB DIMENSIONS ARE:    TOE = 0.900 M         TOTAL BASE WIDTH = 4.360 M
                                 HEEL = 2.950 M         THICK BASE SLAB = 0.610 M
                                      EFF DEPTH BASE SLAB = 0.520 M
 TOE MOMENT =      89.08     KN-M  STEEL REQD = 17.329 SQ CM        % STEEL = 0.0033

HEEL MOMENT =    376.43 KN-M       STEEL REQD = 31.388 SQ CM        % STEEL = 0.0060
```

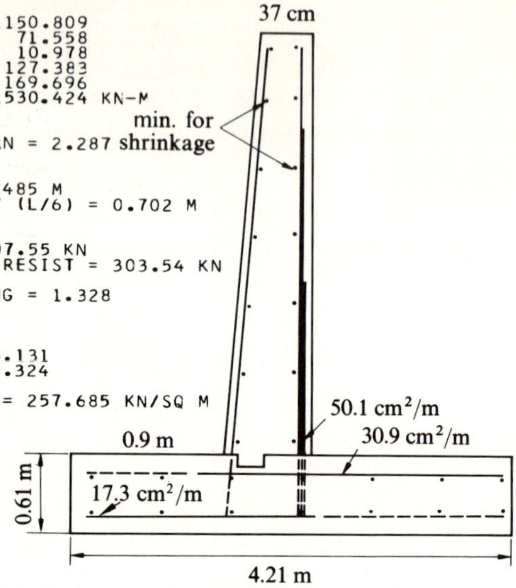

Final design sketch
(no scale)

FIGURE E8-2.2
Remainder of Example 8-2 output and final-design sketch.

A surcharge (SURCHG) on the backfill can be included and soil properties of the backfill different from the base soil. The surcharge is treated as a horizontal projection and must be adjusted if it is known in terms of ground cover. If this is not done, the weight of soil on the heel may not be computed correctly for large β angles. Backfill must not be a cohesive material.

The program computes the stem pressure and finds the minimum stem-base thickness for shear plus 3.5 in (or 9 cm) to allow up to no. 8 rebars and 3 in (7.5 cm) of concrete cover. The top-of-stem thickness is specified by the designer in integer values of, say, 10 in (25 cm), etc. (TOP).

The program computes shear, moments, and steel requirements at the 0.1 points along the stem so that the designer may select cutoff points. The designer must select the rebar sizes and bars to satisfy the 50 percent of P_a indicated in Sec. 8-3.

Next the program computes overturning stability. If the initial base-slab proportions are inadequate to place the resultant in the middle third of base, the heel is increased by 5.0 ft (0.15 m) and the overturning stability repeated up to 3 times; then the toe is increased 0.5 ft (0.15 m). This is repeated until the resultant base pressure is in the middle third of base. The resulting SF is computed. Note that the soil weight over the heel part of the base is placed at heel/2 from the back face of the wall, which results in a small error on the conservative side when $\beta > 0$. The surcharge is treated as the horizontal projection of uniform load.

The sliding SF is investigated and will include passive pressure if the designer specifies. The friction-resistance computation utilizes 0.67ϕ and $0.67c$ for a condition of concrete to soil. The designer must inspect the computed SFs in the output to see if they are adequate. If the SF is inadequate, the designer must provide a base key to improve the sliding SF or increase the heel and reprogram the problem. Similar adjustments will have to be made if the overturning SF is too low since the computer only checks to ensure a resultant within the middle third of the base.

The allowable bearing pressure is computed next, using Eq. (2-27) and factors from Table 2-7.

The actual soil pressure is computed and compared, and the base slab is increased in increments of 0.5 ft (0.15 m) and recycled to the overturning-stability section of the program as many times as necessary to obtain satisfactory bearing pressure.

The base-slab shears and moments are computed. Shear is computed at the stem faces for the toe and heel, and the maximum value is used for design. Depth is made adequate for the largest (critical) shear value found by increasing the base-slab depth in increments of 3 in (7.5 cm) and recycling to the overturning-stability section of the program until the critical shear stresses are satisfied. Once the depth for shear is satisfied, the area of steel for bending is computed. The designer must select the rebars and bar spacing.

If an AREA (or AASHO) design is being made, the programmer requires the program to activate the loss-of-heel-pressure subroutine (HLOSS > 0.). Here the program computes the alternate shear and moment values as described earlier. If these unfactored values are larger, they are used in the design. This may require increasing the footing depth again in increments of 3 in (7.5 cm). Those values used are printed, the critical moments are divided by the LF prior to printing so a ready comparison of critical design moments can be made.

The output identifies all computations so that the designer can check the design at any step. This program will work either metric or fps units but requires three extra data cards containing UNITS, FU1 through FU15.

Line	Operation
4	READ TITLE, UNITS
6–7	READ FU1–FU15 (see data entries in Examples 8-1 and 8-2)
9	READ (6F10.4,15) H = wall height*; G1 = ν backfill soil; G2 = ν base soil; BETA = backfill slope angle; PHI1 = backfill ϕ; PHI2 = base soil ϕ; KPP = activates passive pressure if KPP > 0
10	READ (2F10.4) F1C, FY = concrete and steel stresses (psi or kg/sq cm); TOE = trial toe distance; HEEL = trial heel distance; DC = thickness of base slab; D = depth of base slab; COH = cohesion of base soil
11	READ (3F10.4) FAC = load factor USD; SURCHG = surcharge; HLOSS = check of loss of heel pressure if >0; TOP = minimum top thickness (inches or centimeters)
43–52	Computes stem thickness and checks shear stress
65–79	Computes stem moments and required steel area
80–117	Computes overturning stability
132–142	Computes sliding stability
145–169	Computes allowable capacity and actual soil pressure
171–195	Computes toe and heel shears and moments and designs base-slab thickness
197–215	Computes alternate (AREA) stresses based on loss of heel pressure
216–227	Computes required area of steel for toe and heel moments

* Units not given are feet or meters, kips or kilonewtons, and combinations of these units.

```
        C    J E BOWLES RETAINING WALL DESIGN --ACI 318-71  STRENGTH DESIGN
        C      D = DEPTH OF FTG; COH = COHES CF BASE SOIL; PHI2 = PHI-ANGLE BASE
0001           DIMENSION TITLE(20)
0002           DOUBLE PRECISION UT5,UT6,UT7,UT8
0003           REAL*4 KA,KP1,KPP,NC,NQ,NGAM,IC,IQ,IGAM
0004      6000 READ(1,1000,END=150)TITLE,UT1,UT2,UT3,UT4,UT5,UT6,UT7,UT8
0005      1000 FORMAT(20A4/(4(A4,6X),4(A8,2X)))
0006           READ(1,5)FU1,FU2,FU3,FU4,FU5,FU6,FU7,FU8
0007           READ(1,5)FU9,FU10,FU11,FU12,FU13,FU14,FU15
0008         5 FORMAT(8F10.4)
0009           READ(1,5) H,G1,G2,BETA,PHI1,PHI2,KPP
0010       502 READ(1,5)F1C,FY,TOE,HEEL,DC,D,CCH
0011           READ(1,5)FAC,SURCHG,HLOSS,TOP
0012           WRITE(3,1001)TITLE
0013      1001 FORMAT('1',//,T5,20A4)
0014           IF(HLOSS.LE.0.)WRITE(3,200)
0015       200 FORMAT(T5, 'RETAINING WALL DESIGN BY USD ACI 318-71',// T10, 'GENE
              1RAL INPUT DATA',//)
0016           IF(HLOSS.GT.0.)WRITE(3,208)
0017       208 FORMAT(T5, 'RETAINING WALL DESIGN BY USD ACI 318-71 AND CONSIDERIN
              1G LOSS OF HEEL PRESSURE',//,T10, 'GENERAL INPUT DATA',//)
0018       206 WRITE(3,6)G1,UT6,G2,UT6,BETA,PHI1,PHI2,COH,UT5
0019         6 FORMAT(T5,'SOIL DATA:',T21,'UNIT WT BACKFILL =',F8.3,1X,A7,/,T19,
              1'UNIT WT. BASE SOIL =',F8.3,1X,A7,/,T23,'SLOPE BACKFILL =',F7.3,'
              2DEG',/,T17,'INT FRICT BACKFILL =',F7.3,' DEG',/,T16, 'INT FRICT BA
              3SE SOIL =',F7.3,' DEG',/,T16,'COHES OF BASE SOIL =',F7.2,1X,A7,//)
0020           WRITE(3,7)H,UT1,F1C,UT7,FY,UT7,FAC,SURCHG,UT5
```

```
      7 FORMAT(T7,'WALL HEIGHT ABOVE FTG =',F7.3,1X,A2,/,T9,'28-DAY CONCRE
     1TE STR =',F7.1,1X,A8,/,T5,'YIELD STENGTH OF STEEL =',F8.1,1X,A8,/,
     1 T22,'LOAD FACTOR =',F5.2,/,T25,'SURCHARGE =',F7.3,1X,A7,//)
        WRITE(3,201)TOE,UT1,HEEL,UT1,DC,UT1,D,UT1
    201 FORMAT(T5,'TRIAL FOOTING DIMENSIONS:',T35,'TOE =',F6.3,1X,A2,/,T34
     1,'HEEL =',F6.3,1X,A2,/,T36,'DC =',F6.3,1X,A2,/,T33,'DEPTH =',F6.3,
     2 1X,A2,/,T5,'********  WALL COMPUTATIONS FOLLOW:',//)
C       CALCULATE ACTIVE EARTH PRESSURE BY RANKINE METHOD
        I = 1
        BETA=BETA/57.296
        PHI1 = PHI1/57.296
        PHI2 = PHI2/57.296
        A=COS (BETA)
        B=A**2
        C = (COS(PHI1))**2
        KA = A*(A-SQRT(B-C))/(A+SQRT(B-C))
        WRITE(3,420)KA
    420 FORMAT(T5,'THE RANKINE EARTH PRESSURE COEFF,KA =', F6.5,//)
C       COMPUTE STEM THICK. TOP THICK AS READ USING 'TOP'--EFF UNIT WT >.0
        IF(G1*KA.LT.FU3)G1 = FU3/KA
        QTOP = SURCHG*KA*COS(BETA)
        QBOTW = (SURCHG*KA + G1*H*KA)*COS(BETA)
        PWALL = (QTOP+QBOTW)*H/2.
        WRITE(3,422)PWALL,UT3,G1,UT6
    422 FORMAT(T5,'THE HOR SHEAR FORCE BASE OF WALL =',F7.3,1X,A7, /,T5,
     1'REVISED UNIT WEIGHT OF BACKFILL =',F8.3,1X,A7,//)
        UVC = FU4*.85*SQRT(F1C)
        UVCK = UVC*FU14
        TBC = (PWALL*FAC/UVCK + FU5/FU1)*FU1
C       CALC STEM THICKNESS AT TOP USING BATTER 1:48 ON FRONT FACE
C       TOP AND BASE STEM THICKNESS ROUNDED TO LARGER INTEGER
        TT = TBC - H*FU1/48.
        ITOP = TT
        ATOP = ITOP
        IF(ATOP.LT.TT)TT = ATOP+1.
        IF(TT.LE.TOP)TT = TOP
        TB = TT + H*FU1/48.
        ITB = TB
        ATB = ITB
        IF(ATB.LT.TB)TB = ATB+1.
C       CALCULATE ACTUAL SHEAR STRESS AT BASE OF WALL = ULC
        STRESS = (FAC*PWALL/((TB-FU5)*FU1))*FU9
        WRITE(3,315)STRESS,UT7,UVC,UT7,TT,UT2,TB,UT2
    315 FORMAT(T5,'THE ACTUAL CONCRETE SHEAR STRESS =',F7.2,1X,A8,/,T7,'TH
     2E ALLOWABLE SHEAR STRESS =',F7.2,1X,A8,/,T5,'CALC STEM THICK:', T2
     27,'TOP =',F6.2,1X,A2,/, T24,'BOTTOM =',F6.2,1X,A2,//)
C       COMPUTE ULT MOMENTS IN STEM, STEEL AREA REQD AND % STEEL
        WRITE(3,220)UT1,UT4,UT8,UT2
    220 FORMAT(T5,'ULTIMATE MOMENTS IN STEM AT 0.1 POINTS AND STEEL REQUIR
     1EMENTS PER ',A2,' OF WIDTH',//,T7, 'POINT FROM',T25,'MOMENT,',T41,
     2 'AS',T50, 'PERCENT',T60,'WALL THICK.',T74,'MAX. CODE %',/,T10,'TO
     3P',T26,A7,T39,A5,T51,'STEEL',T64,A2,/)
C       *** CALCULATE MAX , STEEL PER ACI ART 10.2.7 & ART 10.3.2--.75*PB
        IF(F1C-FU7)301,301,302
    302 IDEL = (F1C-FU7)/FU6
        ADEL = IDEL
        AKI = 0.85-ADEL*0.05
    301 IF(F1C.LE.FU7)AKI = 0.85
    305 PB = .6375*AKI*F1C*(FU8/(FY*(FU8+FY)))
C       CALC MOM AND REQ'D STEEL @ 0.1 PTS; ASSUME 3.5 IN OR 9 CM TO CGS
        DY = H/10.
        G = .5*FY/(.85*F1C*FU2)
        DO 32  K = 1,11
        A = K-1
        M = A
        UMOM = FAC*KA*(G1*(A*DY)**3 + 3.*SURCHG*(A*DY)**2)/6.*COS(BETA)
        STEMT = TT + ((TB-TT)/H)*A*DY
        EDCM = (STEMT-FU5)/FU1
        F1 = UMOM/(.9*FY*FU10)
        AS = (EDCM-SQRT(EDCM**2-4.*G*F1))/(2.*G)
        PER = AS/(EDCM*FU1*FU1)
C       MIN % STEEL BASED ON ACI 318-71 ART. 10.5.1
        SMIN = FU11/FY
        IF(PER-SMIN)304,304,32
    304 PER = SMIN
        AS = PER*EDCM*FU1*FU1
     32 WRITE(3,34)M,UMOM,AS,PER,STEMT,PB
     34 FORMAT(T12,I2,T23,F7.3,T39,F7.3,T51,F4.3,T63,F5.2,T77,F4.3)
C       CALCULATE OVERTURNING STABILITY
    350 XHT = HEEL*TAN (BETA)
        HTOT = H+XHT+DC
        W1=((2.*H)+XHT)/2.*HEEL*G1 + SURCHG*HEEL
        W2 = TT*H*FU12/FU1
        W3 = (TB-TT)*H*FU12/(2.*FU1)
        W4 = (TOE+HEEL+TB/FU1)*DC*FU12
        QBASE = SURCHG*KA+G1*HTOT*KA
        PA = (QTOP/COS(BETA) + QBASE)*HTOT/2.
        PAV = PA*SIN (BETA)
        X1 = TOE + TB/FU1 + HEEL/2.
        X2 = TOE + (2.*TB-TT)/(2.*FU1)
        X3 = TOE + (TB-TT)*.667/FU1
        X4 =(TOE + HEEL + TB/FU1)/2.
        X5 = 2.*X4
        RM1 = W1*X1
        RM2= W2*X2
```

```
          RM3=W3*X3
          RM4=W4*X4
          RM5=PAV*X5
          REMO = RM1+RM2+RM3+RM4+RM5
          SUMV=W1+W2+W3+W4+PAV
          WRITE(3,36)UT3,UT1,UT4
       36 FORMAT(//,T10,'WALL STABILITY',//,T7,'PART',T16,'WT OF PART,',1X,A
         14,T35,'ARM',1X,A2,T46,'MOMENT',1X,A4,//)
          WRITE (3,37)W1,X1,RM1,W2,X2,RM2,W3,X3,RM3,W4,X4,RM4,PAV,X5,RM5
       37 FORMAT(T9, '1', T20, F8.3, T36, F6.3, T48, F9.3,/,
         1        T9, '2', T22, F6.3, T36, F6.3, T50, F7.3,/,
         2        T9, '3', T22, F6.3, T36, F6.3, T50, F7.3,/,
         3        T9, '4', T22, F6.3, T36, F6.3, T50, F7.3,/,
         4       T8,'PAV', T22, F6.3, T36, F6.3, T50, F7.3)
          WRITE(3,38)SUMV,REMO,UT4
       38 FORMAT(T7,'SUM OF WTS =',F9.3, 3X,'SUM RESIST MOM =',F10.3,1X,A4/)
C         COMPUTE SF FOR OVERTURNING
          YBAR = ((QTOP/A)*HTOT**2/2. + (QBASE-QTOP/A)*HTOT**2/6.)/PA
          YBAR = (QTOP*HTOT**2/2. + (QBASE-QTOP)*HTOT**2/6.)/PA
          OTMOM = YBAR*PA*COS(BETA)
          FSOT = REMO/OTMOM
C         COMPUTE ECCENTRICITY OF RESULTANT ON BASE
          XNETM=REMO-OTMOM
          XBAR=XNETM/SUMV
          ECCEN = X4-XBAR
          BASE = TOE + HEEL + TB/FU1
          EMAX = BASE/6.
          IF(ECCEN.LE.EMAX)GO TO 46
      306 CONTINUE
          I = I+1
          IF(I/4*4.EQ.I)GO TO 355
          WRITE(3,47)FU13,UT1
       47 FORMAT(/,T5,'WALL UNSTABLE--HEEL INCREASED ',F4.2,1X,A2,' ***')
          HEEL = HEEL + FU13
          GO TO 357
      355 WRITE(3,356)FU13,UT1
      356 FORMAT(/,T5,'WALL UNSTABLE--TOE INCREASED ',F4.2,1X,A2,' *****')
          TOE = TOE + FU13
      357 GO TO 350
       46 WRITE(3,41)YBAR,UT1,OTMOM,UT4,FSOT
       41 FORMAT(T5,'YBAR =',F7.3,1X,A2,/,T5,'OVERTURN MOMENT =',F8.3,1X,A4,
         1 5X,'SF OVERTURN =',F6.3,//)
          WRITE(3,42)XNETM,UT4,XBAR,UT1,ECCEN,UT1,EMAX,UT1
       42 FORMAT(T5,'NET RESIST MOM =',F9.3,1X,A4,5X,'X-BAR =',F6.3,1X,A2,/,
         1T6,'ECCENTRICITY =',F6.3,1X,A2,5X,'MAX ECCENTRICITY (L/6) ='F6.3,1
         2X,A2,//)
C         COMPUTE SLIDING S.F. USING 2/3 FRICT. ANGLE (NOT 2/3TAN (ANGLE))
C        ***NOTE THAT PASSIVE PRESSURE MAY OR MAY NOT BE INCL.
C         *****IF INCL PASSIVE PRESSURE PHI-ANGLE IS BASE SOIL
          IF(KPP.LE.0.)GO TO 35
          KP1 = TAN(45./57.2958 + PHI2/2.)
          KP = KP1**2
          PP = 0.5*G2*KP*D**2 + 2.*COH*D*KP1
       35 PHIP = PHI2*0.667
          FRIC1 = SUMV*TAN (PHIP)
          FRIC2 = COH*BASE*0.667
          IF(KPP.LE.0.)PP = 0.
          TOFR = FRIC1+FRIC2+PP
          SLIDF = PA*COS (BETA)
          FSS = TOFR/SLIDF
          WRITE(3,43)FRIC1,UT3,FRIC2,UT3,PP,UT3,TOFR,UT3,SLIDF,UT3,FSS
       43 FORMAT(T5,'FRIC RESIST =',F6.2,1X,A4,5X,'COHES RESIST =',F6.2,1X,A
         14,/,T5,'PASSIVE RESIS =',F6.2,1X,A4,5X,'TOTAL SLIDING RESIST =',F7
         2.2,1X,A4,//,T5,'DRIVING FORCE =',F7.2, 1X,A4,5X,'*** SF SLIDING ='
         3,F6.3,//)
C         COMPUTATION OF BEARING PRESSURE USING SHAPE, DEPTH AND INCLIN
C            FACTORS OF BRINCH HANSEN (LATEST (1970) REVISIONS)
C          WITH 1-FT STRIP ALL SHAPE FACTORS = 1.
          IF(PHI2.GT.0.)GO TO 91
       92 NC = 5.14
          NQ = 1.
          NGAM = 0.
          GO TO 500
       91 PI = 3.1416
          NQ = EXP(PI*TAN(PHI2))*(TAN(PI/4.+PHI2/2.))**2
          NGAM = 1.8*(NQ-1.)*TAN(PHI2)
          NC = (NQ-1.)*COTAN(PHI2)
      500 BPR = BASE -2.*ECCEN
          IQ = (1.-.5*SLIDF/(SUMV+BPR*COTAN(PHI2)*COH))**5
          IC = IQ - (1.-IQ)/(NQ-1.)
          IF(PHI2.EQ.0.)IC=1.5-0.5*SQRT(1.-SLIDF/(BPR*COH))
          IGAM = (1.-.7*SLIDF/(SUMV+BPR*COH*COTAN(PHI2)))**5
          FDC = 1. + .4*D/BPR
          DQ = 1. + 2.*TAN(PHI2)*(1.-SIN(PHI2))**2*ATAN(D/BPR)
          QULT = COH*NC*FDC*IC + D*G2*NQ*DQ*IQ + .5*BPR*G2*NGAM*IGAM
       61 QALL = QULT/2.
          IF(COH.GT.0.)QALL = QULT/3.
       62 WRITE(3,17)NC,NQ,NGAM,IC,IQ,IGAM,FDC,DQ,QULT,QALL,UT5
       17 FORMAT(T5, 'THE BEARING CAPACITY AND OTHER FACTORS ARE:',/,T6, 'NC
         1 =',F8.3,5X, 'NQ =',F8.3,5X,'NGAM ='F8.3,/,T6,'IC =',F8.3,5X, 'IQ
         3=',F8.3,5X, 'IGAM =',F8.3,/, T6,'FDC=',F8.3, 5X, 'DQ =',F8.3,/, T5
         4,'ULT. SOIL PRESS =',F8.3,3X,'ALLOW SOIL PRESS =',F8.3,1X,A7,//)
      126 QTOE=SUMV*(1.+6.*ECCEN/BASE)/(BASE*1.)
```

```
      QHEEL=SUMV*(1.-6.*ECCEN/BASE)/(BASE*1.)
      WRITE(3,48)QTOE,UT5,QHEEL,UT5
   48 FORMAT(T5, 'ACTUAL SOIL PRESSURES:', T29, 'TOE =', F7.2,1X,A7,/,T2
     18, 'HEEL =',F7.2,1X,A7,//)
      IF(QTOE.GT.QALL)GO TO 306
C     COMPUTE WIDE BEAM SHEAR IN TCE & HEEL; COMPUTE TOE & HEEL MOMENTS
C     COMPUTE SHEAR AT FACE OF STEM; CCMPUTE HEEL MOM @ 3.5 IN (9 CM) IN
C     BACKFACE OF STEM (APPROX CGS); EFFECTIVE D = DC-3.5 IN (9 CM)
  316 DCE = DC - FU5/FU1
      V1 = QTOE - FU12*DC
      V2 = (QTOE-QHEEL)/BASE
      VTOE = V1*TOE - V2*(TOE**2)/2.
      TOEMO = (V1*(TOE**2))/2. - (V2*(TCE**3))/6.
      HEELP = HEEL + FU5/FU1
      AVEHT = (2.*H +XHT)/2.
      SOILP = AVEHT*G1 + DC*FU12 + SURCHG
      V4 = SOILP - QHEEL
      VHEEL = V4*HEEL - V2*(HEEL**2)/2.
      HEELM = V4*(HEELP**2)/2. - V2*(HEELP**3)/6.
      WRITE(3,155)VTOE,UT3,VHEEL,UT3
  155 FORMAT(T5,'SHEAR AT STEM FACES:'/,T10,'TOE SHEAR =',F7.2,1X,A4,/,
     1T9, 'HEEL SHEAR =',F7.2,1X,A4,//)
      VCC = VTOE
      IF(VHEEL.GE.VCC)VCC = VHEEL
      VACT = FAC*VCC/(DCE*FU14)
      IF(UVC-VACT)124,125,125
C     IF SHEAR STRESS EXCEEDS ALLCW FCCT DEPTH INCR 3-IN (7.5 CM)
  124 WRITE(3,131)FU15,UT2
  131 FORMAT(T10,'*** SHEAR STRESS EXCEECED--FTG DEPTH INCR',F5.2,1X,A2
     1,//)
      DC = DC + FU15/FU1
      I = 1
      GO TO 350
  125 WRITE(3,129)VCC,UT3,VACT,UT7,UVC,UT7
  129 FORMAT(T5, 'THE ACTUAL SHEAR STRESS IN FTG. BASED ON  A SHEAR OF',
     1F8.3,1X,A4,' IS',F8.3,1X,A8,/,T10,'THE ALLOWABLE SHEAR STRESS IS',
     1F8.3,1X,A8,//)
      WRITE(3,104)TOEMO,UT4,HEELM,UT4
  104 FORMAT(T5, 'CONVENTIONAL TOE & HEEL MOMENTS',/,T10,'TOE MOMENT =',
     1F8.2,1X,A4,/,T9,'HEEL MOMENT ='F8.2,1X,A4,//)
C     ***** ALTERNATIVE CHECK FOR LOSS OF HEEL PRESSURE--ASSUME TOE CARR
C         OF SUMV AS RECTANGULAR PRESSURE BLOCK
      IF(HLOSS.LE.0.)GO TO 203
      AVTOE = SUMV*.667
      AMTOE = AVTOE*TOE/2.
      AVHEL = (AVEHT*G1 +FU12*DC +SURCHG)*HEEL
      AMHEL = AVHEL*(FU5/FU1 + HEEL/2.)
      AVCC = AVTOE
      IF(AVHEL.GT.AVTOE)AVCC = AVHEL
      AVACT = FAC*AVCC/(DCE*FU14)
C     ALTERNATE CONDITION CONSIDERS APPROX 10% SF ON ULT SHEAR
      IF(AVACT-1.55*UVC)52,52,51
   51 WRITE(3, 107)
  107 FORMAT(T5,'**** ALTERNATE SHEAR CCNDITIONS CCNTROL***',//)
      GO TO 124
   52 CONTINUE
      WRITE(3,108)AVTOE,UT3,AMTOE,UT4,AVHEL,UT3,AMHEL,UT4
  108 FORMAT(T5, 'ALTERNATE TOE, HEEL SHEAR AND MOMENTS',/, T10, 'TOE SH
     1EAR =',F7.2,1X,A4,/,T9,'TOE MOMENT =',F9.2,1X,A7,/,T9,'HEEL SHEAR
     1=',F7.2,1X,A4,/,T8,'HEEL MOMENT =',F9.2,1X,A7,//)
      WRITE(3,112)AVCC,UT3,AVACT,UT7
  112 FORMAT(T5, 'ALTERNATE SHEAR STRESS IN FTG BASED ON A SHEAR OF', F8
     1.3,1X,A4,' IS',F9.3,1X,A8,//)
C     IF ALTERNATE MOMENTS ARE ''FAC'' GREATER THAN CONVENTIONAL MOMENT,
C         THEN ALT. MOMENT CONTROLS FOR STEEL REQUIREMENTS
      IF(AMHEL.GT.FAC*HEELM)HEELM = AMHEL/FAC
      IF(AMTOE.GT.FAC*TOEMO)TOEMO = AMTOE/FAC
C     CALCULATE TOE, HEEL STEEL REQUIREMENTS (SQ IN/FT OR SQ CM/M)
  203 F1 = FAC*TOEMO/(.9*FY*FU10)
      F2 = FAC*HEELM/(.9*FY*FU10)
      ASTOE = (DCE - SQRT (DCE**2 -4.*F1*G))/(2.*G)
      ASHEL = (DCE - SQRT (DCE**2 - 4.*F2*G))/(2.*G)
      PERTO = ASTOE/(DCE*FU1*FU1)
      IF(PERTO-SMIN)307,307,308
  307 PERTO = SMIN
      ASTOE = DCE*SMIN*FU1*FU1
  308 PERHE = ASHEL/(DCE*FU1*FU1)
      IF(PERHE-SMIN)309,309,310
  309 PERHE = SMIN
      ASHEL = SMIN*DCE*FU1*FU1
  310 WRITE(3,145)TOE,UT1,BASE,UT1,HEEL,UT1,DC,UT1,DCE,UT1,TOEMO,UT4,AST
     1OE,UT8,PERTO,HEELM,UT4,ASHEL,UT8,PERHE
  145 FORMAT(T5, 'THE FINAL SLAB DIMENSICNS ARE:', T38,'TOE =',F6.3,1X,A
     12,T58,'TOTAL BASE WIDTH =',F6.3,1X,A2,/,T37,'HEEL =',F6.3,1X,A2,T5
     28,'THICK BASE SLAB =',F6.3,1X,A2,/,T43,'EFF DEPTH BASE SLAB =',F6.
     33,1X,A2,/,T6,'TOE MOMENT =',F9.2,1X,A7,T37,'STEEL REQD =',F7.3,1X,
     4A7,T68,'% STEEL =',F7.4,//,T5,'HEEL MOMENT =', F9.2,1X,A4,T37,'STE
     5EL REQD =',F7.3,1X,A5,T68,'% STEEL =',F7.4,//)
      GO TO 6000
  150 STOP
      END
```

8-7 OTHER DESIGN CONSIDERATIONS

Drainage

The backfill should be drained by placing weep holes at periodic intervals, say, 20 to 30 ft (6 to 10 m) along the wall. These should be protected against plugging by placing a graded filter in the zone. Some concern has been expressed that discharging water onto the toe is poor practice as it will wet the underlying soil. It is difficult to conceive of this as a problem, however, since drainage will generally occur after the base soil is pretty wet.

Ice Thrust

Poorly drained backfill may freeze in the winter. Confined ice thrusts on the order of 1 to 9 ksf [Laba (1970)] have been measured in the laboratory. Certainly thrusts against a retaining wall may be larger than the design active earth pressure but probably not nearly as large as the laboratory values since the retaining wall should deflect under the pressure, thus relieving the stress. Generally, ice would tend to give excessive displacements rather than failure because the wall is somewhat flexible.

Joints

Expansion joints should be provided at intervals of not more than 90 ft (28 m). Contraction joints should be provided at intervals not more than 30 ft (9 m).

Backfill

Backfill quality and density should be specified and job-controlled. Density should be on the order of 90 percent upward of standard compaction procedures (85 percent upward of relative density). The fill should be placed in lifts of not over 4 in for hand (mechanical) tamping and not over 8 in for regular compaction equipment.

Water flooding should not be used for compacting granular material (sand) due to a possible long-term adverse effect on the base soil.

Backfill which is cohesive (roadway abutments) will require both density and moisture control since it has been established [see Bowles (1970), chap. 9] that soils compacted at very low or very high water contents possess lower shear strength than those compacted near optimum. This is especially true if the compacted soil later becomes saturated, a highly likely event in locations such as this.

Frozen soil should not be used since it will not compact well; frozen lumps may not break up, and when they thaw, they are likely to become mud.

Large vibrating equipment used to compact granular backfill should be kept at least 5 to 10 ft from the wall to avoid large wall deflections during backfilling. This safety zone can be compacted with small mechanical hand tamping machines.

8-8 OTHER CAUSES OF RETAINING-WALL INSTABILITY

As indicated earlier, if a retaining wall is located where scour or undercutting is a problem, this must be taken into account. Scour may be 3 or more times the rise in water level.

The AASHO specification [(1969), sec. 1-4.6A] specifies that all footings of the category including retaining walls and abutments shall be at least 4 ft below the stream bed, further stating that this depth is to be increased if site conditions warrant.

Another instability to consider (Fig. 8-11) is where the retaining wall supports a rather large fill and a stratum of soil underlying the base is saturated and cohesive. The weight of the fill will induce consolidation, which in turn will cause the *heel* to move downward. The overall effect is to rotate the retaining wall *into* the fill. If the retaining wall is a bridge abutment, this rotation could displace the bridge seat out from under the superstructure with a sudden discontinuity in the roadway. This settlement type failure should be investigated, as well as the possibility that the system of wall and retained earth will fail in a slope-stability failure mode.

8-9 COUNTERFORT RETAINING WALLS

The counterfort wall shown in Fig. 8-12*a* is similar in design to the cantilever retaining wall. Total wall pressures for overturning and sliding stability and the toe portion are identical in design to the cantilever retaining wall. The same safety factors for sliding and overturning are used.

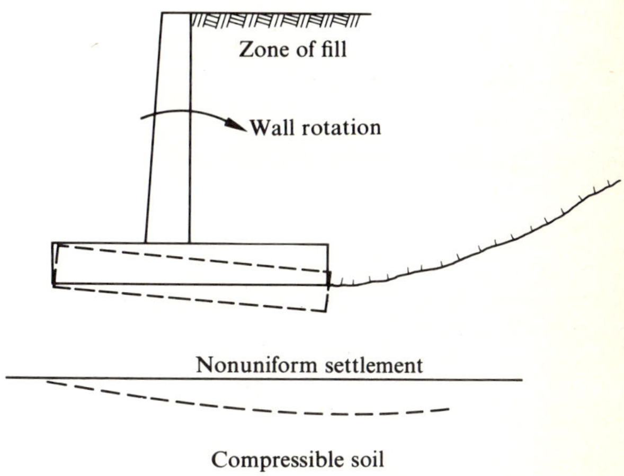

FIGURE 8-11
Deep-seated failure due to differential settlement or excessive shear stresses.

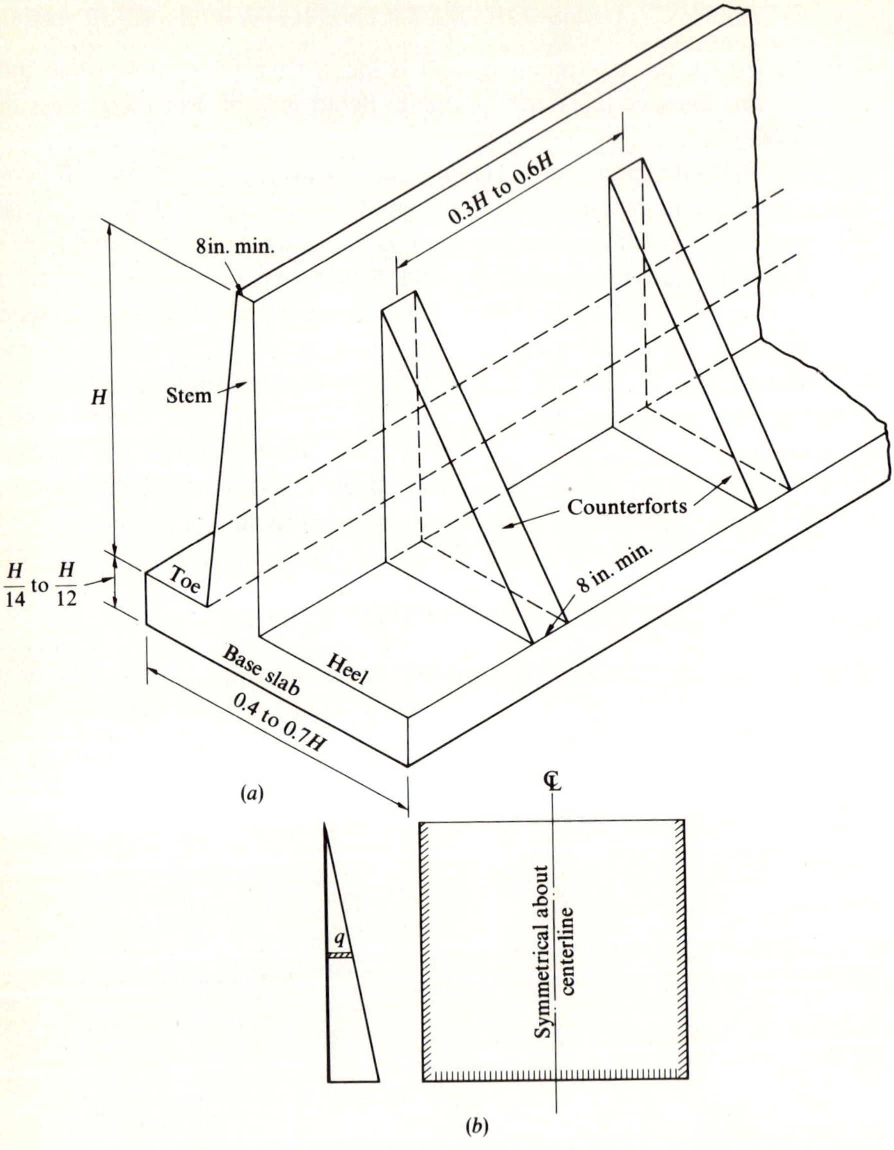

FIGURE 8-12
Counterfort wall: (*a*) trial dimensions; (*b*) stem as plate fixed on three edges and pressure distribution as shown. If counterforts do not extend to wall, top adjustments must be made.

The stem and heel slab differ in design and, strictly speaking, are plates fixed or partly fixed on three edges (Fig. 8-12*b*), depending on the geometry of the counterfort, i.e., whether it spans the entire heel slab or is full wall height. The stem as a plate fixed on three edges must resist a hydrostatic (triangular) pressure. The base slab contains the soil overlying it as a downward pressure and resists the upward footing pressure to give a linear but nonuniform net design pressure.

Approximate procedures are available [Bowles (1968), Huntington (1957)], as well as tables computed using finite differences [Bowles (1968)] to treat the stem in a more theoretical fashion.

Since counterfort walls are rarely used, the design will not be considered here. The reader should consult the given references should the need arise.

PROBLEMS

8-1 Refer to the figure and do the problems assigned, including or excluding passive pressure as instructed. Use the computer output, select the rebars, and sketch the final design. Use f_c' and f_y as assigned. You must estimate initial base-slab dimensions.

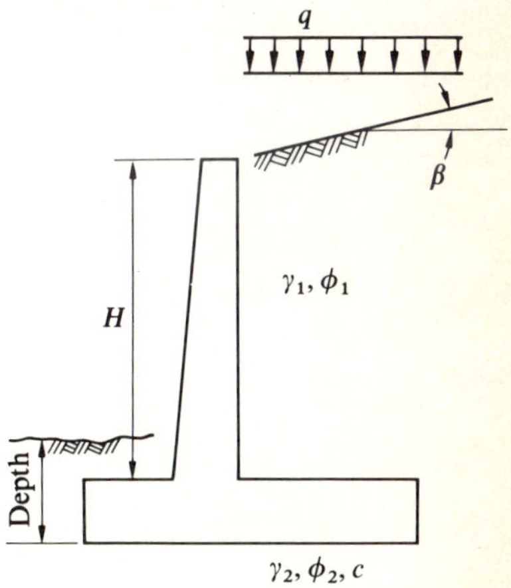

Problem	H	γ_1	γ_2	ϕ_1, deg	ϕ_2, deg	Depth
(*a*)	12 ft	110 pcf	112 pcf	32	34	3 ft
(*b*)	14 ft	110	112	32	34	3 ft
(*c*)	4.9 m	17.3 kN	17.5 kN	32	34	0.9 m
(*d*)	18 ft	110	112	32	34	4 ft
(*e*)	6.1 m	17.3	17.5	32	34	1.2 m
(*f*)	22 ft	115	112	32	34	4 ft
(*g*)	7.3 m	18.9	17.5	32	34	1.2 m
(*h*)	26 ft	120	112	32	34	5 ft
(*i*)	8.5 m	18.9	17.5	32	34	1.5 m
(*j*)	30 ft	120	112	32	34	5 ft

8-2 Repeat the assigned part of Prob. 8-1 if

ϕ_2, deg	c, psf
20	1,200
30	600
30	400
10	1,500
0	4,000

8-3 Repeat the assigned part of Prob. 8-1 if $\beta = 5$, 10, and 15°.

8-4 Repeat the assigned part of Prob. 8-2 if $\beta = 5$, 10, and 15°.

8-5 Repeat the assigned part of Prob. 8-1 if the surcharge pressure is 0.6, 1.0, 1.5, and 2.0 ksf.

8-6 Repeat the assigned part of Prob. 8-1 if the surcharge is 28.7, 48, 72, and 96 kN/sq m.

REFERENCES

AMERICAN ASSOCIATION OF STATE HIGHWAY OFFICIALS (1969): Standard Specifications for Highway Bridges, 10th ed., Washington.

AMERICAN RAILWAY ENGINEERING ASSOCIATION (1958): Manual of Recommended Practice, Chicago.

BOWLES, J. E. (1968): "Foundation Analysis and Design," chap. 7, McGraw-Hill, New York.

——— (1970): "Engineering Properties of Soils and Their Measurement," chap. 9, McGraw-Hill, New York.

HUNTINGTON, W. C. (1957): "Earth Pressures and Retaining Walls," chaps. 2 and 6, Wiley, New York.

LABA, J. T. (1970): Lateral Thrust in Frozen Granular Soils Caused by Temperature Change, *Highw. Res. Rec.* 304, pp. 27–37.

PECK, R. B., H. O. IRELAND, and C. Y. TENG (1948): A Study of Retaining Wall Failures, *Proc. 2d Int. Conf. Soil Mech. Found. Eng., Rotterdam*, vol. 3, pp. 296–299.

9

LATERAL PILES

9-1 LATERAL-PILE CONCEPTS

Early engineers did not design vertical piles to carry lateral loads. Batter piles were provided for this purpose. In fact the Culmann graphical solution [Terzaghi (1943)] widely used up through the early 1950s as a means of analyzing pile groups carrying both vertical and horizontal loads could be used only if batter piles were included to carry the horizontal loads.

The need of a rational method of analysis and design of piles subject to lateral loads has been a matter of concern for some time. Among the earliest large-scale tests are those of Feagin (1937). These tests provided data for the design of the low-sill dam for which they were undertaken. The test data were reasonably adequate since the dam over a period of some 25 years shifted laterally (downstream) only some 1 to 3 in. The shift was probably due to an accumulation of effects such as erosion, vibration, creep, etc., and possibly some of the survey markers may have shifted.

Later Hrennikoff (1950) proposed an analysis of pile groups using the lateral capacity of the pile as a parameter. In the early 1950s others [Palmer and Thompson (1948), Texas A and M (1952), Mason and Bishop (1955), Palmer and Brown (1955),

Howe (1955), Reese and Matlock (1956), Mason (1957), Matlock and Reese (1960)] measured and/or used the finite-difference method of analyzing the lateral pile. Reese and Matlock provided curves as aids since it appeared many people were wary of this analytical method. Bowles (1968) included the finite-difference method along with a computer program in a textbook.

In spite of certain advantages of the finite-difference method over many analysis procedures, some writers preferred alternative methods of analysis [Broms (1964*a*, 1964*b*, 1965), with extensive bibliography].

The author [Bowles (1972)] has developed a method (Sec. 9-2) which has the advantage of including almost any lateral-pile loading scheme and accounting for any type of soil (if the user can describe it), holes, changes in pile section, etc. Since this method is more adaptable to the problem than any other method now available, it is the only method presented in this chapter.

9-2 THE LATERAL PILE BY MATRIX (OR FINITE-ELEMENT) METHODS

This procedure uses the procedure used in Chap. 5 for the beam on an elastic foundation but rotated 90°. Refer to Fig. 9-1 and again recall the three fundamental equations presented in Chap. 5,

$$P = AF$$

$$e = A^T X$$

$$F = Se$$

from which by substitution we have

$$P = ASA^T X \qquad \text{and} \qquad X = ASA^{T-1} P$$

We are ready to proceed with the method.

Figure 9-2 displays the problem of Example 9-1 to illustrate the solution of a lateral pile using this method. No really new or different concepts are introduced from Chap. 5.

The matrix solution here requires inversion of a matrix of size $NP \times NP$, which is twice as large for the same number of nodes as the finite-difference solution. This is offset by ease of programming the loads, nonlinearity of the soil or changes in soil properties, partial embedment, changes in pile-section properties, etc. Also, at most 20 divisions (segments) are required for a solution, thus permitting the use of short segments [since the pile length over 50 ft can generally be neglected (see Sec.

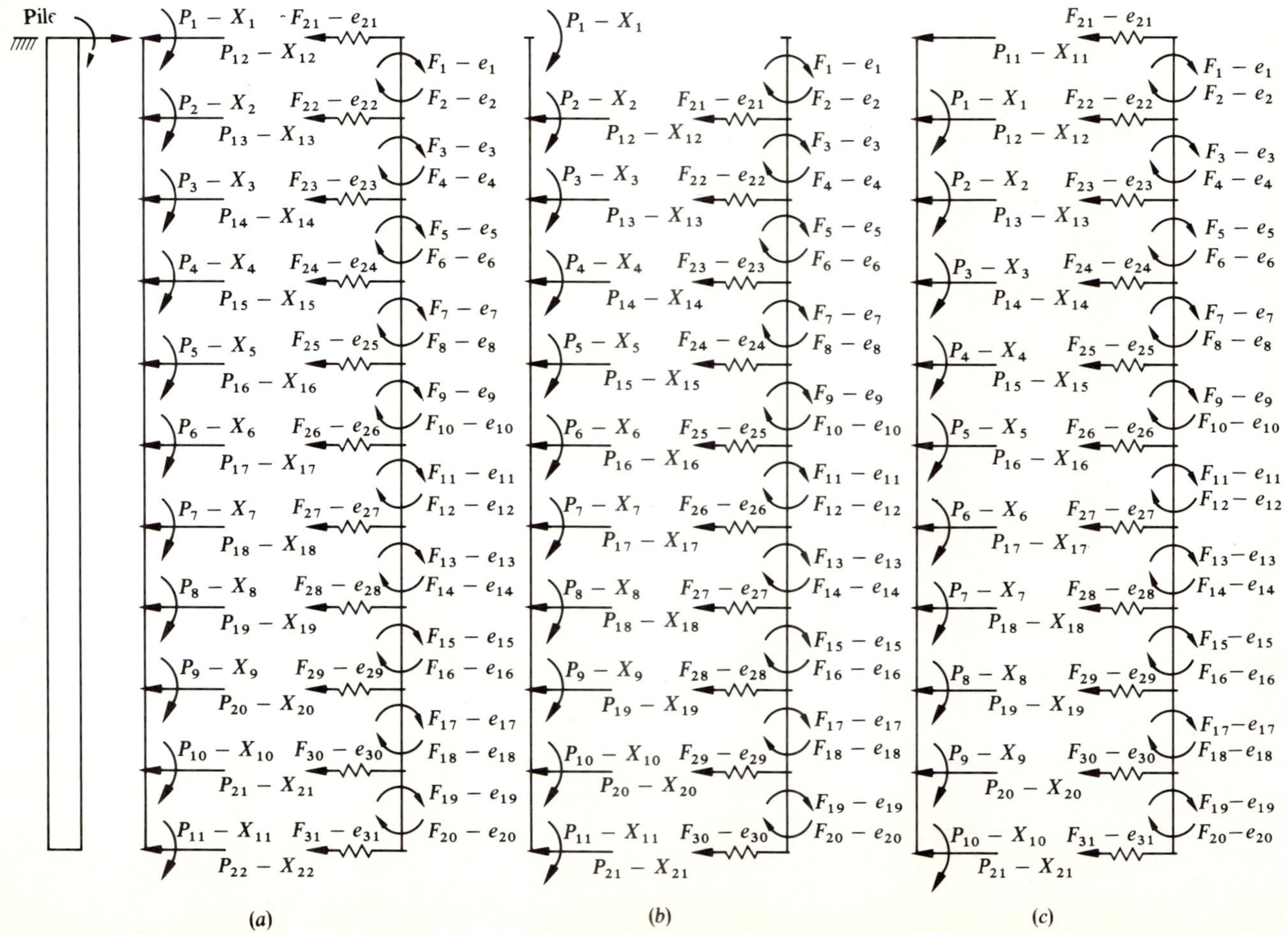

FIGURE 9-1
Coding for various conditions of pile-head fixity: (*a*) free-head condition (rotation and translation) used in the included computer program; (*b*) rotation without translation; (*c*) translation without rotation.

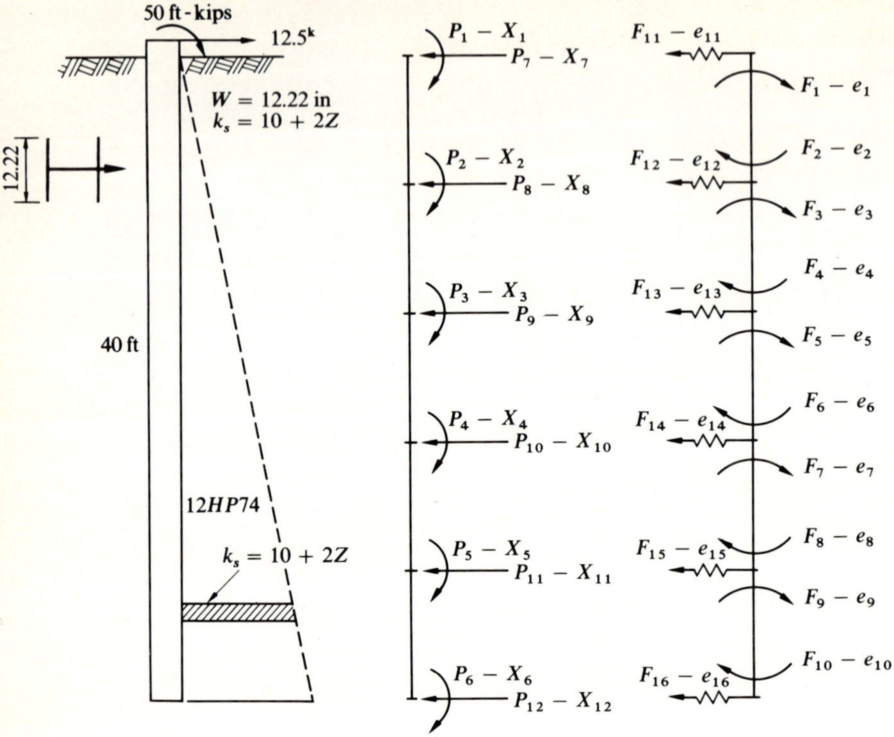

FIGURE 9-2
Coding of pile using five segments (Example 9-1) and soil modulus (kcf) varying with depth as shown.

9-4)]; 10 divisions results in $NP = 22$, and 20 divisions gives $NP = 42$, which is well within the capacity of all but the smallest computers.

9-3 EXAMPLES

Three examples will be used to illustrate the finite-element method of lateral-pile analysis. One example will use five segments to illustrate all the matrices. The same problem will be reworked using ten segments. A third example will illustrate both a metric solution and the method of obtaining pile-head response curves used later in Chap. 13.

EXAMPLE 9-1 Determine the lateral response of a 12HP74 loaded with a lateral force of -12.5 kips (the minus sign indicates that the direction is opposite the basic coding of Fig. 9-1); a moment of $+50$ ft-kips is also applied as shown in Fig. 9-2. From steel property tables the pile width is 12.22 in and has $I = 566.5 \text{ in}^4$. The soil modulus is assumed proportional to depth as $k_s = 10 + 2.0Z$. The pile is 40 ft long. The load is applied at the ground surface.

Five pile segments have been used so that the required matrices can be illustrated in a minimum of space.

SOLUTION It will be necessary to convert the pile width and I to foot units.

$$B = 12.22 \text{ in} = 1.0181 \text{ ft} \qquad I = 566.5 \text{ in}^4 = 0.02731964 \text{ ft}^4$$

$$E_s = 30{,}000 \text{ ksi} = 4{,}320{,}000 \text{ ksf}$$

Computer input is as follows:

Card	Data
1	TITLE
2	UNITS (UT1 – UT6 and FU1 = 12.)
3	5 0 1 1 2 1 1
4	40. 1.0181 0.00 4320000. 2.00
5	10. 2. 1.
6	.02731964
7	1 50.
8	7 −12.50

These cards represent the input data. The output follows in Figs. E9-1.1 and E9-1.2.

Check the computer output:

F(1) = 49.999 from output sheet (50.00 ft-kips given)

−F(2) = F(3) (−110.19 versus +110.189)

F(10) = 0.00

From Fig. E9-1.3

$$\sum M_2 = ?$$

$$50 + 8(12.5) - 8(4.9762) - F(2) = ?$$

$$F(2) = 50 + 100 - 39.84 = 110.16 \ (110.18)$$

```
  J E BOWLES   EXAMPLE 9-1 (12BP74) W/5 DIV.    FREE-HEAD

****** LATERALLY LOADED PILE BY FINITE ELEMENT METHOD

PILE LENGTH = 40.00 FT
PILE WIDTH (IF SQ) =1.0181 FT      PILE DIAM (IF ROUND) =0.0     FT
PILE MOD OF ELAS =  4320000. K/SQ FT
NO OF NODES REQUIRING CORRECT = 0
NODE SOIL STARTS =  1      NO OF LOAD CONDITIONS =  1

   SUBGRADE MODULUS =          10.00 +       2.00*Z**1.000

     MAX LINEAR SOIL DEFORM, XMAX = 2.00 IN
     THE MOMENT OF INERTIA OF THE PILE = 0.0273196 FT**4

     PILE SEGMENT LENGTHS =    8.000 FT



                          THE STATICS MATRIX IS

ROW 1    1.0000   0.0      0.0      0.0      0.0      0.0      0.0      0.0      0.0      0.0
         0.0      0.0      0.0      0.0      0.0      0.0
ROW 2    0.0      1.0000   1.0000   0.0      0.0      0.0      0.0      0.0      0.0      0.0
         0.0      0.0      0.0      0.0      0.0      0.0
ROW 3    0.0      0.0      0.0      1.0000   1.0000   0.0      0.0      0.0      0.0      0.0
         0.0      0.0      0.0      0.0      0.0      0.0
ROW 4    0.0      0.0      0.0      0.0      0.0      1.0000   1.0000   0.0      0.0      0.0
         0.0      0.0      0.0      0.0      0.0      0.0
ROW 5    0.0      0.0      0.0      0.0      0.0      0.0      0.0      1.0000   1.0000   0.0
         0.0      0.0      0.0      0.0      0.0      0.0
ROW 6    0.0      0.0      0.0      0.0      0.0      0.0      0.0      0.0      0.0      1.0000
         0.0      0.0      0.0      0.0      0.0      0.0
ROW 7    0.1250   0.1250   0.0      0.0      0.0      0.0      0.0      0.0      0.0      0.0
        -1.0000   0.0      0.0      0.0      0.0      0.0
ROW 8   -0.1250  -0.1250   0.1250   0.1250   0.0      0.0      0.0      0.0      0.0      0.0
         0.0     -1.0000   0.0      0.0      0.0      0.0
ROW 9    0.0      0.0     -0.1250  -0.1250   0.1250   0.1250   0.0      0.0      0.0      0.0
         0.0      0.0     -1.0000   0.0      0.0      0.0
ROW10    0.0      0.0      0.0      0.0     -0.1250  -0.1250   0.1250   0.1250   0.0      0.0
         0.0      0.0      0.0     -1.0000   0.0      0.0
ROW11    0.0      0.0      0.0      0.0      0.0      0.0     -0.1250  -0.1250   0.1250   0.1250
         0.0      0.0      0.0      0.0     -1.0000   0.0
ROW12    0.0      0.0      0.0      0.0      0.0      0.0      0.0      0.0     -0.1250  -0.1250
         0.0      0.0      0.0      0.0      0.0     -1.0000

  THE SOIL MODULUS AT NODE
        1         10.00000
        2         25.99998
        3         42.00000
        4         58.00000
        5         73.99997
        6         89.99997

  THE P-MATRIX IS AS FOLLOWS     THE NON-LINEAR SOIL FORCES, G(I), ARE

        1          50.0000                0.0
        2           0.0                   0.0
        3           0.0                   0.0
        4           0.0                   0.0
        5           0.0                   0.0
        6           0.0                   0.0
        7         -12.5000                0.0
        8           0.0                   0.0
        9           0.0                   0.0
       10           0.0                   0.0
       11           0.0                   0.0
       12           0.0                   0.0
```

FIGURE E9-1.1
Input data, statics matrix, nodal soil-modulus values, and *P* matrix.

THE S-MATRIX IN TWO COLUMNS

1	59010.42	29505.21
2	29505.21	59010.42
3	59010.42	29505.21
4	29505.21	59010.42
5	59010.42	29505.21
6	29505.21	59010.42
7	59010.42	29505.21
8	29505.21	59010.42
9	59010.42	29505.21
10	29505.21	59010.42
11	31.22	0.0
12	211.76	0.0
13	342.08	0.0
14	472.40	0.0
15	602.71	0.0
16	344.80	0.0

THE ASAT MATRIX WITH 10**3 FACTORED

1	59.010	29.505	0.0	0.0	0.0	0.0	11.064	-11.064	0.0	0.0	0.0	0.0
2	29.505	118.021	29.505	0.0	0.0	0.0	11.064	0.0	-11.064	0.0	0.0	0.0
3	0.0	29.505	118.021	29.505	0.0	0.0	0.0	11.064	0.0	-11.064	0.0	0.0
4	0.0	0.0	29.505	118.021	29.505	0.0	0.0	0.0	11.064	0.0	-11.064	0.0
5	0.0	0.0	0.0	29.505	118.021	29.505	0.0	0.0	0.0	11.064	0.0	-11.064
6	0.0	0.0	0.0	0.0	29.505	59.010	0.0	0.0	0.0	0.0	11.064	-11.064
7	11.064	11.064	0.0	0.0	0.0	0.0	2.797	-2.766	0.0	0.0	0.0	0.0
8	-11.064	0.000	11.064	0.0	0.0	0.0	-2.766	5.744	-2.766	0.0	0.0	0.0
9	0.0	-11.064	0.000	11.064	0.0	0.0	0.0	-2.766	5.874	-2.766	0.0	0.0
10	0.0	0.0	-11.064	0.000	11.064	0.0	0.0	0.0	-2.766	6.005	-2.766	0.0
11	0.0	0.0	0.0	-11.064	0.000	11.064	0.0	0.0	0.0	-2.766	6.135	-2.766
12	0.0	0.0	0.0	0.0	-11.064	-11.064	0.0	0.0	0.0	0.0	-2.766	3.111

THE P-MATRIX
(KIPS OR FT-K) IS

LOAD DIR. 1	50.CCOO
LOAD DIR. 2	0.0
LOAD DIR. 3	0.0
LOAD DIR. 4	0.0
LOAD DIR. 5	0.C
LOAD DIR. 6	0.0
LOAD DIR. 7	-12.5COO
LOAD DIR. 8	0.C
LOAD DIR. 9	0.0
LOAD DIR. 10	0.0
LOAD DIR. 11	0.C
LOAD DIR. 12	0.0

THE JOINT DEFLECTICNS
(RADIANS OR IN) ARE

JOINT DIR. 1	0.015192
JOINT DIR. 2	0.009763
JOINT DIR. 3	0.C03518
JOINT DIR. 4	0.0C0215
JOINT DIR. 5	-0.CC0583
JOINT DIR. 6	-C.CCC590
JOINT DIR. 7	-1.912576
JOINT DIR. 8	-0.682060
JOINT DIR. 9	-0.C64153
JOINT DIR. 10	0.087487
JOINT DIR. 11	0.057219
JOINT DIR. 12	0.00C808

THE BEND. MOMENTS
(KIPS OR FT-K) ARE

MOMENT 1	49.999
MOMENT 2	-110.190
MOMENT 3	110.189
MOMENT 4	-74.092
MOMENT 5	74.091
MOMENT 6	-23.362
MOMENT 7	23.362
MOMENT 8	-0.186
MOMENT 9	0.186
MOMENT 10	0.000
FORCE 11	4.9762
FORCE 12	12.0363
FORCE 13	1.8288
FORCE 14	-3.4440
FORCE 15	-2.8739
FORCE 16	-0.0232

SHEAR AT EACH SEGMENT,KIPS		BENC. MOMENT AT EACH ORDINATE FT-K		SOIL REACTION AT EA. ORD.KIPS
1	7.5238	1	49.9990	-4.9762
2	-4.5125	2	110.1892	-12.0363
3	-6.3413	3	74.0907	-1.8288
4	-2.8972	4	23.3622	3.4440
5	-0.0234	5	0.1857	2.8739
		6	-0.0001	0.0232
			SUM OF SOIL REACTIONS	-12.5001 (-12.500)

FIGURE E9-1.2
S matrix in two columns, ASA^T matrix with 1,000 factored (multiply tabulated values by 1,000), nodal deflection, and moments. Note that at node 1 the F value is 50 ft-kips, which is the applied value. Moment F_{10} should be zero. The sum of the soil reactions considering signs is 12.5 kips (applied value).

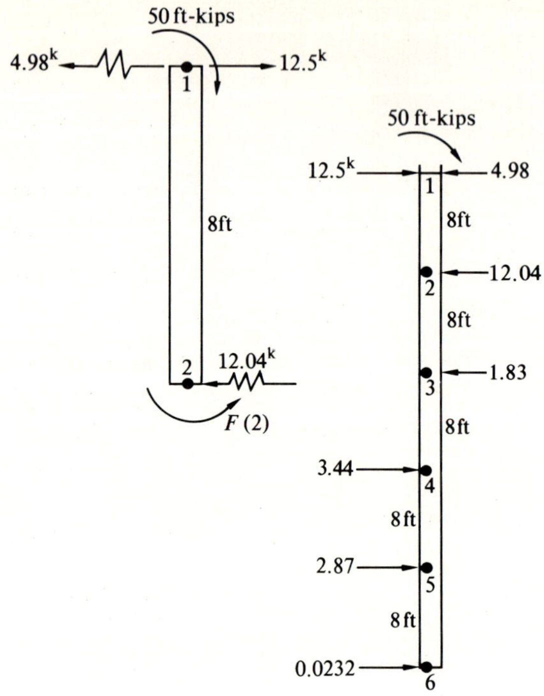

FIGURE E9-1.3

For the pile as a whole

$$\sum F_H = ? \qquad \text{positive to right}$$

$$12.5 + 3.44 + 2.87 + 0.02 - 4.98 - 12.04 - 1.83 \cong 0$$

$$\sum M_{\text{base}} = ? \qquad \text{clockwise} = \text{positive}$$

$$50 + 8[(2.87) + 2(3.44) - 3(1.83) - 4(12.04) - 5(4.98)] + 40(12.5) = ?$$

Solving gives

$$50 + 8(-68.80) + 500 = ?$$

$$-550.4 + 550 \approx 0$$

////

EXAMPLE 9-2 Repeat Example 9-1 using 10 pile segments.

SOLUTION The only card-input change is data card number 3. Change KL to 10:

10	0	1	1	2	1	1

The partial computer output is shown on Fig. E9-2.1. ////

```
J E BOWLES   EXAMPLE 9-2 (12BP74) W/10 DIV -- FREE-HEAD FOR TEXT

******  LATERALLY LOADED PILE BY FINITE ELEMENT METHOD

PILE LENGTH = 40.00 FT
PILE WIDTH (IF SQ) =1.0181 FT        PILE DIAM (IF ROUND) =0.0      FT
PILE MOD OF ELAS =    4320000. K/SQ FT
NO OF NODES REQUIRING CORRECT = 0
NODE SOIL STARTS =   1        NO OF LOAD CONDITIONS =   1

SUBGRADE MODULUS =            10.00 +          2.00*Z**1.000

   MAX LINEAR SOIL DEFORM, XMAX = 2.00 IN
   THE MOMENT OF INERTIA OF THE PILE = 0.0273196 FT**4

   PILE SEGMENT LENGTHS =     4.000 FT


SHEAR AT EACH      BEND. MOMENT AT EACH ORDINATE        SOIL REACTION AT
 SEGMENT,KIPS                 FT-K                        EA. ORD.KIPS

 1    10.4887         1           49.9961 ✓                   -2.0113
 2     3.2171         2           91.9583                     -7.2717
 3    -2.5267         3          104.8083                     -5.7438
 4    -5.7092         4           94.7030                     -3.1825
 5    -6.4402         5           71.8715                     -0.7310
 6    -5.4711         6           46.1162                      0.9691
 7    -3.7109         7           24.2385                      1.7603
 8    -1.9099         8            9.4025                      1.8010
 9    -0.5503         9            1.7706                      1.3595   Checks
10     0.1039        10           -0.4230                      0.6543
                     11           -0.0000 ✓                   -0.1057
                                  SUM OF SOIL REACTIONS      -12.5018   (   -12.500)
```

FIGURE E9-2.1
Computer output for the pile of Example 9-1 using 10 divisions. Note that the moment at node 3 is 104.8 ft-kips compared with 110.2 ft-kips at node 2 (same location) in Example 9-1. More than about eight divisions does not improve the bending-moment computations. In both cases the sum of the soil reactions satisfies statics (= 12.5 kips).

EXAMPLE 9-3 Make a curve of pile-head response for a 12BP74 using metric units. Pile and soil data are:

Item	Metric	fps
B	0.3705 m	1.214 ft
I_x	3.0512×10^{-4} m^4	733.1 in^4
L	15 m	~50 ft
E	204,091,800 kN/sq m	30,000 ksi
XMAX	7.5 cm	3 in

$$k_s = 1{,}570 + 3{,}500Z^{0.67} \text{ kN/cu m} \qquad (10 + 10Z^{0.67} \text{ kips/cu ft})$$

SOLUTION Data cards are:

Card	Data
1	TITLE
	UNITS (UT1–UT6, FU1 = 100.)
2	M CM KN KN-M KN/SQ M KN/CU M 100.
3	10 0 7 1 1 1 0
	Use number of load conditions (NLC) = 7 to read extra loads using only one ASA^T inversion
4	15. .3705 0.0 204091800. 7.5
5	1570. 3500. .67
6	.00030512
7	12 50. (50 kN)
8	12 100.
↓	
13	12 300.

Output for the first lateral load of 50 kN is shown in Fig. E9-3.1. Figure E9-3.2 is a plot of P versus θ and P versus Δ (used in Chap. 13). ////

```
      J E BOWLES   EXAMPLE 9-3  (14BP73) W/10 DIV  USING METRIC UNITS

   ******  LATERALLY LOADED PILE BY FINITE ELEMENT METHOD

   PILE LENGTH = 15.00 M
   PILE WIDTH (IF SQ) =0.3705 M        PILE DIAM (IF ROUND) =C.C      M
   PILE MOD OF ELAS =204091808. KN/SQ M
   NO OF NODES REQUIRING CORRECT = 0
   NODE SOIL STARTS =  1      NO OF LOAD CONDITIONS =  1

      SUBGRADE MODULUS =        1570.00 +    3500.00*Z**0.670

         MAX LINEAR SOIL DEFORM, XMAX = 7.50 CM
         THE MOMENT OF INERTIA OF THE PILE = 0.0003051 M **4

         PILE SEGMENT LENGTHS =     1.500 M


   THE SOIL MODULUS AT NODE
         1      1570.00000
         2      6162.49609
         3      8876.99609
         4     11157.82813
         5     13195.96875
         6     15070.77344
         7     16824.93750
         8     18484.71875
         9     20067.75391
        10     21586.64063
        11     23050.72266

   SHEAR AT EACH      BEND. MOMENT AT EACH ORDINATE       SOIL REACTION AT
    SEGMENT,KN                 KN-M                          EA. ORD.KN

   1  -41.8826         1         -0.0090                        8.1174
   2   -8.2525         2        -62.8188                       33.6301
   3   12.2875         3        -75.1922                       20.5400
   4   17.3433         4        -56.7507                        5.0558
   5   13.3723         5        -30.7297                       -3.9710
   6    7.0557         6        -10.6666                       -6.3167
   7    2.1502         7         -0.0788                       -4.9055
   8   -0.3957         8          3.1506                       -2.5459
   9   -1.0847         9          2.5612                       -0.6889
  10   -0.6287        10          0.9386                        0.4559
                      11          0.0                           0.6257
                                  SUM OF SOIL REACTIONS         49.9970   (     50.000)
```

FIGURE E9-3.1
Input data and partial output data for pile using 10 divisions and metric units. Data are plotted in Fig. E9-3.2.

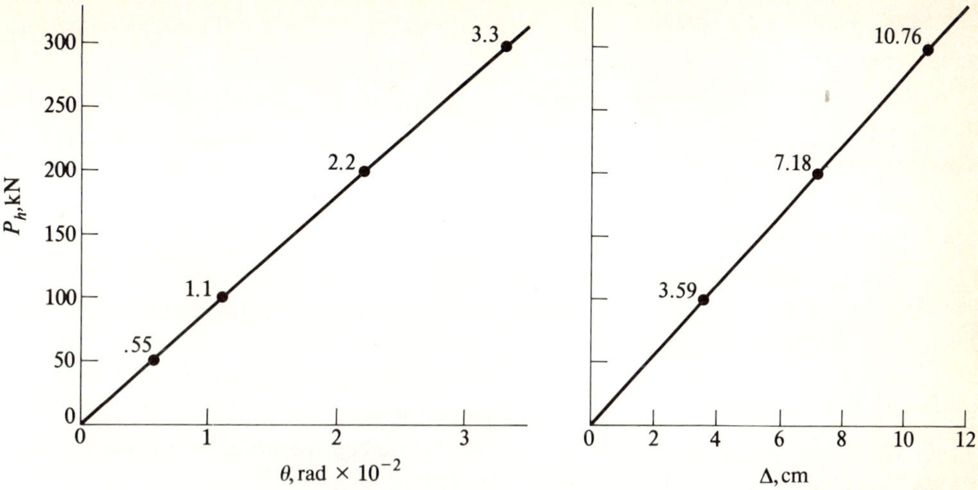

FIGURE E9-3.2
Pile-head response curves (metric units). Curves obtained by varying P_h and plotting the rotation and deflection of node 1.

9-4 SOIL MODULUS AND NONLINEARITY

It is generally conceded that in most soils the modulus-of-subgrade reaction increases with depth according to

$$k_s = A_s + B_s Z^n \tag{9-1}$$

although sometimes

$$k_s = A_s + B\left(\frac{Z}{L}\right)^n \tag{9-1a}$$

is used. However, both equations are made identical by modifying B_s in Eq. (9-1):

$$B_s = \frac{B}{L^n}$$

For sand it appears that $A_s = 0$ and $Z = 1$ is a reasonable approximation. In clay

$$k_s = A_s + B_s Z^n$$

where $n = 0.4$ to 0.8 has been used.

For A_s one may use

$$A_s \approx \begin{cases} 72q_u & \text{kips/cu ft} \\ 2.4q_u & \text{kg/cu cm}^* \end{cases}$$

* To change A_s from kilograms per cubic centimeter to kilonewtons, multiply by 9,807.

One may use Eq. (2-25) by doubling it (although 1.70 to 1.80 might be better, as indicated in the next chapter) to obtain

$$k_s' = 1.30 \sqrt[12]{\frac{E_s B^4}{E_p I_p}} \frac{E_s}{1 - \mu^2} \tag{9-2}$$

Note, however, that the units of k_s' are units of FL^{-2} and include pile width.

The borehole pressure meter may become a particularly effective means of obtaining E_s to convert to k_s. The value of E_s obtained from this device is for the horizontal direction and thus is directly applicable to the lateral modulus. There is also some question whether the lateral modulus obtained from soil tests is different after the pile is driven due to pile-volume displacement. One could, of course, drive a pile, extract it, and determine the lateral modulus using the borehole pressure meter to observe whether a significant change in k_s has occurred.

If lateral deflections are critical, one must use a "good" value of k_s. If only bending moment is important, almost any reasonable value of k_s may be used in the solution. In this respect, this solution is similar to the finite-difference solution. The exponent n in Eq. (9-1) will tend to move the maximum bending moment vertically.

The concept of nonlinear soil response was considered in Sec. 2-9, Fig. 2-8, and Sec. 5-10. One can correct lateral piles for excessive deformation in the same way as for the beam on an elastic foundation, i.e., remove the soil "spring" and apply a negative force of magnitude $F = KX_{\max}$ in the P matrix.

The k_s concentration factor (or method of building the soil spring constant) at a node point is slightly different when the exponent n is greater than zero. The method chosen is based on a parabola given by Newmark (1942):

$$K_1 = \frac{BL}{24}(7k_{s(1)} + 6k_{s(2)} - k_{s(3)})$$

$$K_n = \frac{BL}{24}(7k_{s(n)} + 6k_{s(n-1)} - k_{s(n-2)})$$

any other K

$$K_i = \frac{BL}{12}(k_{s(i-1)} + 10k_{s(i)} + k_{s(i+1)})$$

This is shown in Fig. 9-3 and is valid for the soil modulus varying linearly or as a second-degree parabola. Little error is introduced (less than associated with obtaining k_s) for other values of exponent n by using small segment lengths.

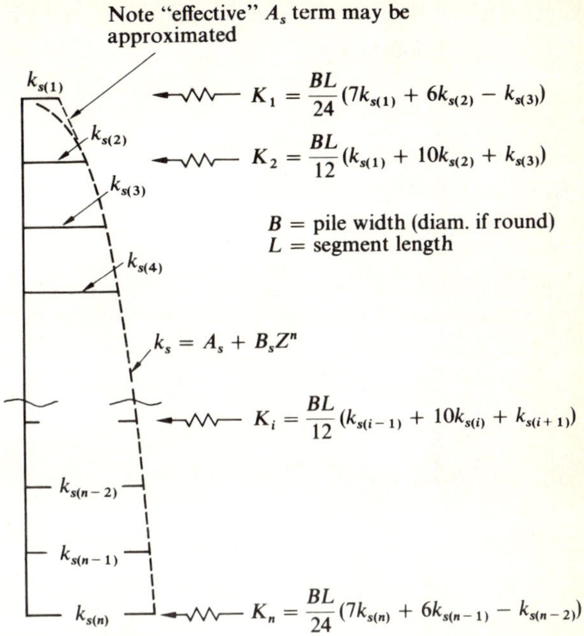

FIGURE 9-3
Soil-modulus variation with depth and method of concentration at nodes to build soil "springs."

9-5 PILE LENGTH AND/OR PARTIAL EMBEDMENT

Pile length is not a critical factor. The author hesitates to state a length which is mainly effective in resisting lateral load since effective length depends on load, soil modulus, and pile stiffness. However, for most piles of reasonable flexural rigidity (EI) and load and length of 30 ft or more the primary effectiveness is in the upper 30 to 40 percent of length. Thus, long piles, say more than 50 to 60 ft in length, can be analyzed on the basis of the top 50 ft.

The soil modulus in depths lower than this critical zone may be anything; hence, in Eq. (9-1) it may be easier to establish the modulus from the top down than the reverse. (The computer program included here requires that the modulus vary from the top down.)

For a pile in an offshore structure or railroad bent, etc., where it is partially freestanding and partially embedded, this solution is quite simple. One simply makes the soil springs in the S matrix zero in the freestanding part or, as in the computer program, starts the soil (JTSS) at the node where embedment begins.

Bending moments and lateral forces may be applied at any location along the pile using this method.

9-6 PILE-HEAD FIXITY

Piles generally terminate in a pile cap of some sort. The terminus may permit both translation or rotation (Fig. 9-1*a*); alternatively the head fixity may allow rotation but no translation. This case (Fig. 9-1*b*) requires modification of the computer program since inspection of Fig. 9-1*a* and *b* shows that P_{12}-X_{12} has been shifted down the pile one node. With this adjustment made, the output will be out of balance, so that for the pile

$$\sum F_H \neq 0.0$$

by the amount of external force required at node 1 to cause no translation.

Lateral movement of the pile head can be treated by computing fixed-end moments based on a value of translation as

$$M_{\text{FEM}} = \frac{6EI\Delta}{L^2}$$

where Δ = assumed value of translation

L = pile segment length

These fixed-end moments including the shear effect are inserted in the P matrix at the appropriate nodes. The F matrix must be corrected for the fixed-end moments in the output.

If translation but no rotation can occur, the coding is as in Fig. 9-1*c*, which also requires modification of the included computer program. This is necessary because P_1-X_1 (the first rotational P-X) is moved down the pile one node. This adjustment will result in the moment

$$F_1 > 0.0$$

in the amount necessary to inhibit rotation. It may be noted in passing that one may allow the pile head to rotate some radians (fixity less than absolute) and compute the resulting fixed-end moments. These values are computed as

$$M_{\text{FEM}} = \begin{cases} \dfrac{4E}{L}\,\theta & \text{top at } F_1 \\ \\ \dfrac{2EI}{L}\,\theta & \text{second node at } F_2 \end{cases}$$

These fixed-end moments at F_1 would subtract from the final computed value of F_1. The fixed-end moment at F_2 would go into the P matrix for P_1. Additionally the fixed-end-moment shear would be entered in P_{11} and P_{12}.

9-7 VALIDITY OF RESULTS

There is a reasonable amount of published literature from which to test the computer solution. It is immediately evident this method agrees with analytical solutions such as in Bowles (1968). It is desirable to check field work and preferably not models.

Results of a series of piles tested by Fruco and Associates (1964) provided the comparison shown in Fig. 9-4. This represents four piles randomly chosen from the

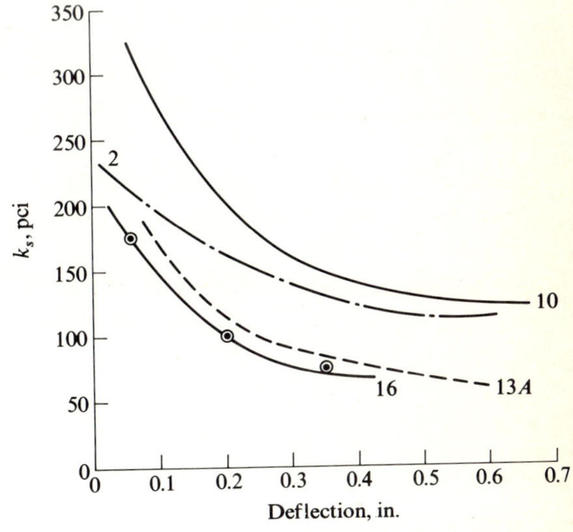

(a)

FIGURE 9-4
Comparison of pile test data and analytical method: (*a*) soil modulus for tests shown in parts (*b*) through (*e*). Measured bending moments taken from graphs. The author has used two variations of $k_s = f$ (depth) to illustrate possibilities of using computer program to establish k_s from measured data. [*Data from Fruco and Associates* (1964).]

Pile load test No. 2

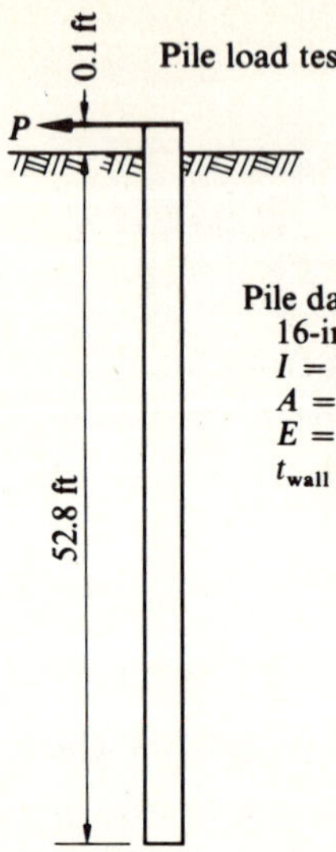

Pile data:
16-in. OD pipe
$I = 838.2$ in.^{4}
$A = 23.86$ in.^{2}
$E = 29{,}000$ ksi
$t_{wall} = 0.312$ in.

$P = 10.5^k$

Dist. from top, ft	Trial 1	Trial 2	Meas. bend. mom., ft-kips
0	1.05	1.05	1.05
5.28	31.56	30.12	31.00
10.56	10.58	9.77	10.00
15.74	−1.18	−0.94	0.0
21.02	−0.96	−1.03	—
26.30	+0.02	−0.07	—

#1 $k_s = 150 + 120Z^1$

#2 $k_s = 150 + 230Z^{0.67}$

(*b*)

Pile load test No. 16

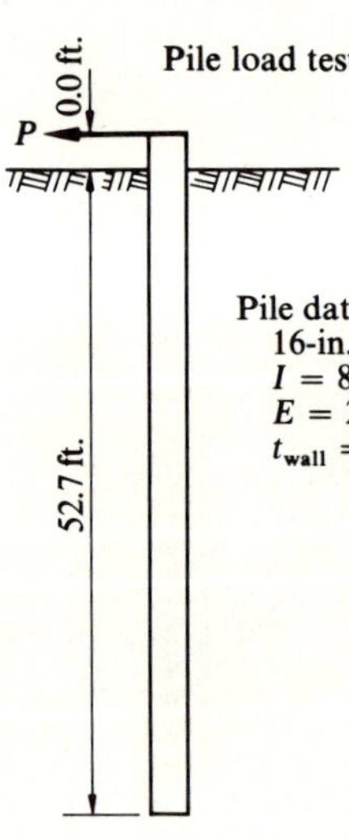

Pile data:
16-in. OD pipe
$I = 829.2$ in.^{4}
$E = 29{,}000$ ksi
$t_{wall} = 0.312$ in.

$P = 19.8^k$

Dist. from top, ft	Trial 1	Trial 2	Meas. bend. mom., ft-kips
0	0	0	0
5.27	61.6	62.4	61.5
10.54	21.7	28.2	26.5
15.81	−2.15	1.19	+0.5
21.08	−2.46	−2.59	—

#1 $k_s = 90 + 125Z^1$

#2 $k_s = 90 + 150Z^{0.67}$

(*c*)

FIGURE 9-4 (*Continued*)

Pile test No. 10

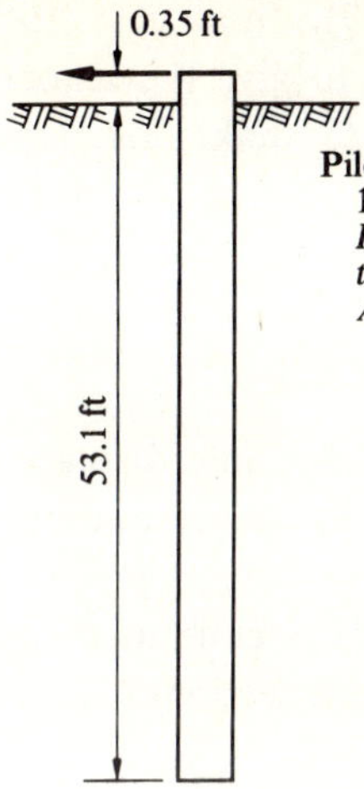

$P = 38.8^k$

Pile data:
16-in. OD pipe
$I = 838.2$ in.^{4}
$t_{wall} = 0.312$ in.
$A = 23.86$ in.^{2}

Dist. from top, ft	Trial 1	Trial 2	Meas. bend. mom., ft-kips
0	13.58	13.58	13.58
5.31	117.89	119.34	120.00
10.62	66.23	61.11	60.00
15.93	14.15	7.43	5.00
21.24	−4.24	−5.83	—

100 ft-kips @ −2.5 ft
100 ft-kips @ −7.5 ft

#1 $k_s = 230 + 30Z^{0.67}$

#2 $k_s = 230 + 30Z^1$

(*d*)

Pile test No. 13*A*

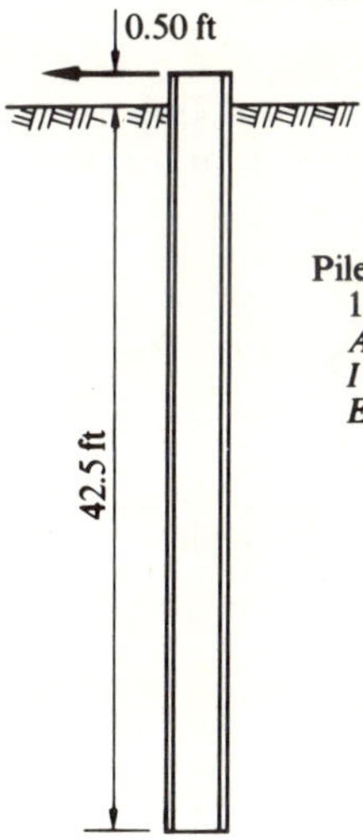

$P = 76.0^k$

Pile data:
14 BP 73
$A = 32.97$ in.^{2}
$I = 855.9$ in.^{4}
$E = 29{,}000$ ksi

Dist. from top, ft	Trial 1	Trial 2	Meas. bend. mom., ft-kips
0	38.00	38.00	38.00
4.25	243.47	255.05	240.00
8.50	210.12	214.87	255.00
12.75	110.05	103.71	160.00
17.00	32.14	23.81	40.00
21.25	—	−8.07	—

180 ft-kips @ −2.5 ft
260 ft-kips @ −7.5 ft
245 ft-kips @ −10 ft
95 ft-kips @ −15 ft

#1 $k_s = 200 + 10Z^1$

#2 $k_s = 150 + 50Z^{0.67}$

(*e*)

FIGURE 9-4 (*Continued*)

large number tested in the cited report. Figure 9-5 displays a comparison of computed versus measured deflections on a fixed-head pile from Mason (1957). In all cases the agreement is quite reasonable. These copious data are included to give the reader some real data to use with the included computer program to make his own comparisons.

9-8 COMPUTER PROGRAM FOR LATERAL PILES

This computer program can be used to compute internal forces and pile displacements for a laterally loaded pile with head free to rotate and translate and for either a nodal moment or lateral force, or both. The pile may be square, as precast-concrete H-piles, or round, as pipe piles. The moment of inertia is computed for round or square solid pile sections and is to be read for all other pile sections.

Note that EXPO is the exponent to compute k_s as

k_s = AS + BS*Z**EXPO

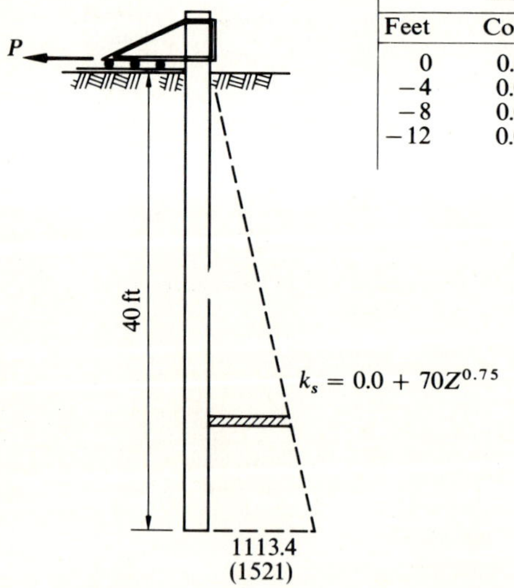

	$P = 5^k$		$P = 15^k$	
	Defl., in		Defl., in	
Feet	Comp.	Meas.	Comp.	Meas.
0	0.09	0.09	0.27	0.27
−4	0.05	0.03	0.15	0.12
−8	0.01	0.0	0.02	−0.02
−12	0.0	0.0	−0.01	−0.01
			0.0	0.0

Pile data:
16 WF 36 about y axis
$b = 15.9$ in (depth)
$E = 29{,}600$ ksi
$EI = 29.2 \times 10^8$ psi (given)

FIGURE 9-5
Fixed-head pile test for comparison of computer program. [*Data from Mason* (1957).] Deflections compare very well when using the variation of soil modulus shown. Mason's value is shown in brackets for the pile base (880 pci varying to the 0.67 exponent). He obtained this value on solving back from the deflections using the finite-difference method available when data were obtained. Solution obtained from computer programmed as Fig. 9-1*c*.

Do not read both BS and EXPO as 0.0 since zero raised to zero power is not defined. The program will compute in either fps or metric units using the UNITS data card and appropriate data entry. FU1 = 12. or 100. (metric).

Line	Operation
1–5	Bookkeeping operations; note that the ASA^T matrix [E(I,J)] is stored over the *A* matrix; therefore, for nonlinear problems it is necessary to rebuild the *A* matrix
6	READ TITLE, UNITS (two cards) Note that FU1 is on the UNIT card
8	READ KL = number of segments; JJS = number of nodes requiring adjustment in the *S* matrix to allow for changes in I or soil spring K; NLC = number of loading conditions. Do not use if XMAX is tested, as program will not do both simultaneously; NI = counter to read moment of inertia; NP = number of nonzero *P*-matrix entries; JTSS = joint soil starts; LIST = counter to list extra output if >1
12	READ* XL = pile length, feet; BX = pile width, feet (if round use 0.); DX = pile diameter, feet (if square use 0.); ELAS = pile modulus of elasticity, ksf; XMAX = maximum value of linear soil deformation
16	READ AS = k_s constant term (kcf); BS = k_s variable with units to result in (kcf); EXPO = exponent
19–34	Forms computation constants
31	READ XI (F10.4) (only if NI > 0)
37–72	Builds flexural (EI) part of *S* matrix in two columns
85–101	Computes subgrade modulus at each node and, using concentration equations, finds K(I,1) at each node (note that first node is *reduced 50 percent*). Also stores soil "springs" for XMAX check [SM(I)]
103	READ I,J, S(I,J) (2I5,F10.4) Reads revised values of *S* matrix according to (and only if) JJS > 0
105–121	Computes and/or reads *P* matrix. The values G(I) are the nonlinear soil values if X > XMAX
140–147	Builds ASA^T over *A* matrix
154–166	Inverts ASA^T
167–181	Computes *X* and *F* matrices. The *F* matrix is corrected for any G(I) soil entries
199–220	Computes shear and bending moment at each segment and writes values
222–239	Nonlinear check and stops if XMAX is exceeded in more than half the nodes

* Substitute meters and kilonewtons for metric problems.

```
      C       JE BOWLES MATRIX DISPLACEMENT ANALYSIS OF A LATERALLY LOADED PILE
      C       *PILE MAY BE FULLY OR PARTIALLY EMBEDDED--JTSS = NODE SOIL BEGINS
      C       UNITS = KIPS (KN); KSF (KN/SQ M); KCF (KN/CU M); FT OR M EXCEPT XM
      C       XL = PILE LENGTH; BX = WIDTH SQ PILE; DX = DIAM IF ROUND; BX OR DX
      C       XMAX = NON-LINEAR SOIL DEFL, IN OR CM;  NP = NO OF NON-ZERO P-ENTR
      C       NI > 0 TO READ MOM OF INERTIA; LIST = A,ASAT IF > 0; JJS = NO NODE
      C        S-MATRIX TO CORRECT;  JTSS = NODE SOIL STARTS = 1 IF FULLY EMBEDD
      C       EXPO = EXPONENT FOR SOIL MOD. VARIATION WITH DEPTH
      C       NLC = NO OF LOADING CONDITIONS--ONLY FOR LARGE VALUES OF XMAX
      C       IF ANY NODE SPRING = 0 FOR XMAX--ADJACENT NODE SPRING REDUCED 25%
      C       (-) P(I) FORCE IS TO RIGHT
0001          DIMENSION X(47),P(47),F(70),F1(70),SOIR(70),SMOD(70),G(67),PM(67)
0002          DIMENSION A(46,67),S(67,2),C(67,46),E(46,46)
0003          DIMENSION LSUM(46),V(46),EE(46),SM(70),PC(46),TITLE(20)
0004          EQUIVALENCE (A(1,1),E(1,1))
0005          DOUBLE PRECISION UT5,UT6
0006     6000 READ (1,1000,END=150) TITLE,UT1,UT2,UT3,UT4,UT5,UT6,FU1,FU2
0007     1000 FORMAT (20A4/4(A4,6X),A8,2X,A8,2X,2F10.2)
0008          READ (1,105)KL,JJS,NLC,NI,NP,JTSS,LIST
0009      105 FORMAT(7I5)
0010          WRITE (3,1001)TITLE
0011     1001 FORMAT('1', T5,20A4)
0012          READ(1,4)XL,BX,DX,ELAS,XMAX
0013        4 FORMAT(6F10.4)
0014      151 WRITE(3,93)XL,UT1,BX,UT1,DX,UT1,ELAS,UT5,JJS,JTSS,NLC
```

```
   93 FORMAT(//,T5,'******  LATERALLY LCADED PILE BY FINITE ELEMENT METH
     1OD',//,T5, 'PILE LENGTH =',F6.2,1X,A2/T5,'PILE WIDTH (IF SQ) =',F6
     2.4,1X,A2,5X,'PILE DIAM (IF ROUND) =',F6.4,1X,A2/T5, 'PILE MOD OF E
     3LAS =',F10.0,1X,A7/T5,'NO OF NODES REQUIRING CORRECT =',I2/T5,'NOD
     4E SOIL STARTS =',I3,5X,'NO OF LOAD CCNDITIONS =',I3//)
      READ(1,4)AS,BS,EXPO
      WRITE(3,214)AS,BS,EXPO,XMAX,UT2
  214 FORMAT(T7,'SUBGRADE MODULUS =',1X,F12.2,' + ',F9.2,'*Z**',F5.3,//,
     1T10,'MAX LINEAR SOIL DEFORM, XMAX =',F5.2,1X,A2)
      M = 3*KL + 1
      N = 2*KL + 2
      KD = M-N
      KP = KL + 1
      KK = 2*KL
      KM = N-1
      K1 = KL+2
      JS = 2
      LSUM(JS-1) = 0
      INDEX2 = 0
      IF(BX.GT.0..AND.NI.LE.0)XI = BX**4/12.
      IF(DX.GT.0..AND.NI.LE.0)XI = .0491*D**4
      IF(NI.GT.0)READ(1,4)XI
      IF(DX.GT.0.)BX = DX
  206 AKL = KL
      H = XL/AKL
      WRITE(3,98)XI,UT1,H,UT1
   98 FORMAT(T10,'THE MOMENT OF INERTIA OF THE PILE =',F10.7,1X,A2,'**4'
     1//T10,'PILE SEGMENT LENGTHS =',F8.3,1X,A2//)
C*** BUILD A-MATRIX ** NON-LINEAR LOOP BEGINS SINCE A & ASAT COMMON
      DO 308  I = 1,M
      SMOD(I) = 0.
      DO 308  J = 1,2
  308 S(I,J) = 0.
   55 DO 508 I=1,N
      DO 508 J =1,M
  508 A(I,J)=0.
      A(KP,KK) = 1.
      A(1,1)=1.0
      NN=2
      DO 501  J = 2,KL
      DO 501 I=2,3
      A(J,NN)=1.0
  501 NN=NN+1
      K = KM
      DO 503  J = K1,N
      A(J,K)=-1.0
  503 K=K+1
      NN=1
      DO 504  J = K1,KM
      DO 504 I=2,3
      A(J,NN)=1./H
  504 NN=NN+1
      K3 = KL+3
      NN=1
      DO 506  J = K3,N
      DO 506 I=2,3
      A(J,NN)=-1./H
  506 NN=NN+1
      IF(JS.GT.2)GO TO 300
      IF(LIST.LE.0)GO TO 9544
      WRITE(3,50)
   50 FORMAT(////,T30,'THE STATICS MATRIX IS',//)
      DO 9151  I = 1,N
 9151 WRITE(3,38)(I,(A(I,J),J=1,M))
   38 FORMAT(1X,'ROW',I2,1X,10F9.4,/,(7X,10F9.4))
C***  BUILD S-MATRIX IN 2-COLUMNS
 9544 DO 107  I= 1,KK
      DO 107  J = 1,2
      S(I,1) = 4.*ELAS*XI/H
      S(I,2) = 2.*ELAS*XI/H
      IF(I/2*2.EQ.I)GO TO 103
      GO TO 107
  103 SAVE = S(I,1)
      S(I,1) = S(I,2)
      S(I,2) = SAVE
  107 CONTINUE
C***BUILD SOIL K--1ST K IS REDUCED 50% ARBITRARILY
      WRITE(3,322)
  322 FORMAT(//,T5, 'THE SOIL MODULUS AT NODE')
      DO 205  I = JTSS,KP
      AM = I-JTSS
      SMOD(I)= AS +BS*(AM*H)**EXPC
  205 WRITE(3,323)I, SMOD(I)
  323 FORMAT(T8, I2, 3X, F12.5)
      S(KK+JTSS,1)=(7.*SMOD(JTSS)+6.*SMOD(JTSS+1)-SMOD(JTSS+2))*H*BX/48.
      S(KK+JTSS,2) = 0.
      S(M,1) = (7.*SMOD(KP)+6.*SMOD(KP-1)-SMOD(KP-2))*H*BX/24.
      S(M,2) = 0.
      KJTS = KK+JTSS+1
      MM = M-1
      MK = N-3
      DO 209  I = KJTS,MM
      S(I,2) = 0.
  209 S(I,1) = H*BX*(SMOD(I-KM)+10.*SMOD(I-KK)+SMOD(I-MK))/12.
      DO 115  I = KM,M
  115 SM(I) = S(I,1)
```

```
C      MODIFICATION OF S-MATRIX--READ AS MANY CARDS AS JJS
       IF(JJS.LE.0)GO TO 99
       READ (1,1008)  (I,J,S(I,J),II=1,JJS)
  1008 FORMAT(2I5,F10.2)
C***   FORM P-MATRIX --EXTERNAL LOADS--NON-LINEAR SOIL EFFECTS
    99 LSUM(JS) = 0
       INDEX2 = INDEX2 + 1
C****  EXTERNAL LOADS INTO P-MATRIX READ NON-ZERO VALUES TO NP
       SUML = 0.
       DO 241  I = 1,N
       G(I) = 0.
   241 P(I) = 0.
       DO 211  JJ = 1,NP
       READ (1,212)I,P(I)
   212 FORMAT(I5, F10.4)
   211 IF(I.GT.KL+1)SUML = SUML + P(I)
       IF(INDEX2.GT.1.AND.NLC.GT.1)WRITE(3,8490)INDEX2
  8490 FORMAT(//,T5,'COMPUTATIONS FOR LOAD CONDITION',I2,//)
C**    NON-LINEAR SOIL ENTRIES--G(I)
   300 DO 213  I =1,N
   213 PM(I) = P(I) + G(I)
       IF(JS-3)215,216,216
   215 WRITE(3,223)
   223 FORMAT(//,T5,'THE P-MATRIX IS AS FOLLOWS', T35, 'THE NON-LINEAR SO
      1IL FORCES, G(I), ARE',//)
       GO TO 225
   216 WRITE(3,48)
    48 FORMAT(//,T5, 'THE MODIFIED P-MATRIX IS',  T35, 'THE NON-LINEAR SO
      1IL FORCES, G(I), ARE',//)
   225 WRITE (3,47)(I,PM(I),G(I),I=1,N)
    47 FORMAT (T8,I2,F15.4, T37, F15.4)
       WRITE(3,226)
   226 FORMAT(//,T5, 'THE S-MATRIX IN TWO COLUMNS')
       WRITE(3,227) (I,S(I,1),S(I,2), I=1,M)
   227 FORMAT(T5,I3,2X,F12.2,2X,F12.2)
       IF(INDEX2.GT.1.AND.NLC.GT.1)GO TO 232
C      *** BUILD THE SAT-MATRIX
       DO 12  I = 1,KK
       DO 12 J=1,N
       KA = I
       IF(I/2*2.EQ.I)KA = KA-1
    12 C(I,J) = S(I,1)*A(J,KA)+S(I,2)*A(J,KA+1)
       DO 530  I = KM,M
       DO 530  J = 1,N
   530 C(I,J) = S(I,1)*A(J,I)
C      ***BUILD THE ASAT-MATRIX
       DO 10   I = 1,N
       DO 15 J=1,N
       EE(J) = 0.
       DO 15 K=1,M
       EE(J) = EE(J)+A(I,K)*C(K,J)
    15 CONTINUE
       DO 10  L = 1,N
    10 E(I,L) = EE(L)
       IF(LIST.LE.0)GO TO 9168
       WRITE(3,9164)
  9164 FORMAT(//,T5,'THE ASAT MATRIX WITH 10**3 FACTORED',//)
       DO 9165  I = 1,N
  9165 WRITE(3,9166)(I,(E(I,J),J=1,N))
  9166 FORMAT(T2,I2,2X,-3P14F8.3)
C****  END OF ASAT FORMATION--INVERT ASAT MATRIX--USE GAUSS-JORDAN METHOD
  9168 DO 25 K=1,N
       DO 20 J=1,N
    20 IF(J.NE.K)E(K,J)=E(K,J)/E(K,K)
       DO 21 I=1,N
       IF(I.EQ.K) GO TO 21
       DO 21 J=1,N
       IF(J.EQ.K) GO TO 21
       E(I,J)=E(I,J)-E(K,J)*E(I,K)
    21 CONTINUE
       DO 22 I=1,N
    22 IF(I.NE.K)E(I,K)=-E(I,K)/E(K,K)
       E(K,K)=1./E(K,K)
    25 CONTINUE
C***   END OF ASAT INVERSION--COMPUTE X-MATRIX
   232 DO 18 I=1,N
       X(I)=0.
       DO 18 K=1,N
       PC(K) = P(K)
       IF(JS.GT.2)PC(K) = PM(K)
    18 X(I) = X(I) + E(I,K)*PC(K)
C*****COMPUTE F-MATRIX--F = SAT*X
       DO 35 I=1,M
       F(I)=0.
       DO 35 K=1,N
    35 F(I)=F(I)+C(I,K)*X(K)
       IF(JS.LE.2)GO TO 39
       DO 37  I = KM,M
    37 F(I) = F(I) + G(I-KD)
    39 DO 233  I = K1,N
   233 X(I) = X(I)*FU1
       WRITE(3,41)UT3,UT4,UT2,UT3,UT4
    41 FORMAT(///T11,'THE P-MATRIX',18X,'THE JOINT DEFLECTIONS',17X,'THE
      1BEND. MOMENTS'/T8,'(',A4,' OR ',A4,') IS',18X,'(RADIANS OR ',A2,
      2') ARE',18X,'(',A4,' OR ',A4,') ARE'/)
C      NOTE THAT ORIGINAL P-MATRIX IS ALWAYS WRITTEN HERE
```

```
      WRITE(3,44)(I,P(I),I,X(I),I,F(I),I=1,KK)
   44 FORMAT(T7,'LOAD DIR.',I3,F10.4,T39,'JOINT DIR.',I3,F13.6,T79,'MOME
     1NT',I3,F12.3)
      DO 49 I=KM,N
      NUM=I+1
   49 WRITE(3,51) I,P(I),I,X(I),I,F(I)
   51 FORMAT(T7,'LOAD DIR.',I3,F10.4,T39,'JOINT DIR.',I3,F13.6,T79,'FORC
     1E ',I3,F11.4)
      IF(NUM.EQ.(M+1)) GO TO 43
      DO 42 I=NUM,M
   42 WRITE(3,52) I,F(I)
   52 FORMAT(T79,'FORCE ',I3,F11.4)
   43 CONTINUE
C***  COMPUTE SEGMENT SHEARS AND MOMENTS
      WRITE(3,234)
  234 FORMAT('1',///,T5,'SHEAR AT EACH',T22,'BEND. MOMENT AT EACH ORDINA
     1TE', T55, 'SOIL REACTION AT')
      WRITE(3,235)UT3,UT4,UT3
  235 FORMAT(T6,'SEGMENT,',A4,T31,A4,T58,'EA. ORD.',A4//)
      VI = 0.
      K = KP
      J = KK
      DO 236 I=1,KL
      K=K+1
      J=J+1
      V(I) = VI - P(K) - F(J)
  236 VI = V(I)
      SUM = 0.
      DO 237 I = 1,KP
      SOIR(I) = -F(I+KK)
  237 SUM = SUM + SOIR(I)
      L = -1
      DO 238  I = 1,KL
      L=L+2
  238 WRITE (3,239)I,V(I), I, F(L), SOIR(I)
  239 FORMAT(T5,I2,F10.4, T24, I2, F15.4, T55, F15.4)
      FX=-F(KK)
      WRITE(3,242)KP,FX,SOIR(KP)
  242 FORMAT(T24,I2,F15.4,T55,F15.4)
      WRITE (3,243)SUM,SUML
  243 FORMAT(T34,'SUM OF SOIL REACTIONS =',T55,F15.4,T72,'(',F10.3,')')
C***  TEST NON-LINEAR--ZERO S-MATRIX AND F = K*XMAX = NEG. P-MATRIX ENTR
      IF(NLC.GT.1)GO TO 149
      LSUN = 0
      K2 = KP+JTSS
      DO 245  I = K2,N
      IF(ABS(X(I)).LT.XMAX)GO TO 245
  244 G(I) = SM(I+KD)*XMAX/FU1
      S(I+KD,1) = 0.
      S(I+KL,1) = 0.75*S(I+KL,1)
      LSUN = LSUN + 1
  245 CONTINUE
      LSUM(JS) = LSUN
      IF(LSUM(JS)-1)149,246,246
  246 IF(JS.EQ.7.OR.LSUM(JS).LE.LSUM(JS-1))GO TO 149
      IF(JS.LE.2)GO TO 54
      IF(LSUM(JS).GE.KL/2+1)GO TO 148
   54 JS = JS+1
      INDEX1 = JS-2
      WRITE(3,1048)INDEX1,LSUM(JS-1)
 1048 FORMAT(//,T10, '****COMPUTATION CYCLE',I3,' XMAX EXCEEDED AT'I3,'
     1NODES',//)
      GO TO 55
  148 WRITE(3,1010)
 1010 FORMAT(//,T5,'**** PILE UNSTABLE--1/2 OR MORE NODES EXCEED XMAX')
  149 IF(INDEX2.LT.NLC)GO TO 99
      GO TO 6000
  150 STOP
      END
```

PROBLEMS

9-1 Modify the included computer program for pile-head rotation without translation.

9-2 Modify the included computer program for pile-head translation without rotation.

9-3 Using the included computer program, what is the effect of reading into the S matrix for $4EI/L$ and $2EI/L$ at F_1 a very large number?

9-4 Using the included computer program, what is the effect of reading into the S matrix a very large number for the soil spring K at the ground-level node?

9-5 Using the modified program of Prob. 9-1, make a plot of M versus θ and R versus θ, where

θ = rotation at node 1

M = various values of external moment applied at node 1 to induce θ

R = amount of unbalanced force resulting from applied moment

What is the significance of these two plots?

9-6 Using the modified program of Prob. 9-2, make a plot of P versus Δ and M versus Δ, where

Δ = deflection of node 1

P = various values of applied lateral force at node 1 to induce Δ

M = unbalanced end moment resulting from applied lateral force

What is the significance of the slopes of these two plots?

9-7 Make a study of the effect of pile stiffness and k_s on maximum bending moment for a selected pile section. Show results graphically.

9-8 Make a study and show results graphically of effective depth of pile (depth of significant deflection or bending moment) as a $f(E1,k_s,L,\ldots)$.

9-9 What is the effect if the laterally loaded pile is battered? Does the computer program require modification to solve this problem?

REFERENCES

BOWLES, J. E. (1968): "Foundation Analysis and Design," chap. 9, McGraw-Hill, New York.

——— (1972): Finite Element Analysis of Lateral Piles, *Proc. Conf. Finite Element Method Civ. Eng., McGill Univ., Montreal*, pp. 677–700.

BROMS, B. B. (1964*a*): Lateral Resistance of Piles in Cohesive Soils, *J. Soil Mech. Found. Div., ASCE*, vol. 90, SM2, March, pp. 27–64.

——— (1964*b*): Lateral Resistance of Piles in Cohesionless Soils, *J. Soil Mech. Found. Div., ASCE*, vol. 90, SM3, May, pp. 123–156 (with extensive bibliography).

——— (1965): Design of Laterally Loaded Piles, *J. Soil Mech. Found. Div., ASCE*, vol. 91, SM3, May, pp. 79–99 (with extensive bibliography).

FEAGIN, L. B. (1937): Lateral Pile Loading Tests, *Trans. ASCE*, vol. 102, pp. 236–288 (including extensive discussion).

FRUCO AND ASSOCIATES (1964): Pile Driving and Loading Tests: Lock and Dam No. 4, Arkansas River and Tributaries, Arkansas and Oklahoma, for U.S. Army Engineer District, Little Rock, Ark., September (see also *J. Soil Mech. Found. Div., ASCE*, vol. 96, SM5, September, 1970).

HOWE, R. J. (1955): A Numerical Method for Predicting the Behavior of Laterally Loaded Piling, *Shell Oil EPR Publ.* 412, Houston, Tex.

HRENNIKOFF, A. (1950): Analysis of Pile Foundations with Batter Piles, *Trans. ASCE*, vol. 115, pp. 351–381.

MASON, HAROLD G. (1957): Field Tests on Laterally Loaded Piles in Sand and Clay, *ASTM STP* 206, pp. 133–149.

———, and J. A. BISHOP (1955): Measurement of Earth Pressure and Deflection along the Embedded Portion of a 40-ft Steel Pile, *ASTM STP* 154-A, pp. 1–22.

MATLOCK, HUDSON, and L. C. REESE (1960): Generalized Solutions for Laterally Loaded Piles, *J. Soil Mech. Found. Div., ASCE*, vol. 86, SM5, October, pp. 63–91.

MCCLELLAND, B., and J. A. FOCHT, JR. (1958): Soil Modulus for Laterally Loaded Piles, *Trans. ASCE*, vol. 123, pp. 1049–1086.

NEWMARK, N. M. (1942): Numerical Procedure for Computing Deflections, Moments, and Buckling Loads, *Proc. ASCE*, vol. 68, no. 5, May, p. 697.

PALMER, L. A., and P. P. BROWN (1955): Piles Subject to Lateral Thrust, *ASTM STP* 154-A, pp. 22–32.

———, and J. B. THOMPSON (1948): The Earth Pressure and Deflection along the Embedded Lengths of Piles Subjected to Lateral Thrust, *Proc. 2d Int. Conf. Soil Mech. Found. Eng., Rotterdam*, vol. 5, pp. 156–161.

REESE, L. C., and H. MATLOCK (1956): Non-dimensional Solutions for Laterally Loaded Piles, *Proc. 8th Tex. Conf. Soil Mech. Found. Eng., Univ. Tex., Austin*, September.

TERZAGHI, K. (1943): "Theoretical Soil Mechanics," p. 364, Wiley, New York.

TEXAS A AND M RESEARCH FOUNDATION (1952): An Investigation of Lateral Loads on a Test Pile, Test Pile Report, *Res. Found. Proj.* 31, College Station, Tex.

10

SHEET-PILE STRUCTURES

10-1 TYPES OF SHEET-PILE STRUCTURES

Sheet piling, for which typical sections[1] are shown in Fig. 10-1, is rather widely used to construct retaining structures, generally waterfront structures, where the pile-section flexibility and resulting deformation are not a major factor.

Sheet piling is also used in many excavations for temporary retaining structures. If the line of piling closes upon itself, the structures are cellular, a topic beyond the scope of this text.

The sheet-pile structure is termed a *cantilever* wall if the wall is laterally unsupported above the dredge line and an *anchored* wall (also anchored bulkhead) if lateral support is provided above the dredge line. *Braced sheeting* describes the structure formed in such a manner that bracing rather than piling embedment provides lateral stability.

Figure 10-2 illustrates typical structural configurations considered in this chapter.

[1] Appendix tables contain more complete listings of sheet-pile sections.

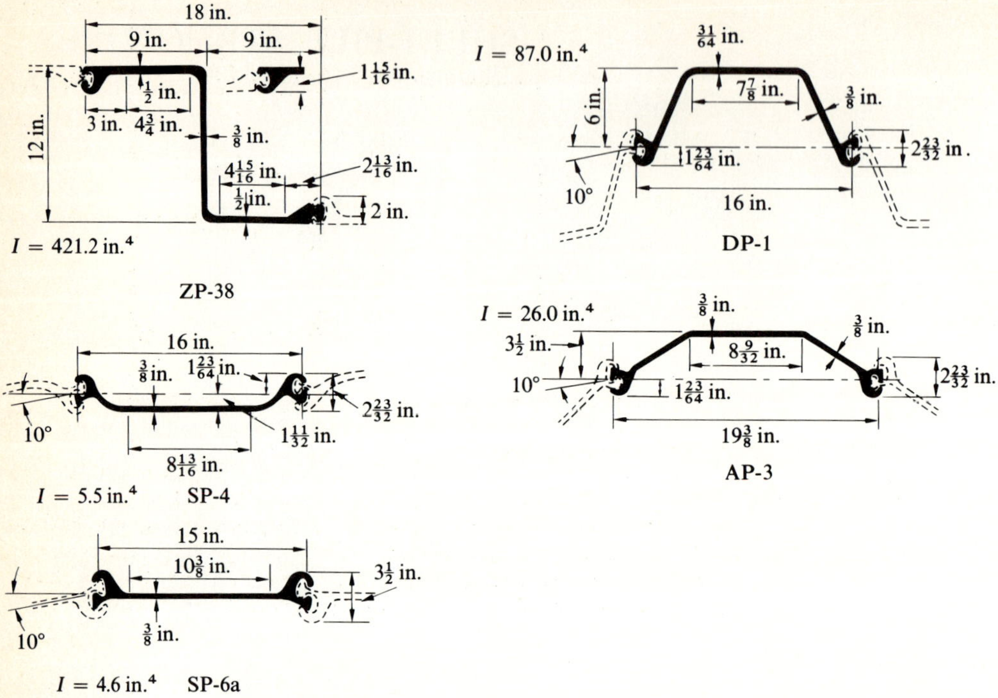

FIGURE 10-1
Typical sheet-pile sections and moments of inertia. (*Bethlehem Steel Corp.*)

10-2 DESIGN METHODS FOR SHEET-PILE WALLS

It is evident (refer to section properties in Appendix B) that with the relatively small section modulus furnished by sheet piling, cantilever walls (Fig. 10-2*a*) can be only of modest height. Heights of anchored piling can be much larger due to the lateral support provided by the anchor rod, which may be located at one or more levels.

A factor to consider where alternative walls are possible is that the cantilever wall is ready for service when the line of piling is driven, whereas the anchorage system for anchored sheet piling is an additional installation.

Several design-method alternatives for sheet-pile walls have been proposed [Ayers and Stokes (1954), Richart (1957), Turabi and Balla (1968), and Haliburton[1] (1968)] which do not seem to have been very widely accepted. Presently, most sheet-pile walls are designed on the basis of *free-earth* or *fixed-earth support*. The methods of analysis are shown in Fig. 10-3 (cantilever walls) and Fig. 10-4 (anchored walls).

[1] See also discussion by Rauhut (1969).

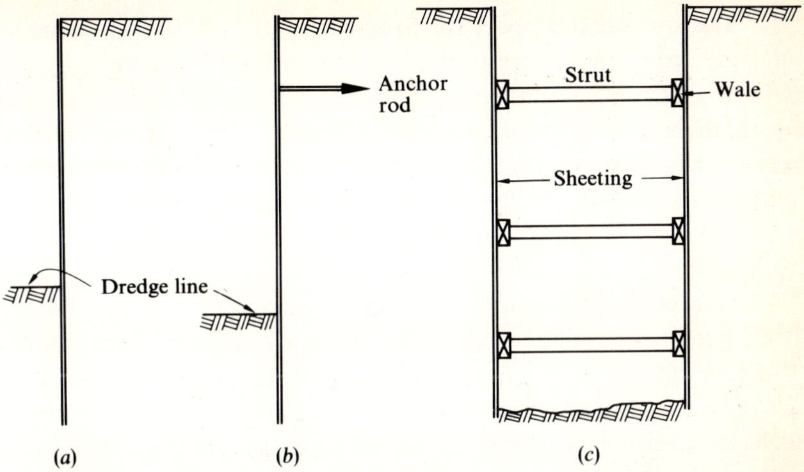

FIGURE 10-2
Sheet-pile installations. The (*a*) cantilever and (*c*) braced sheeting installations are commonly used to temporarily retain excavations for buildings, trenches, etc. (*b*) Anchored sheet-pile installations are more commonly used for waterfront structures (along with cantilevered sheet piling when wall height is low).

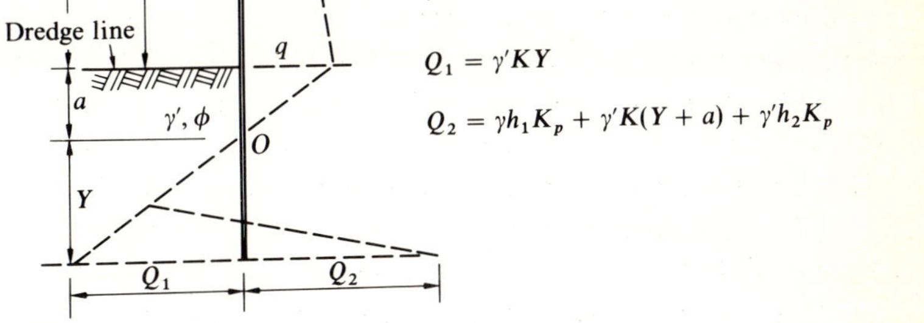

FIGURE 10-3
Assumption of earth pressure in the free-earth-support method of analysis for a cantilever sheet-pile wall in cohesionless soil. In general use Coulomb K_a and Rankine K_p

The free-earth support (or fixed-earth) of Fig. 10-4 is sometimes modified by applying reduction factors to the bending moments computed by this method. The reduction factors, which were proposed by Rowe (1952, 1957), attempt to account for (1) the reduced bending moments actually obtained from anchor-rod and dredge-line deformations and (2) the fact that the pressure (passive) resultant in front of the wall is closer to the dredge line than the distance $X/3$ from the pile base shown in Fig. 10-4. The moment-reduction concept is illustrated in Fig. 10-5 and in Example 10-3.

Conventional design practice uses the concept of active earth pressure on the back face of the wall and passive pressures developed on the soil as shown in Figs. 10-3 and 10-4. For purely cohesive ($\phi = 0$) or cohesionless soils ($c = 0$) solutions are relatively easy to obtain. Solutions are in a fourth-degree equation for cantilever walls in cohesionless soils, third-degree equations for anchored walls in cohesionless soils, and second-degree equations for both type walls for cohesive soils. For ϕ-c soils (a very common occurrence) the solutions are much more difficult to obtain by the free-earth support methods.

Safety Factors

Safety factors are commonly applied by: (1) dividing the passive-earth-pressure coefficients and soil cohesion by a safety factor or (2) arbitrarily increasing the computed embedment depth by 20 to 40 percent.

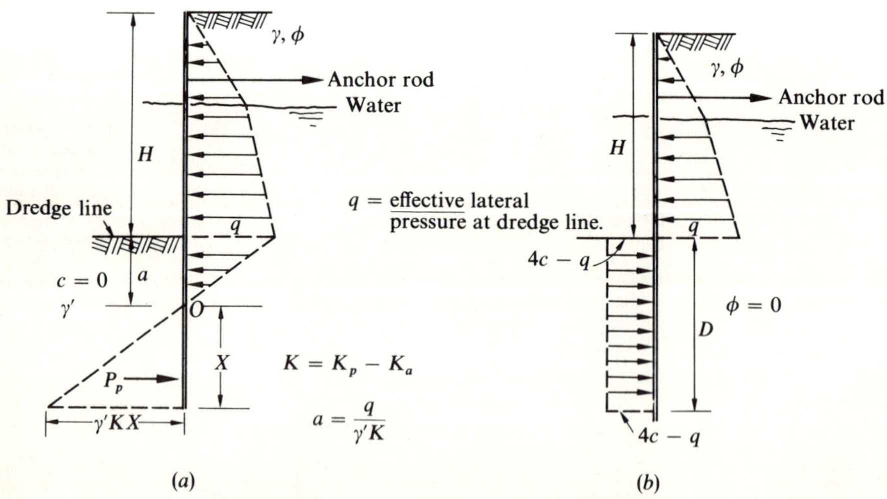

FIGURE 10-4
Soil-pressure assumptions in the free-earth method of analysis and determination of embedment depth of anchored sheet-pile walls: (*a*) cohesionless soil; (*b*) cohesive soil.

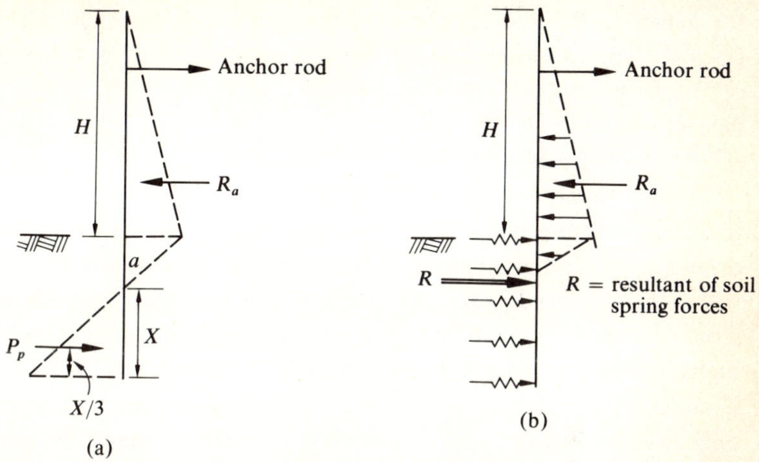

FIGURE 10-5
Moment-reduction concept of free-earth-designed sheet-pile walls explained by the finite-element method. Note that summing moments about the dredge line of free earth (*a*) will give a larger dredge-line moment than the finite element (*b*) because of the larger moment arm. Effectively, the finite-element solution moves the passive wall resistance vertically closer to the dredge line.

Safety factors commonly vary from 1.2 to 2.0 depending on the effect of a failure or large lateral earth movement. A safety factor of 1.3 to 1.5 is recommended for most sheet-pile work.

Steel Design Stresses

Design stresses may vary from $0.6F_y$, commonly used for steel structures (and implying SF = 1.67), to full guaranteed yield stress F_y. The lower values of allowable stresses should be used where uncertainty of soil parameters exist or where excessive earth deformations cannot be tolerated. If one is using a lateral-earth-pressure coefficient of 1.00, if there is little likelihood of water-level fluctuation, and if the soil unit weight is reasonably certain, a design stress of 90 to 100 percent of F_y can be used if wall deformation is not critical.

10-3 SHEET-PILE EARTH-PRESSURE COEFFICIENTS AND WALL FRICTION

Lateral earth pressure against a sheet-pile wall depends on the method of construction, backfilling, dredging, and yielding of the wall and anchor rod. Simplified design procedures have made use of the Rankine earth-pressure coefficients [Eq. (8-2)] for active earth pressure [Ayers and Stokes (1954), Richart (1957), and Anderson (1956)]. The Coulomb earth-pressure coefficient [Eq. (8-1)] is more correct because of the large wall deformations which usually occur, involving slip along the back face of the wall. In the unlikely event of very little wall deformation the Rankine earth-pressure coefficient may be acceptable.

Considerable judgment is required to obtain the angle of internal friction for the earth-pressure coefficient. Cantilever walls are mostly used for excavations by driving a line and excavating on one side. In this case the ϕ angle can be determined within reasonable limits. For anchored walls which may be part of a waterfront structure, the fill may be dredged and deposited through water or, at the least, in a highly fluid state. Friction angles may be zero or nearly so for a short period of time if materials with large amounts of silt or clay are used. The friction angle may be only 25 to 30° for sand and silty sand placed in this manner. In any case, it is important to use a reliable angle of internal friction since a 3 or 4° change in ϕ may increase the lateral pressure 10 to 25 percent.

The angle of wall friction δ will depend on the structural shape used. For straight and very shallow web sheet-pile sections the wall slip surface will develop at the interface of the two materials. When the wall consists of deep arch-web or Z-pile sections, Fig. 10-6 indicates the probable slip surface. With the slip partly soil to soil and partly soil to pile one may use an average angle δ obtained as

$$\tan^{-1} \delta_{av} = \frac{\tan \phi + \tan \delta}{2}$$

Table 10-1 lists some values of skin-friction angle δ for use in the Coulomb equation. Others have used $\delta = m\phi$, where m ranges from 0.5 to 1.00 depending on the material and engineer. The Coulomb earth-pressure coefficient K_a is not highly sensitive to a few degrees' variation in the wall-friction angle.

The Coulomb active-earth-pressure equation provides no ready means of evaluation with a cohesive-soil backfill. For cohesive soils, to make an allowance for wall

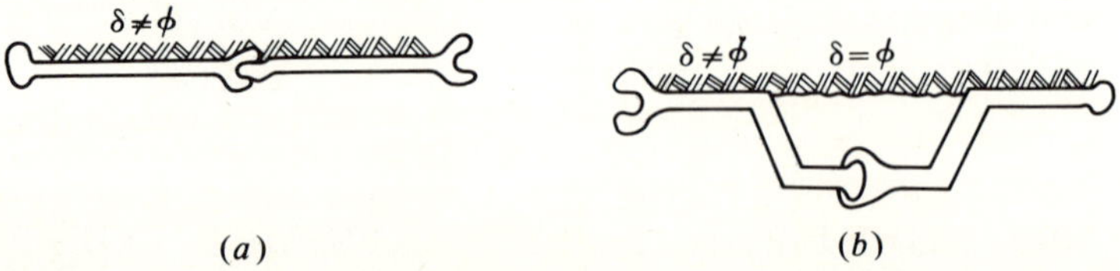

FIGURE 10-6
The shear-plane interface of soil-pile system when considering the Coulomb earth-pressure coefficient including the effect of wall friction: (*a*) straight and shallow web sheet piling; (*b*) deep arch-web and Z-sheet piling.

Table 10-1 SKIN-FRICTION COEFFICIENTS FOR SOIL AND VARIOUS CONSTRUCTION MATERIALS

ϕ = angle of internal friction; δ = friction angle of soil on material; for cohesive soil, c_a = adhesion; c = cohesion, using consistent units

Material	Surface finish	Dense sand, $0.06 < D < 2.0$*		Cohesionless silt, $0.002 < D < 0.06$*			Cohesive granular soil, 50% clay, 50% sand, I_c† = 0.5 to 1.0, w = 13 to 17%		Clay, remolded $D \leq 0.06$*, I_c† = 0.73 to 1.0, w = 22 to 26%	
					Saturated					
		Dry	Saturated	Dry, dense	Loose	Dense				
		δ/ϕ	δ/ϕ	δ/ϕ	δ/ϕ	δ/ϕ	δ/ϕ§	c_a/c	δ/ϕ§	c_a/c
Steel	Polished	0.54	0.64	0.79	0.40	0.68	0.40		0.50	0.25
	Rusted	0.76	0.80	0.95	0.48	0.75	0.65	0.35	0.50	0.50
Wood (pine)	Parallel to grain	0.76	0.85	0.92	0.55	0.87	0.30	0.20	0.60	0.4
	At right angles to grain	0.88	0.89	0.98	0.63	0.95	0.90	0.40	0.70	0.50
Concrete	Metal-formed	0.76	0.80	0.92	0.50	0.87	0.84	0.42	0.68	0.40
(1 to 3 in	Wood-formed	0.88	0.83	0.98	0.62	0.96	0.90	0.58	0.80	0.50
aggregate)	On compacted ground	0.98	0.90	1.00	0.79	1.00	0.95	0.80	0.95	0.60

* Grain size in millimeters. † I_c = consistency index. § Clay may have an angle of internal friction.

SOURCE: Potyondy (1961).

adhesion one must use the method of wedges, as shown in Fig. 10-7. Table 10-1 can be used as a guide for the ratio of cohesion to adhesion in the absence of laboratory tests. The trial wedges can be computer-programmed since the force polygon involves only two unknown vectors (directions are known), one of which is P_a. All weight and cohesion vectors are known, and by incrementing ρ in 1° intervals the maximum value of P_a can be found. Excess pore pressure in the backfill will make computation of the weight vector considerably more difficult.

10-4 DESIGN OF SHEET-PILE WALLS BY MATRIX METHODS

The matrix (finite-element) method of sheet-pile wall analysis and design is by far the most efficient means currently available and can be made to include many special analysis problems such as extra pull on the anchor rod, initial wall deflection, etc. The matrix method directly gives the moment reduction proposed by Rowe (1952, 1954, 1957) by directly considering the soil-pile interaction, the flexibility EI of the pile, and its height. The deformation of the soil at and below the dredge line is also automatically considered, as well as the anchor-rod force and deformation if the pile system is anchored. The same computer program can be used for both cantilever and anchored walls.

The modulus-of-subgrade reaction is again used in this analysis to provide a Winkler lateral support of the sheet pile below the dredge line. For a check on the reliability (and validity) of this concept, the soil pressure resulting from the computed deflections is computed and compared to see whether the soil pressures are reasonable or possible. In passing, it should be noted that the soil pressures obtained are those required for stability and are nearly independent of the value of subgrade modulus used. The final soil pressures are somewhat influenced by the flexural stiffness of the sheet piling and wall height (lateral force to be resisted). The real advantage of this method is that one can now inspect the soil pressures to see if they are reasonable, thus eliminating the uncertainty of the free-earth method, where a passive resistance is computed whether the soil can carry that resistance or not. The soil pressures using the finite-element method are larger nearer the dredge line (which is the cause of Rowe's moment reduction) than those assumed using the free-earth type of analyses.

If, in the opinion of the designer, the soil cannot carry the computed lateral pressures without a soil failure, the wall system must be changed. It may require stiffer pile sections to transfer the lateral resistance deeper, modification of soil, relocation of the anchor rod, reduction of the wall height, etc. Generally just increasing embedment depth (beyond a minimum) for a given pile section does very little to reduce the soil pressure, bending moments, or the lateral deflections. In fact, there is a

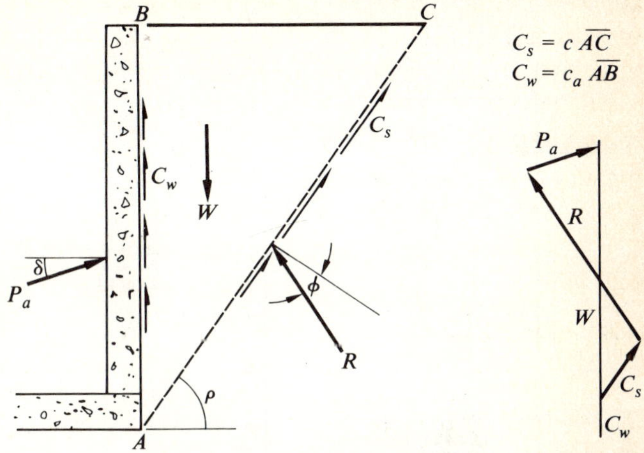

FIGURE 10-7
The use of a trial-wedge solution for backfills with cohesion. The force polygon shown can be programmed on the computer to vary the ρ angle and find the equivalent earth-pressure coefficient.

serious theoretical question of what numerical improvement in safety factor is gained by increasing the embedment beyond that for which lateral deflection and soil pressures are tolerable. Probably a 10 to 30 percent increase in embedment should be made over the required depth, however, to allow for accidental overdredging.

Referring to Fig. 10-8, essentially the same procedure as introduced in Chaps. 5 and 8 is used here. That is,

$$P = AF$$

$$e = A^T X$$

$$F = Se$$

$$X = [ASA^T]^{-1} P$$

and finally the section shear and bending moments are computed as

$$F = SA^T X$$

Coding the problem is shown on the *P-X* and *F-e* diagrams of Fig. 10-8. Note that the soil springs are the end *F*'s with the anchor rod as the very last *F* value; also therefore the last entry in the *S* matrix. This coding scheme allows easy access to the *S* matrix if one needs to modify the anchor rod or treat the soil as nonlinear. This coding also makes it easy to modify the computer program for additional anchor rods.

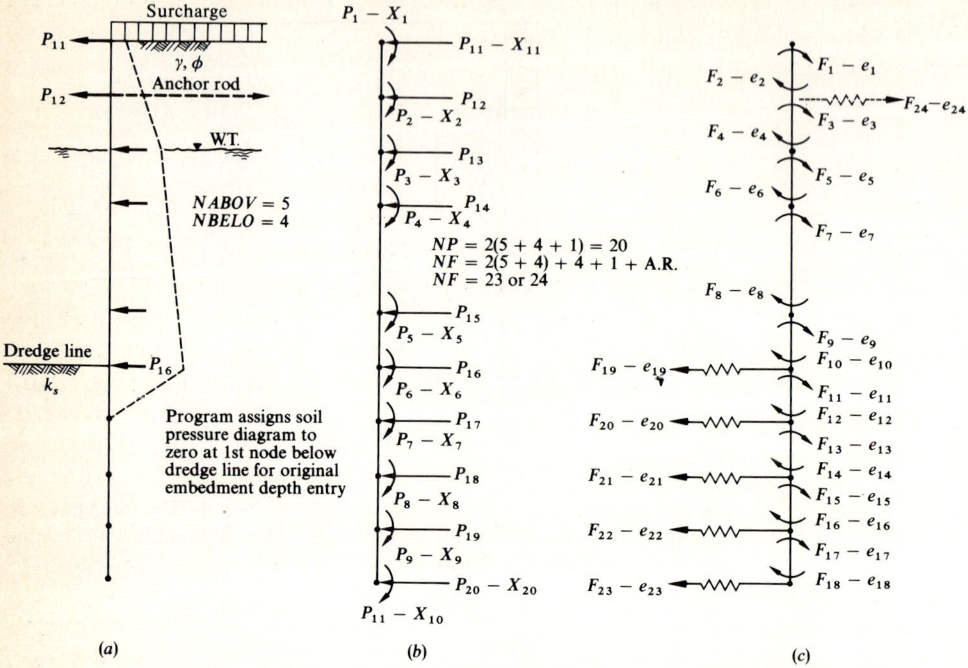

FIGURE 10-8
Coding the sheet-pile wall for the finite-element solution. We may have a surcharge q, as shown, and the wall may be cantilevered (no anchor rod) or anchored (if anchored, NF = 24 as shown). Coding location of soil "springs" and anchor rod allows easy adjustment for wall type or nonlinear soil behavior. Note that the soil "springs" are nodal forces, not element forces. The pressure diagram in (*a*) is converted to equivalent lateral forces as *P*-matrix entries with subscripts shown.

By coding the anchor rod as the last S-matrix entry, the same computer program can be used for designing a cantilever wall by reducing the S matrix by the anchor-rod entry and not entering an anchor-rod value in the A matrix.

10-5 EXAMPLES

This method of analysis will be illustrated by several examples. Example 10-1 illustrates the coding and partial output for a cantilever wall. Example 10-2 illustrates the conventional analysis of an anchored sheet-pile wall to obtain an embedment depth to use in Example 10-3, where the finite-element method is used and compared to Rowe's moment reduction. The sensitivity of the solution to EI and k_s is also illustrated.

EXAMPLE 10-1 Analyze the cantilever sheet-pile wall shown in Fig. E10-1.1 (first page of computer output). Also show a partial check of the output. Note in the checking of results that $F_1 = 0.00$ within computer roundoff; likewise, $F_2 = -F_3$, etc. *Use metric units.*

SOLUTION For the wall shown it is assumed $k_s = 25 + 0.0Z^1$ (kcf). One can determine this value using the procedure of Example 10-3 or methods given in Chap. 2. To reduce deflection (max $\Delta = 1.50$ in) we will use an MZ38 section with a tabulated I of 421.1 in^4 for an 18-in width. The modulus of elasticity is 29,600 ksi. Wall and soil properties are shown on Fig. E10-1.1. Converting to metric units (use Table 2-8) gives

$$I = 421.1 \frac{3.283}{1.5}(41.62)10^{-8} = 0.0003837 \text{ m}^4 \qquad \text{(per meter of wall width)}$$

$$\text{XMAX} = 1.5(2.54) = 3.81 \text{ cm}$$

$$H = 15(0.3048) = 4.572 \text{ m} = \text{depth to water} \qquad \text{Segment length} = 0.910 \text{ m}$$

$$\text{Embedment depth} = 9(0.3048) = 2.743 \text{ m} \qquad \text{Segment length} = 0.550 \text{ m}$$

$$\gamma_{sat} = 132.5(0.15709) = 20.81 \text{ kN/cu m} \qquad 112.0(0.15709) = 17.59 \text{ kN/cu m}$$

$$k_s = 25(157.09) = 3{,}927.5 \text{ kN/cu m}$$

$$E = 29.600(144)(47.882) = 204{,}084{,}000 \text{ kN/sq m}$$

There will be no external *S*-matrix entries (JJS = 0), and we will cycle as required (KSTOP = 0) up to five cycles. The input data cards are as follows:

Card	Data
1	TITLE (see Fig. E10-1.1)
2	UNITS (UT1 – UT6, FU1 – FU4)
	UT1 = M, UT2 = CM, etc., FU1 = 100., FU2 = .3, FU3 = 10., FU4 = 9.807
	NABOV NBELO JJS KSTOP NCYCC
3	5 5 0 0 5
	HWALL HROD ERN ELAS DEMB FAC
4	4.572 0.0 3.837×10^{-4} 20.4×10^{7} 2.743 1.0
	HWAT GSAT GWET PHI DELTA SCHGE XMAX ARODK
	(Note that 10^{-4} and 10^{7} above are not punched on data cards)
5	4.572 20.81 17.59 32. 25. 0.0 3.81 0.0
6	3927.5 0 1.0
7–8	Pile segments H(I), I = 1,MM1 MM1 = 10
	.910 .910 .910 .910 .910 .550 .550 .550
	.550 .550
	NODWAT NODAR (nodes locating water and anchor rod)
9	6 0

These nine cards represent the problem-data input. In this example the computer incremented the embedment depth from 2.743 to 3.34 m. The computer always increments once (0.6 m for cantilever and 0.3 m for anchored sheet piles). Both the 2.743- and 3.34-m embedment-depth output is given so that the user may select the depth most suitable. Note on Fig. E10-1.2 that the top lateral deflection is -0.122 m (12) and the dredge line (17) is $-0.0357 < 0.0381$ m. The 0.6-m increase in depth reduced the dredge-line deflection from 0.0465 to 0.0357 m. Not all the computer output is shown. Also see Fig. 10-1.3.

```
J E BOWLES   EXAMPLE 10-1 <<CANTILEVER SHEET PILE  MZ-38 SECTION>>  ** METRIC

  NO SEGS:    ABOVE D.L. =  5
              BELOW C.L. =  5
  NO NODES REQ. S-MATRIX CORRECT =  0

  WALL HT. ABOVE CREDGE LINE =  4.572 M
  DEPTH GRCUND LINE TC ANC. ROD = 0.0    M
  MOMENT OF INERTIA OF PILE = 0.38370E-03 M **4
  MOCULUS OF ELAST =204084000. KN/SQ M
  INIT. ASSUMED EMBEC DEPTH =  2.743 M
  PRESS RED FAC =1.00

  DEPTH TC WATER= 4.572 M
  SAT. UNIT WT OF SOIL = 20.8000 KN/CU M
  WET UNIT WT OF SOIL = 17.5950 KN/CU M
  ANGLE CF INTERNAL FRICT. = 32.00 DEG.
  MAX SOIL DEFL = 3.81 CM
  SURCHARGE PRESS =    C.0    KN/SQ M    ANGLE WALL FRIC =  25.00

     SCIL MCDULUS:   KS = 3927.50 +    0.0 *Z**1.000

  THE SEGMENT NOS ANC LENGTHS ARE:
   1  C.910    2  0.910    3  0.910    4  0.910    5  0.910
   6  0.550    7  C.550    8  0.550    9  0.550   10  0.550

  NO ANCHOR ROD USED*****
  WATER LEVEL LOCATED AT NODE  6

  NP = 22  NF = 26  NFM = 20  NEND = 26  NDEL =  4  KABCV = 17
  NPCL = 17  NABP1 =  6  M = 11  MM1 = 10  MP1 = 12
```

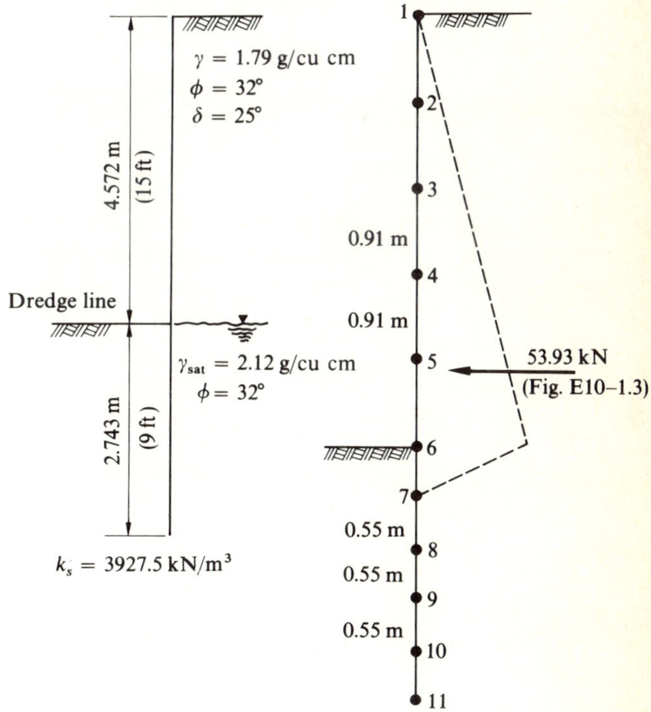

FIGURE E10-1.1
General input data written back as an input check. The lower two lines are computation counters. A sketch of conditions has been placed on the sheet for checking convenience.

```
THE S-MATRIX IN 2-COLS--INCL F-NUMBER
 1   344206.50   172103.19   14   284752.63   569505.38
 2   172103.19   344206.50   15   569505.38   284752.63
 3   344206.50   172103.19   16   284752.63   569505.38
 4   172103.19   344206.50   17   569505.38   284752.63
 5   344206.50   172103.19   18   284752.63   569505.38
 6   172103.19   344206.50   19   569505.38   284752.63
 7   344206.50   172103.19   20   284752.63   569505.38
 8   172103.19   344206.50   21      656.52         0.0
 9   344206.50   172103.19   22     1969.56         0.0
10   172103.19   344206.50   23     2626.08         0.0
11   569505.38   284752.63   24     2626.08         0.0
12   284752.63   569505.38   25     2626.08         0.0
13   569505.38   284752.63   26     1313.04         0.0

THE TOTAL LOAD MATRIX    THE JOINT DEFL               THE FORCE MATRIX
KN    OR KN-M IS         M   OR RADS ARE                 KN    OR KN-M

LOAD DIR.  1  0.0        DIR =  1  -0.0192689     MOMENT  1    -0.0352
LOAD DIR.  2  0.0        DIR =  2  -0.0192655     MOMENT  2     0.5469
LOAD DIR.  3  0.0        DIR =  3  -0.0192324     MOMENT  3    -0.7773
LOAD DIR.  4  0.0        DIR =  4  -0.0191054     MOMENT  4     4.9063
LOAD DIR.  5  0.0        DIR =  5  -0.0187793     MOMENT  5    -5.1484
LOAD DIR.  6  0.0        DIR =  6  -0.0181072     MOMENT  6    16.7227
LOAD DIR.  7  0.0        DIR =  7  -0.0174995     MOMENT  7   -16.8125
LOAD DIR.  8  0.0        DIR =  8  -0.0168594     MOMENT  8    39.3047
LOAD DIR.  9  0.0        DIR =  9  -0.0163735     MOMENT  9   -39.3633
LOAD DIR. 10  0.0        DIR = 10  -0.0161253     MOMENT 10    76.3086
LOAD DIR. 11  0.0        DIR = 11  -0.0160624     MOMENT 11   -76.3398
LOAD DIR. 12  0.6666     DIR = 12   0.1223034     MOMENT 12    96.7188
LOAD DIR. 13  3.9995     DIR = 13   0.1047698     MOMENT 13   -96.5352
LOAD DIR. 14  7.9990     DIR = 14   0.0872496     MOMENT 14    85.7109
LOAD DIR. 15 11.9984     DIR = 15   0.0697957     MOMENT 15   -85.4873
LOAD DIR. 16 15.9979     DIR = 16   0.0525383     MOMENT 16    52.8877
LOAD DIR. 17 13.3609     DIR = 17   0.0357224     MOMENT 17   -52.7305
LOAD DIR. 18  0.0        DIR = 18   0.0238104     MOMENT 18    17.9453
LOAD DIR. 19  0.0        DIR = 19   0.0123278     MOMENT 19   -17.8867
LOAD DIR. 20  0.0        DIR = 20   0.0012301     MOMENT 20     0.0313
LOAD DIR. 21  0.0        DIR = 21  -0.0096213   **FORCE 21   -23.4525
LOAD DIR. 22  0.0        DIR = 22  -0.0203753   **FORCE 22   -46.8960
                                                **FORCE 23   -32.3737
                                                **FORCE 24    -3.2303
                                                **FORCE 25    25.2663
                                                **FORCE 26    26.7536
```

FIGURE E10-1.2

General computer computations for checking. Note that the soil pressure converted to nodal forces is shown, as are the nodal rotations and deflections. Each F force is shown; note those which should be zero or equal and opposite.

```
SHEAR IN EACH   NODAL BEND. MOMENT     SOIL FORCE AND PRESS
 NODE,KN        NODE       KN-M          KN
 1     0.6666     1     -0.0352
 2     4.6661     2     -0.7773
 3    12.6650     3     -5.1484
 4    24.6634     4    -16.8125
 5    40.6614     5    -39.3633
 6    30.5698     6    -76.3398       6   23.4525       140.300 KN/SQ M
 7   -16.3262     7    -96.5352       7   46.8960        93.515 KN/SQ M
 8   -48.7000     8    -85.4873       8   32.3737        48.417 KN/SQ M
 9   -51.9303     9    -52.7305       9    3.2303         4.831 KN/SQ M
10   -26.6640    10    -17.8867      10  -25.2663       -37.788 KN/SQ M
11     0.0895    11      0.0313      11  -26.7536       -80.024 KN/SQ M
   SUM SOIL REACT INCL ANCH. ROD =        53.9327( 54.0223)
```

Actual soil pressure

Checks ΣF_H

```
*****FINAL EMBEDMENT LENGTH OF SHEET PILE =  3.34 M
     OUTPUT ABOVE FOR THIS EMBEDMENT VALUE
```

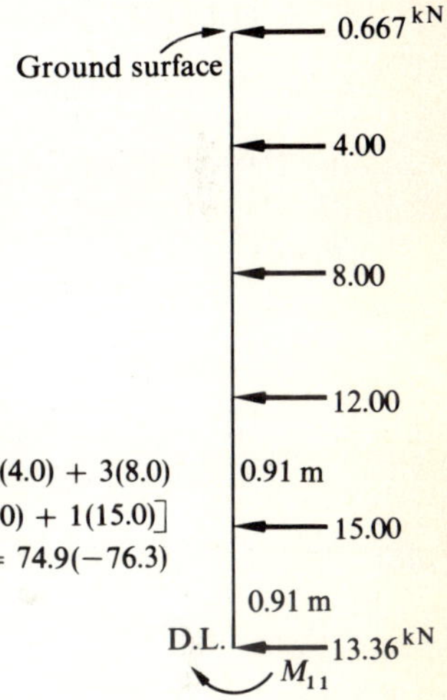

Checking $\Sigma M_{\text{d.l.}}$

$$M_{11} = 0.91[5(0.667) + 4(4.0) + 3(8.0) + 2(12.0) + 1(15.0)]$$
$$= 0.91(82.33) = 74.9(-76.3)$$

FIGURE E10-1.3
Final output identified as shown. Note that embedment depth has been in-increased 0.6 m. Partial static check of moment at dredge line. Program checks $\Sigma F_h = 0$. Soil pressure shown is nearly independent of the modulus-of-subgrade reaction. This problem was run in single precision, and small roundoff errors are present.

////

EXAMPLE 10-2 Determine the depth of embedment, the anchor-rod force, and maximum bending moment in the anchored sheet-pile wall shown (1-ft strip) in Fig. E10-2.1.

SOLUTION The solution will be based on the free-earth-support method. Pressure diagrams and critical dimensions are shown in Fig. E10-2.2. Coulomb active and Rankine passive pressure coefficients will be used.

Find wall pressures and distance a:

$$K_a = 0.2645 \qquad K_p = 3.3921$$

$$\sigma_1 = 6(0.112)(0.2645) = 0.178 \text{ ksf}$$

$$\sigma_2 = \sigma_1 + 14(0.066)(0.2645) = 0.422 \text{ ksf}$$

$$K = K_p - K_a = 3.127$$

$$a = \frac{\sigma_2}{\gamma' K} = \frac{0.422}{0.066(3.127)} = 2.045 \text{ ft}$$

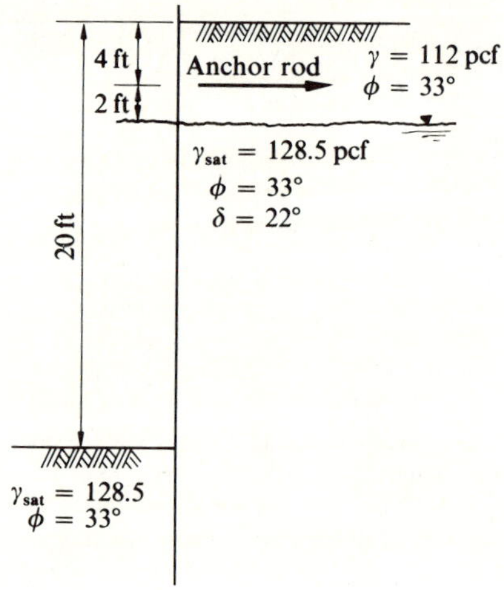

FIGURE E10-2.1

Find R_a and $\bar{y}$

$$R_a = 3(0.178) + 7(0.178 + 0.422) + 0.422(1.022) = 5.165 \text{ kips}$$

Summing moments about 0, we find

$$\bar{y} = 9.482 \text{ ft}$$

Find X as $\sum M_{ar} = 0$:

$$y'R_p - \bar{y}R_a = 0$$

$$\gamma' \frac{KX^2}{2}(18.045 + 0.67X) - 9.482(5.165) = 0$$

$$0.0688X^3 + 1.8621X^2 - 48.974 = 0$$

Solving, we have

$$X \cong 4.80 \text{ ft} \qquad D = X + a = 4.80 + 2.05 = 6.85 \text{ ft}$$

Find anchor-rod force:

$$R_p = 0.066(3.127)(4.80)^2(\tfrac{1}{2}) = 2.38 \text{ kips}$$

$$\sum F_H = 0$$

$$F_a + R_p - R_a = 0$$

$$F_a = 5.165 - 2.38 = 2.78 \text{ kips}$$

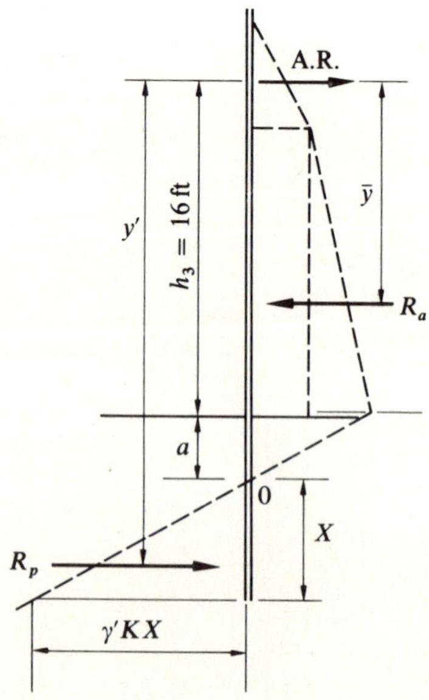

FIGURE E10-2.2

This completes computations of critical wall dimensions for SF = 1. The following is a summary for several safety factors:

SF	X, ft	a, ft	D, ft	F_a, kips
1	4.80	2.05	6.85	2.78
1.2	5.30	2.50	7.80	2.88
1.3	5.50	2.73	8.23	2.97
1.4	5.70	2.96	8.66	3.04

Find the maximum moment for SF = 1.3 (see Fig. E10-2.3). Find location of zero shear ($\sum F_h = 0$):

$$0.534 - 2.97 + 0.178y + 0.066(0.2645)y\,\frac{y}{2} = 0$$

$$0.00873y^2 + 0.178y = 2.436$$

$$y = 9.375 \text{ ft below W.T.}$$

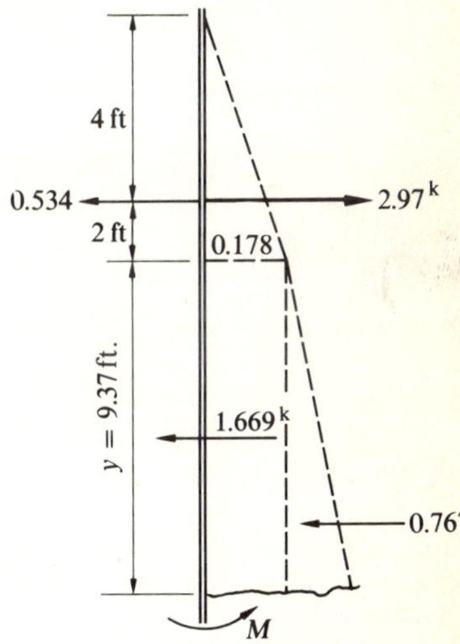

FIGURE E10-2.3

The maximum bending moment is

$$M = 0.534(15.375 - 4) - 2.97(11.375) + 1.669\,\frac{9.375}{2} + 0.767\,\frac{9.375}{3}$$

$$= -27.71 + 7.828 + 2.40$$

$$= -17.482 \text{ ft-kips}$$

////

EXAMPLE 10-3 Analyze and compare the finite-element output with the bending moment obtained in Example 10-2. Also compare with Rowe's moment-reduction concept. Rowe's moment-reduction curves are in Bowles (1968) and Rowe (1952, 1954, 1957). Show partial problem statics check on computer output sheets.

SOLUTION (using a MP110 sheet-pile section). Take $D = 8$ ft as average of SF = 1.2 and 1.3 of Example 10-2. Take anchor rod as 1.825-in-diam cable 20 ft long and spaced 16 ft on centers.

$$K_{ar} = \frac{AE}{L} = \frac{0.7854(1.825)^2(29{,}600)}{20(16)} = 242.0 \text{ kips/ft}$$

The soil modulus can be estimated as follows:

$$q_{ult} = qN_q + \tfrac{1}{2}\gamma BN_\gamma$$

$$= 0.066(26.1)Z + \tfrac{1}{2}(0.066)(1)(29.3)$$

$$= 1.7Z + 1$$

$$k = \frac{q}{\Delta} = 12q$$

but double for soil on both sides

$$k_s = 24 + 1.7(24)Z \rightarrow 24 + 40Z^1$$

Also try

$$k_s = 12 + 40Z^1 \qquad \text{and} \qquad k_s = 48 + 40Z^1$$

The problem is entered on nine data cards, as follows (refer to Fig. E10-3.1 for nodes, elements, etc.):

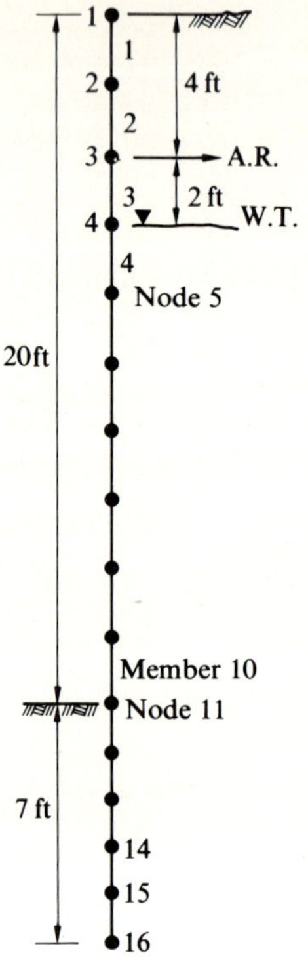

$NP = 2(10 + 5 + 1) = 32$

$NF = 2(10 + 5) + 6 + 1 = 37$

FIGURE E10-3.1

Card	Data
1	TITLE
2	UNITS (in this example fps units) UT1–UT6, FU1 = 12., FU2 = 1., FU3 = 144., FU4 = .0625
3	NABOV NBELO JJS KSTOP NCYCL 10 5 0 0 5
4	HWALL HROD ERN ELAS DEMB FAC 20. 4.0 .00315 4262400. 7.0* 1.
5	HWAT GSAT GWET PHI DELTA SCHGE XMAX ARODK 6.0 .1285 .112 33. 22. 0.0 1.5 242.0
6	12. 40. 1.00
7–8	Pile-segment lengths H(I), I = 1,MM1 MM1 = 15 2.0 2.0 2.0 2.0 2.0 2.0 2.0 2.0 2.0 2.0 1.4 1.4 1.4 1.4 1.4
9	4 3 (water node and anchor-rod node)

* Note that embedment depth is initially 7.0 ft since the computer program automatically increments depth of anchored walls by 1 ft.

Partial output and output checks are illustrated in Figs. E10-3.2, E10-3.3, and E10-3.4 following.

```
J E BOWLES    EXAMPLE 10-3 <<ANCHOR SHEET PILE OF EXAMPLE 10-2  MP110>>

   NO SEGS:   ABOVE D.L. = 10
              BELOW D.L. =  5
   NO NODES REQ. S-MATRIX CORRECT =  0

   WALL HT. ABOVE DREDGE LINE = 20.000 FT
   DEPTH GROUND LINE TO ANC. ROD = 4.000 FT
   MOMENT OF INERTIA OF PILE = 0.31500E-02 FT**4
   MODULUS OF ELAST =  4262400. K/SQ FT
   INIT. ASSUMED EMBED DEPTH =   7.000 FT
   PRESS RED FAC =1.00

   DEPTH TO WATER= 6.000 FT
   SAT. UNIT WT OF SOIL =   0.1285 K/CU FT
   WET UNIT WT OF SOIL =   0.1120 K/CU FT
   ANGLE OF INTERNAL FRICT. = 33.00 DEG.
   MAX SOIL DEFL = 1.50 IN
   SURCHARGE PRESS =    0.0    K/SQ FT    ANGLE WALL FRIC =   22.00

       SOIL MODULUS:   KS =    24.00 +   40.00*Z**1.000

   THE SEGMENT NOS AND LENGTHS ARE:
    1   2.000     2   2.000     3   2.000     4   2.000     5   2.000
    6   2.000     7   2.000     8   2.000     9   2.000    10   2.000
   11   1.400    12   1.400    13   1.400    14   1.400    15   1.400


   ANCHOR ROD LOCATED AT NODE   3
     WATER LEVEL LOCATED AT NODE   4


   NP = 32   NF = 37   NFM = 30   NEND = 36   NDEL =   4   KABOV = 27
   NPDL = 27   NABP1 = 11   M = 16   MM1 = 15   MP1 = 17
```

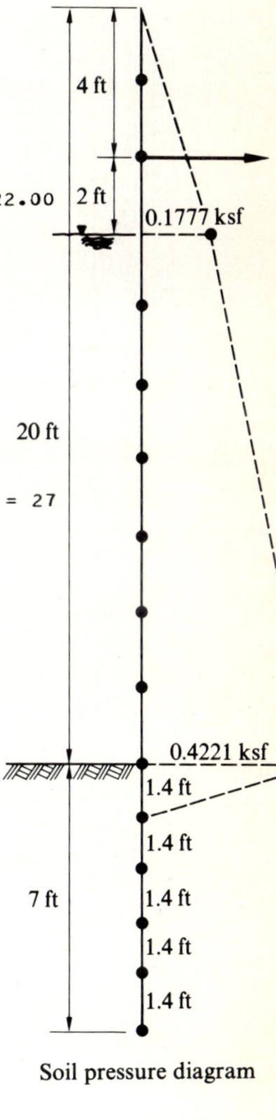

FIGURE E10-3.2

Soil pressure diagram

THE TOTAL LOAD MATRIX KIPS OR FT-K IS		THE JOINT DEFL FT OR RADS ARE		THE FORCE MATRIX KIPS OR FT-K	
LOAD DIR. 1	0.0	DIR = 1	0.0039698	MOMENT 1	-0.0011
LOAD DIR. 2	0.0	DIR = 2	0.0039728	MOMENT 2	0.0387
LOAD DIR. 3	0.0	DIR = 3	0.0039993	MOMENT 3	-0.0376
LOAD DIR. 4	0.0	DIR = 4	0.0037780	MOMENT 4	0.3192
LOAD DIR. 5	0.0	DIR = 5	0.0030712	MOMENT 5	-0.3176
LOAD DIR. 6	0.0	DIR = 6	0.0019942	MOMENT 6	-3.2896
LOAD DIR. 7	0.0	DIR = 7	0.0006841	MOMENT 7	3.2893
LOAD DIR. 8	0.0	DIR = 8	-0.0007009	MOMENT 8	-6.2004
LOAD DIR. 9	0.0	DIR = 9	-0.0019819	MOMENT 9	6.2017
LOAD DIR. 10	0.0	DIR = 10	-0.0029593	MOMENT 10	-8.2593
LOAD DIR. 11	0.0	DIR = 11	-0.0034126	MOMENT 11	8.2605
LOAD DIR. 12	0.0	DIR = 12	-0.0033026	MOMENT 12	-9.3289
LOAD DIR. 13	0.0	DIR = 13	-0.0028908	MOMENT 13	9.3293
LOAD DIR. 14	0.0	DIR = 14	-0.0024239	MOMENT 14	-9.2661
LOAD DIR. 15	0.0	DIR = 15	-0.0021224	MOMENT 15	9.2668
LOAD DIR. 16	0.0	DIR = 16	-0.0020326	MOMENT 16	-7.9331
LOAD DIR. 17	0.0197	DIR = 17	-0.0069091	MOMENT 17	7.9341
LOAD DIR. 18	0.1185	DIR = 18	0.0010325	MOMENT 18	-5.1895
LOAD DIR. 19	0.2370	DIR = 19	0.0089976	MOMENT 19	5.1904
LOAD DIR. 20	0.3473	DIR = 20	0.0168645	MOMENT 20	-0.8958
LOAD DIR. 21	0.4253	DIR = 21	0.0237860	MOMENT 21	0.8965
LOAD DIR. 22	0.4951	DIR = 22	0.0289024	MOMENT 22	3.0071
LOAD DIR. 23	0.5649	DIR = 23	0.0316072	MOMENT 23	-3.0070
LOAD DIR. 24	0.6347	DIR = 24	0.0315888	MOMENT 24	4.8918
LOAD DIR. 25	0.7045	DIR = 25	0.0288730	MOMENT 25	-4.8921
LOAD DIR. 26	0.7743	DIR = 26	0.0238636	MOMENT 26	4.0627
LOAD DIR. 27	0.6074	DIR = 27	0.0173851	MOMENT 27	-4.0632
LOAD DIR. 28	0.0	DIR = 28	0.0119586	MOMENT 28	1.7209
LOAD DIR. 29	0.0	DIR = 29	0.0069777	MOMENT 29	-1.7210
LOAD DIR. 30	0.0	DIR = 30	0.0027375	MOMENT 30	-0.0001
LOAD DIR. 31	0.0	DIR = 31	-0.0008669	**FORCE 31	-0.3152
LOAD DIR. 32	0.0	DIR = 32	-0.0041670	**FORCE 32	-1.2628
				**FORCE 33	-1.6970
				**FORCE 34	-0.9461
				**FORCE 35	0.3884
				**FORCE 36	1.0756
				ANCHOR ROD FORCE 37	2.1792

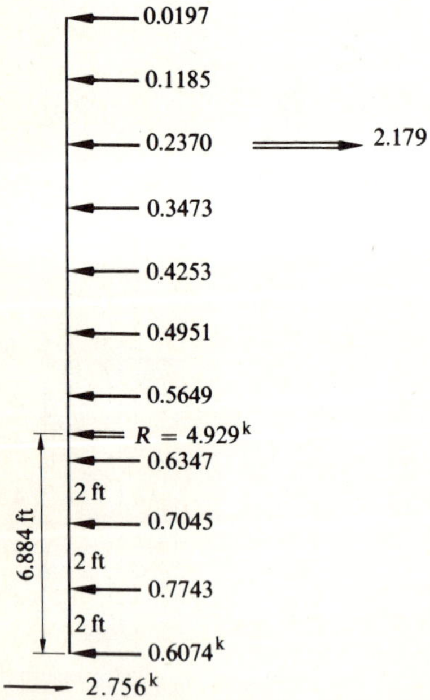

Check $\Sigma F_H = 0$

$$\Sigma F_H = 2.179 + 2.756 - 4.929$$
$$= 4.935 - 4.929 \cong 0$$

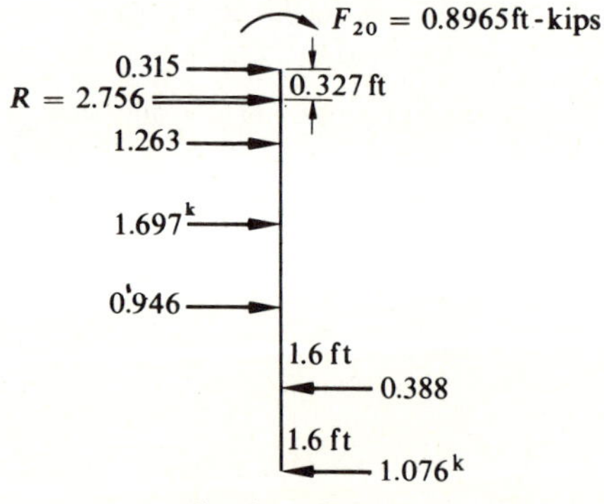

Statics check

FIGURE E10-3.3

```
SHEAR IN EACH  NODAL BEND. MOMENT    SOIL FORCE AND PRESS
NODE,KIPS       NODE       FT-K          KIPS
1     0.0197      1    -0.0011
2     0.1382      2    -0.0376
3    -1.8040      3    -0.3176
4    -1.4567      4     3.2893
5    -1.0315      5     6.2017
6    -0.5364      6     8.2605
7     0.0285      7     9.3293
8     0.6632      8     9.2668
9     1.3677      9     7.9341
10    2.1420     10     5.1904
11    2.4342     11     0.8965     11    0.3152       0.417 K/SQ FT
12    1.1714     12    -3.0070     12    1.2628       1.052 K/SQ FT
13   -0.5256     13    -4.8921     13    1.6970       1.061 K/SQ FT
14   -1.4717     14    -4.0632     14    0.9461       0.591 K/SQ FT
15   -1.0833     15    -1.7210     15   -0.3884      -0.243 K/SQ FT
16   -0.0077     16    -0.0001     16   -1.0756      -1.433 K/SQ FT
   SUM SOIL REACT INCL ANCH. ROD =       4.9363(    4.9287)

*****FINAL EMBEDMENT LENGTH OF SHEET PILE =  8.00 FT
      OUTPUT ABOVE FOR THIS EMBEDMENT VALUE
```

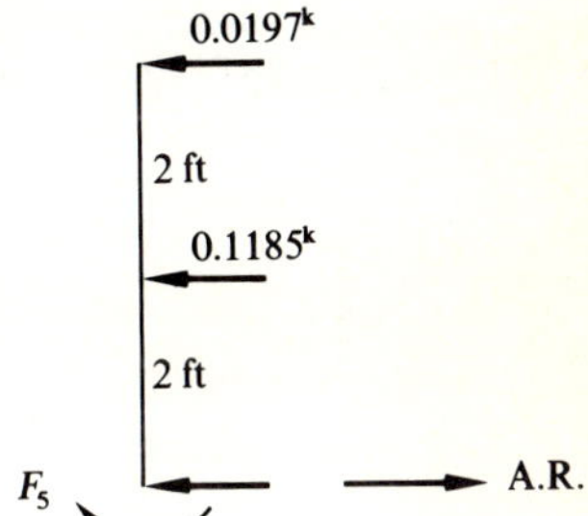

$F_5 = 4(0.0197) - 2(0.1185) = 0$

$F_5 = 0.3158\ (-0.3176)$

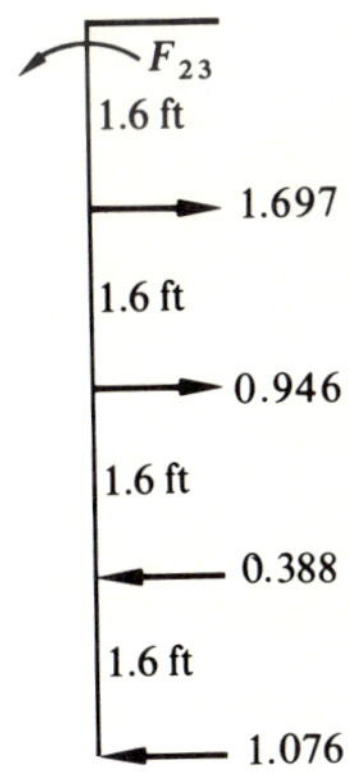

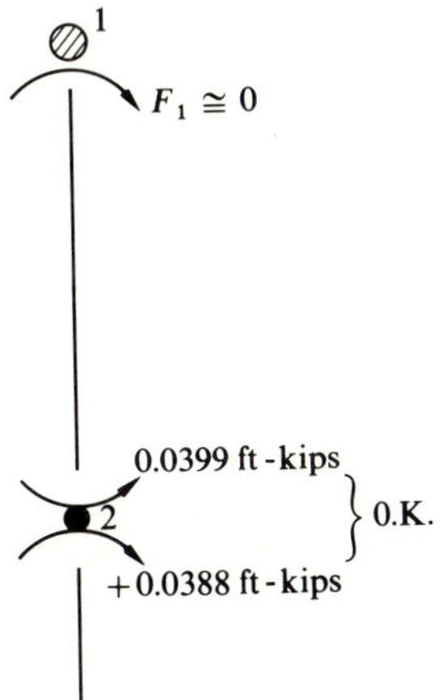

Check moments at nodes 1 and 2

$F_{23} + 1.6(1.697) + 3.2(0.946)$

$-4.8(0.388) - 6.4(1.076) = 0$

$F_{23} = -2.715 - 3.027 + 1.862 + 6.886$

$F_{23} = 3.0128\ (+3.007)$

FIGURE E10-3.4

A tabulation of computations follows (based on 8-ft final embedment).

k_s	Section	Force in anchor rod, kips	Bending moment, ft-kips at node		Max soil pressure	
			At 7 ft	At 8 ft	ksf	Node
$12 + 40Z^1$	MP110	2.20	9.55	9.55	1.461	16
	MP116	2.12	8.87	8.70	1.589	16
	MZ27	2.41	11.17	11.56	0.910	16
$24 + 40Z^1$	MP110	2.14	9.33	9.27	1.433	16
	MP116	2.10	8.67	8.45	1.528	16
	MZ27	2.38	10.92	11.25	0.937	16
$48 + 40Z^1$	MP110	2.10	8.97	8.81	1.383	16
	MP116	2.06	8.36	8.07	1.423	16
	MZ27	2.33	10.49	10.72	0.979	16

This tabulation illustrates that the solution is relatively insensitive to k_s in terms of either bending moment or soil pressure.

Check Rowe's moment-reduction method:

$$\rho = \frac{H^4}{EI} \qquad H \text{ in feet, } EI \text{ in psi}$$

For MP116

$$I = 39.75 \text{ in}^4 \qquad \text{(per foot of wall)}$$

$$\rho = \frac{28^4}{29.6 \times 10^6 I} = \frac{0.02076}{I} = 0.000522$$

$$\log \rho = -3.282$$

For MP110

$$I = 65.4 \text{ in}^4 \qquad \rho = 0.000317$$

$$\log \rho = -3.498$$

For MZ27

$$I = 184.20 \qquad \rho = 0.000113$$

$$\log \rho = -3.948$$

$$\alpha = \frac{H_{\mathrm{dL}}}{H} = \frac{20}{28} = 0.71 \qquad \beta = \frac{H_{\mathrm{ar}}}{H} = \frac{4}{28} = 0.14$$

From moment-reduction curves at α, β, $\log \rho$ one obtains M/M_0 = reduction factor as follows:

Sheet pile	Loose sand	Dense sand	Average	Finite element*
MP116	0.63	0.47	0.55	0.51†
MP110	1.10	0.58	0.84	0.55
MZ27	>1	0.78	0.90	0.66

* Using $k_s = 12 + 40Z^1$.
† Value found as 8.87/17.48, etc.

////

10-6 VALIDITY OF THE MATRIX SOLUTION AND GENERAL COMMENTS

The validity of the solution has been checked using results reported by Rowe (1952) in Fig. 10-9. Data shown on the figure should enable the user to verify the data also. Since Rowe provided stress-strain data on the soil used in the tests, Eq. (9-2) was used to establish the k_s values shown on Fig. 10-9.

This method of solution was further compared to selected test data from Tschebotarioff's (1949*b*) Princeton University tests. Some of these data were also made available by Tschebotarioff (1948, 1949*a*). Typical tests and comparisons are shown in Fig. 10-10*a*.

A full-scale field-test comparison [Matich et al. (1964)] is shown in Fig. 10-10*b*.

The matrix method should use enough pile segments both above and below the dredge line to reasonably define the elastic curve of the pile. This can generally be accomplished using five to eight segments above the dredge line and four to six segments below, depending on the location of the anchor rod.

Most analytical methods do not consider the wall construction method or sequence. Matching measured values with an analytical solution is very difficult except for Rowe's model tests. Since Rowe filled both sides of the wall equally with dry sand, then excavated one side, the backfill method and sequence of attaching the anchor rod were not problem parameters. Tschebotarioff's tests were far more realistic of on-site construction since they included both saturated and wet soils, anchor-rod sequence, and backfilling operation. As a consequence it is considerably more difficult to obtain an analytical comparison.

The field test (Fig. 10-10*b*) is even more difficult to match since the backfill varied and the backfill operation undoubtedly influenced the measured stresses. It should be noted, however, that in both the Tschebotarioff tests and the field test the matrix method satisfactorily computes bending moments, although in the field test the location of maximum moment differs somewhat from the analytical location.

Sheet pile data: *Sheet pile (no anchor yield) Fig. 7 of Rowe reference.*

Moment of inertia,

$I = 0.11\ \text{in}^4$

Mod. of elasticity,

$E = 29{,}000$ ksi

Unit weight of soil:

$\gamma_{sat} = $ —

$\gamma_{wet} = 0.090$ kcf

Angle of internal friction,

$\phi = 30°$

Mod.-of-subgrade reaction,

$k_s = 40$ kcf

Note: Pressure reduction factor = 0.8.
Assume anchor rod K = 600 kips/ft.

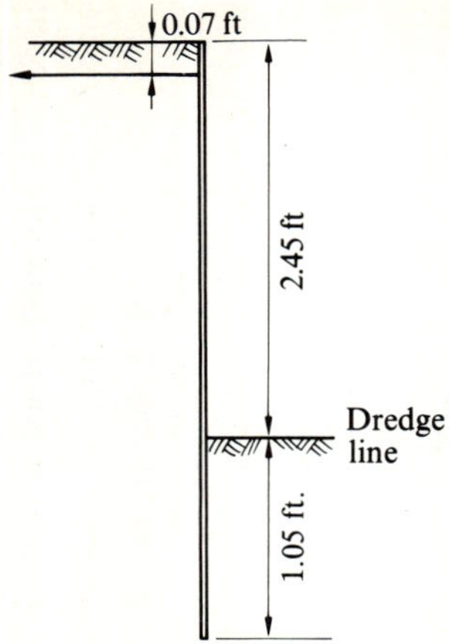

(*a*)

Dist. from top, ft	Node no.	Bending moment, ft-kips	
		Computed	Measured
0	1	0	0
0.07	2	0	0
0.408	3	0.0109	0.009
0.816	4	0.0223	0.020
1.22	5	0.0301	0.028
1.63	6	0.0328	0.033
2.04	7	0.0297	0.030
2.45	8	0.0198	0.022
2.71	9	0.0102	0.015
2.97	10	0.0043	0.009
3.24	11	0.0011	0.004
3.50	12	0	0

Sheet pile data: *Anchored sheet (no yield of anchor road) pile Fig. 8 of Rowe reference.*

Moment of inertia,

$I = 0.11\ \text{in}^4$

Mod. of elasticity,

$E = 29{,}000$ ksi

Unit weight of soil:

$\gamma_{sat} = $ —

$\gamma_{wet} = 0.090$ kcf

Angle of internal friction,

$\phi = 30°$

Mod.-of-subgrade reaction,

$k_s = 40$ kcf

Note: Method of no-yield not specified; therefore, may affect solution to extent increasing measured moments in vicinity of anchor rod.

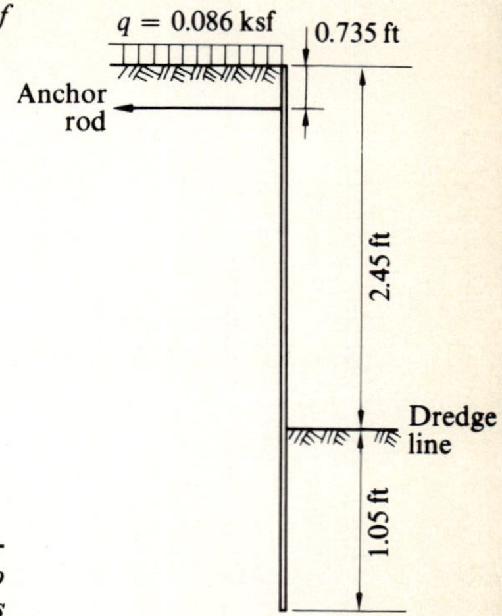

(*b*)

Dist. from top, ft	Node no.	Bending moment, ft-kips	
		Computed	Measured
0	1	0	0
0.408	2	−0.002	−0.004
0.735	3	−0.009	−0.017 (A.R.)
0.816	4	−0.011	−0.013
1.22	5	+0.014	+0.009
1.63	6	+0.30	+0.026
2.04	7	+0.34	+0.30 (+13%)
2.45	8	+0.25	+0.026 (D.L.)
2.71	9	+0.014	+0.020
2.97	10	+0.006	+0.015
3.24	11	+0.002	+0.007
3.50	12	0	0 4.2 in

FIGURE 10-9
Sheet-pile model tests. Typical data given by Rowe (1952) are shown in (*a*) and (*b*) versus computed data.

Sheet pile data: *Test no. 42 stage IV Tschebotarioff's Princeton tests.*
Moment of inertia,
$I = 0.0156$ in^4
Mod. of elasticity,
$E = 30{,}000$ ksi
Unit weight of soil:
$\gamma_{sat} = 0.1284$ kcf
$\gamma_{wet} = 0.110$ kcf
Angle of internal friction,
$\phi = 36°$
Mod.-of-subgrade reaction,
$k_s = 20$ kcf
Note: With these parameters anchor rod force computed to be 165 *lb vs.* 114 *lb measured.* ϕ *increased slightly due to increased soil density. Values below use factor* = 1.00.

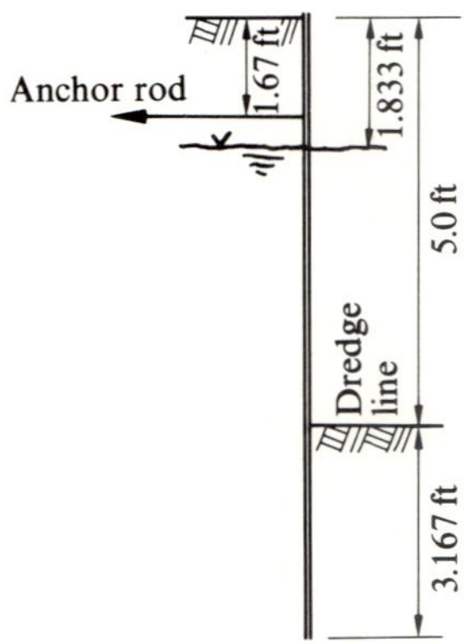

(*a*)

Dist. from top, ft	Node no.	Bending moment, ft-kips	
		Computed	Measured
0	1	0	0
0.5	2	−0.0005	−0.001
1.0	3	−0.004	−0.0037
1.5	4	−0.0145	−0.010
1.67	5	−0.020	−0.018 (A.R.)
2.0	6	−0.034	0
2.5	7	+0.0156	+0.027
3.0	8	+0.0508	+0.046
3.5	9	+0.0692	+0.051 (+36%)
4.0	10	+0.0688	+0.043
4.5	11	+0.0478	+0.025
5.0	12	+0.004	−0.004 (D.L.)
5.352	13	−0.031	−0.027
5.704	14	−0.040	−0.030
6.056	15	−0.036	−0.027

Sheet pile data: *Ship channel extension, Toronto, location no.* 1 (*Matich et al.*, 1964).
Moment of inertia,
$I = 217.5$ in^4
Mod. of elasticity,
$E = 30{,}000$ ksi
Unit weight of soil:
$\gamma_{sat} = 0.125$ kcf
$\gamma_{wet} = 0.107$ kcf
Angle of internal friction,
$\phi = 36°$
Mod.-of-subgrade reaction,
$k_s = 40$ kcf
Note: Anchor rod K = 630 *kips/ft. Values taken from small scale graph for bending moments.*

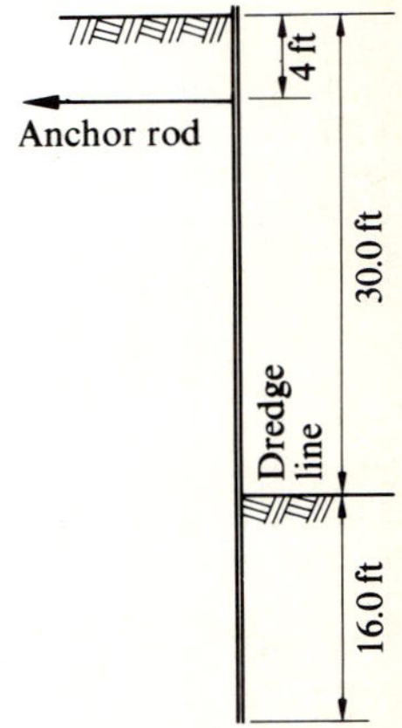

(*b*)

Dist. from top, ft	Node no.	Bending moment, ft-kips	
		Computed	Measured
0	1	0	0
4.0	2	−0.268	0 (A.R.)
4.857	3	+3.50	0
9.71	4	+22.71	25.00
14.57	5	36.35	35.00
19.43	6	42.55	42.00
24.29	7	39.62	37.00
29.14	8	25.88	23.00
34.00	9	−0.365	−4.00
38.00	10	−21.31	−29.00
42.00	11	−19.55	−28.00
46.00	12	−9.13	−4.00
50.00	13	0	0

FIGURE 10-10
(*a*) Data for a model sheet pile as reported by Tschebotarioff (1949*b*). Bending moments obtained from small-scale graph. (*b*) Comparison of computed and measured values of bending moment on a full-scale (field) anchored bulkhead. [*Data from Matich et al.* (1964).]

Other Comments

As stated earlier, one may allow deformation of the piling and arbitrarily change the anchor-rod pull as a *P*-matrix entry.

The pressure diagram can be modified to reflect increased anchor-rod pull, but a problem of what earth-pressure coefficient to use will arise. The program can be easily modified to allow more than one anchor-rod location.

Nonlinear soil deformation may be allowed by removing soil springs where the deformation is too large, replacing their effect with a *P*-matrix entry, and recycling the computation, as in Chap. 5 (also in computer program).

10-7 BRACED SHEETING

The problem of braced sheeting can be analyzed using the same form of solution. Referring to Fig. 10-11, we note the sequence of coding. Here the struts are the springs holding the wall in place rather than soil springs. Due to space limitation the computer program is not included.

The amount of embedment of the sheeting at the bottom of the excavation will determine whether to consider fixity, use a soil spring, or assume a free end, as at the top.

10-8 COMPUTER PROGRAM FOR CANTILEVERED AND ANCHORED SHEET-PILE WALLS

This program will analyze any cantilever or anchored (one anchor rod) sheet-pile wall. Provision for nonlinear soil deformation is included. Figure 10-8 illustrates the method of applying the active Coulomb earth pressure. A factor (FAC) can be applied to reduce (or increase) the pressure as

$$q_h = \text{FAC(Coulomb computed pressure)}$$

A surcharge can be applied on the backfill. The modulus-of-subgrade reaction is in the general form of $k_s = A + Bz^n$. The first (dredge-line) node soil spring is reduced 50 percent and the next lower node 25 percent for poor soil conditions usually encountered at these locations. The user may change these factors if desired.

If KSTOP = 0, the program computes through, then increases embedment depth 1 ft (30 cm) for anchored and 2 ft for cantilevered walls, and continues to

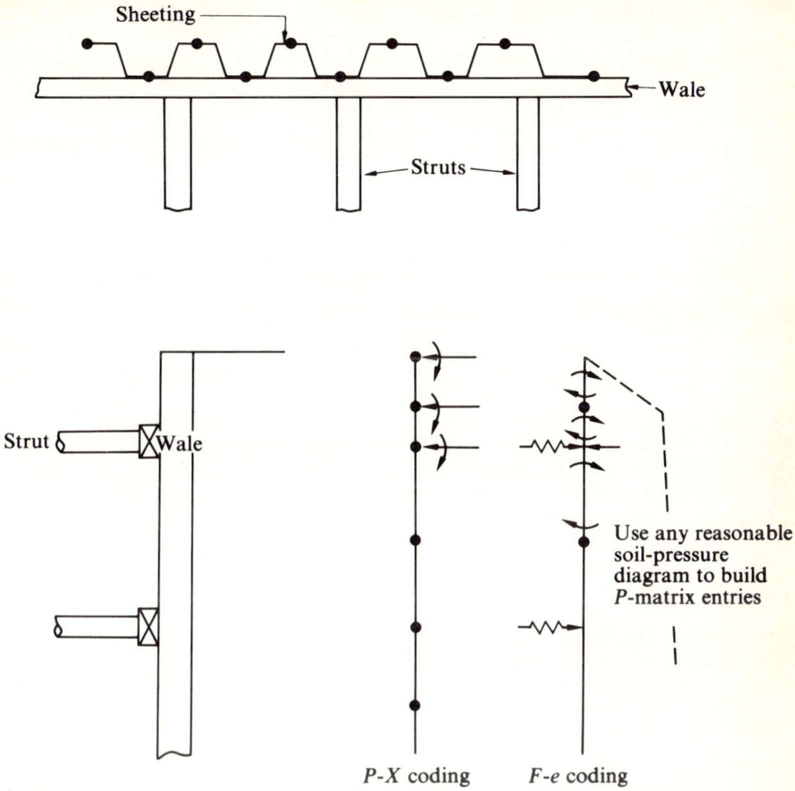

FIGURE 10-11
Finite-element analysis of braced excavation. Use as many nodes as desired; preferably use strut "springs" at a node.

recycle until the dredge-line deflections no longer increase and/or decrease. The program then checks for nonlinear soil conditions, with embedment depth again increased until the wall is stable. The program halts on one computation if KSTOP $>$ 0. Wall computations are normally for a unit width of wall, so that the anchor-rod spring (and moment of inertia of pile) is prorated as illustrated in Example 10-3.

Increased program flexibility is obtained by reading the wall-element lengths, but those below the dredge line must all be of the same length since incrementing the embedment depth results in dividing the new depth by the number of segments below the dredge line (NBELO) on succeeding cycles.

Line	Operation
1–5	Bookkeeping, note ASA^T (or E) and A matrix use common core
6	READ TITLE, UNITS (two cards)
8	READ NABOV = number of nodes above dredge line; NBELO = number of nodes below dredge line; KSTOP = means of doing problem once (no depth increase); NCYCL = number of iterations unless instability occurs
14	READ HWALL = wall height above dredge line; HROD = depth to anchor rod from surface; ERN = moment of inertia of unit width of wall; ELAS = modulus of elasticity of pile; DEMB = initial embedment depth; FAC = lateral-earth-pressure reduction coefficient
18	READ HWAT = depth to water surface; GSAT = saturated unit weight of backfill; GWET = wet unit weight of backfill; PHI = ϕ angle of backfill; SCHGE = surcharge if there is one (use 0. if none); XMAX = maximum linear soil deflection; ARODK = spring constant of anchor rod
21–56	Forms computation constants and counters and writes out critical values
61–109	Forms A matrix and writes out on first cycle
111–158	Forms S matrix in two columns
162–189	Computes Coulomb earth pressure and builds P matrix
190–197	Forms SA^T matrix
198–204	Forms ASA^T matrix and stores in A matrix
205–215	Inverts ASA^T (E matrix)
216–219	Computes X matrix
220–228	Computes F matrix
246–267	Computes shear, bending moments, soil reactions at nodes and actual soil pressure on embedded part of pile. Writes out values for checking and final design
268	Checks value of KSTOP
269–283	Checks deflections for embedment increment and checks nonlinear effects using XMAX

```
          C   J E BOWLES   CANTILEVER AND ANCHORED SHEETPILE WALL ANALYSIS PROGRAM
          C      NABOV,NBELO = NO OF SEGMENTS ABOVE & BELOW DREDGE LINE,RESPECTIVEL
          C      HROD,HWAT = DEPTH TO ANCH ROD OR WATER, FT OR M.
          C      AS,BS,EXPO PERTAIN TO MODULUS OF SUBGRADE REACTION,KCF OR T/CU M.
          C      FOR CANTILEVER SHEET-PILE WALL READ NODAR = -1 OR 0 ***
          C      UNIT WIDTH = 1 FT OR 1 METER;     **  SCHGE = SURCHARGE, KSF OR TSM
          C      KSTOP USED TO COMPUTE 1-CYCLE WITHOUT TEST FOR DEFL. OR XMAX IF >1
          C      NCYCL = NO OF ITERATIONS--STOPS PRCG REGARDLESS OF STABILITY
          C      ELAS = MOD. OF ELAST K/SQ FT OR KN/SQ M
          C      ERN = MOMENT OF INERTIA FT**4 OR M**4
          C      FU1 = 12. OR 100.; FU2 = 1. OR .3; FU3 = 144. OR 10.; FU4 = .0625
0001             DIMENSION G(50),PT(50),PRESS(50),P(50),H(30),X(50),V(50)
0002             DIMENSION XMIN(15),SS(50),D(50),SOIR(50),SOILP(50),LS(50),SMOD(25)
CC03             DIMENSION A(44,55),S(60,2),C(54,44),E(44,44),F(55),TITLE(20)
CC04             EQUIVALENCE (E(1,1), A(1,1))
0005             DOUBLE PRECISION UT5,UT6
0006        5C00 READ(1,1000,END=150)TITLE,UT1,UT2,UT3,UT4,UT5,UT6,FU1,FU2,FU3,FU4
0007        1CC0 FORMAT(20A4/4(A4,6X),A8,2X,A8,2X,4F5.1)
0008          10 READ(1,1001)NABOV,NBELO,JJS,KSTCP,NCYCL
0009        1001 FORMAT(5I5)
0010             WRITE(3,2001)TITLE
0011        2C01 FORMAT('1',T5,//,20A4,/)
0012             WRITE(3,2002)NABOV,NBELO,JJS
0013        2C02 FORMAT(T5,'NO SEGS:  ABOVE D.L. =',I3,/,T15,'BELOW D.L. =',I3,/,T5
                1, 'NO NODES REQ. S-MATRIX CCRRECT =',I3,//)
0014             READ(1,1004)HWALL,HROD,ERN,ELAS,DEMB,FAC
0015        1C04 FORMAT(8F10.4)
0016             WRITE(3,2004)HWALL,UT1,HROD,UT1,ERN,UT1,ELAS,UT5,DEMB,UT1,FAC
0017        2004 FORMAT(T5,'WALL HT. ABOVE DREDGE LINE =',F7.3,1X,A2/T5,'DEPTH GROU
                1ND LINE TO ANC. ROD =',F6.3,1X,A2/T5,'MOMENT OF INERTIA OF PILE ='
                2,G12.5,1X,A2,'**4',/T5, 'MODULUS OF ELAST =',F10.0,1X,A7/T5, 'INIT
                3. ASSUMED EMBED DEPTH =',F7.3,1X,A2/T5,'PRESS RED FAC =',F4.2,//)
0018             READ(1,1004)HWAT,GSAT,GWET,PHI,DELTA,SCHGE,XMAX,ARODK
C019             WRITE(3,2007)HWAT,UT1,GSAT,UT6,GWET,UT6,PHI,XMAX,UT2,SCHGE,UT5,
                1DELTA
0020        2007 FORMAT(T5,'DEPTH TO WATER=',F6.3,1X,A2 / T5,'SAT. UNIT WT OF SOIL
                1=',F8.4,1X,A7 / T5,'WET UNIT WT OF SOIL =',F8.4,1X,A7 / T5,'ANGLE
                2OF INTERNAL FRICT. =',F6.2,' DEG.' / T5,'MAX SOIL DEFL =',F5.2,1X,
                3A2 / T5,'SURCHARGE PRESS =',F8.3,1X,A7,3X,'ANGLE WALL FRIC ='F7.2)
0021             XMA = XMAX/FU1
```

```
      M = NABOV + NBELO + 1
      MM1 = M-1
      MP1 = M+1
      READ(1,1004)AS,BS,EXPO
      WRITE(3,2003)AS,BS,EXPO
 2003 FORMAT(//,T8,'SOIL MODULUS:  KS =',F8.2,' +',F7.2,'*Z**',F5.3)
      READ(1,1004)(H(I),I=1,MM1)
      READ(1,1008)NODWAT,NODAR
 1008 FORMAT(2I5)
      WRITE(3,2005)
 2005 FORMAT(//,T5,'THE SEGMENT NOS AND LENGTHS ARE:')
      WRITE(3,2009)(I,H(I),I=1,MM1)
 2009 FORMAT(5(T3,5(2X,I2,F7.3),/))
      NAR = 0
      IF(NODAR.LE.0)WRITE(3,2015)NODWAT
 2015 FORMAT(//,T5,'NO ANCHOR ROD USED*****',/, T5,'WATER LEVEL LOCATED
     1AT NODE',I3)
      IF(NODAR.GT.0)WRITE(3,2012)NODAR,NODWAT
 2012 FORMAT(//,T5,'ANCHOR ROD LOCATED AT NODE',I3,/,T7,'WATER LEVEL LOC
     1ATED AT NODE',I3,//)
  200 IF(NODAR.GT.0)NAR = 1
C     ***COMPUTATION CONSTANTS FORMED HERE***
      NP = 2*M
      NPM1 = NP-1
      NPP1 = NP+1
      NFM = 2*(M-1)
      NF = NFM+NBELO+NAR+1
      NFM1 = NF-1
      NFMM1 = NFM-1
      NFMP1 = NFM+1
      KABOV = M + NABOV + 1
      NEND = NF
      IF(NODAR.GT.0)NEND = NEND-1
      NDEL = NEND-NP
      NPDL = NP-NBELO
      NABP1 = NABOV + 1
      XMIN(1) = 0.00
      JK = 2
      ANBELO = NBELO
      WRITE(3,2017)NP,NF,NFM,NEND,NDEL,KABOV,NPDL,NABP1,M,MM1,MP1
 2017 FORMAT(//,T5,'NP =',I3,2X,'NF =',I3,2X,'NFM =',I3,2X,'NEND =',I3,
     12X, 'NDEL =',I3,2X,'KABOV =',I3,/,T5,'NPDL =',I3,2X,'NABP1 =',I3,2
     2X,'M =',I3,2X,'MM1 =',I3,2X,'MP1 =',I3)
      DO 600  I = 1,NP
      P(I) = 0.
      PT(I) = 0.
  600 G(I)=0.0
C*****FORM A-MATRIX---BEGIN LOOP FOR STABILITY ***
  100 DO 480  I = 1,NP
      DO 480  J = 1,NF
  480 A(I,J) = 0.
      LCOUN = 0
      IF(JK.EQ.2)GO TO 101
      DO 121  I = NABP1,MM1
  121 H(I) = DEMB/ANBELO
      WRITE(3,2009)(I,H(I),I = 1,MM1)
  101 A(1,1) = 1.
      A(M,NFM) = 1.
      NN = 2
      K = 2
      DO 481  J = 2,MM1
      L = K+1
      DO 481  I = K,L
      A(J,I) = 1.
  481 K = L+1
      A(NP,NFM-1) = -1./H(MM1)
      A(NP,NFM) = -1./H(MM1)
      L = NFM + 1
      DO 482  J = KABOV,NP
      A(J,L) = -1.
  482 L = L+1
      IF(NODAR.GT.0)A(M+NODAR,NF)=-1.0
  485 A(M+1,1) = 1./H(1)
      A(M+1,2) = 1./H(1)
      MP2 = M+2
      K = 0
      I2 = 0
      DO 486  J = MP2,NPM1
      I2 = I2+1
      DO 486  I = 1,2
      K = K+1
      A(J,K) = -1./H(I2)
  486 A(J,K+2)=1.0/H(I2+1)
      IF(JK.GT.2)GO TO 1100
      WRITE(3,2024)
 2024 FORMAT('1',//,T20, 'THE STATICS MATRIX',//)
      M2 = 0
  818 M1 = M2+1
      M2 = MIN0(M1+14,NF)
      DO 819  I = 1,NP
  819 WRITE(3,1898)I, (A(I,J), J = M1,M2)
 1898 FORMAT(T3,I3,2X,15F6.3)
      IF(M2.LT.NF)WRITE(3,1921)
 1921 FORMAT('1',//,T5, 'THE ADDITIONAL PART OF THE STATICS MATRIX',//)
      IF(M2.LT.NF)GO TO 818
```

```
C*****FORM S-MATRIX IN 2 COLUMNS
      J = 0
      DO 510  I = 1,NFM
      IF(I/2*2.NE.I)J = J+1
      S(I,1) = 4.*ERN*ELAS/H(J)
      S(I,2) = 2.*ERN*ELAS/H(J)
      IF(I/2*2.NE.I)GO TO 510
      SAVES = S(I,1)
      S(I,1) = S(I,2)
      S(I,2) = SAVES
  510 CONTINUE
C     FORM REMAINDER OF S-MATRIX DUE TO SOIL
 1100 J = NABOV
      IF(LS(JK-1).GT.0.AND.JK.GT.2)GO TO 3060
      JJ = NBELO+1
      DH = 0.
      DO 495  I = 1,JJ
      SMOD(I) = AS + (BS*DH)**EXPO
      J = J+1
  495 DH=DH+H(J)
      J = NABOV
      IF(NODAR.LE.0)NFM1 = NF
      IJ = 1
      DO 511  I = NFMP1,NFM1
      J = J+1
      IF(I.GT.NFMP1)GO TO 496
      S(I,1) = (7.*SMOD(IJ)+6.*SMOD(IJ+1)-SMOD(IJ+2))*H(J)/24.
      GO TO 499
  496 IF(I.EQ.NFM1)GO TO 497
      S(I,1) = (3.*SMOD(IJ-1)+10.*SMOD(IJ)-SMOD(IJ+1))*H(J-1)/24. + (3.*
     1SMOD(IJ+1)+10.*SMOD(IJ)-SMOD(IJ-1))*H(J)/24.
      GO TO 499
  497 S(I,1) = (7.*SMOD(IJ)+6.*SMOD(IJ-1)-SMOD(IJ-2))*H(J-1)/24.
  499 IJ = IJ+1
      S(I,2) = 0.
C     REDUCE 1ST SOIL SPRING 50% & 2ND SPRING 25%
      IF(I.EQ.NFMP1)S(I,1) = .5*S(I,1)
      IF(I.EQ.NFMP1+1)S(I,1) = .75*S(I,1)
  511 SS(I)=S(I,1)
      IF(NODAR.GT.0)S(NF,1) = AROCK
      S(NF,2) = 0.
 3060 WRITE(3,2061)
 2061 FORMAT('1',///,T5,'THE S-MATRIX IN 2-COLS--INCL F-NUMBER')
      NFDB2 = NF/2
      IF(NF/2*2.NE.NF)NFDB2 = NFDB2+1
      DO 507  I = 1,NFDB2
      LL = I+NFDB2
      IF(NFDB2.NE.NF/2.AND.I.EQ.NFDB2)WRITE(3,2051)I,(S(I,J),J=1,2)
      IF(NFDB2.NE.NF/2.AND.I.EQ.NFDB2)GO TO 507
      WRITE(3,2051)I,(S(I,J),J=1,2),LL,(S(LL,J),J=1,2)
 2051 FORMAT(T5,I3,2F12.2,2X,I3,2F12.2)
  507 CONTINUE
      IF(JK.GT.2)GO TO 1111
      DO 515  I = 1,NP
  515 PRESS(I) = 0.
C*****COMPUTE COULOMB EARTH PRESSURE COEFFICIENT
      DA = DELTA/57.2958
      PHE = PHI/57.2958
      ROOT = SQRT(SIN(PHE+DA)*SIN(PHE)/COS(DA))
      CKA = (COS(PHE)**2/(COS(DA)*(1.+ROOT)**2))
      WRITE(3,2025)CKA
 2025 FORMAT(//,T7,'THE COULOMB PRESS. COEFF =',F8.6)
      TH = 0.
      DO 505 I=1,NABP1
      IF(TH.LE.HWAT)J=I
      IF(TH.LE.HWAT)PRESS(I) = SCHGE*CKA + GWET*TH*CKA
      IF(TH.GT.HWAT)PRESS(I)=PRESS(J)+(GSAT - FU4)*(TH-HWAT)*CKA
  505 TH = TH + H(I)
C     FORM P-MATRIX DUE TO WALL PRESSURE
      WRITE(3,2031) UT6
 2031 FORMAT(T5,'SOIL MODULUS,',A7,T27,'NODAL SOIL PRESSURE AND P-MATRIX
     1 VALUES')
      J = 0
      SUML = 0.
      DO 512  I = MP1,KABOV
      IF(I.EQ.KABOV)PRESS(J+2) = 0.
      IF(I.NE.MP1)P(I)=H(J)*(2.*PRESS(J+1)+PRESS(J))/6.+H(J+1)*(2.*PRESS
     A(J+1)+PRESS(J+2))/6.
      IF(I.EQ.MP1)P(I)=H(1)*(2.*PRESS(1)+PRESS(2))/6.
  508 SUML = SUML+P(I)
      J = J+1
      IF(J.LE.JJ)WRITE(3,2030)J,SMOD(J),J,PRESS(J),I,P(I)
  512 IF(J.GT.JJ)WRITE(3,2032)J,PRESS(J),I,P(I)
 2030 FORMAT(T9,I3,2X,F10.4,2(2X,I3,2X,F10.4))
 2032 FORMAT(T26,I3,2X,F10.4,2X,I3,2X,F10.4)
C     COMPLETE P-MATRIX
 1111 DO 524  I = 1,NP
  524 PT(I) =  P(I) + G(I)
C******FORM SAT-MATRIX
      DO 552  I = 1,NFM
      DO 552  J = 1,NP
      KA = I
      IF(I/2*2.EQ.I)KA = I-1
  552 C(I,J) = S(I,1)*A(J,KA) + S(I,2)*A(J,KA+1)
      DO 553  I = NFMP1,NF
      DO  553  J = 1, NP
```

```
  553 C(I,J) = S(I,1)*A(J,I)
C*******END OF SAT MATRIX---BUILD ASAT MATRIX AND STORE OVER A-MATRIX
      DO 557 I = 1,NP
      DO 559  J = 1,NP
      D(J) = 0.
      DO 559  K = 1,NF
  559 D(J) = D(J) + A(I,K)*C(K,J)
      DO 557  L = 1,NP
  557 E(I,L) = D(L)
C     INVERT ASAT-MATRIX USING GAUSS-JORDAN METHOD
      DO 562  K = 1,NP
      DO 563  J = 1,NP
  563 IF(J.NE.K)E(K,J)=E(K,J)/E(K,K)
      DO 565  I = 1,NP
      IF(I.EQ.K)GO TO 565
      DO 565  J = 1,NP
      IF(J.NE.K)E(I,J)=E(I,J)-E(K,J)*E(I,K)
  565 CONTINUE
      DO 567  I = 1,NP
  567 IF(I.NE.K)E(I,K)=-E(I,K)/E(K,K)
  562 E(K,K)=1.0/E(K,K)
C     ***END OF ASAT INVERSION--COMPUTE X-MATRIX
      DO 575  I = 1,NP
      X(I) = 0.
      DO 575  K = 1,NP
  575 X(I) = X(I) + E(I,K)*PT(K)
C     ******COMPUTE F-MATRIX USING F = SAT*X
      DO 580  I = 1,NF
      F(I) = 0.
      DO 580  K = 1,NP
  580 F(I)=F(I)+C(I,K)*X(K)
      IF(JK.LE.2)GO TO 582
      J = 0
      DO 581  I = NFMP1,NEND
      F(I) = F(I) + G(NPDL+J)
  581 J=J+1
      WRITE(3,2101)
 2101 FORMAT('1',//)
  582 WRITE(3,2026) UT3,UT4,UT1,UT3,UT4
 2026 FORMAT(//T6,'THE TOTAL LOAD MATRIX',4X,'THE JOINT DEFL',8X,'THE FO
     1RCE MATRIX'/T6,A4,' OR ',A4,' IS',10X,A2,' OR RADS ARE',11X,A4,
     2' OR ',A4//)
   30 WRITE(3,1021)(I,PT(I),I,X(I),I,F(I),I=1,NFM)
 1021 FORMAT(T4,'LOAD DIR.',I3,F8.4,T27,'DIR =',I3,2X,F10.7,T50,'MOMENT'
     1,I3,F10.4)
   36 WRITE(3,2027)(I,PT(I),I,X(I),I,F(I),I=NFMP1,NP)
 2027 FORMAT(T4,'LOAD DIR.',I3,F8.4,T27,'DIR =',I3,2X,F10.7,T49,'**FORCE
     1',I3,F10.4)
      WRITE(3,2033)(I,F(I),I=NPP1,NEND)
 2033 FORMAT(T49,'**FORCE',I3,F10.4)
      IF(NODAR.GT.0)WRITE(3,2035)NF,F(NF)
 2035 FORMAT(T40,'ANCHOR ROD FORCE',I3,F10.4)
C*****CALCULATION OF SHEARS AND FINAL FIXED-END MOMENTS       ****
      WRITE(3,2034) UT3,UT4,UT3
 2034 FORMAT('1'//T5,'SHEAR AT EACH',2X,'NODAL BEND. MOMENT',3X,'SOIL FO
     1RCE AND PRESS' / T6,'NODE,',A4,4X,'NODE',6X,A4,10X,A4)
      VI = 0.
      SUM = 0.
      NSOIL = NFMP1-NABOV-1
      DO 53  I = 1,M
      IF(I.GT.NABOV)GO TO 51
      V(I) = VI+PT(I+M)
      IF(I.EQ.NODAR)V(I) = V(I)-F(NF)
      GO TO 53
   51 V(I) = VI+PT(I+M)+F(I+NSOIL)
      SOIR(I) = -F(I+NSOIL)
      SOILP(I) = X(I+M)*SMOD(I-NABOV)
      IF(X(I+M).GT.XMA)SOILP(I) = XMA*SMOD(I-NABOV)
      SUM = SUM+SOIR(I)
      IF(NODAR.GT.0.AND.I.EQ.M)SUM = SUM+F(NF)
   53 VI = V(I)
      L = -1
      DO 54  I = 1,M
      L = L+2
      IF(L.GT.NFM)L = NFM
      IF(I.GT.NABOV)WRITE(3,2038)I,V(I),I,F(L),I,SOIR(I),SOILP(I),UT5
 2038 FORMAT(T5,I2,F10.4,4X,I2,F10.4, 6X,I2,F9.4,2X,F9.3,1X,A7)
   54 IF(I.LE.NABOV)WRITE(3,2037)I,V(I),I,F(L)
 2037 FORMAT(T5,I2,F10.4,4X,I2,F10.4)
      WRITE(3,2039)SUM,SUML
 2039 FORMAT(T9,'SUM SOIL REACT INCL ANCH. ROD =',F10.4,'(',F8.4,')',//)
C*****TEST FOR STABILITY--CURRENT X(JK)<X(JK-1)--ALSO NON-LINEAR TEST
      IF(KSTOP.GT.0) GO TO 538
  996 LS(1) = 0
      LS(JK) = 0
      XMIN(JK) = X(KABOV)
      DO 56  I = KABOV,NPM1
   56 IF(X(I+1).LT.XMIN(JK))XMIN(JK)=X(I+1)
      IF(LS(JK).GT.0.OR.XMIN(JK).GE.(XMIN(JK-1)-.0002).OR.

     1 JK.GE.4.AND.ABS(XMIN(JK)).GT.XMA)GO TO 3000
      GO TO 70
 3000 DO 59  I = NPM1,NEND
      IF(X(I-NDEL).GT.0.)GO TO 59
      IF(ABS(X(I-NDEL))-XMA)59,60,60
```

```
   60 G(I-NDEL) = -SS(I)*XMA
      S(I,1) = 0.
      LS(JK) = LS(JK)+1
   59 CONTINUE
   70 JKM1 = JK-1
      IF(JK.EQ.2)GO TO 75
      IF(LS(JK).LE.LS(JKM1))GO TO 538
   75 WRITE(3,2102)JK,LS(JK),LS(JKM1),XMA,XMIN(JK),XMIN(JKM1)
 2102 FORMAT(T5,'JK =',I3,2X,'LS(JK) =',I3,2X,'LS(JKM1) =',I3,2X,'XMA ='
     1,F9.6,/,T5, 'XMIN(JK) =',F9.6,3X,'XMIN(JKM1) =',F9.6)
      IF(JK.EQ.NCYCL)GO TO 538
      WRITE(3,2041)JKM1
 2041 FORMAT(//,T10,'COMPUTATIONS FOR CYCLE',I3,' FOLLOW',//)
C**** EMBED DEPTH INCR U-FT OR .3-M ANCH SHEETPILE;  2-FT OR .6 M CANTIL
      IF(LS(JK).GT.0.AND.JK.GT.2.AND.LS(JK).LT.(NBELO+1)/2+1)GO TO 250
      IF(LS(JK).GE.((NBELO+1)/2+1))GO TO 538
   79 DEMB = DEMB+FU2
      IF(NODAR.LE.0)DEMB = DEMB+FU2
      WRITE(3,2042)DEMB,UT1
 2042 FORMAT(// T5,'PILE EMBEDMENT INCREASED TO',F6.2,1X,A2//)
  250 JK = JK+1
      GO TO 100
  538 WRITE(3,2044)DEMB,UT1
 2044 FORMAT(///T5,'*****FINAL EMBEDMENT LENGTH OF SHEET PILE =',F6.2,
     11X,A2 / T10,'OUTPUT ABOVE FOR THIS EMBEDMENT VALUE')
      IF(LS(JK).GE.((NBELO+1)/2+1))WRITE(3,2054)LS(JK)
 2054 FORMAT(T5,'*********PILE UNSTABLE',I3,' NODES ZEROED IN S-MATRIX')
      IF(LS(JK).GE.((NBELO+1)/2+1))LCOUN = 1
      IF(LCOUN.NE.1.OR.(JK.EQ.NCYCL).OR.(KSTOP.GT.0))GO TO 80
      LS(JK) = 0
      GO TO 79
   80 GO TO 5000
  150 STOP
      END
```

PROBLEMS

10-1 Design a sheet-pile wall to be embedded in a soil with $q_u = 0.4$ ksf, $\phi = 15°$. Wall height is to be 15 ft as measured from the dredge line. Backfill is $\gamma = 110$ pcf, $e = 0.55$, $\phi = 32°$. Water level will be 5 ft below ground surface. Find the embedment depth and sheet-pile section required. Use $F_y = 36$ ksi for steel. State your SF and design stress.

10-2 Repeat Prob. 10-1 if the wall is 25 ft and water level is 5 ft below ground surface. Use an anchor rod at a spacing of 10 ft on centers. Required:

Pile section
Embedment depth
Reasonably optimum anchor-rod location
Wale section using a pair of channels back to back

10-3 Referring to Example 10-3, how can you move the pile bending moment vertically? Use the computer program and check your plan.

Problems 10-4 through 10-7 should be partial class and partial subgroup projects, where each group of three to five students program selected parts to reduce computer congestion. Certain of plots (a) to (g) may be omitted at the discretion of the instructor. Each group should submit a set of graphs and discussion of data using its output data and the output data from the other groups.

10-4 Make a sheet-pile study by graphing the following:
(*a*) Anchor-rod location versus maximum bending moment
(*b*) Anchor-rod location versus dredge-line deflection
(*c*) Dredge-line deflection versus pile moment of inertia

(*d*) Dredge-line deflection versus depth of embedment
(*e*) Maximum soil pressure versus depth of embedment
(*f*) Dredge-line deflection versus modulus-of-subgrade reaction
(*g*) Maximum soil pressure versus modulus-of-subgrade reaction

Make the study using:

Wall height = 18 ft
No water above dredge line
$\gamma = 110$ pcf, $\gamma_{sat} = 132.4$ pcf
Embedment depth from 6 to 18 ft in increments of 3 ft
Anchor rod from 0 to 16 ft in 2-ft increments

10-5 Do the appropriate parts of Prob. 10-4 for *no anchor rod.*

10-6 Repeat Prob. 10-4 if wall height is 24 ft. Adjust embedment depth and increments and anchor-rod increments.

10-7 Repeat Prob. 10-4 if water is 6 ft below the ground surface.

10-8 Make a comparison of bending moment obtained for any given anchored wall holding the moment of inertia, wall height, and anchor rod constant and varying k_s (same as varying density). How does this result compare to Rowe's moment-reduction curves for sand?[1]

REFERENCES

ANDERSON, PAUL (1956): "Substructure Analysis," 2d ed., chap. 2, Ronald, New York.

AYERS, JAMES R., and R. C. STOKES (1954): The Design of Flexible Bulkheads, *Trans. ASCE*, vol. 119, pp. 373–383.

BOWLES, J. E. (1968): "Foundation Analysis and Design," chap. 8, McGraw-Hill, New York.

HALIBURTON, T. A. (1968): Numerical Analysis of Flexible Retaining Structures, *J. Soil Mech. Found. Div., ASCE*, vol. 94, SM6, November, pp. 1233–1252.

MATICH, M. A. J., R. P. HENDERSON, and D. B. OATES (1964): Performance Measurements on Two New Anchored Bulkheads, *Can. Geotech. J.*, vol. 1, no. 3, July, pp. 167–178.

POTYONDY, J. G. (1961): Skin Friction between Various Soils and Construction Materials, *Geotech. (Lond.)*, vol. 11, no. 4, December, pp. 339–353.

RAUHUT, JAMES B. (1969): Discussion: Numerical Analysis of Flexible Retaining Structures, *J. Soil Mech. Found. Div., ASCE*, vol. 95, SM6, November, pp. 1553–1564.

RICHART, FRANK E., JR. (1957): Analysis for Sheet-Pile Retaining Walls, *Trans. ASCE*, vol. 122, pp. 1113–1135.

ROWE, P. W. (1952): Anchored Sheet-Pile Walls, *Proc. Inst. Civ. Eng. (Lond.)*, vol. 1, pp. 27–70.

——— (1954): Discussion: The Design of Flexible Bulkheads, *Trans. ASCE*, vol. 119, pp. 388–390.

[1] Rowe's curves may be found in Rowe (1952, p. 59); Terzaghi (1954; p. 1261); Bowles (1968, p. 396).

——— (1957): Sheet Pile Walls in Clay, *Proc. Inst. Civ. Eng.* (*Lond.*), vol. 7, July, pp. 629–654.

TERZAGHI, KARL (1954): Anchored Bulkheads, *Trans. ASCE*, vol. 119, pp. 1243–1280.

TSCHEBOTARIOFF, G. P. (1948): Determination from Bending Strain Measurements of the Distribution of Lateral Earth Pressures against Model Flexible Bulkheads, *Geotech.* (*Lond.*), vol. 1, no. 2, pp. 98–111.

——— (1949*a*): Large-Scale Model Earth Pressure Tests on Flexible Bulkheads, *Trans. ASCE*, vol. 114, pp. 415–454 (and discussion pp. 507–524).

——— (1949*b*): Final Report: Large-Scale Earth Pressure Tests with Model Flexible Bulkheads, Princeton University.

TURABI, DAFALLA A., and ARPAD BALLA (1968): Sheet Pile Analysis by Distribution Theory, *J. Soil Mech. Found. Div., ASCE*, vol. 94, SM1, January, pp. 291–322 (see also discussion May 1969, closure January 1970).

11

PILE STRESSES: WAVE EQUATION

11-1 INTRODUCTION

This chapter and Chap. 12 consider two methods of evaluating pile stresses (or predicting pile performance). Dynamic stresses and performance will be evaluated, using the wave equation of the next articles. This equation can be used to determine: (1) whether the pile can be driven using the proposed pile–pile-hammer combination, (2) whether the pile will reach the desired ultimate load capacity using an estimate of the ultimate capacity based on set (blows per inch), and (3) what the values of the driving stresses (and tensile stresses in the case of concrete piles) will be. Wave-equation correlations with pile-load-test data and the limited number of piles reported in the literature instrumented for dynamic response indicate that the above claims are valid [Samson et al. (1963), Mosley (1967), Raamot (1967)].

11-2 THE WAVE EQUATION

Historical

The impact on the end of a longitudinal rod to describe pile driving has been considered historically by several writers. For those interested in the historical development Smith (1962) and Samson et al. (1963) provide several references.

An analytical method using the wave-equation concept was proposed by Smith (1955, 1962), a mechanical engineer with Raymond International Inc. (Concrete Pile Division). Smith's method is a finite-difference solution which can be done by hand but is really practical only with a digital computer. Since the publication by Smith others have made considerable contributions to the application of this means of analysis [Samson et al. (1963), Forehand and Reese (1964), Graff (1965), Mosley (1967), Raamot (1967), Bowles (1968, 1970), Davisson (1970), Mosley and Raamot (1970), and Hirsch et al. (1970)]. The author [Bowles (1968)] included a modest program to output pile forces at corresponding time intervals as an early aid for people unfamiliar with the analytical method. The current program (included) is considerably more sophisticated.

The Mathematical Model and Symbols

The pile and the corresponding finite-element model is as shown in Fig. 11-1. Certain driving equipment and/or methods may require slight to major modification of this model, but these modifications are not considered here in order not to obscure the analysis.

This method replaces the differential equation describing transmission of a shock wave along the pile with a numerical (or difference) equivalent. Figure 11-2 basically describes the method.

1 At the beginning of $t = 1$ (DT = 1)[1] the pile-driver ram [W(2)] impacts on the spring with an initial velocity V(2,1) or the v_1 of Fig. 11-2.

2 This velocity V(2,1) displaces the cap-block spring SP(2) at the end of the first time interval an amount [D(2,2)] according to the equation

$$y_1 = v_1 \, \Delta t$$

3 This displacement produces a cap-block force [F(2,2)] computed as

$$F = Ky_1$$

[1] These are the computer program variables as a further reader aid.

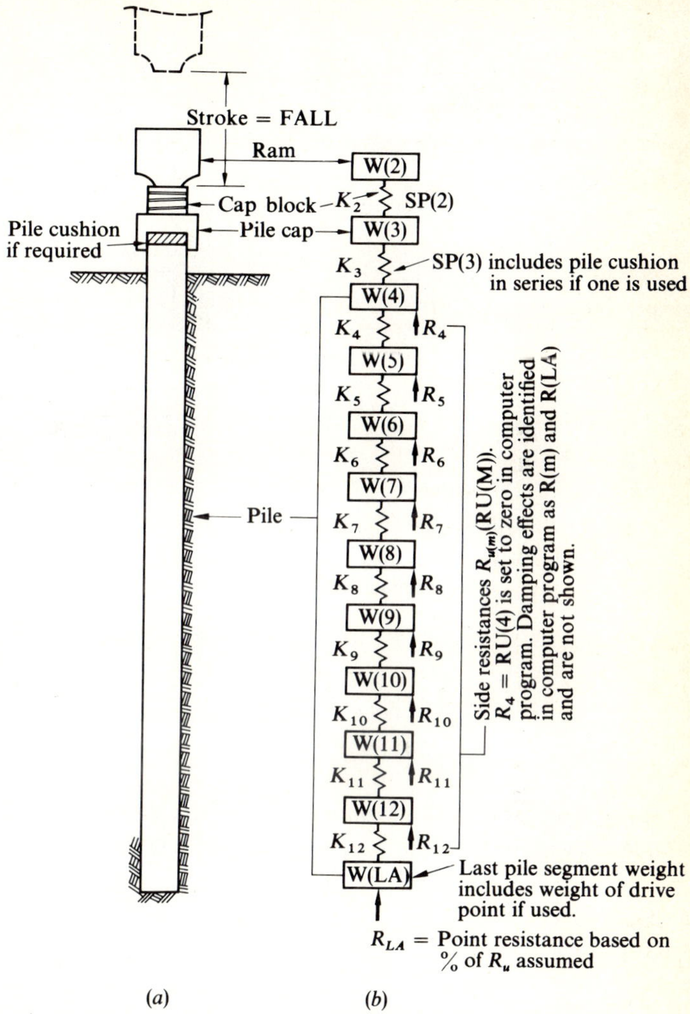

FIGURE 11-1
(*a*) Pile at approximate embedment depth. (*b*) The finite-difference model of a pile for the wave-equation solution. The pile-segment and spring subscripts correspond to a 12-element system (10 pile segments) as used in the included computer program. The weight of the segment is concentrated at the bottom of the spring. Note that R_{LA} includes the side resistance of the bottommost pile element.

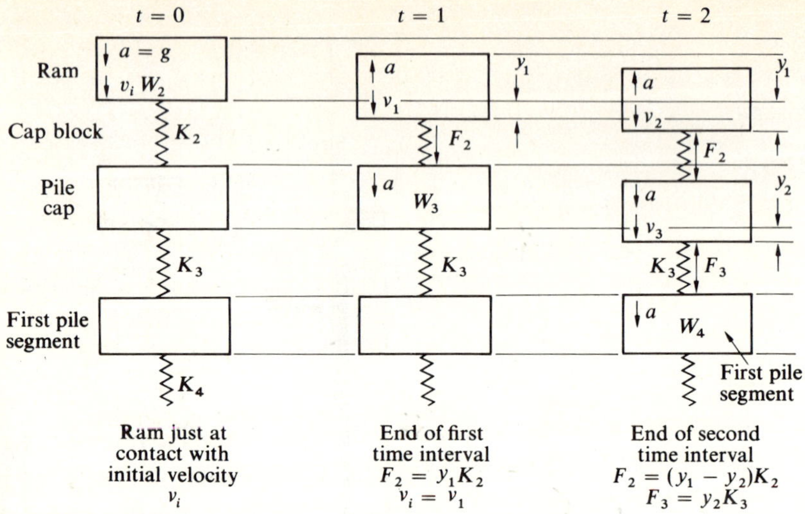

FIGURE 11-2
The wave-equation analysis examined through time increments from the instant of ram contact with the cap block to the end of two time intervals.

4 The force F(2,2) accelerates the pile cap [W(3)] downward according to

$$F = ma$$

No other event of importance occurs during this time interval.

5 At the end of the next time interval (DT = 2):

a The cap block has moved a distance with a velocity v_2 of

$$y_2 = v_2 \, \Delta t$$

b The pile cap has moved a distance based on the acceleration of the pile cap due to F(2,2) to obtain a velocity of

$$v_3 = at$$

and a resulting displacement [D(3,2)] of

$$y_3 = v_3 \, \Delta t$$

c Resulting in a new cap-block force based on the net compression of the cap block of $y_1 - y_2$ to obtain the force as

$$\mathrm{F(2,2)} = K_2(y_1 - y_2)$$

d And also resulting in a force F(3,2) between the pile cap and the first pile segment based on the displacement y_2 and the spring constant of the first pile segment K_3 [SP(3)] as force F_3 of Fig. 11-2

$$F(3,2) = K_3 y_2$$

This process is continued for all the spring elements and the point and repeated for the number of time intervals required to obtain the desired output information.

The following paragraphs will develop this method in detail. To avoid breaking the discussion the necessary symbols are grouped and identified at this point.

List of Symbols

The terms used in the following discussion are as follows:

A = cross-sectional area of pile, sq in (sq cm)

C_m = spring compression of element m at $t = i$, in (cm)

C'_m = spring compression at $t = i - 1$ time intervals, in (cm)

D_m = displacement of element m at $t = t$, in (cm)

D'_m = displacement of element m at $t - 1$ time intervals, in (cm)

D''_m = displacement at $t = t - 2$ time intervals, in (cm)

D_g = ground plastic displacement at $t = i$, in (cm)

D'_g = ground plastic displacement at $t = t - 1$ time intervals, in (cm)

E = modulus of elasticity of pile materials, ksi (kN/sq cm)

E_f = pile-hammer efficiency

g = acceleration of gravity, in/sec^2 (32.2 ft/sec^2 or 980.7 cm/sec^2)

$t - 1$ = one time interval before current time

J_p = damping constant used with pile-point resistance, sec/ft (sec/m)

J_s = damping constant used with side resistance, sec/ft (sec/m)

K_m = pile-element spring constants including cap, cap block, cushion, etc., kips/in (kN/cm)

K'_m = soil spring constant, kips/in (kN/cm)

ΔL = length of pile element, ft (m)

M, m = mass = weight/g (also used as element counter—note context)

Q = quake or maximum elastic ground deformation, in (cm)

T = time interval used in computations, sec

$\ddot{u}$ = second derivation of displacement u with respect to time

v_m = velocity of element m, in/sec (cm/sec)

v'_m = velocity of element m at $t = i - 1$

From inspection of Fig. 11-3*a*, the instantaneous displacement D_m of any element is the sum of the displacement one unit of time back (D'_m) plus the product of instantaneous element velocity and time interval

$$D_m = D'_m + v_m T \tag{11-1}$$

and the net element compression (Fig. 11-3*b*) is computed as

$$B - C$$

or in terms of displacements (and because we compute all element displacements before computing spring compressions)

$$C_m = D_m - D_{m+1} \tag{11-2}$$

The resulting force in the element spring K_m is simply

$$F_m = K_m C_m \tag{11-3}$$

The accelerating force F_{am} on any element m is (Fig. 11-3*c*)

$$-F_{m-1} + F_{am} + F_m + R_m = 0$$

or

$$F_{am} = F_{m-1} - F_m - R_m \tag{11-4}$$

The velocity of element m is computed from the conventional velocity equation as

$$v = v_0 + at$$

which becomes in this case (since $a = F_{am}/m$ or $a = F_{am}g/W_m$)

$$v_m = v'_m + \frac{F_{am}g}{W_m} T \tag{11-5}$$

The final general element-displacement equation can be obtained by multiplying Eq. (11-5) by T, to obtain

$$v_m T = v'_m T + \frac{F_{am}g}{W_m} T^2 \tag{a}$$

Rearranging Eq. (11-1), we find that

$$v_m T = D_m - D'_m$$

and by analogy

$$v'_m T = D'_m - D''_m$$

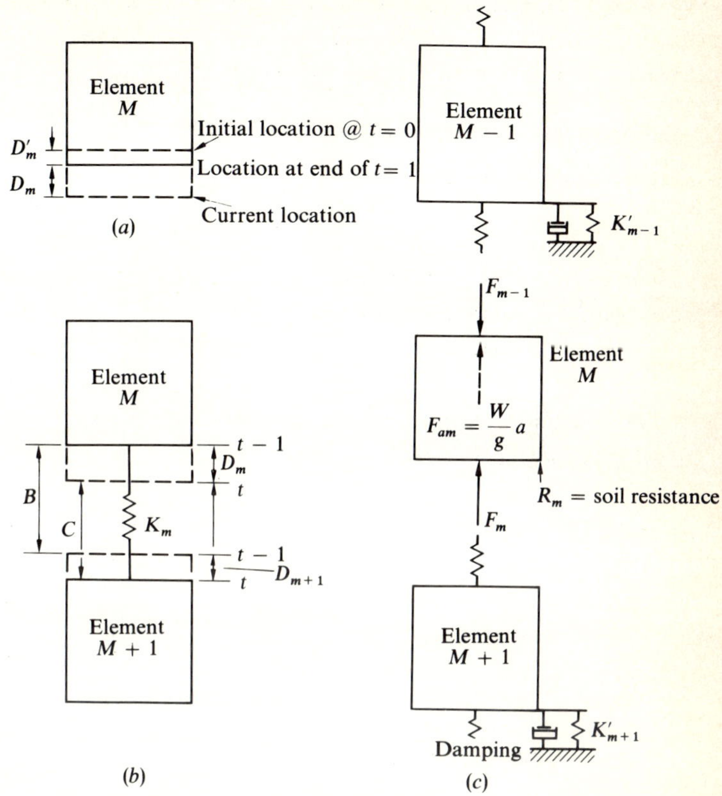

FIGURE 11-3
Element forces and displacements. The terminology closely matches that of the computer program. Note that the soil resistance is shown as R_m in c on element M whereas on element $M + 1$ it is shown as a soil spring and dashpot in parallel.

Therefore Eq. (*a*) becomes

$$D_m = 2D'_m - D''_m + \frac{F_{am}g}{W_m} T^2 \tag{11-6}$$

Now let us pause and consider the differential equation of impact of a long slender rod subjected to side resistance R, as in Fig. 11-4. The unit strain is

$$\varepsilon = \frac{\partial u}{\partial y}$$

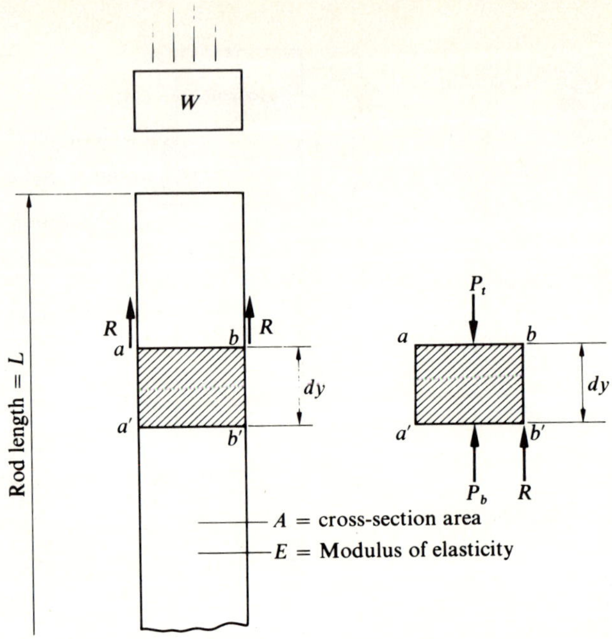

FIGURE 11-4
Transmission of strain (and stress) in a long rod tapped on one end with a moving mass.

The force is

$$P_t = EA\varepsilon = EA\frac{\partial u}{\partial y}$$

The force on the bottom of the rod element is

$$P_b = P_t - \Delta P \pm R$$

And the net force (which produces acceleration of the element) on the element of length dy is

$$P_{net} = P_t - P_b \pm R$$

or

$$P_{net} = AE\frac{\partial u}{\partial y} - \left(AE\frac{\partial u}{\partial y} - AE\frac{\partial^2 u}{\partial y^2}\,dy\right) = AE\frac{\partial^2 u}{\partial y^2}\,dy \pm R$$

but the unbalanced force P_{net} is also

$$P_{\text{net}} = Ma = \frac{W}{g}\ddot{u} = \frac{W}{g}\frac{\partial^2 u}{\partial t^2}$$

Equating values of P_{net} and introducing $\rho = W/gA\,dy$, we have

$$\frac{\partial^2 u}{\partial t^2} - \frac{E}{\rho}\frac{\partial^2 u}{\partial y^2} \pm R = 0 \tag{11-7}$$

We can express $\partial^2 u/\partial t^2$ in finite-difference form using the *first-backward*-difference equation of Table 4-1 (since we cannot travel forward in time, we cannot use forward or central differences). This equation is

$$y''_n = \frac{y_n - 2y_{n-1} + y_{n-2}}{(\Delta x)^2}$$

and converting to terminology consistent with this problem, we obtain

$$\frac{D_m - 2D'_m + D''_m}{T^2} \approx \frac{\partial^2 u}{\partial t^2} \tag{b}$$

Multiplying both sides of (*b*) by T^2 and noting that $(\partial^2 u/\partial t^2)T^2$ is $F_{am}gT^2/W_m$ of Eq. (11-6), we have obtained this equation in two ways.

11-3 OTHER FACTORS IN SOLUTION

Let us next investigate the remaining details of obtaining the wave-equation solution. We need pile-element velocities, found from Eqs. (11-4) and (11-5), to obtain

$$v_m = v'_m + (F_{m-1} - F_m - R_m)\frac{Tg}{W_m} \tag{11-8}$$

and the force F_m is from Eq. (11-3)

$$F_m = (D_m - D_{m+1})K_m \tag{c}$$

The ultimate pile resistance R_u can be distributed in some manner to the pile elements so that the sum of pile-element resistances totals R_u. The distribution may be even[1] (common) and based on the percentage of load *estimated* to be carried by side friction and point resistance. The soil "spring" constant is computed as

$$K'_m = \frac{R_{um}}{Q}$$

[1] Some people are of the opinion (and the included computer program assumes) that the first in-the-ground pile segment may not have a soil resistance due to driving and other surface disturbances.

and the instantaneous pile-element resistances are computed as follows (refer to Fig. 11-5). Let the amount of soil deformation in excess of the quake Q be D_{sm}, defined as

$$D_{sm} = \pm D_m \mp Q$$

or, using the computer approach,

$$D_m - Q \leq D_{sm} \leq D_m + Q$$

since D_m may be either + or −. This requires using a computer subroutine [SUBROUTINE NO. 1 for DE(M,2)] to find D_{sm}, the plastic soil deformation.

With the plastic soil deformation evaluated, the elastic soil deformation is the total deformation less the plastic deformation, or

$$D_m - D_{sm}$$

and the resulting soil resistance (elastic) is

$$R_m = (D_m - D_{sm})K'_m \qquad (d)$$

But with damping present (Fig. 11-5*d*) we must modify Eq. (*d*). This is accomplished by assuming that R_m is the sum of two resistances, elastic and damping, which can be written as

$$R_m = R_e + R_d \qquad (e)$$

and assuming further that damping resistance can be written

$$R_d = R_e J_s v_m \qquad (f)$$

i.e., a function of the element velocity, elastic resistance, and damping factor. Combining terms, we have[1]

$$R_m = (D_m - D_{sm})K'_m(1 + J_s v_m) \qquad (11\text{-}9)$$

[1] Note that for the pile with 100 percent point load $K'_m = 0$ and no damping occurs according to Eq. (11-9) even if $J_s \neq 0$, a situation which may require serious consideration of Eq. (11-10).

FIGURE 11-5
Idealized development of pile-element soil resistance. Parts (*a*), (*b*), and (*c*) do not include damping. (*a*) D(M,2) < Q; element displacement is less than the quake. (*b*) Pile-point displacement is larger than Q, as shown. When point displacement D(LA,2) is less than Q, use (*a*). (*c*) General idealized displacement condition for the *m*th element. At the beginning of the record both displacement and t are zero. Displacement builds up to D(M,2); force builds up to $R_m = QK'_m$; then plastic deformation [DE(M,2)] occurs; the quake is recovered from B to C at end of time increment. Next interval of time begins, and the cycle repeats as *CBDE* at the end of t_2. A negative element displacement is shown as occurring next, through *EFGH* ($t = t_3$). The fourth time element starts from an initial displacement of H. (*d*) The general situation of (*a*), (*b*), and (*c*) when damping is included.

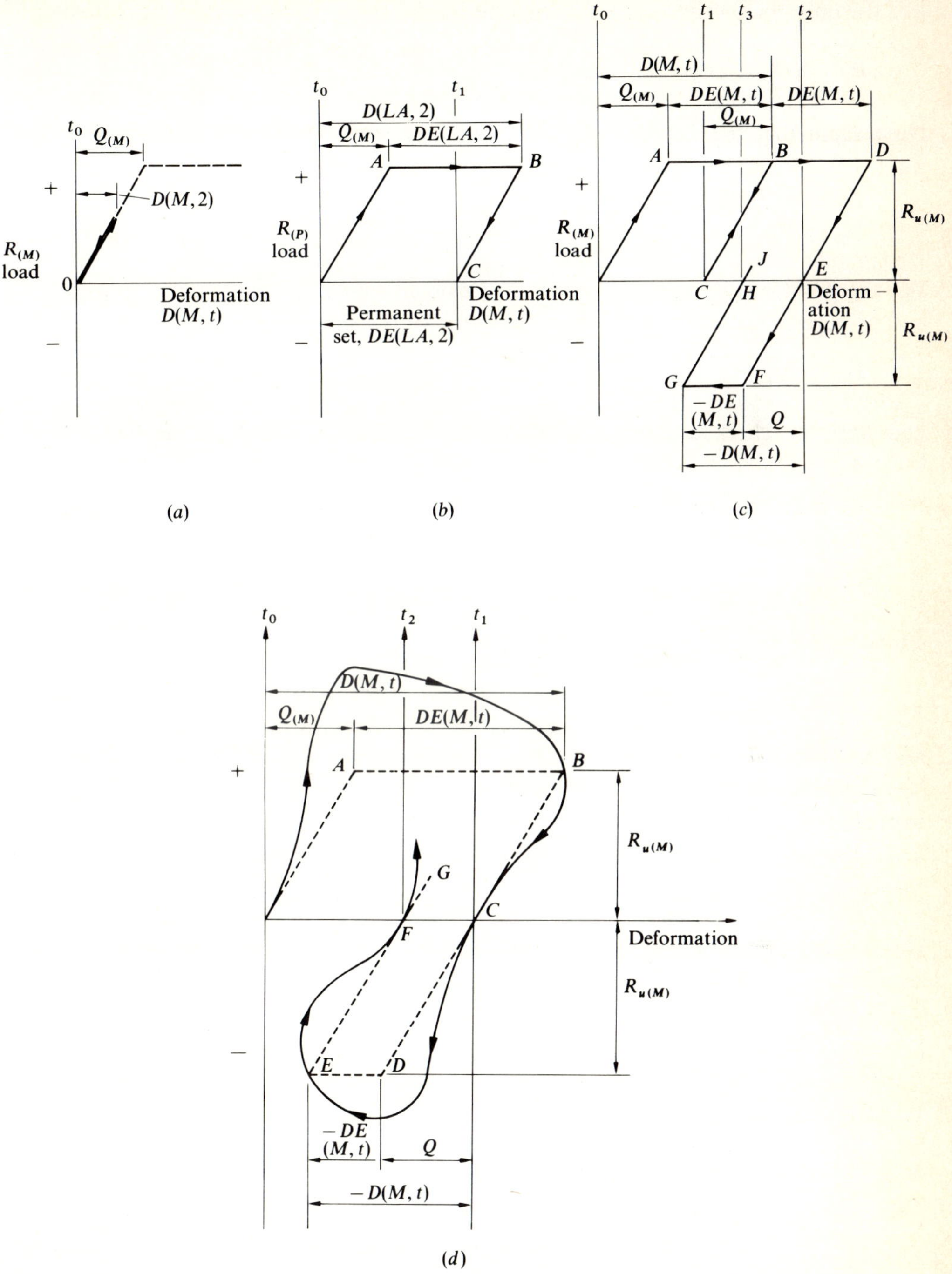

t0
Q(M)
+
D(M, 2)
R(M)
load
0
Deformation
D(M, t)
−
(a)
t0
t1
D(LA, 2)
Q(M)
DE(LA, 2)
A
B
+
R(P)
load
C
Deformation
D(M, t)
Permanent
set, DE(LA, 2)
−
(b)
t0
t1
t3
t2
D(M, t)
Q(M)
DE(M, t)
DE(M, t)
Q(M)
A
B
D
+
R(M)
load
Ru(M)
J
C
H
E
Deform-
ation
D(M, t)
−
Ru(M)
G
F
− DE
(M, t)
Q
− D(M, t)
(c)
t0
t2
t1
D(M, t)
Q(M)
DE(M, t)
+
A
B
Ru(M)
G
C
F
Deformation
Ru(M)
−
E
D
− DE
(M, t)
Q
− D(M, t)
(d)

For the point (subscripts p) by analogy we have

$$R_p = (D_p - D_{sp})K_p'(1 + J_p v_p) \tag{11-9a}$$

Note again that the definition of D_{sm} (and D_{sp}) limits the elastic displacements in Eq. (11-9) to

$$D_m - D_{sm} \leq Q$$

For the point of the pile (Fig. 11-5b) it is evident that D_{sp} = permanent pile set(s), and no reversal of sign is possible, as was true of the side resistances; thus

$$D_{sp} \geq D_p - Q$$

and must be checked using a computer subroutine [SUBROUTINE NO. 2 for DE(LA,2)]. If we plot displacement versus time, the displacements D_{sm} or D_{sp} will lag D_m (or D_p) by the amount of the quake Q.

11-4 PILE-HEAD ATTACHMENTS

To avoid pile damage, expected both on the basis of practical experience and analytical results from the wave equation, a pile cap block is generally inserted between the ram and anvil or pile-cap assemblage. This device, which may be of wood, Micarta, Micarta and aluminum plates sandwiched, or other materials, avoids impacting the metal ram directly on a metal–pile-hammer interface, thus increasing the hammer life. This element is considered to be weightless although it may weigh as much as 100 lb (generally 20 to 40).

A pile cushion placed between the pile cap (or anvil) and the pile head, with which it is in direct contact, may be used to even out the contact surface and reduce driving stresses in the pile. The pile cushion is soft material such as wood planking 3 to 6 in thick.

Generally the cap block, pile cap (or anvil), and pile cushion rest on top of each other and are attached in such a manner that they transmit only compressive forces (a situation idealized in Fig. 11-6b).

Since the pile cap block and cushion are different materials than the ram or pile cap, we are concerned with the loss of energy on ram impact. The energy loss can be depicted by the force-displacement diagram of Fig. 11-7, where the area DBC is

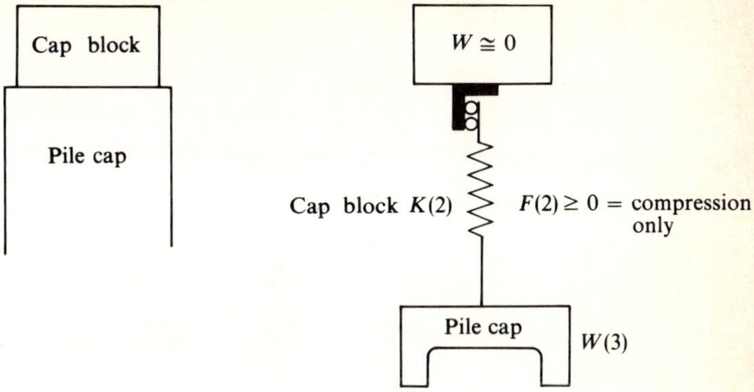

(a)

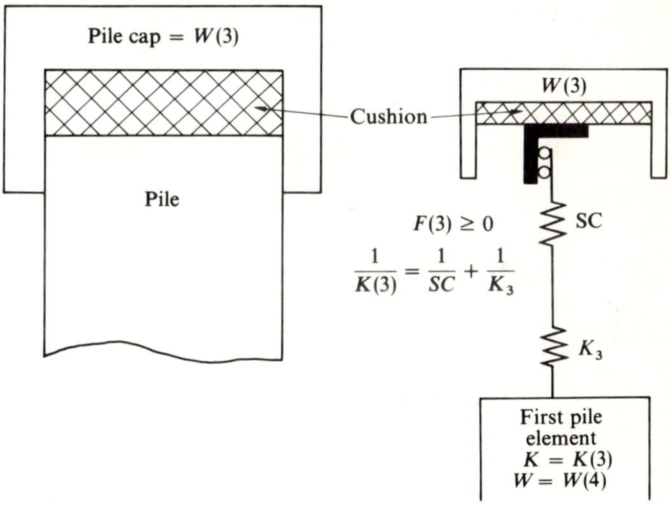

(b)

FIGURE 11-6
Pile-hammer interfacing accessories. (*a*) Cap-block and pile-cap system for compression force only. (*b*) Pile-cap and pile-cushion system for compression force only. Note that both pile cap block and pile cap (or pile cushion if used) are idealized, with connections which can transmit only compression.

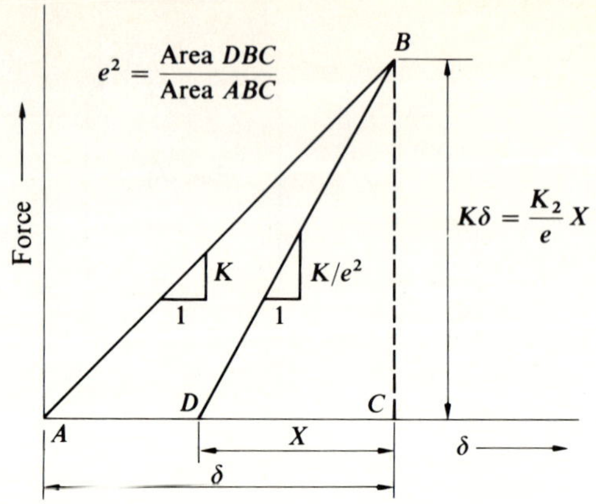

FIGURE 11-7
Pile cap-block (and pile-cushion) compression and coefficient-of-restitution characteristics.

output energy and the area ABC is input energy. From the energy equation (impulse type)

$$e(M_1 v_i + M_2 v_i') = M_1 v_f + M_2 v_f'$$

where M_1 = mass of ram

M_2 = mass of cap block or pile cushion ≈ 0

v_i, v_i' = initial velocities

($v_i = 0$ for M_2); therefore,

$$e = \frac{v_f}{v_i} \qquad \text{and} \qquad e^2 = \frac{v_f^2}{v_i^2}$$

The kinetic-energy equation is

$$\text{KE} = \tfrac{1}{2}Mv^2$$

Therefore, it is evident that in Fig. 11-7

$$e = \frac{v_f^2}{v_i^2} = \frac{\text{area } DBC}{\text{area } ABC}$$

Referring to Fig. 11-8, which represents the force deformation of springs with restitution, we compute the cap-block and cushion spring forces as

$$F = KC_2 \tag{g}$$

until the change in compression $(C_2 - C_2')$ between two time intervals is negative. At this time a term C_{max} is introduced

$$C_{max} = C_2'$$

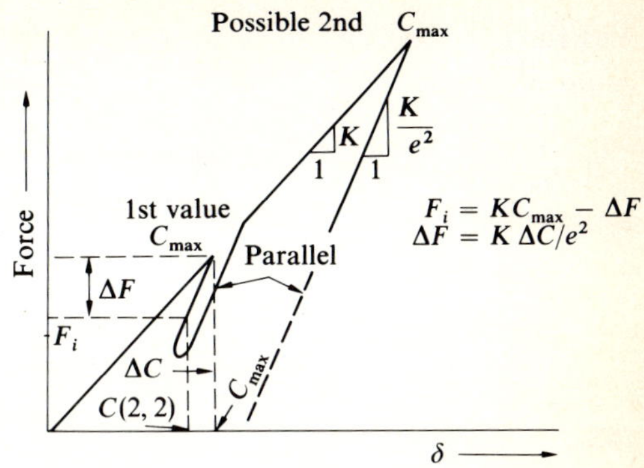

FIGURE 11-8
Force-displacement diagram for cap block (and pile cushion) with restitution. This concept is used in computer subroutine programs 3 and 4.

and the instantaneous cap-block (and/or cushion) force is computed as

$$KC_{max} + \frac{K(C - C_{max})}{e^2} \tag{11-10}$$

Equation (11-10) is used until $C_2 - C_2' \geq 0$; then Eq. (g) is used again, the cycles being repeated as required.

Subroutines

Computer subroutines are required for both the pile cap block [forces and deformations in cap-block spring SP(2)] and the pile cap and pile cushion [forces and deformations in spring SP(3)]. Note that when a pile cushion is used, spring SP(3) is obtained as follows:

$$\frac{1}{\text{SP}'(3)} = \frac{1}{\text{SC}} + \frac{1}{\text{SP}(3)} \tag{h}$$

This equation is obtained considering that with two springs in series, the force is the same in both springs and the total deformation is the sum of the individual spring deformations. Here SP′(3) is the equivalent spring combining the effects of the pile-cushion spring SC and the computed first-pile-element spring SP(3) $= AE/\Delta L$.

Computer subroutines (SUBROUTINE NO. 3 and SUBROUTINE NO. 4) check that the forces in springs 2 and 3 are always zero or compression

F(2,2) ≥ 0

F(3,2) ≥ 0

and further check when restitution occurs as follows (using SUBROUTINE NO. 3 as an example):

1 Check current compression and compression one time interval earlier as (C(2,1) = C'_m)

$$\mathrm{C(2,2) - C(2,1) \geq 0}$$

and as long as the difference is as shown,

$$\mathrm{F(2,2) = C(2,2)*SP(2)}$$

2 If C(2,2) − C(2,1) < 0, then

$$\mathrm{C(2,1) = CMAX}$$

and Eq. (11-10) is used.

3 Continue using Eq. (11-10) with this value of CMAX until the situation in Eq. (*a*) obtains. Now use

$$\mathrm{F(2,2) = C(2,2)*SP(2)}$$

until the spring-compression difference is again negative, etc.

4 Check that F(2,2) ≥ 0. If F(2,2) is a tension value (−), it is set to zero for reasons previously stated.

5 Repeat steps 1 through 4 as required.

The wave equation requires the following computation steps:

1 Compute displacements D_2 to D_p using Eq. (11-1) and using consistent units. Current values are D(M,2); previous values D'_m are D(M,1).

2 Compute the plastic ground displacements [DE(M,2)] using SUBROUTINE NO. 1 for elements other than pile point and using SUBROUTINE NO. 2 for the point.

3 Compute side and point ground resistances R_m and R_p; note in computer program M = 4 to LA these are R(M) and R(LA).

4 Compute the spring compression in each spring C_m (M = 2,L) using Eq. (11-2). These are C(M,2) for current values, and the previous values C'_m are C(M,1).

5 Compute force F_2 in spring SP(2) (the cap block) using SUBROUTINE 3 [force F(2,2)].

6 Compute the force F_3 in spring SP(3) (the first pile segment with or without a cushion) using SUBROUTINE NO. 4 [force F(3,2)].

7 Compute the remainder of pile-element forces using Eq. (11-3) [F(M,2)].

8 Compute the velocity of each element using Eq. (11-5) in expanded form. These are V(M,2).

The computer should be programmed to stop when the following two conditions are reached:

1 $D_{sp} - D'_{sp} \leq 0$ (DE(LA,2) − DE(LA,1)) ≤ 0.
2 All element velocities are simultaneously negative or zero.

The logic of condition 1 is based on the permanent set reaching a maximum value from which it should not decrease. Because of computer roundoff, the ≤ sign is used rather than an equality. Condition 2 ensures (by analysis) that the pile does not come out of the ground.

11-5 INPUT PARAMETERS

The wave equation requires certain input data. These are:

1 Weight of:
  a Pile
  b Pile-cap or driving helmet (pile cap block and pile cushion generally assumed weightless)
  c Ram
  d Pile tip or driving shoe (if used)
2 Material properties of spring constant and coefficient of restitution for:
  a Cap block
  b Cushion
3 Soil properties of:
  a Quake
  b Damping constants: side = J_s, point = J_p
4 Pile properties of:
  a Modulus of elasticity
  b Cross-sectional area
  c Length of pile elements and estimates of ultimate pile resistance R_u and amount and distribution of side resistance (amount of R_u carried by skin friction)

Weights

Manufacturers' catalogs can be consulted for weight data needed as input. Currently steel H-piles run from about 36 to 117 lb/ft. Pile-cap or driving helmets commonly range from 300 to 6,000 lb, depending on hammer size and pile to be driven. Ram weights vary from around 3,000 to 20,000 lb, and the pile point (if used) may range from 50 to 500 lb. Tables in Appendixes A and B provide some useful data on piles, pile hammers, etc.

Cap-Block and Cushion Properties

Smith (1962) and Hirsch et al. (1970) have provided data for several cushion materials. These data enable one to compute the cap block or pile-cushion spring as

$$K = \frac{AE}{L}$$

where A = cross-sectional area of cap block or cushion

L = length

E = elastic modulus from Table 11-1

Soil Properties

The maximum amount of elastic soil deformation is the quake. Smith (1962) proposed a value of 0.1. Forehand and Reese (1964) indicated that the values should be as follows:

	Quake		Damping constant J_p	
Soil	in	cm	sec/ft	sec/m
Sand	0.05–0.20	0.13–0.51	0.10–0.20	0.33–0.66
Clay	0.05–0.30	0.13–0.76	0.40–1.00	1.31–3.3

Table 11-1 VALUES OF SECANT MODULUS OF ELASTICITY AND COEFFICIENTS OF RESTITUTION FOR SEVERAL CAP-BLOCK AND PILE-CUSHION MATERIALS*

Material	E, ksi	E, kN/sq cm	Coefficient of restitution e
Micarta	450	310.2	0.80
Hardwood, oak	45	31.02	0.50
Asbestos disks	45	31.02	0.50
Plywood, fir	35	24.1	0.40
Pine	25	17.2	0.30
Softwood, gum	30	20.7	0.25
Steel on steel, using pipe piles			0.55
Using H-piles or concrete piles			0.50

* Data from Smith (1962) and Hirsch et al. (1970).

The side damping constant J_s is usually taken as

$$J_s = \frac{J_p}{3}$$

Bowles (1970) illustrates that quake and damping are not very critical; thus, as long as "reasonable" values are selected, say, Q around 0.1 to 0.15 and $J = 0.15$ to 0.30 (fps units), the computed results will be satisfactory most of the time. Obviously if the correct values are known, they should be used.

Pile Properties and Time Increment

The modulus of elasticity E, pile cross-sectional area A, and element lengths ΔL are needed to compute the element "spring" values as

$$K_m = \frac{AE}{\Delta L} \qquad \text{lb/in} \qquad \text{or} \qquad \text{kN/cm} \tag{11-11}$$

It is necessary to select a time increment T for use in the computations. Theoretically compression waves in an elastic material travel at a velocity of

$$v^2 = \frac{E}{\rho}$$

where ρ is the mass density (unit weight/gravity constant). Thus the time is

$$T = \frac{\Delta L}{v} = \frac{\Delta L}{\sqrt{E/\rho}}$$

which simplifies to

$$T = \sqrt{\frac{W_m}{K_m g}} \tag{11-12}$$

If the time interval is larger than this value, termed T_{cr}, the computations become unstable.[1] The pile cap (or helmet) and the first pile element (with a cushion) may have a different T than the pile segments; therefore, T in general should be somewhat less than T_{cr}. If T is too small, it takes many iterations to complete the computations. Smith (1962) proposed using

$$T \approx \frac{T_{cr}}{2}$$

[1] Generally determined by getting an overflow-error message on the computer.

and recommended in general that

Pile material	ΔL ft	m	T, sec
Steel	8–10	2.4–3.1	$\frac{1}{4000}$
Concrete	8–10	2.4–3.1	$\frac{1}{4000}$
Wood	8–10	2.4–3.1	$\frac{1}{3000}$

The author recommends $T_{cr}/2 < T < T_{cr}$ as a compromise value.

One should be aware that if the correct value of T for the system is used, the minimum number of iterations is obtained; if T is too big, the computations diverge; if it is too small, it will take many iterations. One must use a constant time interval since a variable time interval to match each element produces a rapid convergence but the results do not inspire confidence. One should try to use a time interval that will give results in 35 to 60 iterations.

Work by the author and also by Samson et al. (1963) indicates that the solution is only very slightly sensitive to T if the value is less than T_{cr} and well within the accuracy of the rest of the problem.

Pile-segment length is not too critical. Generally the lengths should be kept to between 5 to 10 ft (2 to $3\frac{1}{2}$ m). Too few elements may not yield the desired computational accuracy, but more than 10 to 12 segments do not improve the accuracy (an exception being a very long pile, where more than 12 segments may be required). Note, however, that a short pile with, say, 10 segments (small ΔL) may be troublesome due to increased sensitivity to T. For example, using a 30-ft concrete pile and 10 pile segments 3 ft in length requires the time interval T to be between 0.00023 and 0.00033 sec for a solution and uses many iterations. Much easier and just as satisfactory is the solution obtained using five 6-ft pile segments.

Initial Velocity

The displacement of the first element requires an initial velocity [V(M,1) in computer program] computed from the kinetic-energy equation

$$\tfrac{1}{2}Mv^2 = E_f \times \text{ram energy}$$

or

$$v = \sqrt{\frac{E_f \times \text{rated ram energy} \times 2g}{\text{weight of ram}}}$$

which further simplifies (as in computer program) to

$$v = \sqrt{2g(E_f) \times \text{ht of fall}} \qquad (11\text{-}13)$$

and requires the programmer to determine the height (or equivalent height) of ram fall if it is not given in the manufacturer's data sheets. Typical pile-hammer efficiencies are given in Table 11–2.

11-6 GENERAL APPLICATION OF THE WAVE EQUATION

The wave equation can be adapted to any pile configuration and impact driving method.

PIPE PILES

Without mandrel Obtain area and estimate total and segment weights, and total length and weight of point if driven closed-end. Manufacturers' catalogs can be consulted for pile-cap (or follower) weight. The approximate cap-block area can be estimated and its spring constant computed. A pile cushion is generally not used.

With mandrel Treat the mandrel as the pile but increase the weight for the pipe shell.

H-piles Treat like pipe pile. A pile cushion is generally not used because of pipe-cap (or follower) configuration.

Wood piles Same as H-piles.

Concrete piles Same as pipe or H-piles, but generally a pile cushion (wooden blocks) is placed between the pile cap and concrete to reduce driving stresses and prevent spalling.

Step-taper piles Piles are mandrel-driven. Mandrel has collars to fit onto the shoulders formed at step locations of pile sections. Treat as for pipe piles; i.e., mandrel is analyzed as pile. Segment weights include the contribution of pile shell segment. Springs, however, are obtained using average mandrel cross section (not at shoulder) within the segments. It is not necessary to modify the side resistance as

$$R_u = f\,(\text{diam})$$

as computed results are within data accuracy using either procedure.

Table 11-2 PILE-HAMMER EFFICIENCIES*

Hammer	Efficiency
Single-acting steam (or air)	0.75–0.85
Double-acting steam (or air)	0.70–0.80
Diesel	0.85–1.00

* Values are for hammers in good condition and ideal operating conditions.

Tapered piles Same as pipe piles if driven without a mandrel except area and weight are not constant from segment to segment. With a mandrel, treat as a step-tapered pile. Again it is not necessary to treat

$$R_u = f(\text{diam})$$

Effect of Soil Parameters

Soil parameters are not very critical in the wave-equation solution. The soil-parameter effect is approximately as follows (with the percent of load applied to the pile point held constant). The usual range of quake values of 0.1 to 0.3 will make maximum computed pile forces vary by not over 1 to 2 percent [see Bowles (1970)]. Increasing quake by 100 percent may decrease the set as much as 8 to 15 percent.

Damping constant J_p (and using $J_s = J_p/3$) will vary the maximum computed pile forces on the order of 1 to 2 percent for reasonable variations in J_p.

Varying the percent of load applied to the pile point influences both the maximum pile forces and the maximum point set (holding Q constant). The percent increase in maximum pile force for percent of R_u on the point is as follows:

R_u on point, %	Change in segment force, %
0–75	15–20
75–100	45–55

taking in both cases the base force at 0 percent R_u on the point. The pronounced effect on the maximum pile forces at 100 percent point load is actually due to the complete assumed loss of side resistance and damping. Probably very few practical pile problems should be analyzed as 100 percent point bearing when driven through any soil.

Pile set depends heavily on the assumed ultimate pile resistance R_u and the assumed percent point resistance. Varying the percent of point load from 0 to 75 percent can vary the set 50 to 100 percent; i.e., reducing the set to one-half the value at 0 percent point load is a 100 percent reduction. Increasing R_u also reduces the set; i.e., doubling R_u will reduce the set as much as 95 to 100 percent.

Plot of Ultimate Load versus Set in Blows per Inch

A feature of the wave equation is that one can make a plot of R_u versus $1/s$ (vary R_u from say 50 to 500 kips and obtain s at these loads). Mosley (1967) and Hirsch et al. (1970) present several load tests versus wave-equation analysis and consistently the wave equation predicts ultimate pile capacity within ±25 percent. This curve may be used to obtain another very useful bit of information. It is well known that piles may increase in ultimate resistance after driving has stopped. When splicing piles in the field this problem requires that the operation does not take too long. The phenomenon is termed *freezing*. Sometimes a pile loses resistance after driving halts, termed *relaxation*. This can be detected by redriving (also *retapping*) the pile and recording the new set obtained, which is then plotted on the curve (see Fig. 11-9). As an approximation the new ultimate resistance is the $1/s$ intercept on the curve projected to the load ordinate.

Incidentally, if the coordinate points (R_u, $1/s$) do not plot a reasonably smooth curve, an instability of some type exists (pile-hammer combination, time increment, or pile-cap spring, etc.).

It should be evident that one may try several input soil parameters, plot the resulting curves of R_u versus set (blows per inch), then plot the load-test value. That curve closest to the load test can be taken as the job curve.

Plot of Driving Stress versus Set in Blows per Inch

For a given R_u one can obtain the driving stresses by selecting enough point set values and the corresponding maximum pile-element forces; this means of course that one must run the curve through the origin and that for part of the analysis no point displacements DE(LA,i) occur. The maximum element force occurring prior to the point having a displacement value (as it usually does) simply fixes the asymptote of the curve.

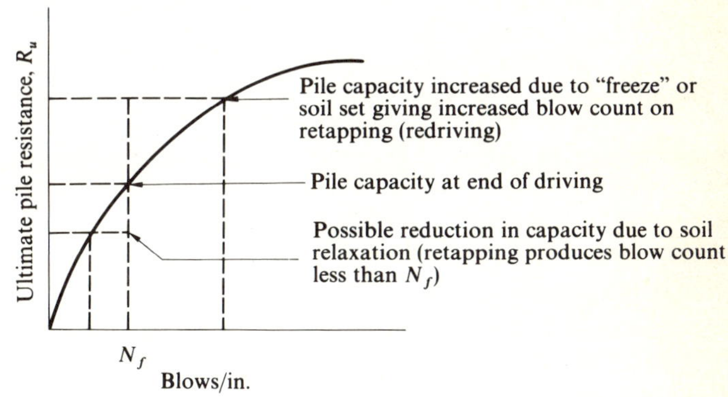

FIGURE 11-9
Using the wave-equation curve of R_u versus blows per inch to evaluate the increase or decrease in pile capacity at some elapsed time after driving the pile to a set of N_f blows per inch.

Negative Forces

It has long been recognized that tension forces sometimes exist in piles during driving. The wave equation is the only method so far available for analyzing the values quantitatively. Tensile (or negative) forces can be important when driving concrete piles. Tensile forces appear in driving friction (percent on point < 100) piles and may be present in driving point-bearing piles if the ultimate resistance and pile-hammer combination is critical. The computer program is set to reenter the computation loop to check negative forces, although so far the author has found that when the velocities are all negative and the program exits the first time, the largest negative force has generally already occurred. These negative forces may compute rather large. The negative forces do not, however, depend on the time increment (T = DT) selected for the computation.

Driving Point-Bearing Piles

Piles driven to refusal can be analyzed for driving stresses but may require an estimation until a pile is driven to evaluate the probable value of set. One can then try various large values of quake, say ≥ 0.5, to compute a compatible set. The quake yielding the best set fit is used for the analysis.

Other Input Parameters

The wave equation is rather sensitive to initial ram velocity. It is not very sensitive to the cap-block spring; i.e., doubling the spring value can increase the set 1 to 10 percent depending on R_u. A typical example with $R_u = 620$ kN and 25 percent R_u on point (other data of Example 11-3) gives the following:

Cap-block spring, kN/cm	s, cm	F_{max}, kN
3,500	1.87	1,581.2
7,000	1.88	1,609.3

The solution is not very sensitive to pile cross-sectional area (reflected in pile-segment springs); however, the weight does influence the problem considerably. This is illustrated [see also Mosley (1967)] in the examples plotted later, where a mandrel-driven step-taper pile (on the order of 200 lb/ft with mandrel) displays as much as

100 percent increase in load capacity for the same set (in blows per inch) over light pipe-pile sections.

Workability of Pile-Hammer Combinations

The wave equation will rapidly show whether a given system is satisfactory or not. Obviously when input of the desired R_u results in segment forces that are too large for the material, either the pile or hammer must be changed. Likewise, if the permanent set is zero to very small (the result is $1/s$ being very large) when the desired R_u is used as input, the hammer is too small or the pile is too heavy. If the computations blow up, it is a signal that all is not well with one of these input parameters or DT.

Gravity Effects

Smith (1962) indicated that gravity effects can be included in the computations. Samson et al. (1963) indicate the method of including gravity on a study which concludes that gravity effects are negligible.

Internal Damping

Smith also indicated that one might include internal damping by modification of Eq. (11-3) to read

$$F = C_m K_m + B K_m \frac{C_m - C'_m}{\Delta t} \tag{11-14}$$

Smith further suggested that B should be a small value such as 0.00016 to 0.00025. Since no data currently exist, the value of B is at present an estimation.

Since damping disappears when $D_m - D_{sm} = 0$ in Eqs. (11-9) and (11-9*a*), it was also recommended to consider the use of

$$R_m = (D_m - D_{sm})K'_m + J'K'_m Q v_m \tag{11-15}$$

and

$$R_p = (D_p - D_{sp})K'_p + JK_p Q v_p \tag{11-16}$$

after $D_m - D_{sm}$ or $D_p - D_{sp}$ first equals Q. This would require additional subroutines. The loss of damping effect is the primary cause of the large increase in the computed internal segment forces obtained when 100 percent of R_u is carried by the pile point. To avoid this loss of damping and still be realistic, few piles should be analyzed as 100 percent point-bearing.

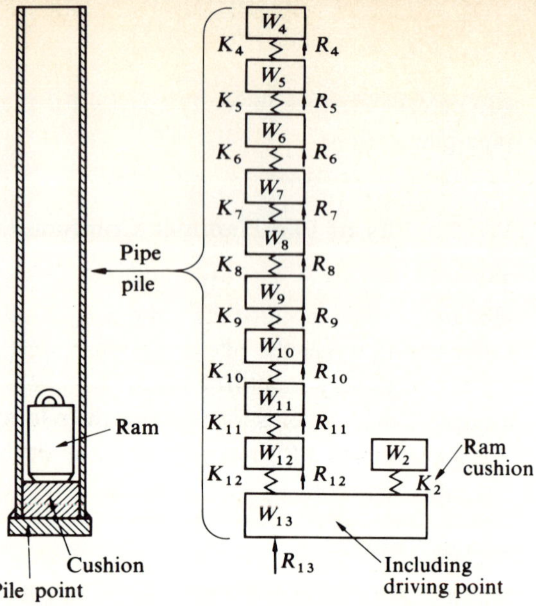

FIGURE 11-10
Pipe pile with point driving.

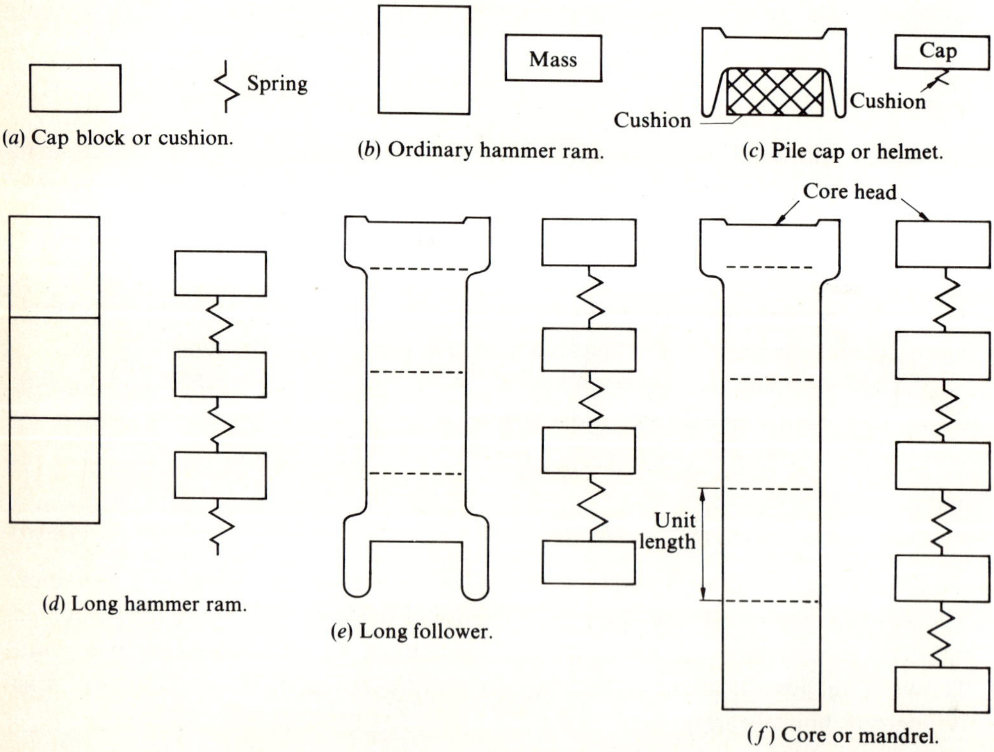

FIGURE 11-11
Method of modeling various types of driving equipment for use in the wave equation. For the mandrel find the average area of the drive stem to compute springs; add the shell weight to the mandrel to obtain element weights and treat the mandrel as a pile.

Driving Methods and Systems

Point driving is illustrated in Fig. 11-10, from which it is evident that the computer program requires modification, as the spring force in K_2 is in general

$$F(2) = K(2)(D(2,2) - D(LA,2))$$

likewise spring K_{12} is stretched rather than compressed, as the pile will be dragged down.

Other possibilities of driving systems are shown in Fig. 11-11. The designer must make necessary modifications in the computer program to take care of these systems, but the use of a mandrel does not require program modification. Treat the mandrel as the pile and estimate its shaft cross section if not accurately known for computing the pile-segment springs. To obtain segment weights add the mandrel to the shell weight.

11-7 WAVE-EQUATION EXAMPLES

EXAMPLE 11-1 Plot a curve of R_u versus set and driving stresses versus set for R_u = 200 kips for the pipe pile shown [pile *B* given in Mosley (1967)]; refer to Fig. E11-1.1. Use the following input data:

10 pile segments at 8 ft (NELEM = 10 + 2 = 12, ELEML = 8.0)
Use linear variation of R_u (JJS = 0)
Vary R_u (NCHECK = 3)
Not all element forces needed (IFWRIT = 0)
Constant pile section (PILTYP = 1.)
Wall thickness of pipe pile = 0.25 in (use either 0. or 0.25)
Negative forces not needed (COMFM = 0)
Ram weight = 6.5 kips [W(2)]; pile cap = 0.925 kips
Ram fall = 3 ft (FALL); hammer efficiency = 0.8 (EFF)
Weight of drive point = 0.0 lb (DRIVPT); time interval = 0.00025 sec (DT)
J_s = 0.05 sec/ft (SJ); J_p = 0.15 sec/ft (PJ)
Pile cushion = 0.0 (SC)
Modulus of elasticity of pile = 30,000.0 ksi (EMOD)
e cap block = 0.80 (EPCB); *e* pile cushion = 1.00 (EPC)
Weight of pile per foot = 0.0338 kips (WFT)
Pile cross section = 9.817 sq in (AREA)
Q = 0.10 in (Q); $\%R_u$ on point = 0.25% (PER) pile driven open-end
R_u = 100 to 400 kips (RUTOT)
Spring constant of cap block = 4,500 kips/in [SP (2)]

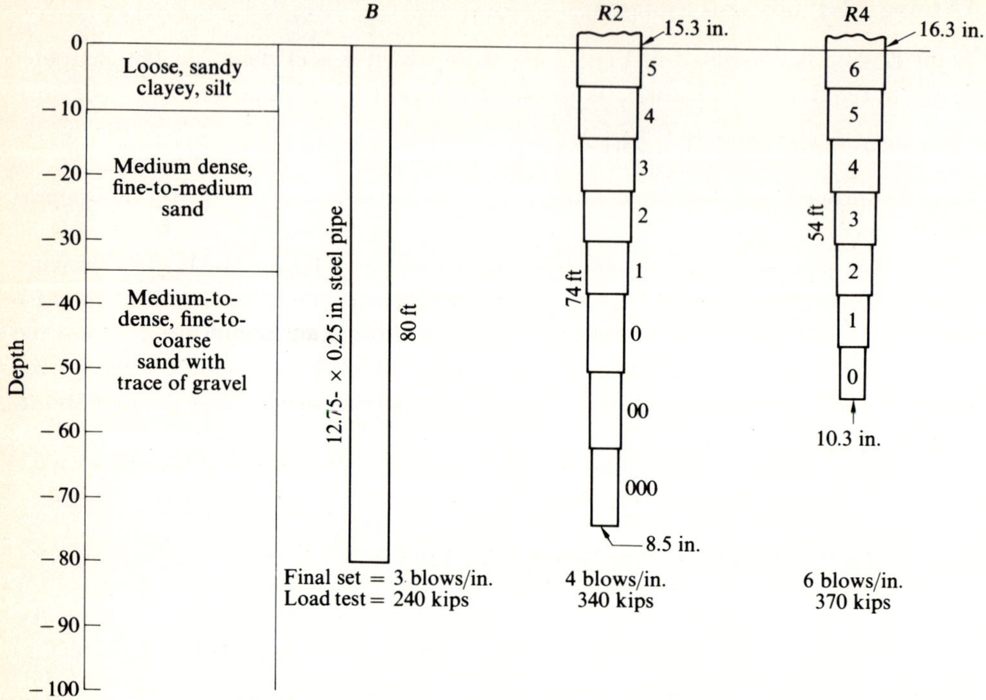

FIGURE E11-1.1
Pile profiles. Pile *B* is used in Example 11-1; pile R4 is used in Example 11-2. [*Mosely* (1967).]

SOLUTION Data cards are as follows:

Card	Data
1	TITLE
	UT1–UT7
2	FT IN KIPS KIPS/IN KIPS/SQ IN FT/SEC SQ IN
	FU1–FU3
3	12. 144. 32.2
4	12 0 3 0 1. 8.00 0.0 0.
	Could use t_{wall} = 0.25 in
5	6.5 .925 3.00 .80 0.0 .00025
6	.050 .150 0.00
7	30000.00 .80 1.00
8	.0338 9.817
9	.100
10	.25
11	100. 4500.
12	200. 4500. NCKECH = 3 recycles to 5150 to read cards 12 to 16
↓	
16	400. 4500.

Computer output is as follows (maximum force occurred in first pile segment in all cases):

R_u, kips	F_{max}, kips	Average set s, in	$1/s$	No. of iterations I
100	269.0	0.914	1.09	117
150	271.7	0.620	1.61	81
200	274.3	0.388	2.58	68
250	276.9	0.198	5.05	63
275	284.3	0.121	8.26	60
300	296.3	0.052	19.2	59
325	308.2	0.009	111.1	57
350				>150

The plot of ultimate resistance R_u versus blows per inch is shown in Fig. E11-1.2.

For $R_u = 200$ kips the following additional output and computations are made:

DT	s, in	F_{max}, kips	Blows/in	σ, ksi
11	0	274.3*	∞	27.94
26	0.025	241.5	40	24.60
29	0.103	227.6	9.7	23.2
39	0.179	223.8	5.6	22.8
47	0.280	191.0	3.6	19.5
54	0.339	186.3	2.9	19.0
64	0.388*	165.7	2.6	16.9

* Maximum.

$$\sigma = \frac{274.3}{9.817} = 27.94 \text{ ksi}$$

A plot of σ versus blows per inch is shown in Fig. E11-1.3. ////

EXAMPLE 11-2 Plot a curve of R_u versus set (blows per inch) for pile R4 (refer to Fig. E11-1.1).

Use the following input data for R4:

7 pile segments at 8 ft (56 versus 54 ft actually in system)
Ram 6,500 lb; height of fall = 3.00 ft; Efficiency = 0.80
$J_s = 0.05$ sec/ft; $J_p = 0.15$ sec/ft; $Q = 0.10$ in
Pile cap block = 14,000 kips/in; $e = 0.80$
No cushion, $\Delta t = 0.00025$ sec
Drive point estimated at 100 lb; % of R_u on point = 50%

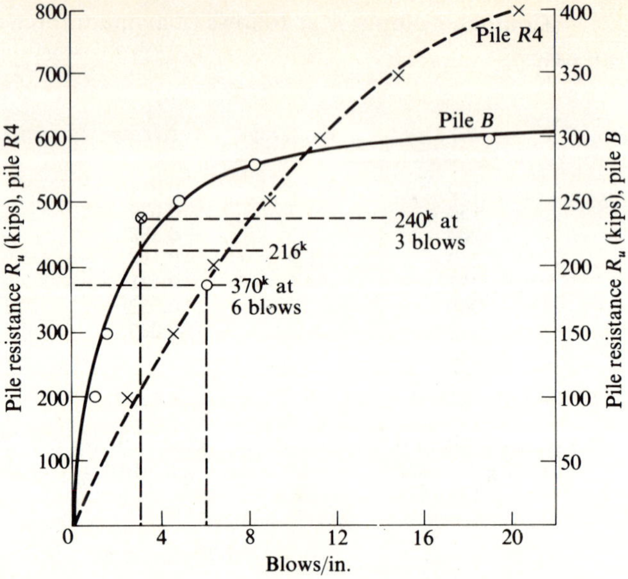

FIGURE E11-1.2
Plot of R_u versus blows per inch. Pile *B* is from Example 11-1; pile R4 is from Example 11-2. Field load-test data also plotted. Note about 10 percent error for pile *B* and apparent correct wave-equation model for pile *R*4.

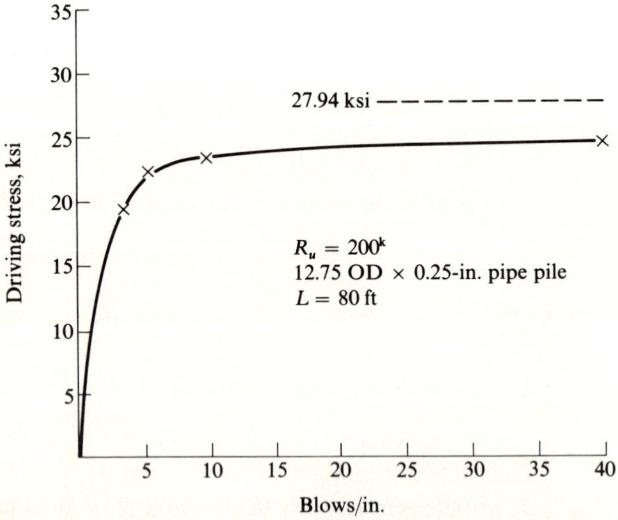

FIGURE E11-1.3
Plot of set versus driving stresses for R_u = 200 kips, percent point = 0.25. Other data in Example 11-1.

SOLUTION The pile consists of step-taper sections numbers 0 to 6 at 1-in increase per section (see Tables B-6*a* and B-6*b* in Appendix B).

0 = 10.375 in

1 = 11.375 in

6 = 16.375 in

An inspection of core weights indicates 13,000 lb for this system, which will be taken as 1,290 lb per foot of shaft and 210 lb concentrated at the shoulders where sections join (inspection of appendix tables indicates that the weight of the core section is 1,500 lb). The pile weight per foot is the weight of the mandrel section divided by 8.0 plus shell

$$\text{Weight per ft} = \tfrac{1500}{8} = 187.50 \text{ lb}$$

$$+ \text{ shell} = \underline{10.00} \text{ est}$$

$$197.50 \text{ lb}$$

The pile cap weighs 13,000 − 7(1,500) = 2,500 lb. The computed cross-sectional area is 197.5(144)/490 = 58.05 sq in. The load-test value is 370 kips; final field resistance is 6 blows per inch. For these input data the following output is obtained (plotted on Fig. E11-1.2):

R_u, kips	$1/s$	s, in	I	F_{max} (element 4),* kips
200	2.64	0.3782	78	1,086.0
300	4.69	0.2132	53	1,064.0
400	6.67	0.1500	40	1,065.2
500	9.12	0.1097	36	1,066.4
600	11.49	0.087	31	1,067.6
700	15.04	0.0665	29	1,068.8
800	20.49	0.0488	27	1,070

* These large values are primarily in the core (or mandrel).

EXAMPLE 11-3 Analyze a 30 × 30 cm × 24 m concrete pile for the effect of the pile cushion on compression and tension driving stresses. Use only two pile-cushion values. Use a British 7b hammer. Use metric UNITS card (equivalent to those used in Example 11-1) and FU1 = 100., FU2 = 10000., and FU3 = 9.807. Input data:

Ram wt = 29.90 kN; fall = 1.37 m; E = 0.75
Pile cap = 5.34 kN; pile cap block = 3,500 kN/cm; e = 0.80
Pile cushion = 1,051 and 2,101 kN/cm
e for both cushions = 0.50

E_c = 2,896 kN/sq cm; element length = 2.4 m
J_s = 0.164 sec/m; J_p = 0.492 sec/m; Q = 0.254 cm
Pile weight per meter = 2.101 kN/m; driving point = 0.0 kN
Δt = 0.000333 sec

SOLUTION Varying the percent of R_u on point and the pile-cushion spring (SC), we have the following typical values:

PER	I	F_{max}	DT	s, cm	F_{min}	DT	Element no.
		Spring cushion = 1,051 kN/cm			R_u = 620 kN		
0	96	1,614.2	32	2.08	−870.5	54	11
0.25	95	1,581.2	31	1.87	−662.1	60	8
0.50	94	1,550.5	31	1.76	−483.2	60	8
0.75	81	1,528.8	33	1.66	−314.1	67	5
1.00	80	1,534.0	33	1.59	−277.1	67	5
		Spring cushion = 2,101 kN/cm			R_u = 620 kN		
0	124	1,921.3	26	2.48	−1,238.1	55	9
0.25	113	1,877.4	25	2.09	−905.1	54	9
0.50	95	1,863.3	30	1.83	−631.1	57	8
0.75	91	1,862.9	30	1.72	−464.6	64	4
1.00	90	1,865.8	32	1.62	−421.6	64	4

These data indicate that decreasing the pile cushion (SC) by one-half reduces the negative (tension) stresses from 30 to 50 percent depending on the percent of R_u applied to the pile point. The effect on the compressive stresses is considerably less.

11-8 WAVE-EQUATION COMPUTER PROGRAM

This program will compute segment weights and spring constants for constant-section piles of any shape. It will also compute properties (average) for round tapered (hollow or solid) piles and allow reading individual segment areas and weights for stepped piles. Several problems or parameters can be studied through use of a computed GO TO statement via use of NCHECK. Output may be pile-segment forces at each time interval, as well as point displacements and maximum forces in a segment at each time interval. Also given is the maximum force ever obtained in any pile segment and, as a check, the last force and element velocity computation. For concrete piles the program will obtain the maximum negative (tension) force if specified via use of COMFM. This program automatically excludes the soil-side resistance on the first pile segment regardless of segment length unless read when JJS > 0.

Program scans the point set values [DE(M,2)] for the five largest values unless values differ by more than 0.005 unit, sums the values, and divides by the number of values used. Average set value and the number of values used are printed for convenience in plotting the curve of $1/s$ versus R_u. This program will solve problems in either fps or metric units, as specified by the user through data cards UT1-UT7 and FU1-FU3. Refer to Examples 11-1 and 11-3 for card entry values.

Line	Operation
3	READ TITLE, UT1–UT7 (use two cards)
7	READ FU1, FU2, FU3
8	READ (4I5, 4F10.2) NELEM = number of elements including ram and cap; JJS = 0 to compute linear distribution of RU on sides of pile segments; JJS = 1 if READ distribution of RU on the sides of pile segments; NCHECK = counter in computed GO TO statement; 1 = vary Q other data to Q = constant; 2 = vary point load % of RU; 3 = vary RU; 4 = change all problem data including TITLE; 5 = stop after 1 run IFWRIT = control to write element forces; 0 = does not write; ≥1 = writes element forces for each time interval in groups of 7 PILTYP = pile type; >0 = constant-area section; 0 = read area and weight of each pile segment; <0 = compute average area and weight of each pile segment ELEML = length of pile segments, feet or meters; TWALL = wall thickness of pipe piles; use 0.00 if PILTYP ≥ 0; use radius if program computes area of solid round piles COMFM = switch to compute negative (tension) element forces; 0 = negative forces not required; >0 = maximum tension force obtained
10	READ (6F10.3) W(2) = ram weight, kips or kilonewtons; W(3) = pile-cap weight, kips or kilonewtons; FALL = ram fall, feet or meters; EFF = hammer efficiency; DRIVPT = weight of drive point, kips or kilonewtons; DT = time interval, seconds
12	READ (6F10.4) leave card blank for spaces not used SJ = J_s, sec/ft or sec/m; PJ = J_p, sec/ft or sec/m; SC = spring constant pile cushion, kips/in or kN/cm
13	READ (6F10.3) EMOD = E pile, ksi or kN/sq cm; EPCB = coefficient of restitution cap block; EPC = coefficient of restitution cushion
17	Checks pile type (PILTYP)
18	READ (6F10.4) constant-section piles WFT = weight per unit length of pile, kips or kilonewtons; AREA = cross-sectional area of pile, sq in or sq cm
22	READ (6F10.3) read element weights and areas of each pile segment
24	READ (6F10.3) DIAT = top outside diameter of pile, inches or centimeters; DIAB = bottom diameter of pile, inches or centimeters; UNITWT = unit weight of pile material, kcf or kN/cu m
25–34	Computes tapered-pile properties including spring constants
36–41	Computes spring constants for all piles and element weights of constant-section piles
43–44	Computes modified spring constant for first pile segment if a pile cushion used
50	READ Q (loop for NCHECK = 1) 0 = quake, inches or centimeters
52	READ PER (loop for NCHECK = 2) PER = % RU on pile point
54	READ (loop for NCHECK = 3) RUTOT = assumed ultimate pile resistance, kips or kilonewtons; SP(2) = spring constant cap block, kips/in or kN/cm
71–73	Computes side resistances excluding first pile segment

Line	Operation
75	READ (IF JJS > 0) Read I, RU(I) at four values per card if other than linear variation of RUTOT is assumed for side resistance
81	Computes soil spring constants FK(M)
80–99	Initializes variables F, V, R, C, D and sets counters
100	Begins DO loop for computing element forces, velocities, etc.
103–104	Computes element displacements
108–150	SUBROUTINE NO. 1 through 4
153–155	Computes pile element forces [F(M,2)]
155	Stores computed forces for writing [B(M,KL)]
159–172	Checks negative (tension) forces to stop computations
173–177	Finds largest segment force, location, and time interval of occurrence
196–204	Writes element forces if required
205–217	Computes new element velocities; checks if all negative; checks current pile point set against value retained from last time interval
218–223	Redefines all variables from subscripts (M,2) to subscripts (M,1)
225	End of iteration loop
226–234	Finds up to five largest point set values to average for the pile set for that R_u value
234–268	Writes various data
269	GO TO statement activated by NCHECK

```
            C      J E BOWLES  WAVE EQUATION FCR PILE RESPONSE TO IMPACT TYPE DRIVING
            C      PILE MAY BE STRAIGHT,TAPERED, STEPPED OR H-SECTIONS
            C****************       *****************      ********************
0001               DIMENSION C(15,2),F(15,2),V(15,2),B(15,7),W(15),SP(15),D(15,2),DE(
                  A15,2),EK(15),R(15),IX(15),RU(15),AF(15,2),DE1(150),D1(150),KKK(20)
                  B,FMAX(150),JX(150),FM(15),A(15),CB(15),DI(15),FMIN(15,2),TITLE(20)
C002               DOUBLE PRECISION UT4,UT5,UT6,UT7
C003          5C00 READ(1,1000,END=150)TITLE,UT1,UT2,UT3,UT4,UT5,UT6,UT7
0004          1000 FORMAT(20A4,/,3(A4,6X),4(A8,2X))
C005               WRITE (3,1001)TITLE
0006          1C01 FORMAT('1',//,T5,20A4)
0007               READ(1,501)FU1,FU2,FU3
0008           100 READ(1,500)NELEM,JJS,NCHECK,IFWRIT,PILTYP,ELEML,TWALL,COMFM
CC09           500 FORMAT(4I5,4F10.2)
0010               READ(1,501)W(2),W(3),FALL,EFF,CRIVPT,DT
0011           501 FORMAT(6F10.3)
0012               READ(1,501)SJ,PJ,SC
0013               READ(1,501)EMOD,EPCB,EPC
0014               LA = NELEM + 1
0015               KO = NELEM-2
0016               X = KO
0017               IF(PILTYP)63,62,61
0018            61 READ(1,501)WFT,AREA
0019               CO 85  I = 3,NELEM
002C            85 A(I) = AREA
0021               GO TO 97
0022            62 READ(1,501)(W(M+1),A(M),M=3,NELEM)
0023               GO TO 97
0024            63 READ(1,501)DIAT,DIAB,UNITWT
0025               DELD = (DIAT-DIAB)/X
0026               BB = 0.5
0027               DO 94  I = 1,KO
0028               DB(I) = DIAT - BB*DELD
C029               DI(I) = DB(I) - 2.*TWALL
C030               A(I) = 0.7854*(DB(I)**2 - DI(I)**2)
0031               SP(I+2) = A(I)*EMOD/(ELEML*FU1)
0032               W(I+3) = A(I)*UNITWT*ELEML/FU2
0033               BB = BB+1.
0034            94 CONTINUE
0035               GO TO 98
0036            97 DO 31 M = 3,NELEM
0037            31 SP(M) = A(M)*EMOD/(ELEML*FU1)
0038               IF(PILTYP.LE.0.)GO TO 98
0039               DO 30  M = 4,LA
0C40            30 W(M) = ELEML*WFT
0041            98 W(LA) = W(LA) + DRIVPT
0042               IF(SC)60,6C,59
0043            59 SPX = (1./SP(3))+(1./SC)
0044               SP(3) = 1./SPX
0045            60 CONTINUE
0046               ECB = 1./EPCB**2
0047               EPC2 = 1./EPC**2
0048               T = DT*FU1
0049               G = DT*FU3
0050          5C50 READ(1,501)Q
```

```
      IF(Q.LE.0.)GO TO 5000
 5100 READ(1,501)PER
      IF(PER.LT.0.)GO TO 5000
 5150 READ(1,501)RUTOT,SP(2)
      IF(RUTOT)150,5000,6000
 6000 WRITE(3,300)PER
  300 FORMAT(///,T20,'**** % OF RUTOT CARRIED BY PILE POINT =',F6.3)
      WRITE(3,601)X,ELEML,UT1,NELEM,EMOD,UT5
  601 FORMAT(//,T5,'** GENERAL INPUT DATA:',/,T6,'NO OF PILE SEGMENTS ='
     1,F5.1,3X,'LENGTH OF PILE ELEM =',F7.3,1X,A2,/,T6,'NO OF ELEMENTS I
     2NCL RAM & CAP =',I4,/,T6,'PILE MCDULUS OF ELAST =',F11.1,1X,A8,//)
      IF(PILTYP.LE.0.)WRITE(3,602)(W(M),M=4,LA),UT3,(A(N),N=1,KO),UT7
  602 FORMAT(T10,'COMPUTED PILE SEGMENT PROPERTIES',/,T5,'SEG WTS =',5F1
     10.4,/,T19,5F10.4,1X,A4,/,T5,'SEG AREAS =',5F10.4,/,T19,5F10.4,1X,A
     35,//)
      IF(PILTYP.GT.0.)WRITE(3,603)UT1,WFT,UT3,AREA,UT7
  603 FORMAT(/,T5, 'WT/',A2, ' OF PILE =',F9.4,1X,A4,3X, 'PILE X-SECT ='
     1,F8.3,1X,A5,/)
      WRITE(3,604)UT3,W(2),W(3),W(LA),DRIVPT,FALL,UT1,EFF,SJ,PJ
  604 FORMAT(T5,'ELEM WTS,',1X,A4,':',3X,'RAM =',F8.3,3X,'PILE CAP =',F8
     A.4,/,T15, 'WT BOT ELEM + DRIVE PT =',F8.4,3X,'WT DRIVE PT =', F8.4
     B,/, T6,'HT OF RAM FALL =',F6.3,1X,A2,5X, 'HAMMER EFF =',F5.2,/,T5,
     C 'SIDE DAMP CONST,SJ =',F6.3,3X,'POINT DAMP CONST,PJ =',F6.3,//)
      WRITE(3,605)UT4,SP(2),SC,SP(3),SP(4),EPCB,EPC,RUTOT,UT3
  605 FORMAT(T5,'SPRING CONST,'1X,A5,':',3X,'CAPBLOCK =',F12.3,3X,'PILE
     1CUSHION =',F12.3,/,T8,'1ST PILE SEG =',F12.3,3X,'2ND PILE SEG =',F
     212.3,/,T5,'COEFF OF RESTIT PILE CAPBLOCK =',F6.3,3X,'COEFF OF REST
     3IT PILE CUSHION =',F6.3, /,T5,'*********** ASSUMED ULTIMATE PILE R
     4ESISTANCE =',F10.2,1X,A4,//)
      WRITE(3,618)UT3
  618 FORMAT(T15,'I',T22,'RU(I),',1X,A4)
      IF(JJS.EQ.1)GO TO 612
  102 DO 4   M = 4,NELEM
    4 RU(M) = (1.-PER)*RUTOT/(X-2.)
      RU(4) = 0.0
      GO TO 616
C     READ SIDE RESISTANCES AT 1-SEG/CARD--USE 0. AS REQ'D  (WHEN JJS>0)
  612 READ(1,615)(I,RU(I),I=4,NELEM)
  615 FORMAT(4(I5,F10.4))
  616 RU(LA) = RUTOT*PER
      WRITE(3,621)(M,RU(M),M=4,LA)
  621 FORMAT(T14,I2,T18,F12.2)
  625 DO 8   M = 3,LA
      IF(M.GT.3)EK(M) = RU(M)/Q
    8 V(M,1) = 0.0
      V(2,1) = SQRT (FALL*EFF*FU3*2.)
      F(1,2) = 0.
      DO 3   M = 2,LA
      R(M) = 0.0
      C(M,1) = 0.
    3 D(M,1) = 0.
      DE(LA,1) = 0.
C     *****SET COMPUTATION CONSTANTS TO ZERO
      KL = 0
      BMIN = 0.
      NMIN = 0
      IMIN = 0
      KCOUN = 0
      KKK(1) = 0
      DEMAX = -1.
      WRITE(3,404)DT
  404 FORMAT(//,T5, 'TIME INTERVAL =',F10.7,' SEC',//)
C     ***  START LOOP TO COMPUTE PILE SEGMENT FORCES
      I = 0
 1898 I = I+1
      MM = I
      LSUM = 0
C     **D(LA,2) IS DP OF SMITH'S PAPER = INSTANT. POINT DISPLACEMENT**
      DO 6   M = 2, LA
    6 D(M,2) = D(M,1) + V(M,1)*T
      DO 41   M = 4,NELEM
      DE(M,2) = 0.
      IF(PER.EQ.1.)GO TO 41
C******* SUBROUTINE NO 1*** TO OBTAIN PLASTIC DEFORM OF TOTAL DEFORM
      IF(Q-D(M,2).GT.0.0)GO TO 40
      DE(M,2) = D(M,2)-Q
      GO TO 41
   40 IF(D(M,2)+Q.GE.0.0)GO TO 41
      DE(M,2) = D(M,2)+Q
   41 CONTINUE
C    ***** SUBROUTINE NO. 2 *****
C     **DE(LA,2) IS D PRIME P OF SMITH'S PAPER = PERMENANT SET OF PILE**
C     **DE(LA,2) CANNOT BE LESS THAN C(LA,2) - Q********
      DE(LA,2) = 0.0
      IF(Q-D(LA,2).GT.0.0)GO TO 15
      DE(LA,2) = D(LA,2) - Q
      IF(PER.EQ.1.)GO TO 39
   15 DO 42 M = 4, NELEM
   42 R(M) = (D(M,2)-DE(M,2))*EK(M)*(1.+SJ*V(M,1))
   39 R(LA) = (D(LA,2) - DE(LA,2))*EK(LA)*(1.+PJ*V(LA,1))
      CALL OVERFL(IV)
      IF(IV.NE.2)GO TO 2596
C     FIND MAX POINT SET VALUE (DE(LA,2)) FOR LATER AVERAGING OF SET VAL
      DE1(I) = DE(LA,2)
      IF(DE1(I).LT.DEMAX)GO TO 628
      DEMAX = DE1(I)
```

```
  628 D1(I) = D(LA,2)
      DO 43 M = 2, NELEM
   43 C(M,2) = D(M,2) - D(M+1,2)
C ******* SUBROUTINE NO. 3   *** CCMPUTE FORCE IN PILE CAP
      IF(I.EQ.1)GO TO 18
      IF(C(2,2).LT.C(2,1).AND.IM.EQ.0)GO TO 20
      IF(C(2,2).LT.CMAX)GO TC 20
   18 IF(C(2,2).GE.C(2,1))IM = 0
      F(2,2) = C(2,2)*SP(2)
      GO TO 17
   20 IM = IM+1
      IF(IM.EQ.1)CMAXH = C(2,1)
      IF(IM.GE.1)CMAX = CMAXH
      F(2,2) = SP(2)*CMAX + SP(2)*ECB*(C(2,2)-CMAX)
   17 IF(F(2,2).LT.0.)F(2,2) = 0.
C ******* SUBROUTINE NO. 4   **   CCMPUTE FORCE IN 1ST PILE SEGMENT
C       SETS F(3,2) NOT LESS THAN ZERC--IF F(3,2) CAN CARRY TENSION
C         REMOVE STATEMENT NUMBERED 24 BELCW
      IF(I.EQ.1)GO TO 19
      IF(C(3,2).LT.C(3,1).AND.KM.EQ.0)GC TC 25
      IF(C(3,2).LT.CMAXX)GO TO 25
   19 IF(C(3,2).GE.C(3,1))KM = 0
      F(3,2) = C(3,2)*SP(3)
      GO TO 24
   25 KM = KM+1
      IF(KM.EQ.1)CMAXXH = C(3,1)
      IF(KM.GE.1)CMAXX = CMAXXH
      F(3,2) = SP(3)*CMAXX + SP(3)*EPC2*(C(3,2)-CMAXX)
   24 IF(F(3,2).LT.0.)F(3,2) = 0.
C ******* COMPUTE FORCES IN REMAINING PILE SEGMENTS*************
   29 KL = KL+1
      IX(KL) = I
      DO 105  M = 4, NELEM
      F(M,2) = C(M,2)*SP(M)
  105 B(M,KL) = F(M,2)
      F(LA,2) = R(LA)
      IF(KCOUN.LE.0)GO TO 2235
 2222 KKK(2) = 0
C     TEST FOR TENSION (NEGATIVE) SEGMENT FORCES --CONC PILES
      DO 2223  M = 2,LA
      FM(M) =-F(M,2)
      IF(F(M,2).GE.FM(M))GO TO 2223
      IF(F(M,2).LT.F(M,1))KKK(2) = KKK(2)+1
 2223 CONTINUE
 2235 B(2,KL) = F(2,2)
      B(3,KL) = F(3,2)
      B(LA,KL) = R(LA)
      IF(I.GT.1) GO TO 109
      DO 115  N = 2,LA
      AF(N,1) = B(N,KL)
      FMIN(N,1) = B(N,KL)
      FMIN(N,2) = 1.
  115 AF(N,2) = 1
C     FIND MAXIMUM ELEMENT FORCE AND CCRRESPONDING DT
  109 DO 108  N = 2,LA
      IF(B(N,KL).LE.AF(N,1))GO TO 107
      AF(N,1) = B(N,KL)
      AF(N,2) = I
  107 IF(COMFM.LE.0.)GO TO 108
      IF(B(N,KL).GT.FMIN(N,1))GC TC 106
      FMIN(N,1) = B(N,KL)
      FMIN(N,2) = I
  106 IF(N.LT.4.OR.N.GT.NELEM)GC TC 108
      IF(B(N,KL).GE.BMIN)GO TO 108
      BMIN = B(N,KL)
      NMIN = N
      IMIN = I
  108 CONTINUE
  283 IF(KCOUN.LE.0)GO TO 2240
      IF(KKK(2).LE.KKK(1))GO TO 2402
      KKK(1) = KKK(2)
 2240 FMAX(I) = B(4,KL)
      JX(I) = 4
      DO 55  K = 5,LA
      IF(B(K,KL).GT.FMAX(I))FMAX(I) = B(K,KL)
      IF(B(K,KL).EQ.FMAX(I))JX(I) = K
   55 CONTINUE
      IF(KL.LT.7)GO TO 131
      IF(I.LE.7.AND.IFWRIT.GT.0)WRITE(3,630)UT3
  630 FORMAT(//,T15,'COMPUTED PILE SEGMENT FORCES IN',1X,A4,//)
      IF(IFWRIT.LE.0)GO TO 130
      WRITE(3,297)(IX(NN),NN=1,7)
  297 FORMAT(//,T7,'DT =',3X,I4,6(7X,I4))
      WRITE(3,298)((B(II,JJ),JJ=1,7),II=2,LA)
  298 FORMAT(9X,7F11.4)
  130 KL = 0
  131 DO 44 M = 2, LA
      IF(M.EQ.LA)GO TO 32
      V(M,2) = V(M,1)+(F(M-1,2)-F(M,2) - R(M))*G/W(M)
      GO TO 46
   32 V(M,2) = V(M,1) + (F(M-1,2)-R(M))*G/W(M)
   46 CALL OVERFL(IV)
      IF(IV.NE.2)GO TO 2596
      IF(V(M,2).GT.0.0)GO TO 44
      LSUM = LSUM + 1
   44 CONTINUE
```

```
      IF((LSUM+1-LA).LT.0)GO TO 54
      IF(DE(LA,2).LT.DE(LA,1))GO TC 48
   54 LSUM = 0
      DO 49  M = 2, LA
      D(M,1) = D(M,2)
      F(M,1) = F(M,2)
      IF(M.NE.LA)C(M,1) = C(M,2)
   49 V(M,1) = V(M,2)
      DE(LA,1) = DE(LA,2)
      IF(KCOUN.EQ.0.AND.I.EQ.150)GC TC 2595
 1111 IF(I.LT.150)GO TO 1898
C     END OF SEGMENT FORCE LOOP AND SET FOR 300 ITERATIONS
C     FIND LARGEST SET VALUES IF NCT LESS THAN 0.005 OF MAX VALUE
   48 DSET = 0.
      LL = 0
      DO 56  L = 1,MM
      DIFF = DEMAX - DE1(L)
      IF(ABS(DIFF).GT.0.005)GO TC 56
      DSET = DSET + DE1(L)
      LL = LL + 1
   56 CONTINUE
      SET = DSET/LL
      WRITE(3,301)MM,Q,UT2,SET,UT2,LL
  301 FORMAT(//,T5,'NUMBER OF ITERATICNS =',I4,3X,'QUAKE =',F5.3,1X,A2,3
     1X,'AVERAGE SET =',F7.4,1X,A2, 3X, 'NO OF VALUES USED =',I2,//)
      IF(KCOUN.GT.0)GO TO 2402
      DO 207  M = 1,MM,11
      M1 = MIN0(M+10,MM)
      WRITE(3,302)(K,K=M,M1)
  302 FORMAT(T3, 'DT=',T7,I3,10(6X,I3))
      WRITE(3,305)(DE1(K),K=M,M1)
  305 FORMAT(T2,'SET=',T6,11F9.5)
      WRITE(3,306)(D1(K),K=M,M1)
  3C6 FORMAT(T3, 'D=',T6,11F9.5)
      WRITE(3,307)(FMAX(K),K=M,M1)
  307 FORMAT(T3,'F=',T6,11F9.3)
  207 WRITE(3,309)(JX(K),K=M,M1)
  309 FORMAT(T3, 'ELEM NO',T10,I3,10(6X,I3))
      WRITE(3,308)UT3,UT6
  308 FORMAT(//,T15, 'THE FCRCES IN PILE SEGMENTS ARE AS FOLLOWS',//,T7,
     1 'MAX ELEM FORCES AT DT',T37,'LAST FORCE COMP,'1X,A4,T65,'LAST V(M
     2,2),'1X,A6,/)
      WRITE(3,304)(N,(AF(N,M),M=1,2),F(N,2),V(N,2),N=2,LA)
  304 FORMAT (I8,2X,F12.2,6X, F5.0, 5X, F12.2, 8X, F12.2)
      KCOUN = KCOUN+1
      IF(COMFM.GT.0..AND.MM.NE.150)GC TO 54
      GO TO 2600
 2402 WRITE(3,320)I,UT3,UT6
  320 FORMAT(///,T5, 'MIN ELEM FORCES (CONC PILES) AT DT =',I5,3X,'LAST
     1FORCE CCMP,',1X,A4,4X,'LAST V(M,2),',1X,A6,//)
      WRITE(3,322)((FMIN(M,L),L=1,2),F(M,2),V(M,2),M=2,LA)
  322 FORMAT(T5,F14.3,5X,F5.0,T45,F14.3,T70,F14.5)
      WRITE(3,631)BMIN,UT3,NMIN,IMIN
  631 FORMAT(//,T10,'MIN NEG. SEGMENT FCRCE =',F12.4,1X,A4, 3X,'IN ELEME
     1NT',I3,' AT ITERATION NO',I3,//)
      IF(KCOUN.GT.0)GO TO 2600
 2595 WRITE(3,325)
  325 FORMAT(////,T5,'***** PROBLEM UNSTABLE OVER 150 ITERATIONS -- DT T
     1OO SMALL *******',//)
      GO TO 2600
 2596 WRITE(3,324)
  324 FORMAT(////,T5,'***** PROBLEM UNSTABLE CHANGE ONE OF FOLLOWING:',/
     A,T8,'PILE HAMMER',/,T8,'DECREASE DT',/,T8,'INCREASE OR DECREASE RU
     B',/,T8, 'CHANGE CAPBLOCK OR CUSHICN SPRING CONSTANT',////)
 2600 GO TO(5050,5100,5150,5000,150),NCHECK
  150 STOP
      END
```

PROBLEMS

11-1 Verify the partial output of Examples 11-1 to 11-3.

11-2 Vary the input parameters and make a set-versus-R_u curve for pile R2 (Fig. E11-1.1) to fit the load test shown.

11-3 Redraw the curve for pile *B* shown on Fig. E11-1.2 for percent of R_u on the point of 0.0, 0.50, and 0.75; is a better fit to load test obtained? Plot a curve of set versus driving stress for each percent for R_u = 200 kips.

11-4 Repeat Fig. E11-1.3 for R_u = 100 kips.

11-5 Repeat Fig. E11-1.3 for percent of R_u = 0.0, 0.25, 0.75, and 1.00 and R_u = 890 kN.

11-6 Make a study of the effect of the spring constant of the pile cap block [SP(2)] on driving a pile.

11-7 A 12.5-in square prestressed concrete pile 48 ft long is driven with a Raymond 65CH pile hammer (same as Example 11-1). A Micarta-aluminum cap block is used. Final driving set is 27 blows per foot; load test is 400 kips. Make a curve of set versus R_u to fit the driving data.

11-8 A 10.75 OD × 0.25 wall pipe pile is driven 61 ft using the same hammer system as Prob. 11-7. Final driving set is 10 blows per foot; load test = 280 kips. Convert data to metric units and make a curve of set versus R_u.

11-9 A 36-in OD × 5-ft wall prestressed concrete pile 80 ft long is driven 62 ft into the ground (remainder is freestanding, partly in water). The lower 4 ft of pipe is filled with concrete as a driving point. The pile is driven with a Raymond 5/0 hammer (fall = 3.25 ft; ram = 17,500 lb). Final driving set = 30 blows per inch, and load test = 1,330 kips. Make a curve of set versus R_u. Assume a cap block of Micarta with an area of 3 sq ft and 9 in long.

11-10 Repeat Prob. 11-9 using equivalent metric units.

11-11 A 14BP74 with a length of 24 m is driven with a DE30 hammer to point bearing on rock. No cushion is used, but a Micarta cap block of K = 4,380 kN/cm is used. Assuming 20 blows for 6-cm set, make a curve of set versus R_u and estimate R_u. If the working load is 6.205 kN/sq cm and the yield point of steel is 26.577 kN/sq cm, is this system satisfactory?

11-12 For Prob. 11-8, what are the smallest and largest commercially available hammers which can be used to drive the pile? Use the hammers listed in Appendix A.

REFERENCES

BOWLES, J. E. (1968): "Foundation Analysis and Design," chap. 9, McGraw-Hill, New York.

——— (1970): The Wave Equation used to Establish Pile Load-Stress Criteria, *Proc. Conf. Des. Install. Pile Found. Cell. Struct., Lehigh Univ.*, pp. 309–317.

DAVISSON, M. T. (1970): Design Pile Capacity, *Proc. Conf. Des. Install. Pile Found. Cell. Struct., Lehigh Univ.*, pp. 75–86.

FOREHAND, P. W., and J. L. REESE, JR. (1964): Prediction of Pile Capacity by the Wave Equation, *J. Soil Mech. Found. Div., ASCE*, vol. 90, SM2, pp. 1–25.

GRAFF, C. RUSS (1965): The Wave Equation and Pile Driving, *Found. Facts* (Raymond Concrete Pile Division, New York), vol. 1, no. 2, pp. 8, 9, 18.

HIRSCH, T. J., L. L. LOWERY, H. M. COYLE, and C. H. SAMSON, JR. (1970): Pile-Driving Analysis by One-Dimensional Wave Theory: State of the Art, *Highw. Res. Rec.* 333, HRB, Washington, pp. 33–54.

MOSLEY, ERNEST T. (1967): Wave Equation Analysis, *Found. Facts* (Raymond Concrete Pile Division, New York), vol. 3, no. 2, pp. 15–17.

———, and TONIS RAAMOT (1970): Pile Driving Formulas, *Highw. Res. Rec.* 333, HRB, Washington, pp. 23–32.

RAAMOT, TONIS (1967): Analysis of Pile Driving by the Wave Equation, *Found. Facts* (Raymond Concrete Pile Division, New York), vol. 3, no. 1, pp. 10–12.

SAMSON, C. H., JR., T. J. HIRSCH, and L. L. LOWERY, JR. (1963): Computer Study of Dynamic Behavior of Piling, *J. Struct. Div., ASCE,* vol. 89, ST4, August, pp. 413–449.

SMITH, E. A. L. (1955): Impact and Longitudinal Wave Transmission, *Trans. ASME,* August, pp. 963–973.

——— (1962): Pile Driving Analysis by the Wave Equation, *Trans. ASCE,* vol. 127, pt. 1, pp. 1145–1193 (includes discussion).

12

PILE STRESSES: STATIC LOADING

12-1 PILE-SOIL INTERACTION

Chapter 11 was concerned with pile stresses from dynamic (driving) loads. This chapter is concerned with the evaluation of pile stresses and deformations under static and/or working loads. This method of analysis can be used for partially or fully embedded piles, with or without a batter. While this method of analysis was not developed to solve the flagpole, signpost, or other partially embedded piles, it can be used for this class of problem.

The part of a pile embedded in the ground will carry part of the vertical load by shear transfer along the pile shaft to the adjacent soil, and the remainder of the load will be carried by the point. Field tests by many researchers [e.g., Tavenas (1971), Vesić (1970), Sherman (1969), D'Appolonia and Romualdi (1963), Mohan et al.

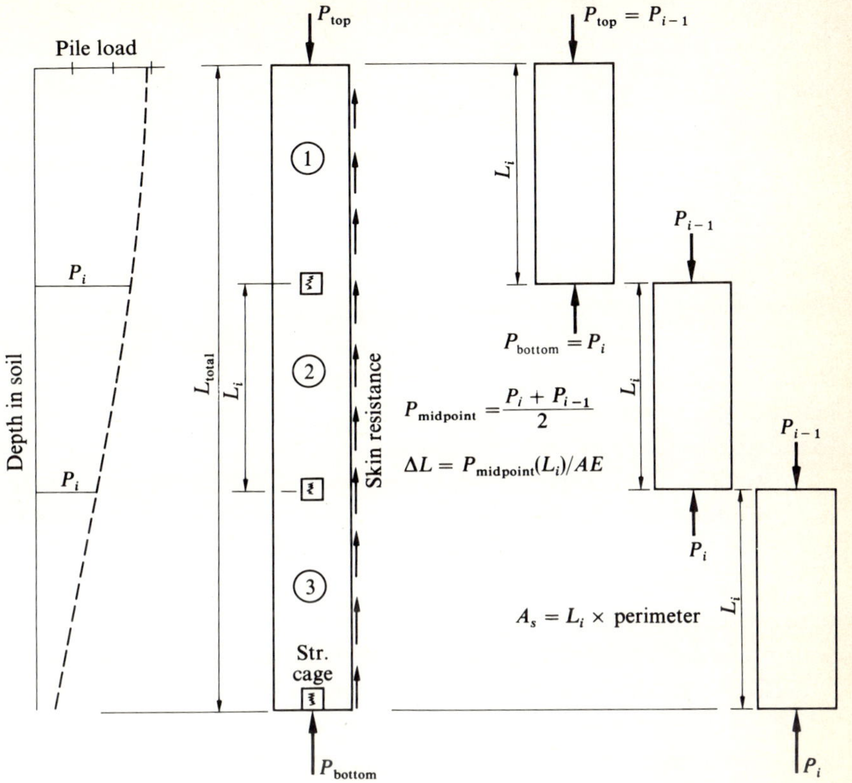

FIGURE 12-1
Pile with load variation with depth, showing use of finite elements of length L_i and loads P_i, P_{i-1} to evaluate the skin resistance of soil on the element.

(1963), Seed and Reese (1957)] have verified this as the in situ state and regardless of whether the pile is a point-bearing or friction pile.

The load transfer to the soil via skin resistance (refer to Fig. 12-1) is computed as

$$\tau_{av} = \frac{P_{top} - P_{bot}}{\text{perimeter} \times L_i} \tag{a}$$

Load tests with instrumentation to measure pile shaft load, usually through strain gages or accumulated deformation with rods to the ground surface, indicate that the average shear resistance across an increment of shaft (element of length L_i) can be related to the element deflection [DEFL(I)] as

$$\tau\Big|_{i-1}^{i} = \frac{\Delta P}{A_s} = \left(\int_0^{L_i} \varepsilon \, dL + \text{point deflection}\right) C \tag{b}$$

and

$$\varepsilon = \frac{P_{\text{midpoint}}}{AE} \tag{c}$$

ΔP is the change in pile forces (P_{i-1}, P_i) at the ends of an element with a surface area of A_s and length L_i. The strain ε is computed from the pile load at the middle of the segment (P_{midpoint}), and AE is the pile-shaft cross section and modulus of elasticity. It is possible that neither the element force nor the point deflection is linear. The load-transfer curves shown in this chapter of actual pile load tests indicate that in general load transfer is nonlinear. The C term of Eq. (*b*) is a compression constant (analogous to the A' constant used in Chap. 13).

Actual soil shear-transfer characteristics at a site can be obtained by instrumenting a pile for a load test to measure the strain ε at various points along the pile shaft. The strain is related to the load carried by the pile at that point; thus, the load carried by the soil between adjacent instrumented points can be found. The load is related to shear strength τ through Eq. (*b*). A plot of τ (or load transfer) versus deformation (or strain) can now be made, as in Fig. 12-2*c* or *d*.

Coyle and Reese (1966) proposed a set of curves of deformation versus load-transfer–shear-strength ratio (Fig. 12-3) based on the analysis of pile responses over a very wide geographic area to be used in conjunction with the measured shear strength of the soil at the given site. The curves of Fig. 12-3 should be used only after their local validity has been established or when no better data are available.

The response of the pile point is more difficult to evaluate than the side resistance. We may, however:

1 Treat the pile point as a bearing-capacity problem and estimate point resistance and movements.

2 Treat the point resistance as a spring, using the coefficient-of-subgrade reaction concept, apply to the point all the pile load not carried in shear, and estimate the point deflection.

3 Apply a percent of the total load to the point.

4 In the case of open pipe piles actually measure the point deflection at various loads. In the case of H- or other solid piles plot the rebound curve and estimate the point deflection under various loads.

The method of analysis used in this chapter incorporates steps 2 and 3.

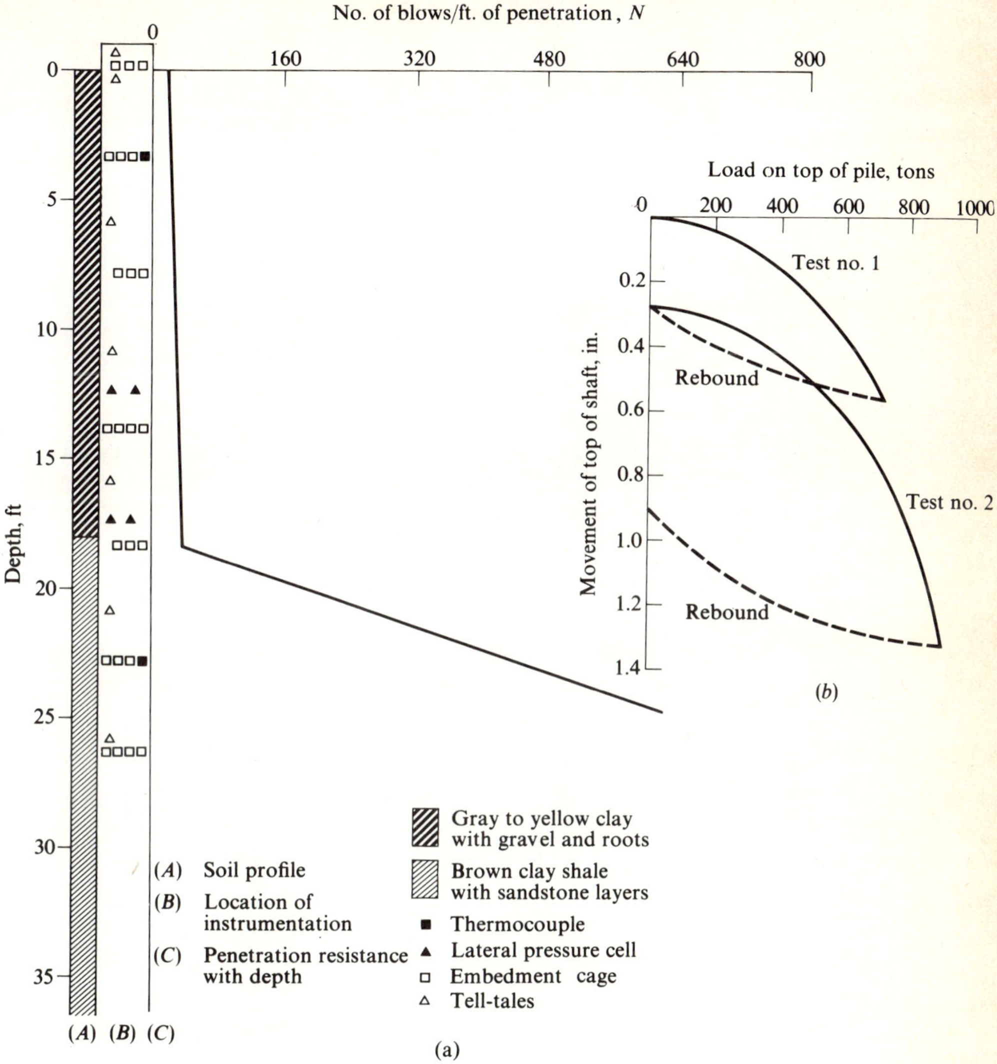

FIGURE 12-2
Pile-load-test data from Reese et al. (1969) (note that tons used here are 2,000 lb). (*a*) soil profile, blow count, and pile-instrumentation locations; (*b*) load-settlement curves for two tests; (*c*) load-transfer curves for two tests; (*d*) plot of load in pile shaft versus depth to obtain the data to plot (*c*).

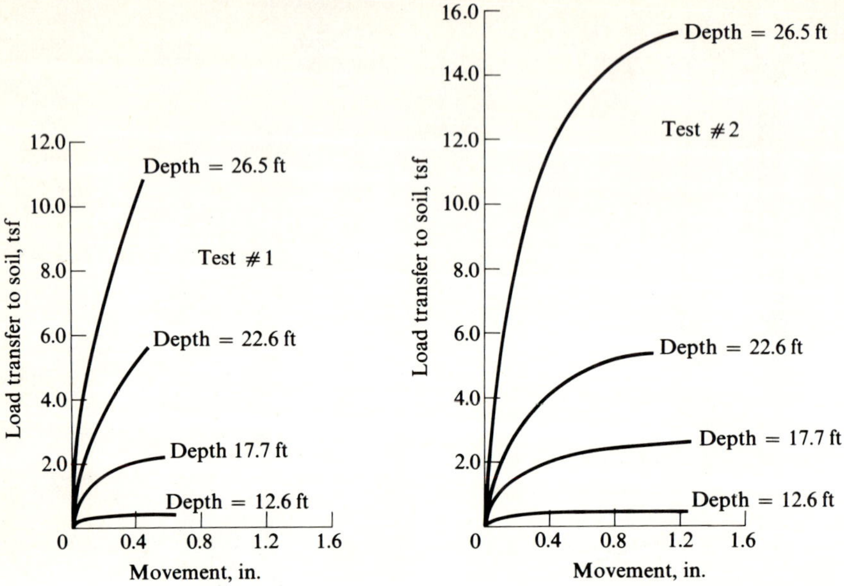

(c)

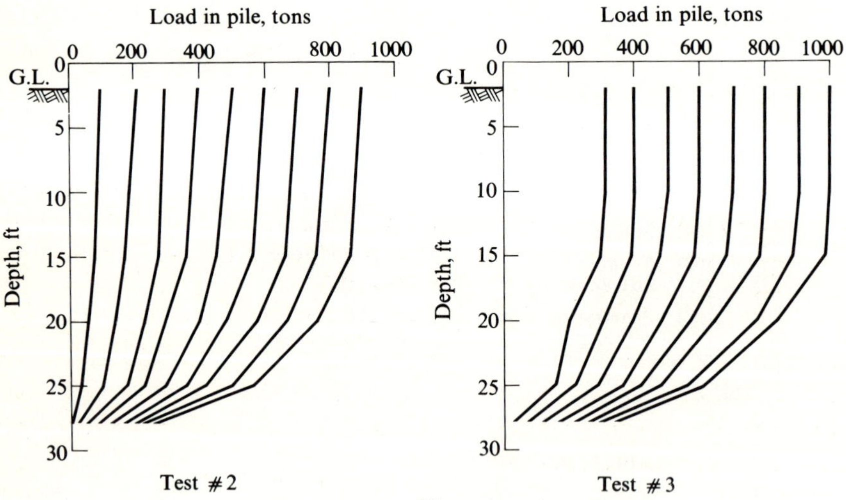

(d)

FIGURE 12-2 (*Continued*)

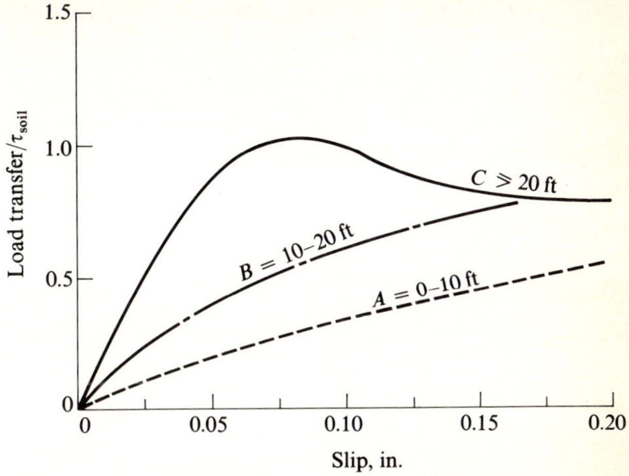

FIGURE 12-3
Approximate ratio of load transfer–soil shear strength versus pile slip. With the ratio and the actual soil shear strength known, one can compute the shear resistance on the element or load transfer. [*After Coyle and Reese* (1966).]

12-2 MATRIX SOLUTION OF THE PROBLEM

The following matrix solution is a completely general solution for any two-dimensional pile loading with three degrees of freedom. The method is applicable to both vertical and batter piles as long as the batter is in the plane of loading. The pile may be fully or partially embedded.

Two modes of failure of a pile are possible, namely, buckling (which is not considered here) and the failure when the combined stresses of compression and bending cause a material rupture. The combined-stress condition can be examined with this computer program including the increased bending stresses due to gravity loads as lateral displacement of partially embedded piles occurs (commonly termed the $P\Delta$ *effect* in tall-building analysis). In passing, it may be noted that the lateral-pile solution of Chap. 9 is a special case of this method of solution.

Referring to Fig. 12-4, we divide a pile into a number of segments (10 are generally sufficient for fully embedded piles) of any length. We set up a local coordinate system for the *i*th element to obtain the local A and S matrices as follows:

$$A = \begin{array}{c|cccccc} P \backslash F & 1 & 2 & 3 & 4 & 5 & 6 \\ \hline 1 & +\dfrac{\cos\alpha}{L} & +\dfrac{\cos\alpha}{L} & +\sin\alpha & 0 & -\cos\alpha & 0 \\ 2 & -\dfrac{\sin\alpha}{L} & -\dfrac{\sin\alpha}{L} & +\cos\alpha & 0 & +\sin\alpha & 0 \\ 3 & +1.0 & 0 & 0 & 0 & 0 & 0 \\ 4 & -\dfrac{\cos\alpha}{L} & -\dfrac{\cos\alpha}{L} & 0 & -\sin\alpha & 0 & -\cos\alpha \\ 5 & +\dfrac{\sin\alpha}{L} & +\dfrac{\sin\alpha}{L} & 0 & -\cos\alpha & 0 & +\sin\alpha \\ 6 & 0 & +1.0 & 0 & 0 & 0 & 0 \end{array}$$

$$S = \begin{array}{c|cccccc} F \backslash e & 1 & 2 & 3 & 4 & 5 & 6 \\ \hline 1 & \dfrac{4EI}{L} & \dfrac{2EI}{L} & 0 & 0 & 0 & 0 \\ 2 & \dfrac{2EI}{L} & \dfrac{4EI}{L} & 0 & 0 & 0 & 0 \\ 3 & 0 & 0 & \dfrac{2EA}{L} & 0 & 0 & 0 \\ 4 & 0 & 0 & 0 & \dfrac{2EA}{L} & 0 & 0 \\ 5 & 0 & 0 & 0 & 0 & K_1 & 0 \\ 6 & 0 & 0 & 0 & 0 & 0 & K_1 \end{array}$$

The SA^T and ASA^T matrix is then computer-generated for each element in turn. The final (sometimes termed *global*) ASA^T matrix is the sum of the member (or *local*) ASA^T matrix values of common PX subscripts. For example (Fig. 12-4), the final ASA^T values for P_4X_4 includes the sums of the values P_4X_4 for the first element and the values found at P_4X_4 for the second element. Likewise

$$P_8X_8|_{\text{final}} = P_8X_8|_{\text{element 2}} + P_8X_8|_{\text{element 3}}$$

As in previous chapters, the following matrix equations are solved

$$P = AF$$

$$e = BX = A^TX$$

$$F = Se$$

$$X = [ASA^T]^{-1}P$$

$$F = SA^TX$$

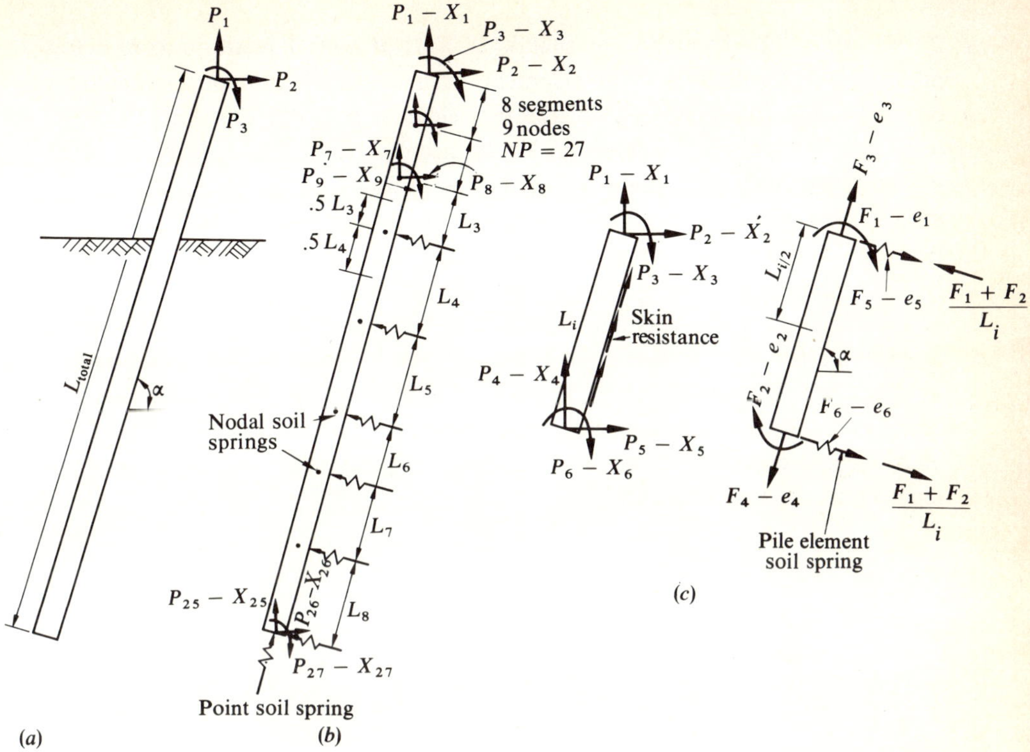

FIGURE 12-4
Pile divided into finite elements and the coding used in the analysis: (*a*) actual pile; (*b*) analytical model; (*c*) element *PX* and *Fe* coding. Note, however, that output spring forces if positive will act on element of (*c*) in the direction opposite that shown. The element springs actually represent element contribution to nodes.

Several features appear in this solution, however, which have not occurred previously.

1 After inverting the ASA^T to obtain a solution of X (displacements) as $X = [ASA^T]^{-1}P$, we must ensure compatibility of the axial components of X and the load-transfer characteristics of the soil. To do this we must read in enough load-transfer curves to describe the actual system load transfer; compute X_i and the corresponding pile-segment movements with respect to the soil; use the curves to obtain load-transfer values; re-form the P matrix; and recompute X_i. This is cycled as many times as necessary until the desired convergence of measured to computed pile slip is obtained.

2 From Fig. 12-4 it can be seen that X(1), X(2) of node 1 relate to axial deformation of the top half of the first element, whereas the X's of node 2 relate to the deformation in the lower half of the first element plus the deformation of the top half of the second pile element. This concept is continued to the point where it is also evident that the point X's [X(NP − 2), X(NP − 1)] represent the axial deformation of the bottom half of the bottom pile segment. To effect the nodal deformations in this sequence, it is necessary to modify the element S matrix [ES(3,3) and ES(4,4)] appropriately when building the ASA^T matrix. It is also necessary to recognize this deformation sequence to compute the pile-element movements for load transfer. In the computer program this is accomplished by dividing S(3,3) and S(4,4) by terms F1 or F2 (also F3). These terms are 1 at end nodes and for interior nodes are computed as follows. Let the ratio of two adjacent pile segments at any node be defined by dividing the smaller segment length into the larger to obtain[1]

$$\text{RATIO} = \frac{\text{LMAX}}{\text{LMIN}}$$

Next define a sum square term (SUMSQ) as

$$\text{SUMSQ} = (1 + \text{RATIO})^2$$

Now

$$\text{F1} = \text{SUMSQ} \qquad \text{F2} = \frac{\text{SUMSQ}}{\text{RATIO}}$$

Obviously this operation must be performed on both ends of interior pile segments to complete the ASA^T matrix adjustment.

3 The element movements are obtained starting with the bottom element (NM), which moves as follows:

$$\begin{aligned}\Delta(\text{NM}) &= \text{point movement} + \text{X(NP} - 3)\sin\alpha + \text{X(NP} - 2)\cos\alpha \\ &\quad + \frac{\text{L(NM)}}{\text{L(NM)} + \text{L(NM} - 1)}\,\text{X(NP} - 5)\sin\alpha \\ &\quad + \frac{\text{L(NM)}}{\text{L(NM)} + \text{L(NM} - 1)}\,\text{X(NP} - 4)\cos\alpha\end{aligned}$$

A similar computation is made for each element, using the deformation of the previous lower element and the additional element deformations. The ratio

[1] Using computer program notation.

$L_i/(L_i + L_{i-1})$ is used to obtain that part of the total nodal movement applicable to the element under consideration. The axial deformation for any node is based on 2AE/L of the adjacent elements. This problem also arose in building the ASA^T matrix in step 2.

4 The kth element deflection or relative pile-soil movement is

$$\text{Deflection} = \text{total element movement} = \text{point deflection} + \sum_{i=NM}^{i=NM-k} \varepsilon L_i$$

and this value is used to enter the appropriate curve of deflection versus shear resistance to find the amount of the pile load carried by shear along that element length. This friction component is

$$\text{Friction load} = (L_k)(\tau) \times \text{perimeter of pile}$$

The element friction load is used to revise the P matrix at each node in turn down the pile from top to bottom, and the inverted ASA^T matrix is used again to compute new X values and recycled until the current and preceding slip values meet some required specification (say, ≤ 0.002).

5 When the X values have been found to the required degree of computational precision, these values are used together with the element SA^T (which in the computer program is recomputed to save storage) to obtain the element forces as

$$F_i = (SA^T)_i X$$

Again, because of the method of problem formulation and definition of the X values, the S matrix must be adjusted to obtain compatibility.

6 Precautions must be taken to avoid tensile forces in the pile subjected to compressive loads if the pile is too long and too much friction or shear resistance[1] is available from the soil. The computer program avoids this by zeroing the P matrix if there is more shear resistance available in a pile segment than is needed to carry the nodal force at the top of that segment (for either a tension or compression pile).

12-3 PROBLEM CHECKING

In complex problems like this it is important to make an adequate number of checks. Parts of the problem output can be readily checked as follows (refer also to examples):

1 Moments F(1) and F(2) of adjacent elements must be equal and opposite in sign.

[1] Also skin resistance.

2 Axial forces F(3) of element $i + 1$ and F(4) of element i must be equal and opposite in sign.

3 Lateral soil springs F(5) and F(6) must be added at interior nodes to obtain the "nodal" spring. Also note that the nodal (joint) effect is *opposite in sign* to the element effect.

4 $\sum M = 0$ of element 1 at the lower end, and $\sum M = 0$ of entire pile. At other points $\sum M$ must include all the pile elements to that point to be zero and must include nodal soil spring effects (not element spring effects).

5 $\sum F_H = 0$, $\sum F_v = 0$ for entire pile and must include skin-resistance effect. In general, the three equations of static equilibrium are satisfied for each node but not for the individual pile segments.

6 A special case of this solution is the laterally loaded pile of Chap. 9. One can check part of the problem by computing the lateral pile by both methods. The answers should check, and the nodal springs of step 3 above should check with those of the lateral pile. The first soil node spring will, of course, depend on both programs' using the same reduction factor (currently 50 percent in Chap. 9 and here).

12-4 THE $P\Delta$ EFFECT

This method can allow for the $P\Delta$ effect in partially embedded piles, i.e., the increase in bending due to lateral deflection. Since X_2 is the translation of the top of the pile, one can obtain a revised P_3 down the pile to the joint where embedment takes place as

$$P_{(3i)} = P_{3(i)} - P_1(X2_1 - X2_i)$$

This value of P_3 in the P matrix can be used to recompute X_i until the current and last values of X_2 are

$$X_i|_i^n \leq X_{i-1}|_i^n \leq \text{accuracy required}$$

The computation for the $P\Delta$ effect can be made in the computer program if NDELT $\geq$ 1. Note that this is applied only up to the joint where soil begins since the $P\Delta$ effect becomes somewhat indeterminate from that point onward.

12-5 EXAMPLES

This method of analysis of pile stresses will be illustrated by several examples.

EXAMPLE 12-1 Use the pile-soil system shown in Figs. 12-5 and E12-1.1 with three

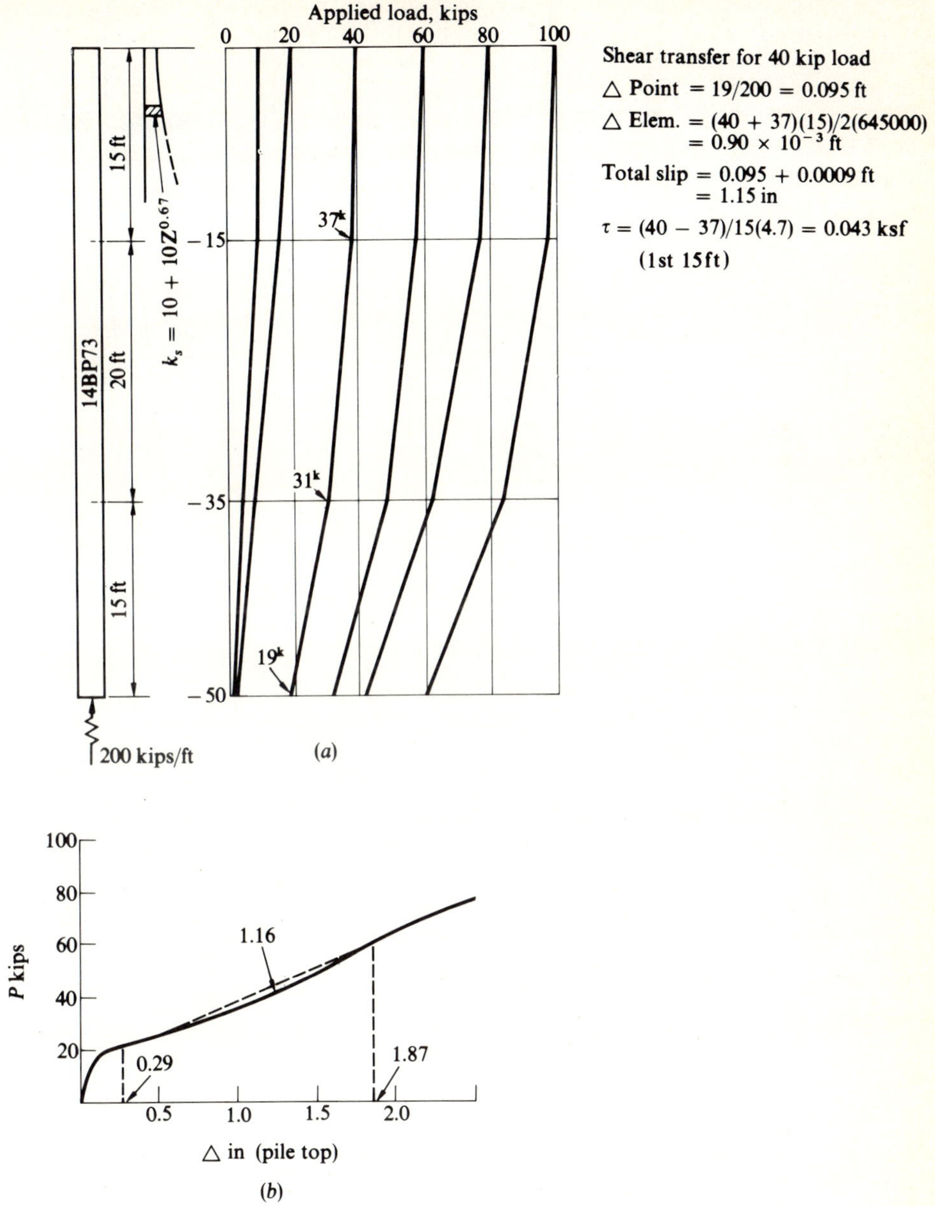

FIGURE 12-5
Load-transfer data, pile and soil profile, and plot of pile-top-deformation versus load (for use in Chap. 13). (*a*) Load-transfer curves. (*b*) Computed pile-response curve.

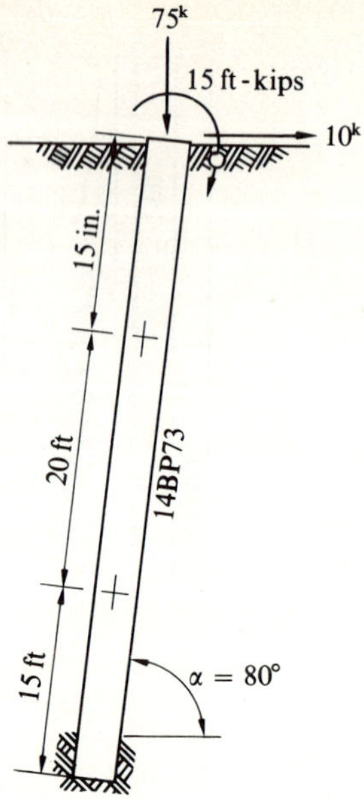

FIGURE E12-1.1

pile segments. Give a complete computer output and show checks to illustrate the method. Data for 14BP73:

$I_{x-x} = 733.1 \text{ in}^4$ Area = 21.5 sq in

$$\text{Perimeter} = \frac{2(13.64 + 14.586)}{12} = 4.70 \text{ ft}$$

Width = 14.586 in $E = 30{,}000$ ksi

$L = 50$ ft $\alpha = 80°$

SOLUTION With three segments

NP = (3 + 1)3 = 12

The $P\Delta$ effect will not be considered; therefore, NDELT = 0. Three loads are used; NNZP = 3. The load is given, not computed by the computer; therefore,

IPRD = 1. Only one computer run is to be made for this pile: NLC = 0 or 1. The soil starts at joint 1: JTSOIL = 1. No pile nodes require separately read-in soil springs: JJS = 0. The point spring (POINTK) is assumed at 200 kips/ft.

For the given load no percent of pile load is assumed to be carried by the point (if some is, the computer will make the determination). Thus, PERPP = 0. A point deflection is estimated from the load-transfer curves:

POINTX = 0.19 ft

We want a complete computer listing, LIST = 1. Also eight load-transfer data points are used; NSTRPT = 8. We can now assemble the input cards:

Card	Data
1	TITLE (see first line of computer output)
2	UNITS
	FT IN KIPS FT-KIPS KIPS/SQ FT KIPS/CU FT KIPS/FT KIPS/SQ IN
	NP NM NNZP NC IPRD NLC JTSOIL NDELT JJS LIST NSTRPT IUNIT
3	12 3 3 3 1 0 1 0 0 1 8 1
	E PIL POINTK AREAP ALPHA PERPP POINTX XMAX
4	30000. 50. 500. 21.46 80. 0. −.19 2.50
5–7	I P(I)
	1 −75.
	2 10.
	3 15.
8	AS BS EXPO ($k_s = a + bz^n$)
	10. 10. 0.67
9	NEC(JJ) Number of segments for first set of load-transfer data
10	X deflections (eight points)
11	Y shear ordinates (eight points)
	Refer to computer output for shear and deflection values
12	NEC(JJ) Number of segments for second set of load-transfer data
13	X
14	Y
15	NEC(JJ) Number of segments for third set of load-transfer data
16	X
17	Y
18–20	XL(I) XIN(I) BMEM(I) PER(I) AREA(I)
	15. 733.1 14.586 4.70 21.46
	20. 733.1 14.586 4.70 21.46
	15. 733.1 14.586 4.70 21.46

These 20 cards make up input; the output from the computer sheets is shown in Fig. E12-1.2 (pages 402–404).

J E BOWLES   EXAMPLE 12-1   USE 3-SEQ.   USE 14BP73 PILE @ 80 DEG BATTER

```
E =    30CCC.COO K/SQ IN        PILE LENGTH = 50.000 FT
POINT SPRING =      20C.000 K/FT      PILE POINT X-SECTICN =  21.500 SQ IN
          INITIAL PCINT DEFL =-0.1880 FT
          ASSUMED PCINT LOAD =   0.0     PERCENT      NC CF PILE SEGMENTS =   3
               JOINT SOIL STARTS =  1     NO OF LOAD CCNCITICNS =  1
  MAX POINT DEFL TO REDUCE POINTK 25% = 2.5000 IN

SOIL MODULUS =     10.0 +    10.00Z**0.670 K/CU FT
```

THE INITIAL P-MATRIX

1	-75.000
2	10.000
3	15.000

NC OF PILE ELEMENTS FOR FOLLOWING CURVE DATA = 1

XC =	0.0	0.06300	0.12500	1.15200	1.93600	2.542C0	3.62700	10.00000	IN
YC =	0.0	0.020CC	0.05700	0.04300	0.04300	0.04300	0.04300	0.04300	K/SQ FT

NC OF PILE ELEMENTS FOR FOLLOWING CURVE DATA = 1

XC =	0.0	0.067C0	0.12400	1.15300	1.94000	2.54600	3.66700	10.00000	IN
YC =	0.0	0.06000	0.08500	0.06400	0.0960C	C.16CCC	0.15C00	0.15000	K/SQ FT

NC OF PILE ELEMENTS FOR FOLLOWING CURVE DATA = 1

XC =	0.C	0.06100	0.12100	1.14700	1.93100	2.53500	3.62700	10.00000	IN
YC =	0.C	0.0570C	0.08600	0.17000	0.23000	0.28400	0.32600	0.32600	K/SQ FT

NO OF SHEAR STR CURVES = 3   NO OF ELEMENTS PER CURVE IN SEQUENCE = 1   1   1

MEMNC	NP1	NP2	NP3	NP4	NP5	NP6	ALPHA	L, FT	I, IN**4	WIDTH, FT	PERIM, FT	A SQ IN
1	1	2	3	4	5	6	80.000	15.00	733.10	1.22	4.70	21.50
2	4	5	6	7	8	9	80.000	20.00	733.10	1.22	4.70	21.50
3	7	8	9	10	11	12	80.000	15.00	733.10	1.22	4.70	21.50

NODE	SCIL MODULUS AT NODE	MEMNC	SOIL SPRING 1	SOIL SPRING 2
1	10.000	1	144.328	475.154
2	71.373	2	1075.472	1265.499
3	118.274	3	1180.460	1269.276
4	147.502			

THE ASAT MATRIX

1	83427.31	14589.33	707.23	-16.37	92.86	707.23	0.0	C.0	0.0	0.0	0.0	0.0
2	14589.32	3259.86	-4010.90	92.86	-526.66	-4010.90	0.0	0.0	0.0	0.0	0.0	0.0
3	707.23	-4010.90	40727.77	-707.23	4010.9C	2C363.88	0.0	C.0	0.0	0.0	0.0	0.0
4	-16.37	92.86	-707.23	35815.77	5905.73	-309.41	-6.91	39.18	397.82	0.0	0.0	0.0
5	92.86	-526.66	4010.90	5905.73	3364.10	1754.77	39.18	-222.19	-2256.13	0.0	0.0	0.0
6	707.23	-4010.90	20363.88	-309.41	1754.77	71273.56	-397.82	2256.13	15272.91	0.0	0.0	0.0
7	0.0	0.0	0.0	-6.91	39.18	-397.82	35842.77	5752.62	309.41	-16.37	92.86	707.23
8	0.0	0.0	0.0	39.18	-222.19	2256.13	5752.62	4232.43	-1754.77	92.86	-526.66	-4010.90
9	0.0	0.0	0.0	397.82	-2256.13	15272.91	309.41	-1754.77	71273.56	-707.23	4010.90	20363.88
10	0.0	0.0	0.0	0.0	0.0	0.0	-16.37	92.86	-707.23	83461.25	14396.95	-707.23
11	0.0	0.0	0.0	0.0	0.0	0.0	92.86	-526.66	4010.90	14396.95	4350.89	4010.90
12	0.0	0.0	0.0	0.0	0.0	0.0	707.23	-4010.90	20363.88	-707.23	4010.90	40727.77

```
THE X-MATRIX FOR CYCLE  1 IS
1 -0.021915   2  0.119458   3  0.009673   4 -0.001229   5  0.006971   6  0.004264
7  0.000483   8 -0.002738   9 -0.000820  10 -0.000160  11  0.000905  12  0.000040

ELEM NO  TOT ELEM DISP,FT   ELEM MOVE,FT      ELEM SLIP, IN    PTDEF1 =-0.1880000 FT
    1        -0.188839        -0.000839         2.266063
    2        -0.188000        -0.000000         2.256000
    3        -0.188000        -0.000000         2.256000

     PRINT COUNTERS USED TO RECOMPUTE P-MATRIX
          NEC(I2)=  1    XC(JJ,I2)=1.93600    XC(JJ+1,I2) =2.54200
     JJJ=  1    JJ=  5    I2=  1    KCC=  1
       CURRENT VALUE PAXIS(  2) =    -69.093 KIPS      ELEM. FRIC =      3.031 KIPS
         CURRENT VALUE PAXIS(I) =    -72.124 KIPS
          NEC(I2)=  1    XC(JJ,I2)=1.94000    XC(JJ+1,I2) =2.54600
     JJJ=  1    JJ=  5    I2=  2    KCC=  2
       CURRENT VALUE PAXIS(  3) =    -56.932 KIPS      ELEM. FRIC =     12.161 KIPS
         CURRENT VALUE PAXIS(I) =    -69.093 KIPS
          NEC(I2)=  1    XC(JJ,I2)=1.93100    XC(JJ+1,I2) =2.53500
     JJJ=  1    JJ=  5    I2=  3    KCC=  3
       CURRENT VALUE PAXIS(  4) =    -38.668 KIPS      ELEM. FRIC =     18.263 KIPS
         CURRENT VALUE PAXIS(I) =    -56.932 KIPS

THE POINT DEFL DUE TO    -37.6000 KIPS IS  -0.1880000 FT


NM  SLIP, IN    SHEAR, K/SQ FT       P-MATRIX IN 3-COLS
 1  2.266063      0.04300      1    -75.000   2    10.000   3     15.000
 2  2.256000      0.12937      4    -68.043   5   -11.998   6      0.0
 3  2.256000      0.25906      7    -56.067   8    -9.886   9      0.0
                              10    -38.081  11    -6.715  12      0.0
```

FIGURE E12-1.2
Partial computer output for Example 12-1.

```
THE X-MATRIX FOR CYCLE  2 IS
1 -0.021915   2  0.119458   3  0.009673   4 -0.003075   5  C.006645   6  0.004264
7 -0.001038   8 -0.003006   9 -0.000820  10 -0.000602  11  0.000827  12  0.000040

ELEM NC   TOT ELEM DISP,FT    ELEM MOVE,FT      ELEM SLIP, IN     PTDEF1 =-0.1933402 FT
    1         -0.198048         -0.001642         2.376573
    2         -0.196406         -0.001954         2.356868
    3         -0.194452         -0.001112         2.333421

     PRINT COUNTERS USED TO RECOMPUTE P-MATRIX
          NEC(I2)=  1     XC(JJ,I2)=1.93600    XC(JJ+1,I2) =2.54200
     JJJ=  2    JJ=  5    I2=  1    KCC=  1
       CURRENT VALUE PAXIS(  2) =    -69.093 KIPS      ELEM. FRIC =     3.031 KIPS
         CURRENT VALUE PAXIS(I) =    -72.124 KIPS
          NEC(I2)=  1     XC(JJ,I2)=1.94000    XC(JJ+1,I2) =2.54600
     JJJ=  2    JJ=  5    I2=  2    KCC=  2
       CURRENT VALUE PAXIS(  3) =    -55.930 KIPS      ELEM. FRIC =    13.162 KIPS
         CURRENT VALUE PAXIS(I) =    -69.093 KIPS
          NEC(I2)=  1     XC(JJ,I2)=1.93100    XC(JJ+1,I2) =2.53500
     JJJ=  2    JJ=  5    I2=  3    KCC=  3
       CURRENT VALUE PAXIS(  4) =    -37.179 KIPS      ELEM. FRIC =    18.751 KIPS
         CURRENT VALUE PAXIS(I) =    -55.930 KIPS

THE POINT CEFL DUE TC    -38.6680 KIPS IS  -0.1933402 FT
```

FIGURE E12-1.2 (*Continued*)

Partial checking of computer output is as follows (refer to Fig. E12-1.3 and note application of element spring effects):

Deflection of element 1 = 2.331 in

Therefore, interpolation of load-transfer data is

$$\tau = 0.043 \text{ ksf}$$

$$P_{ax} = 75 \sin 80° - 10 \cos 80°$$

$$= 72.19 \text{ kips}$$

$$P_{ax} \text{ at base} = 72.19 - 0.043(15)(4.7)$$

$$= 69.16 \text{ kips}$$

$$\sum M \text{ base} = ? \qquad \text{clockwise} = \text{positive}$$

$$-M_b + 15 + 75(15)(0.17365) + 10(15)(0.98481) - 15(17.528) = 0$$

$$M_b = 95.18 \text{ ft-kips} \qquad 95.15 \text{ computer}$$

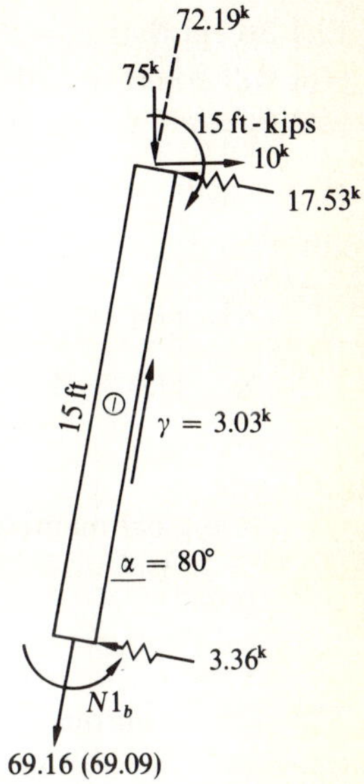

FIGURE E12-1.3

For entire pile (Fig. E12-1.4) and using data from computer output:

$$\text{Shear transfer} = 3.03 + 12.71 + 18.45 = 34.19 \text{ kips}$$

$$\text{Point} = 72.19 - 34.19 = 38.00\ (37.91)$$

$$H \text{ component} = 72.19\ (0.17365) = 12.54$$

$$+ P_2 = \underline{10.00}$$

$$\text{Total } H = 22.54$$

$$\text{Resisting } H = (17.53 + 28.50 + 1.17 - 6.80)(0.9841) = 22.52 \text{ kips}$$

$$\sum M \text{ pile} = ?$$

$$75(50)(0.17365) + 50(10)(0.98481) - 50(17.53) - 35(10.97) + 15(6.80) \approx 0$$

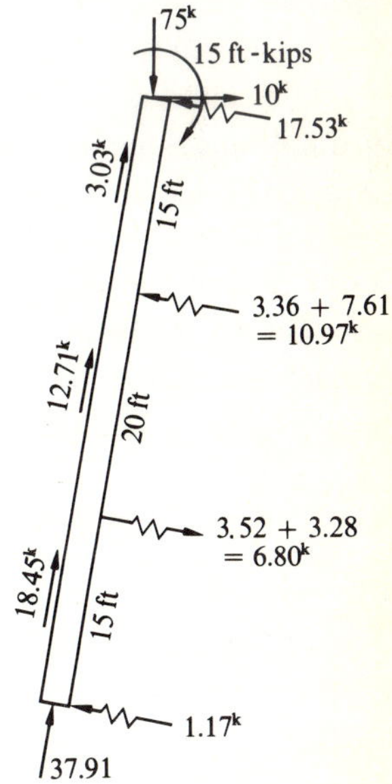

FIGURE E12-1.4

////

EXAMPLE 12-2 Make a curve of load versus deflection for use in the pile group analysis of Chap. 13 for a 14BP73. The applicable data are in Fig. 12-5 and Example 12-1 except that $\alpha = 90°$. Note that if this problem is correctly modeled, the resulting plot will coincide with the field load test.

SOLUTION We will use 10 segments, NP = 33. With one load [P(1)] NNZP = 1. We will increment six loads from −10 to −80 kips; therefore, NLC = 6. As not all output is desired, LIST = 0.

NEC(JJ) = 3 three segments in top 15 ft of pile

NEC(JJ) = 4 four segments in middle 20 ft of pile

NEC(JJ) = 3 three segments in lower 15 ft of pile

A typical member data card contains:

5.	733.1	14.586	4.70	21.5

The partial output for P(1) = −20 kips is plotted in Fig. 12-5*b*. Figure E12-2.1 (pages 407–408) shows the partial computer input. ////

EXAMPLE 12-3 Compare the $P\Delta$ effect on a vertical and batter pile as shown in Fig. 12-6 and use metric units. For this example, we will assume a reasonable soil modulus but read in very large spring values for the first two soil nodes to control the ground-line slope and deflection. The pile is a 14BP73; see Example 12-1 for pile data. The batter is with respect to the strong axis. We will use two shear-transfer curves, one with NEC(JJ) = 5 and all Y = 0 to take care of 25 ft of pile above ground, and the second with NEC(JJ) = 10 for below-ground load-transfer data. Obviously the below-ground shear-transfer data could be such that more curves would be needed. See the partial computer output for values used.

SOLUTION

NP = 16 × 3 = 48	NLC = 1
NM = 15	JTSOIL = 6
NNZP = 2	NDELT = 1 (second run is 0)
NC = 2	RSPR(6,1) = 99999999. = RSPR(6,2)
IPRD = 1	LIST = 1 (not all output for this problem is shown in text)
	NSTRPT = 8

```
J E BOWLES EXAMPLE 12-2  PILE RESPONSE FOR FIG 12-5 USE IN CHAP 13

E =    30000.000 K/SQ IN       PILE LENGTH = 50.COO FT
POINT SPRING =     200.000 K/FT      PILE PCINT X-SECTICN =   21.460 SQ IN
          INITIAL POINT DEFL =-0.0050 FT
          ASSUMED POINT LOAD =   0.0    PERCENT      NC CF PILE SEGMENTS = 10
               JCINT SOIL STARTS =  1   NO OF LOAC CONCITICNS =   3
  MAX POINT DEFL TO REDUCE POINTK 25% = 2.0000 IN

SOIL MODULUS =     10.0 +   10.00Z**0.670 K/CU FT

THE INITIAL P-MATRIX
1         -10.000

     NC OF PILE ELEMENTS FOR FOLLOWING CURVE DATA =   3

XC =   0.0       0.06300   0.12500   1.15200   1.93600   2.54200   3.62700   10.00000   IN
YC =   0.C       C.02CCC   0.05700   C.04300   0.04300   0.C4300   0.04300    0.04300   K/SQ FT

     NC OF PILE ELEMENTS FOR FOLLOWING CURVE DATA =   4

XC =   0.0       0.06700   0.12400   1.15300   1.94000   2.546CC   3.66700   10.00000   IN
YC =   0.C       C.06000   0.08500   0.06400   0.09600   0.16000   0.15000    0.15000   K/SQ FT
     NC CF PILE ELEMENTS FOR FOLLOWING CURVE DATA =   3

XC =   0.0       0.061CC   0.12100   1.14700   1.93100   2.53500   3.62700   10.00000   IN
YC =   0.C       C.057CC   0.08600   0.17000   0.23000   0.28400   0.32600    0.32600   K/SQ FT
 NO OF SHEAR STR CURVES =  3    NC OF ELEMENTS PER CURVE IN SEQUENCE =  3    4    3
```

MEMNC	NP1	NP2	NP3	NP4	NP5	NP6	ALPHA	L, FT	I, IN**4	WIDTH, FT	PERIM, FT	A SQ IN
1	1	2	3	4	5	6	90.000	5.CC	733.10	1.22	4.70	21.46
2	4	5	6	7	8	9	90.000	5.00	733.10	1.22	4.70	21.46
3	7	8	9	10	11	12	90.000	5.00	733.10	1.22	4.70	21.46
4	10	11	12	13	14	15	90.000	5.CC	733.10	1.22	4.70	21.46
5	13	14	15	16	17	18	90.000	5.00	733.10	1.22	4.70	21.46
6	16	17	18	19	20	21	90.000	5.00	733.10	1.22	4.70	21.46
7	19	20	21	22	23	24	90.000	5.00	733.10	1.22	4.70	21.46
8	22	23	24	25	26	27	90.000	5.0C	733.10	1.22	4.70	21.46
9	25	26	27	28	29	30	90.000	5.00	733.10	1.22	4.70	21.46
10	28	29	3C	31	32	33	90.000	5.00	733.10	1.22	4.70	21.46

NODE	SCIL MODULUS AT NODE	MEMNC	SOIL SPRING 1	SOIL SPRING 2
1	10.C00	1	31.604	92.986
2	39.397	2	138.022	155.623
3	56.773	3	187.702	202.491
4	71.373	4	230.366	243.581
5	84.420	5	268.883	281.039
6	96.421	6	304.526	315.900
7	107.650	7	338.007	348.768
8	118.274	8	369.774	380.038
9	128.408	9	400.136	409.984
10	138.130	10	429.325	438.818
11	147.502			

FIGURE E12-2.1
Partial computer input. Note that P(1) is initially −10 kips but output shown after four iterations is for P(1) = −20 kips.

```
THE FINAL COMPUTED PILE ELEMENT FORCES AND OTHER DATA AFTER 4 ITERATIONS FOLLOWS

MEMNO     END MOMENTS              AXIAL FORCE          SOIL REACTIONS       ELEM. DEFL,FT      SOIL SPRINGS
   1      0.000      0.000     -20.000     18.663      0.000      0.000      -0.01088          31.604      92.986
   2     -0.C00     -0.000     -18.663     17.325      0.CCC      C.000      -0.01073         138.022     155.623
   3     -0.C00     -0.000     -17.325     15.987      0.000      0.000      -0.01059         187.702     202.491
   4     -0.C00     -0.000     -15.987     13.991      0.000      0.000      -0.01046         230.366     243.581
   5     -0.000     -0.000     -13.991     11.995      C.CCC      C.000      -0.01035         268.883     281.039
   6     -C.000     -0.000     -11.995      9.998      0.000      0.000      -0.01025         304.526     315.900
   7     -0.C00     -0.000      -9.998      8.000      0.000      0.000      -0.01016         338.007     348.768
   8     -0.000     -0.000      -8.C00      5.967      C.CCC      C.000      -0.01009         369.774     380.038
   9     -C.000     -0.000      -5.967      3.934      0.000      0.000      -0.01004         400.136     409.984
  10     -0.C00     -C.000      -3.934      1.903      0.000      0.000      -0.01000         429.325     438.818

     SUM HOR. SOIL REACTIONS =        0.00 KIPS

          PILE TOP MOVEMENTS:   VERT =   -0.13059 IN
                                HORZ =    0.0     IN
                                 ROT =    0.0     RAD

****** THE ULTIMATE PILE RESISTANCE =     -20.0000 KIPS VERTICAL AND     0.0    KIPS HORIZ
       THE ORIGINAL MOMENT APPLIED =        0.0     FT-K
          POINT FORCE =          1.903 KIPS

     CURRENT PROB NO =   3    POINTX = -0.067000 FT

THE INITIAL P-MATRIX
1        -30.000

THE X-MATRIX FOR CYCLE  1 IS
 1 -0.C00116     2  0.0          3  0.0          4  0.0          5  0.0          6  0.0
 7  0.0          8  0.0          9  0.0         10  0.0         11  0.0         12  0.0
13  0.0         14  0.0         15  0.0         16  0.0         17  0.0         18  0.0
19  0.0         20  0.0         21  0.0         22  0.0         23  0.0         24  0.0
25  0.0         26  0.0         27  0.0         28  0.0         29  0.0         30  0.0

ELEM NO   TCT ELEM DISP,FT    ELEM MCVE,FT      ELEM SLIP, IN     PTDEF1 =-0.0670000 FT
   1          -0.067116        -0.000116         0.805397
   2          -C.067000         0.0              0.804000
   3          -0.067000         0.0              0.804CCC
   4          -0.067C00         0.0              0.804000
   5          -0.067C00         0.0              0.804000
   6          -0.067000         0.0              0.804C00
   7          -0.067000         0.0              0.804000
   8          -0.067C00         C.0              0.804C00
   9          -0.067000         0.0              0.804CC0
  10          -0.067000         0.0              0.804CCC

THE POINT DEFL DUE TO     -13.4000 KIPS IS  -0.0670000 FT
```

FIGURE E12-2.1 *(Continued)*

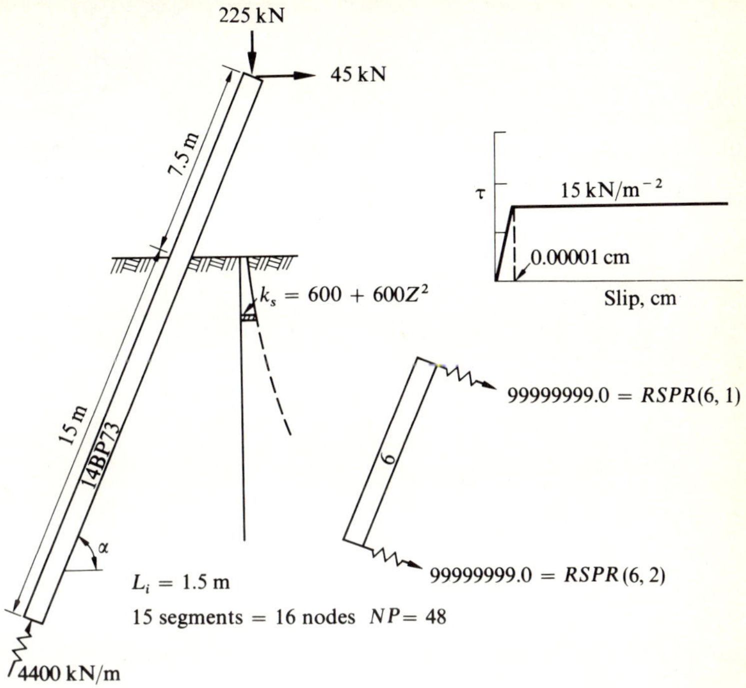

FIGURE 12-6
Pile-soil conditions for Example 12-3.

Other data are straightforward with

ALPHA = 80 and 90°

Partial output (see also Fig. E12-3.1, pages 410–412) follows:

Without $P\Delta$			With $P\Delta$		
α, deg	Δ_{top}, cm	$M_{max(6)}$, kN-m	Δ_{top}, cm	M_{max}, kN-m	Moment increase w/P-Δ, %
90	11.9	−337.5	14.8	370.8	9.9
80	21.7	−625	26.9	685.7	9.7

Checking:

For vertical pile ($\alpha = 90°$)

M_6 should be $45 \times 7.5 = 337.5$ kN-m   checks output

M_6 with $P\Delta$ effect included should be $337.5 + 0.14812(225) = 370.8$

```
J E BOWLES   EXAMPLE 12-3   P-DELTA EFFECT OF PILE FIG 12-6   METRIC UNITS

E = 210930C.C00 KG/SQ CM        PILE LENGTH = 22.500 M
POINT SPRING =      44C0.000 KN/M       PILE POINT X-SECTICN = 138.400 SQ CM
          INITIAL POINT DEFL = 0.0      M
          ASSUMED PCINT LOAD =   0.0     PERCENT      NC OF PILE SEGMENTS = 15
               JOINT SOIL STARTS =  6    NO OF LOAD CONCITICNS =  1
  MAX POINT DEFL TO REDUCE POINTK 25% = 2.0000 CM

  *** P-DELTA EFFECT IS CONSIDERED IN COMPUTATIONS **

SOIL MCDULUS =    600.0 +  600.00Z**2.000 KN/CU M

THE INITIAL P-MATRIX
1        -225.000
2          45.000

     NC OF PILE ELEMENTS FOR FOLLOWING CURVE DATA =  5

XC =    0.0        0.10000   0.20000   0.30000   0.40000   0.50000   0.60000 200.00000  CM
YC =    0.0        0.0       0.0       0.0       0.0       C.0       0.0       0.0      KN/SQ M

     NO OF PILE ELEMENTS FOR FOLLOWING CURVE DATA = 10

XC =    0.0        0.00001   0.25000   0.50000   0.75000   1.CCCCO   2.00000 100.00000  CM
YC =    0.0       15.00000  15.00000  15.00000  15.00000  15.CCCOO  15.00000  15.00000  KN/SQ M

NO OF SHEAR STR CURVES =  2    NO OF ELEMENTS PER CURVE IN SEQUENCE =  5   10
```

MEMNC	NP1	NP2	NP3	NP4	NP5	NP6	ALPHA	L, M	I, CM**4	WIDTH, M	PERIM, M	A SQ CM
1	1	2	3	4	5	6	90.000	1.50	30511.60	0.37	1.43	138.40
2	4	5	6	7	8	9	90.000	1.50	30511.60	0.37	1.43	138.40
3	7	8	9	10	11	12	90.000	1.50	3C511.60	0.37	1.43	138.40
4	10	11	12	13	14	15	90.000	1.50	30511.60	0.37	1.43	138.40
5	13	14	15	16	17	18	90.000	1.50	30511.60	0.37	1.43	138.40
6	16	17	18	19	20	21	90.000	1.50	3C511.60	0.37	1.43	138.40
7	19	20	21	22	23	24	90.000	1.50	30511.60	0.37	1.43	138.40
8	22	23	24	25	26	27	90.000	1.50	30511.60	0.37	1.43	138.40
9	25	26	27	28	29	30	90.000	1.50	3C511.60	0.37	1.43	138.40
10	28	29	30	31	32	33	90.000	1.50	30511.60	0.37	1.43	138.40
11	31	32	33	34	35	36	90.000	1.50	30511.60	0.37	1.43	138.40
12	34	35	36	37	38	39	90.000	1.50	30511.60	0.37	1.43	138.40
13	37	38	39	40	41	42	90.000	1.50	3C511.60	0.37	1.43	138.40
14	40	41	42	43	44	45	90.000	1.50	30511.60	0.37	1.43	138.40
15	43	44	45	46	47	48	90.C00	1.50	30511.60	0.37	1.43	138.40

FIGURE E12-3.1*a*
Input data for partially embedded vertical pile.

```
THE X-MATRIX FOR CYCLE  7 IS
 1 -0.000059    2  0.148120    3  0.027474    4 -0.000118    5  0.107310    6  0.026672
 7 -0.000118    8  0.069070    9  0.024047   10 -0.000118   11  0.036287   12  0.019396
13 -0.000118   14  0.012132   15  0.012543   16 -0.000118   17  0.000004   18  0.003360
19 -0.000101   20 -0.000004   21 -0.001315   22 -0.000084   23  0.001068   24 -0.000227
25 -0.000067   26  0.000984   27  0.000247   28 -0.000050   29  0.000537   30  0.000300
31 -0.000033   32  0.000172   33  0.000177   34 -0.000017   35  0.000005   36  0.000056
37  0.0        38 -0.000028   39 -0.000001   40  0.0        41 -0.000016   42 -0.000011
43  0.0        44 -0.000003   45 -0.000006   46  0.0        47  0.000004   48 -0.000004


THE FINAL COMPUTED PILE ELEMENT FORCES AND OTHER DATA AFTER 7 ITERATIONS FOLLOWS

MEMNO      END MOMENTS              AXIAL FORCE           SOIL REACTIONS     ELEM. DEFL,M      SOIL SPRINGS
  1      -0.035    -67.535   -225.000    224.946      0.0        0.0       -0.00312         0.0           0.0
  2      76.695   -144.157   -224.946    224.965      0.0        0.0       -0.00301         0.0           0.0
  3     161.962   -229.450   -224.965    224.981      0.0        0.0       -0.00289         0.0           0.0
  4     254.613   -322.089   -224.981    224.993      0.0        0.0       -0.00277         0.0           0.0
  5     352.692   -420.175   -224.993    225.000      0.0        0.0       -0.00265         0.0           0.0
  6     453.500     60.095   -225.000    192.757    387.381   -361.460     -0.00254   99999984.00099999984.000
  7     -60.093     31.489   -192.757    160.514     -0.003      1.313     -0.00243       854.464      1229.594
  8     -31.489      8.426   -160.514    128.272      2.382      2.810     -0.00233      2229.941      2855.159
  9      -8.427     -3.990   -128.272     96.030      4.287      2.808     -0.00226      4355.684      5230.992
 10       3.990     -6.370    -96.030     63.787      3.882      1.438     -0.00220      7231.691      8357.086
 11       6.370     -3.793    -63.787     31.545      1.868      0.059     -0.00216     10857.961     12233.438
 12       3.793     -1.017    -31.545      0.000      0.073     -0.478     -0.00213     15234.492     16860.059
 13       1.017      0.176     -0.000      0.000     -0.577     -0.353     -0.00212     20361.262     22236.891
 14      -0.176      0.216     -0.000      0.000     -0.416     -0.079     -0.00212     26238.313     28364.055
 15      -0.216     -0.000     -0.000     -0.000     -0.092      0.144     -0.00212     32865.625     35241.457

    SUM HCR. SOIL REACTIONS =       44.99 KN

         PILE TOP MOVEMENTS:  VERT =  -0.31246 CM
                              HORZ =  14.81204 CM
                               ROT =   0.02747 RAD

****** THE ULTIMATE PILE RESISTANCE =    -225.0000 KN   VERTICAL AND   45.000 KN   HORIZ
       THE ORIGINAL MOMENT APPLIED =        0.0     KN-M
          POINT FORCE =        -0.000 KN
```

FIGURE E12-3.1*b*
Output for vertical pile partially embedded with $P\Delta$ effect.

```
THE X-MATRIX FOR CYCLE  7 IS
 1 -0.047528   2  0.269223   3  0.050726   4 -0.034500   5  0.195012   6  0.049240
 7 -0.022242   8  0.125494   9  0.044386  10 -0.011735  11  0.065909  12  0.035791
13 -0.003996  14  0.022018  15  0.023139  16 -0.000112  17 -0.000012  18  0.006194
19 -0.000093  20 -0.000023  21 -0.002425  22 -0.000419  23  0.001926  24 -0.000418
25 -0.000375  26  0.001776  27  0.000456  28 -0.000216  29  0.000967  30  0.000553
31 -0.000082  32  0.000308  33  0.000326  34 -0.000012  35  0.000007  36  0.000104
37  0.000009  38 -0.000051  39 -0.000002  40  0.000005  41 -0.000029  42 -0.000020
43  0.000001  44 -0.000005  45 -0.000012  46 -0.000001  47  0.000007  48 -0.000007


THE FINAL COMPUTED PILE ELEMENT FORCES AND OTHER DATA AFTER 7 ITERATIONS FOLLOWS

MEMNO     END MOMENTS           AXIAL FORCE          SOIL REACTIONS      ELEM. DEFL,M       SOIL SPRINGS
   1      0.047  -125.008   -214.250   213.715     0.0       0.0        -0.00296          0.0           0.0
   2    141.828  -266.711   -213.715   213.715     0.0       0.0        -0.00284          0.0           0.0
   3    299.107  -424.152   -213.715   213.789     0.0       0.0        -0.00273          0.0           0.0
   4    469.934  -594.825   -213.789   213.750     0.0       0.0        -0.00262          0.0           0.0
   5    650.504  -775.490   -213.750   213.767     0.0       0.0        -0.00251          0.0           0.0
   6    836.068   110.788   -213.767   181.523   714.575  -666.374      -0.00240   99999984.00099999984.000
   7   -110.790    58.052   -181.523   149.281    -0.006     2.421      -0.00229        854.464      1229.594
   8    -58.051    15.536   -149.281   117.037     4.391     5.181      -0.00221       2229.941      2855.159
   9    -15.534    -7.355   -117.037    84.797     7.904     5.177      -0.00214       4355.684      5230.992
  10      7.355   -11.745    -84.797    52.555     7.157     2.650      -0.00208       7231.691      8357.086
  11     11.744    -6.992    -52.555    20.313     3.443     0.109      -0.00205      10857.961     12233.438
  12      6.992    -1.875    -20.313     0.000     0.135    -0.881      -0.00203      15234.492     16860.059
  13      1.875     0.325     -0.000    -0.000    -1.064    -0.650      -0.00202      20361.262     22236.891
  14     -0.325     0.399      0.000     0.0      -0.767    -0.146      -0.00202      26238.313     28364.055
  15     -0.399     0.000      0.0      -0.000    -0.169     0.266      -0.00202      32865.625     35241.457

     SUM HCR. SOIL REACTIONS =      82.09 KN

          PILE TOP MOVEMENTS:  VERT =  -0.29114 CM
                               HORZ =  26.92227 CM
                                ROT =   0.05073 RAD

****** THE ULTIMATE PILE RESISTANCE =   -225.0000 KN   VERTICAL AND   45.000 KN   HORIZ
       THE ORIGINAL MOMENT APPLIED =      0.0     KN-M
          POINT FORCE =       -0.000 KN
```

FIGURE E12-3.1*c*

Output for partially embedded pile with $P\Delta$ effect; $\alpha = 80°$.

The output (refer to Figs. E12-3.1*c* and E12-3.2) corrected for applied external node moments is as follows:

$$\text{Joint moments due to } P\Delta = 225[(0.1481 - 0.1073) + (0.1073 - 0.0690) + (0.0690 - 0.0362) + (0.0362 - 0.0121) + (0.0121 - 0)]$$

$$= 9.16 + 8.60 + 7.38 + 5.42 + 2.72$$

Note that these moments accumulate from node to node down the pile in applying corrections to the output, therefore, obtaining the accumulated sum and subtracting from M_6 given in output, we have

$$M_6 = 453.50 - 9.16 - 17.76 - 25.14 - 30.56$$

$$= 370.9 \qquad (\approx 370.8)$$

For the battered pile with the $P\Delta$ effect

$$M_6 = 45(7.5)(0.9841) = 332.13$$
$$+ 225(7.5)(0.17365) = 293.03$$
$$+ 225(0.269) = 60.52$$
$$685.68$$

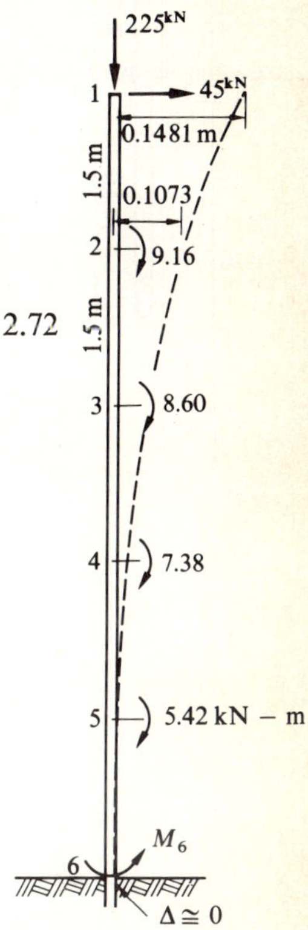

FIGURE E12-3.2

The output corrected for accumulated node moments is

$$836.1 - 16.68 - 32.08 - 45.47 - 55.33 = 686.55 \approx 685.68 \qquad ////$$

EXAMPLE 12-4 Using the data shown in Fig. 12-7, compute a curve of load versus deflection and compare to the field load test. Plot data directly on Fig. 12-7*b*. The pile is a 6-in-OD pipe with a cross section of 1.4 sq in of metal and a point area of 0.196 sq ft.

SOLUTION It was necessary to write a small computer program to convert the load in the pile to shear versus slip data. It was also necessary to estimate the point K, or spring value, the point load from Fig. 12-7*b*, and the resulting point deflection which must be added to each element-deflection value to obtain total slip for an element.

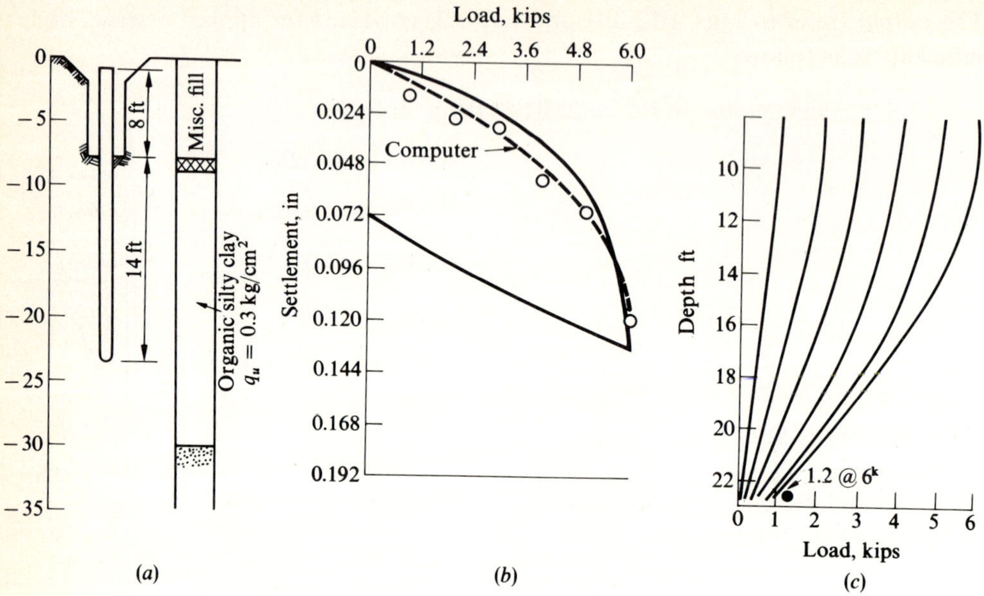

FIGURE 12-7
Pile-load-test data used in Example 12-4. (*a*) Soil-pile system. (*b*) Load-settlement curve with computer output for Example 12-4 added. (*c*) Load-transfer curves. Load-settlement curve of (*b*) is for test 7 given in reference. [*Seed and Reese* (1957).]

Pile load-depth data were read from the curve at even 2-ft increments. The program computed the average force in the pile element. Loads at 8, 10, 16 ft as obtained from the reference graphs are as follows (other data are on first page of computer output):

Depth, ft	Load = 1	2	3	4.1	5.1	6 kips
8	1	2	3	4.1	5.1	6
10	0.930	1.93	2.87	3.9	5.0	5.9
16	0.567	1.29	2.0	2.78	3.74	4.3

The output of the deflection-shear computer program was punched onto cards in 8F10.7 format to use directly as input in the computer program.

The soil modulus at the pile point is estimated as follows:

$$q_{\text{ult}} = q_{\text{u}} + qN_q$$

$$q_{ult} = 300 + 1.8(670)(1)(0.75) \quad \text{using a 25\% depth reduction and } \gamma = 1.8 \text{ g/cu cm}$$

$$= 300 + 6{,}030 = 9{,}345 \text{ g/sq cm}$$

$$k_s = \frac{9.345}{2.54} = 3.68 \text{ kg/cu cm} = 230 \text{ kips/cu ft}$$

$$K = A_p k_s = 0.196(230) = 45 \text{ kips/ft} \qquad \text{use 50 kips/ft}$$

After several sets of computations it appeared (due to load testing the pile several times) that K should be about 225 kips/ft. From Fig. 12-7*b* the point forces and resulting point deflections based on the computed K of 225 ft/kips are approximately

P, kips	P_{bot}, kips	Δ, ft
0	0	0
1	0.10	0.00044
2	0.25	0.00111
3	0.50	0.00222
4.1	0.75	0.00333
5.1	0.95	0.00422
6.0	1.2	0.00530

The Δ's are converted to inches for building-shear-versus-slip curves and are used with a minus sign and units of feet as POINTX in the pile program to obtain computed slip.

The shear-slip program was adjusted to obtain τ at slip $= 0$ and for one point beyond the 6.1-kip load to avoid having the computer search for undefined values. This is shown in the partial computer listing (Fig. E12-4.1), which includes the load-transfer curves.

The first pages of input, as well as output, for the 2-kip load are shown. The computed data are superimposed on Fig. 12-7*b* for comparison of computed versus measured response. It should be evident to the reader that modeling this pile is difficult due both to the small loads and deflections. Actual numerical deflection differences are small, however (0.018 versus 0.027 in computed at 2 kips), and the computed point loads match the values from the graph very well (250 versus 408 lb computed for 2-kip load).

The output could possibly be improved, but the validity of the method and the computer-program possibilities have been demonstrated. Any soil-parameter improvements are left as an exercise for the reader. ////

```
  J E BOWLES   EXAMPLE 12-4  PILE RESPONSE CURVE DATA SEED & REESE (1957)

 E =    30CC0.000 K/SQ IN       PILE LENGTH = 22.000 FT
 POINT SPRING =      225.000 K/FT      PILE POINT X-SECTION =  28.30C SQ IN
           INITIAL POINT DEFL =-0.0004 FT
           ASSUMED POINT LOAD =  0.0     PERCENT      NC CF PILE SEGMENTS =  9
                JOINT SOIL STARTS =  3   NO OF LOAD CONDITIONS =  3
   MAX PCINT DEFL TO REDUCE POINTK 25% = 0.0600 IN

 SOIL MODULUS =       5.0 +     3.00Z**1.000 K/CU FT

 THE INITIAL P-MATRIX
 1          -1.000

      NO OF PILE ELEMENTS FOR FOLLOWING CURVE DATA =  2

XC =   0.0        0.10000    0.20000    0.30000    0.40000    C.5CCCO    0.6C000    1.00000   IN
YC =   0.0        0.0        0.0        0.0        0.0        0.0        0.0        0.0       K/SQ FT

      NO CF PILE ELEMENTS FOR FOLLOWING CURVE DATA =  1

XC =   0.0        0.C0584    0.01446    0.02835    0.04228    0.05355    0.06700    8.06700   IN
YC =   0.01592    0.01592    0.01592    0.03183    0.06366    C.C3183    0.03183    0.03183   K/SQ FT

      NO OF PILE ELEMENTS FOR FOLLOWING CURVE DATA =  1

XC =   0.0        0.00581    0.01443    0.02829    0.04220    C.C5346    0.C6691    8.06691   IN
YC =   0.01592    0.01592    0.01592    0.03183    0.03183    0.06366    0.06366    0.06366   K/SQ FT

      NC OF PILE ELEMENTS FOR FOLLOWING CURVE DATA =  1

XC =   0.C        C.00577    0.01436    0.02815    0.04202    0.05324    0.06669    8.06668   IN
YC =   0.01048    0.03183    0.06366    0.12732    0.15915    C.19C99    0.19099    0.19099   K/SQ FT

      NC OF PILE ELEMENTS FOR FOLLOWING CURVE DATA =  1

XC =   0.0        0.00571    0.01419    0.02792    0.04174    C.C5292    0.06629    8.06628   IN
YC =   0.C        C.03183    0.12732    0.12732    0.15915    0.15915    0.25465    0.35014   K/SQ FT

      NO OF PILE ELEMENTS FOR FOLLOWING CURVE DATA =  1

XC =   0.0        0.00559    0.01397    0.02769    0.04141    0.05254    0.06579    8.06578   IN
YC =   0.08487    0.09549    0.11141    0.12732    0.20690    C.27C56    0.30239    0.33422   K/SQ FT

      NO OF PILE ELEMENTS FOR FOLLOWING CURVE DATA =  1

XC =   0.0        0.00547    0.01377    0.02742    0.04104    C.C52C2    0.06521    8.06521   IN
YC =   0.C0587    C.04775    0.11141    0.17507    0.20690    0.30239    0.33422    0.36606   K/SQ FT

      NC CF PILE ELEMENTS FOR FOLLOWING CURVE DATA =  1

XC =   0.0        0.C0538    0.01357    0.02711    0.04064    0.05148    0.06460    8.06460   IN
YC =   0.00595    0.04775    0.11141    0.17507    0.23873    C.3C239    0.35014    0.39789   K/SQ FT

  NO OF SHEAR STR CURVES =  8   NO OF ELEMENTS PER CURVE IN SEQUENCE =  2    1    1    1    1
 1    1    1
```

```
THE FINAL COMPUTED PILE ELEMENT FORCES AND OTHER DATA AFTER 6 ITERATIONS FOLLOWS

MEMNO     END MOMENTS             AXIAL FORCE          SOIL REACTIONS      ELEM. DEFL,FT      SOIL SPRINGS
   1    -0.000    -0.000      -2.000    2.000       0.0       0.0          -0.00223           0.0        0.0
   2    -0.000    -0.000      -2.000    2.000       0.0       0.0          -0.00204           0.0        0.0
   3    -0.000    -0.000      -2.000    1.945       0.000     0.000        -0.00185           1.750      4.500
   4    -0.000    -0.000      -1.945    1.893       0.000     0.000        -0.00176           6.500      7.500
   5    -0.000    -0.000      -1.893    1.700       0.000     0.000        -0.00167           9.500     10.500
   6    -0.000    -0.000      -1.700    1.347       0.000     0.000        -0.00158          12.500     13.500
   7    -0.000    -0.000      -1.347    1.007       0.000     0.000        -0.00151          15.500     16.500
   8    -0.000    -0.000      -1.007    0.707       0.000     0.000        -0.00145          18.500     19.500
   9    -0.000    -0.000      -0.707    0.408       0.000     0.000        -0.00141          21.500     22.500

     SUM HOR. SOIL REACTIONS =       0.00 KIPS

          PILE TOP MOVEMENTS:  VERT =  -0.02679 IN
                               HORZ =   0.00001 IN
                                ROT =   0.0      RAD

****** THE ULTIMATE PILE RESISTANCE =      -2.0000 KIPS VERTICAL AND     0.0    KIPS HORIZ
       THE ORIGINAL MOMENT APPLIED =        0.0     FT-K
          POINT FORCE =          0.408 KIPS

     CURRENT PROB NO =  3   POINTX = -0.002222 FT

THE INITIAL P-MATRIX
1          -3.000

THE X-MATRIX FOR CYCLE  1 IS
 1 -0.000143    2  0.000001   3  0.0       4  0.0       5  0.000000    6  0.0
 7  0.0         8  0.000000   9  0.0      10  0.0      11  0.0        12  0.0
13  0.0        14  0.0       15  0.0      16  0.0      17  0.0        18  0.0
19  0.0        20  0.0       21  0.0      22  0.0      23  0.0        24  0.0
25  0.0        26  0.0       27  0.0      28  0.0      29  0.0        30  0.0

ELEM NO   TOT ELEM DISP,FT   ELEM MOVE,FT    ELEM SLIP, IN     PTDEF1 =-0.0022220 FT
    1        -0.002365         -0.000143        0.028378
    2        -0.002222          0.000000        0.026664
    3        -0.002222          0.000000        0.026664
    4        -0.002222          0.0             0.026664
    5        -0.002222          0.0             0.026664
    6        -0.002222          0.0             0.026664
    7        -0.002222          0.0             0.026664
    8        -0.002222          0.0             0.026664
    9        -0.002222          0.0             0.026664

THE POINT DEFL DUE TO      -0.4999 KIPS IS  -0.0022220 FT
```

FIGURE E12-4.1
Problem input for 1-kip and partial output for the 2-kip pile load.

12-6 COMPUTER PROGRAM

This computer program will compute for either vertical or battered piles any of the following as specified by the user:

1 Pile ultimate load capacity and response, i.e., deflection-rotation effect and pile element forces, given a point deflection (POINTX < 0), or percent of load carried on the point (PERPP > 0), or for a friction pile. For these latter operations, IPRD $= 0$.

2 Pile load response for a given load (IPRD > 0).

3 Load-response curves (as in Example 12-2) if NLC > 0.

4 Response of partially embedded piles (with or without batter).

5 Pile may be tapered, round, square, or H-piles, and changes in moment of inertia or segment length can be accommodated from element to element.

6 Any number of shear-transfer curves can be used. They must be read in order from top to bottom with instructions to the computer (NEC) of the number of pile elements to be used for a set of curve data.

7 Any reasonable variation of soil modulus with depth can be used.

Any number of elements can be used by changing the DIMENSION statements; an ASA^T matrix of size 33×33 is required for 10 elements.

This program may oscillate (1) if the pile is too long (overdesigned) for the given load when the load is fixed (IPRD > 0) or (2) if pile loads are small and POINTX is specified unless the convergence criterion is relaxed. Convergence can be improved by making a trial run, observing the point deflection, then rerunning the problem with a revised POINTX value. In general, one must use plots of load versus depth to obtain load-transfer curves, as a simple made-up problem may not converge. If the load-transfer curves are

$$\tau = 0 \qquad \text{at slip} = 0$$

as obtained when using load-test data which include a point-deflection value, the program will compute only a minimum pile load; this will be analytically correct but physically incorrect for friction piles unless a reasonably correct value of POINTX is used.

In general, 8 to 10 pile segments should be used, but the segments do not have to be equal in length. This feature makes it very easy to model stratified soils and piles of varying moment of inertia or different materials.

The program is currently set for seven iterations (JJJ) and for a SLIP convergence of 0.02. For a given problem, these criteria may or may not be adequate.

It should be evident that if the computed load-response curve does not plot reasonably well onto the actual load-test curve, the pile-soil system has not been correctly modeled. One must vary one or more parameters, point spring, point load, etc., until the load-response curves match.

One may allow for nonlinear point deflections using XMAX. The program currently reduces the point spring 25 percent when the computed point deflection exceeds XMAX. This reduction must be reflected initially in the load-transfer curves.

Example 12-3 illustrates that pile bending moments in the output may require correcting for the fixed-end moments (which for this particular class of problem is a cumulative effect) and not simply subtracting the fixed-end-moment value at that node from the computed moment, as is commonly done for most structural-analysis problems.

Line	Operation
1	DIMENSION
3	READ TITLE, UT1–UT8
5	READ (1ZI5) NP = number of P's (10 elements NP = 33); NM = number of elements; NNZP = number of nonzero P's in P matrix to be read; NC = number of load-transfer curves; IPRD = switch if pile loads are known (activated if >0); NLC = switch to vary pile loads externally (activated if >0); JTSOIL = joint (or node) soil starts (use 1 if fully embedded); NDELT = switch to include $P\Delta$ effect on partially embedded piles (use 1 to include effect or 0 if not included); JJS = number of nodes where soil springs are read from cards; LIST = listing of all computations if LIST > 1; otherwise lists only P and X matrix of each cycle; NSTRPT = number of points on shear-transfer curves to enable the user to program the curve data [punch them on cards in 8F10.7 FORMAT and use directly as input with NEC(JJ) inserted as required]; IU = specification for fps (1) or metric units (2)
14	READ (8F10.4) E = modulus of elasticity, ksi (kg/sq cm); PIL = pile length, feet (meters); POINTK = point spring constant, kips/ft (kN/m); AREAP = area of pile point, sq in (sq cm); ALPHA = pile batter from horizontal, deg (90 percent = vertical); PERPP = percent of pile load carried by point (decimal); POINTX = point deflection, feet (meters) (known or assumed), generally read 0.00 or a minus value; XMAX = maximum point deflection, inches (centimeters) without a 25 percent point spring reduction
20	READ AS, BS, EXPO (8F10.4) Lateral subgrade modulus as k = AS + BS*Z**EXPO
34–42	Zeros and reads initial P matrix
51–56	Convert P(1), P(2) into equivalent axial load PAXIS(1) and hold P(1) and P(2) for recycle; used if IPRD > 0
57–66	Reads shear-transfer data and writes back: NEC(JJ) = number of elements to be used with data set; XC(I,JJ) = deflection, inches (abscissa of curve data, use up to eight points); YC(I,JJ) = shear strength mobilized for corresponding XC deflection, ksf (ordinates of curve data, use up to eight points) data read at up to eight points, and written back as check
69–85	Sets global P matrix [NPE(I,J)] and reads and stores member data using one card per member READ (5F10.4): XL(I) = pile element length, feet (note it does not have to be a constant); XIN(I) = member moment of inertia, in^4; BMEM(I) = member width, inches (projected width for round piles); PER(I) = member perimeter, feet; AREA(I) = member area, sq in; RSPR(I,1), RSPR(I,2) = soil springs if desired to read a modification here, leave blank if no modification (if metric, use corresponding metric units)

Line	Operation
86–91	Computes soil modulus with depth for each nodal point and writes values for checking
92	READ I, SK(I) if JJS > 0 to modify soil modulus at the *i*th node
95	Begins DO loop for ASA^T matrix. Note IP = pile element number
99–110	Computes member soil springs SSPR1 = value at top of each element
117–162	Computes nonzero values of element *A* and *S* matrices and adjusts ES(3,3), ES(4,4) using F1, F2, F3 as appropriate
163–167	Builds element SA^T matrix
169–178	Builds ASA^T matrix using each member contribution and writes final values if LIST > 0
187–199	Inverts ASA^T matrix
200	Begins loop for obtaining compatibility of *X* and shear-transfer strength
203–208	Computes current *X* matrix using 1 as counter for current value and writes values out
222–242	Computes element deflections as summing process for point to top. Point deflection is computed if necessary. Note that element deflections use the pro rata part of the nodal deflection applicable to that element using XPER1 and XPER2. The current point deflection PTDEF1 is an average on the third and succeeding cycles of the current and preceding point-deflection (PTDEF2) values to improve convergence. All values are written out if LIST > 0
244–258	Computes element slip as an element-deflection summation + point-deflection number
266–301	Using element slip [SLIP(M,JJ)] and the appropriate load-transfer curve computes element shear resistance [FRIC(I)] and corresponding adjustments to the pile axial load PAXIS(I). Writes values, including curve used and coordinates and current values of PAXIS(I) and FRIC(I) if LIST > 0
302–312	Makes required adjustments for point contribution to pile capacity
314–336	Recomputes *P* matrix. Sets P(I) = 0 if the element shear resistance exceeds the amount required for static stability
337–341	Includes $P\Delta$ effect in *P* matrix (moments) if desired (NDELT > 0)
344–350	Writes new *P* matrix for error check
355	Causes at least two cycles of *X* to be computed
356–368	Compares the new element slip value to preceding cycle value so that when all values are within specification, recycling halts, or if seven cycles, computed
375–388	Computes element forces based on last *X* matrix and writes values out; also checks $\sum F_h$ as SUMH
397–400	Computes pile top movements and writes out
409	READ Additional values of POINTX if NLC > 1 value for each new set of *P* data

```
      C      J E BOWLES  STATIC PILE ANALYSIS BY FINITE ELEMENT METHOD
      C      PILE MAY BE BATTERED; FULLY OR PARTIALLY EMBEDDED AND WITH VARIABL
      C        CROSS-SECTION--USE EITHER FPS CR METRIC UNITS
0001         DIMENSION ASAT(48,48),EA(6,6),P(48),X(48,2),XIN(20),XMOD(48,2),
            1SLIP(20,2),SS(20),DELX(20),SK(20),SUMDX(20),NEC(10),XC(10,10),
            2YC(10,10),PF(51),ELEMX(20),XL(20),XI(20),PER(20),BMEM(20),B(20),
            3AREA(20),ES(6,6),ESAT(6,6),EASAT(6,6),F(6),FRIC(20),PAXIS(20),
            4MNO(20),NPE(20,6),EAOL(20),SM(20),RSPR(20,2),TITLE(20)
0002         DOUBLE PRECISION UT5,UT6,UT7,UT8
0003    7C00 READ(1,1000,END=6000)TITLE,UT1,UT2,UT3,UT4,UT5,UT6,UT7,UT8
0004    1000 FORMAT(20A4/4(A4,6X),4(A8,2X))
0005         READ(1,101)NP,NM,NNZP,NC,IPRD,NLC,JTSOIL,NDELT,JJS,LIST,NSTRPT,IU
0006     101 FORMAT(12I5)
0007         WRITE(3,1001) TITLE
0008    1001 FORMAT('1',///,T5,20A4)
CC09         NMP = NM+1
0010         FU1 = 12.
0011         IF(IU.EQ.2)FU1 = 100.
0012         FU2 = 144.
0013         IF(IU.EQ.2)FU2 = 98.07
0014         READ(1,102)E,PIL,POINTK,AREAP,ALPHA,PERPP,POINTX,XMAX
0015     102 FORMAT(8F10.4)
0016         WRITE(3,625)E,UT8,PIL,UT1,PCINTK,UT7,AREAP,UT2,POINTX,UT1,PERPP,NM
            1,JTSOIL,NLC ,XMAX,UT2
0017     625 FORMAT(//,T5,'E =',F12.3,1X,A8,           5X,'PILE LENGTH =',F7.3,1
            1X,A2/T5, 'POINT SPRING =',F12.3,1X,A4,5X, 'PILE POINT X-SECTION ='
            2,F8.3, ' SQ ',A2/ T15,'INITIAL PCINT DEFL =',F7.4,1X,A2/ T15,'ASSU
            3MED POINT LOAD =',F8.4,' PERCENT',5X,'NO OF PILE SEGMENTS =',I3/T2
            40,'JOINT SCIL STARTS =',I3,3X,'NO OF LOAD CONDITIONS =',I3,/,T7, '
            5MAX POINT DEFL TO REDUCE PCINTK 25% =',F7.4,1X,A2)
```

```
      IF(NDELT.GT.0)WRITE(3,632)
  632 FORMAT(/,T7,'*** P-DELTA EFFECT IS CONSIDERED IN COMPUTATIONS **')
      READ(1,102)AS,BS,EXPO
      WRITE(3,1005)AS,BS,EXPO,UT6
 1005 FORMAT(//,T5,'SOIL MODULUS =',F8.1,' + ',F7.2,'Z**',F5.3,1X,A8)
      SAVEK = POINTK
      E = E*FU2
      AREAP = AREAP/(FU1*FU1)
      DO 103 I=1,NP
      DO 103 J=1,NP
  103 ASAT(I,J)=0.0
      JCOUN = 1
 7500 II=0
      JJ=0
      JJJ = 1
      SUMH = 0.
      DO 4 I=1,NP
      X(I,2) = 0.
    4 P(I) = 0.
      WRITE(3,546)
  546 FORMAT(//,T5,'THE INITIAL P-MATRIX')
      DO 5 IP=1,NNZP
      READ(1,6)I,P(I)
    6 FORMAT(I5,F10.4)
    5 WRITE(3,547)I,P(I)
  547 FORMAT(T3,I3,3X,F12.3)
      NMP1 = NM+1
      DO 7510  I = 1,NMP1
      PAXIS(I) = 0.
 7510 FRIC(I) = 0.
      ALPHAR = ALPHA/57.2957731
      XSIN=SIN(ALPHAR)
      XCOS=COS(ALPHAR)
      PAXIS(1) = P(1)*XSIN + P(2)*XCOS
      IPAXIS = PAXIS(1)
      IF(IPRD.LE.0)PAXIS(1) = 0.
      IF(IPRD.GT.0)PHOLDV = P(1)
      IF(IPRD.GT.0)PHOLDH = P(2)
      IF(NLC.GT.0.AND.JCOUN.GT.1)GO TO 2000
C     READ LOAD TRANSFER DATA AT 8 COORDS PER CARD X = SLIP; Y = SHEAR S
C     NEC(JJ) = NO OF ELEMENTS FOR CORRESPONDING LOAD TRANSFER CURVE DAT
      DO 1012  JJ = 1,NC
      READ(1,6)NEC(JJ)
      READ(1,102)(XC(I,JJ),I=1,NSTRPT)
      READ(1,102)(YC(I,JJ),I=1,NSTRPT)
      WRITE(3,548)NEC(JJ),(XC(I,JJ),I=1,NSTRPT),UT2
  548 FORMAT(//T10,'NO OF PILE ELEMENTS FOR FOLLOWING CURVE DATA =',I3//
     1T2,'XC =',8F10.5,2X,A2)
 1012 WRITE(3,549)(YC(I,JJ),I=1,NSTRPT),UT5
  549 FORMAT(T2,'YC =',8F10.5,2X,A7)
      WRITE(3,177)NC,(NEC(JJ),JJ=1,NC)
  177 FORMAT(//,T5,'NO OF SHEAR STR CURVES =',I3,3X, 'NO OF ELEMENTS PER
     1 CURVE IN SEQUENCE =', 5(I3,2X))
      WRITE(3,104)UT1,UT2,UT1,UT1,UT2
  104 FORMAT(     ///,T7, 'MEMNO',2X,'NP1', 2X,'NP2',2X,'NP3',2X,'NP4',2X
     2,'NP5',2X,   'NP6', 3X, 'ALPHA',4X,'L, ',A2,2X,'I, 'A2,'**4',2X,'WI
     3DTH, 'A2,2X,'PERIM, ',A2,2X,'A SQ ',A2//)
      K = 1
      DO 520  I = 1,NM
      MNO(I) = I
      NPE(I,1) = K
      NPE(I,2) = K+1
      NPE(I,3) = K+2
      NPE(I,4) = K+3
      NPE(I,5) = K+4
      NPE(I,6) = K+5
      K = K+3
      READ(1,102)XL(I),XIN(I),BMEM(I),PER(I),AREA(I),RSPR(I,1),RSPR(I,2)
      B(I) = BMEM(I)/FU1
      WRITE(3,108)MNO(I),(NPE(I,IZ),IZ=1,6),ALPHA,XL(I),XIN(I),B(I),PER(
     AI),AREA(I)
  108 FORMAT(5X,7I5,3X,F7.3,2X,F6.2,2X,F8.2,3(4X,F6.2))
      AREA(I) = AREA(I)/(FU1*FU1)
      EAOL(I) = E*AREA(I)/XL(I)*2.
  520 XI(I) = XIN(I)/(FU1**4)
      AL = 0.
      DO 522 I = 1,NMP
      SK(I) = AS + BS*AL**EXPO
      IF(I.LT.JTSOIL)SK(I) = 0.0
  522 IF(I.GE.JTSOIL.AND.I.LE.NM)AL=AL+XL(I)
C*****MODIFY SUBGRADE MODULUS IF JJS>0
  212 FORMAT(I5,F10.1)
      IF(JJS.GT.0)READ(1,212)(JTNO,SK(JTNO),M=1,JJS)
  517 WRITE(3,887)
  887 FORMAT(//,T5,'NODE',3X, 'SOIL MODULUS AT NODE',3X,'MEMNO',5X,'SOIL
     1 SPRING 1',3X, 'SOIL SPRING 2')
 8000 DO 205 IP = 1,NM
 8050 EIOL = E*XI(IP)/XL(IP)
      SINOL = XSIN/XL(IP)
      COSOL = XCOS/XL(IP)
      SSPR1 = 0.0
      SSPR2 = 0.0
      IF(IP.LE.JTSOIL-1)GO TO 528
      IF(IP.EQ.NM)GO TO 523
      SSPR1 = XL(IP)*B(IP)*(7.*SK(IP)+6.*SK(IP+1)-SK(IP+2))/24.
```

```
      SSPR2 = XL(IP)*B(IP)*(3.*SK(IP)+10.*SK(IP+1)-SK(IP+2))/24.
      IF(RSPR(IP,1).GT.0.)SSPR1 = RSPR(IP,1)
      IF(RSPR(IP,2).GT.0.)SSPR2 = RSPR(IP,2)
      IF(IP.EQ.JTSOIL.AND.RSPR(IP,1).LE.0.)SSPR1 = .5*SSPR1
      GO TO 528
  523 SSPR1 = XL(IP)*B(IP)*(3.*SK(IP+1)+10.*SK(IP)-SK(IP-1))/24.
      SSPR2 = XL(IP)*B(IP)*(7.*SK(IP+1)+6.*SK(IP)-SK(IP-1))/24.
  528 IF(II.LE.0)WRITE(3,888)IP,SK(IP),IP,SSPR1,SSPR2
  888 FORMAT(T5,I4,10X,F12.3,T35,I4,5X,F12.3,5X,F12.3)
      DO 1010 IJ = 1,6
      DO 1010 JJ = 1,6
      EA(IJ,JJ) = 0.0
 1010 ES(IJ,JJ) = 0.0
      EA(1,1)=+COSOL
      EA(1,2)=+COSOL
      EA(1,3)=+XSIN
      EA(1,5)=-XCOS
      EA(2,1)=-SINOL
      EA(2,2)=-SINOL
      EA(2,3)=+XCOS
      EA(2,5)=+XSIN
      EA(3,1)=1.0
      EA(4,1)=-COSOL
      EA(4,2)=-CCSOL
      EA(4,4)=-XSIN
      EA(4,6)=-XCOS
      EA(5,1)=+SINOL
      EA(5,2)=+SINOL
      EA(5,4)=-XCOS
      EA(5,6) = +XSIN
      EA(6,2) = 1.0
      ES(1,1) = 4.*EIOL
      ES(1,2) = 2.*EIOL
      ES(2,1) = 2.*EIOL
      ES(2,2) = 4.*EIOL
      IF(IP.LT.NM)XLMAX = MAX1(XL(IP),XL(IP+1))
      IF(IP.LT.NM)XLMIN = MIN1(XL(IP),XL(IP+1))
      IF(IP.GT.1)XLMAX2 = MAX1(XL(IP-1),XL(IP))
      IF(IP.GT.1)XLMIN2 = MIN1(XL(IP-1),XL(IP))
      RATIO = XLMAX/XLMIN
      SUMSQ = (1.+RATIO)**2
      FAC2 = SUMSQ/(RATIO**2)
      F1 = SUMSQ
      IF(IP.EQ.1)GO TO 756
      IF(XL(IP).GT.XL(IP+1))F1 = FAC2
      RATIO2 = XLMAX2/XLMIN2
      SUMSQ2 = (1.+RATIO2)**2
      FAC4 = SUMSQ2/(RATIO2**2)
      F3 = SUMSQ2
      IF(XL(IP).GT.XL(IP-1))F3 = FAC4
  756 IF(IP.EQ.1)F3 = 1.
      ES(3,3) = EAOL(IP)/F3
      IF(II.GT.0)ES(3,3) = EAOL(IP)
      IF(IP.EQ.1)ES(3,3) = EAOL(IP)
      ES(4,4) = EAOL(IP)/F1
      IF(II.GT.0)ES(4,4) = EAOL(IP)
      IF(IP.EQ.NM)ES(4,4) = EAOL(IP)
      ES(5,5) = SSPR1
      ES(6,6) = SSPR2
      DO 202  I = 1,6
      DO 202 J=1,6
      ESAT(I,J)=0.0
      DO 202  K = 1,6
  202 ESAT(I,J)=ESAT(I,J)+ES(I,K)*EA(J,K)
      IF(II.GT.0) GO TO 605
  203 DO 204 I=1,6
      DO 204 J=1,6
      EASAT(I,J)=0.0
      DO 204  K = 1,6
  204 EASAT(I,J)=EASAT(I,J)+EA(I,K)*ESAT(K,J)
      DO 205 I=1,6
      NS1 = NPE(IP,I)
      DO 205 J=1,6
      NS2 = NPE(IP,J)
  205 ASAT(NS1,NS2)=ASAT(NS1,NS2)+EASAT(I,J)
  521 WRITE(3,889)NMP,SK(NMP)
  889 FORMAT(T5,I4,10X,F12.3)
      IF(LIST.LE.0)GO TO 518
      WRITE(3,304)
  304 FORMAT(//,T5,'THE ASAT MATRIX',/)
      DO 302 I=1,NP
  302 WRITE(3,303)I,(ASAT(I,J),J=1,NP)
  303 FORMAT(I3,2X,12F10.2)
C     ASAT INVERSION
  518 M = NP
      DO 25 K=1,M
      DO 20 J=1,M
   20 IF(J.NE.K)ASAT(K,J)=ASAT(K,J)/ASAT(K,K)
      DO 21 I=1,M
      IF(I.EQ.K) GO TO 21
      DO 21 J=1,M
      IF(J.EQ.K) GO TO 21
      ASAT(I,J)=ASAT(I,J)-ASAT(K,J)*ASAT(I,K)
   21 CONTINUE
      DO 22 I=1,M
   22 IF(I.NE.K)ASAT(I,K)=-ASAT(I,K)/ASAT(K,K)
```

```
   25 ASAT(K,K)=1.0/ASAT(K,K)
C     *******END OF ASAT INVERSION--COMPUTE X-MATRIX *****
 2000 P11B = P(1)
      P21B = P(2)
      P31B = P(3)
      DO 7 I=1,NP
      X(I,1)=0.0
      DO 7 J=1,NP
      X(I,1)=X(I,1)+ASAT(I,J)*P(J)
      IF(ABS(X(I,1)).LT..0000001)X(I,1) = 0.0
    7 XMOD(I,1) = X(I,1)
      XPDEL = X(2,2)
      WRITE(3,890)JJJ
  890 FORMAT(//,T5,'THE X-MATRIX FOR CYCLE',I3,' IS')
      ANP = NP/6
      INP = ANP
      IF(INP.LT.ANP)INP=INP+1
      IZ = 1
      DO 188  I = 1,INP
      IZP = IZ+5
      IF(IZP.GT.NP)IZP = NP
      WRITE(3,623)(NN,X(NN,1),NN=IZ,IZP)
  188 IZ = IZP+1
  623 FORMAT(T3,I3,F10.6,5(I4,F10.6))
C     ****** COMPUTE ELEMENT DEFLECTICNS *****
      MM = 1
      DO 26 I = 1,NM
      XPER1 = XL(I)/(XL(I)+XL(I+1))
      IF(I.GT.1)GO TO 27
      ELEMX(I) = X(MM,1)*XSIN + X(MM+1,1)*XCOS + X(MM+3,1)*XPER1*XSIN +
     AX(MM+4,1)*XPER1*XCOS
      GO TO 26
   27 XPER2 = XL(I)/(XL(I-1)+XL(I))
      IF(I.EQ.NM) GO TO 28
      ELEMX(I) = X(MM,1)*XPER2*XSIN + X(MM+1,1)*XPER2*XCOS + X(MM+3,1)*X
     >PER1*XSIN + X(MM+4,1)*XPER1*XCCS
      GO TO 26
   28 ELEMX(NM) = X(NP-2,1)*XSIN + X(NP-1,1)*XCOS + X(MM,1)*XPER2*XSIN +
     A X(MM+1,1)*XPER2*XCOS
      IF(P(NP-2).GT.0.0.OR.POINTK.LE.C.)GO TO 26
      POINTF = P(NP-2)/XSIN
C     NOTE PTDEF1 IS APPROX VALUE CF PRECED CYCLE IF P(NP-2)<0.
      POINTK = SAVEK
  311 PTDEF1 = POINTF/POINTK
      IF(POINTK.LE..75*SAVEK)GO TC 26
      IF(ABS(PTDEF1*FU1).GT.XMAX)PCINTK = .75*SAVEK
      IF(POINTK.LE..75*SAVEK)GO TC 311
   26 MM=MM+3
      IF(JJJ.LT.2.AND.PTDEF1.GT.PCINTX)PTCEF1 = POINTX
      IF(JJJ.GT.2)PTDEF1 = (PTDEF1+PTDEF2)/2.
      MM=NM
      DO 9 I=1,NM
      IF(I.GT.1) GO TO 10
      SUMDX(MM) = ELEMX(MM)+PTDEF1
      DELX(MM)=ELEMX(MM)
      SLIP(MM,1) = (DELX(MM) + PTCEF1)*FU1
      GO TO 11
   10 DELX(MM)=DELX(MM+1)+ELEMX(MM)
      SUMDX(MM) = SUMDX(MM+1) + ELEMX(MM)
      SLIP(MM,1) = (DELX(MM) + PTCEF1)*FU1
   11 IF(SLIP(MM,1).LT.0.)SLIP(MM,1) = -SLIP(MM,1)
    9 MM=MM-1
      WRITE(3,30)UT1,UT1,UT2,PTDEF1,UT1
   30 FORMAT(//T5,'ELEM NO',2X,'TOT ELEM DISP,'A2,3X,'ELEM MOVE,',A2,5X,
     1'ELEM SLIP, ',A2,4X,'PTDEF1 =',F10.7,1X,A2)
      DO 32 I=1,NM
   32 WRITE(3,31)I,SUMDX(I),ELEMX(I),SLIP(I,1)
   31 FORMAT(T7,I3,5X,F12.6,5X,F12.6,3X,F12.6)
      IF(LIST.GT.0)WRITE(3,626)
  626 FORMAT(//,T10,'PRINT COUNTERS USED TO RECOMPUTE P-MATRIX')
 3060 K = 1
      KN = 1
      I2 = 1
      KCC = NEC(KN)
      DO 501  KZ = 1,NP
  501 IF(KZ/3*3.NE.KZ)P(KZ)=0.
      DO 52  I = 1,NM
      IF(I.GT.KCC)I2=KN+1
      IF(I2.NE.KN)KCC = KCC+NEC(I2)
      DO 54  JJ = 1,8
      IF(XC(JJ,I2).GT.SLIP(I,1))GC TO 43
      IF(XC(JJ,I2).EQ.SLIP(I,1))GO TO 44
   54 CONTINUE
   43 JJ = JJ-1
   44 C1 = ABS(YC(JJ+1,I2)-YC(JJ,I2))
      DELYC = 0.
      C2 = XC(JJ+1,I2)-XC(JJ,I2)
      IF(YC(JJ+1,I2).LT.YC(JJ,I2))C1 = -C1
      DELTAX = SLIP(I,1) - XC(JJ,I2)
      IF(ABS(C2).LE..0000001)C2 = 0.
      IF(C2.EQ.0.)GO TO 261
      DELYC = DELTAX*C1/C2
  261 SS(I) = YC(JJ,I2)+DELYC
      FRIC(I) = XL(I)*PER(I)*SS(I)
      JPLUS = I+1
      IF(IPRD.GT.0)GO TO 45
      IF(IPAXIS)234,235,235
```

```
  234 PAXIS(I+1) = PAXIS(I) - FRIC(I)
      GO TO 46
  235 PAXIS(I+1) = PAXIS(I)+FRIC(I)
      GO TO 46
   45 IF(PAXIS(I))81,80,80
   80 PAXIS(I+1) = PAXIS(I)-FRIC(I)
      IF(PAXIS(I+1).LT.0..AND.FRIC(I).NE.0.)PAXIS(I+1) = 0.0
      GO TO 46
   81 PAXIS(I+1) = PAXIS(I)+FRIC(I)
      IF(PAXIS(I+1).GT.0..AND.FRIC(I).NE.0.)PAXIS(I+1) = 0.0
   46 IF(LIST.GT.0)WRITE(3,154)NEC(I2),XC(JJ,I2),XC(JJ+1,I2),JJJ,JJ,I2,
     1KCC,JPLUS,PAXIS(I+1),UT3,FRIC(I),UT3,PAXIS(I),UT3
  154 FORMAT(T15,'NEC(I2)=',I3,3X,'XC(JJ,I2)=',F7.5,3X,'XC(JJ+1,I2) =',
     1F7.5/ T10,'JJJ=',I3,3X,'JJ=',I3,3X,'I2=',I3,3X,'KCC=',I3/ T12,
     2'CURRENT VALUE PAXIS(',I3,') =',F10.3,1X,A4,5X,'ELEM. FRIC =',
     3F10.3,1X,A4/ T14,'CURRENT VALUE PAXIS(I) =',F10.3,1X,A4)
   52 KN = I2
C     POINT EFFECTS INCLUDED IN P-MATRIX--***--NOTE POINTX OR PERPP ARE
C          READ AS DATA;   PTDEFL IS CCMPUTED POINT DEFLECTION
      POINTF = PTDEF1*POINTK
      IF(PTDEF1.GT.0.)POINTF = 0.0
      IF(IPRD.GT.0)GO TO 151
      IF(PERPP.GT.0.)GO TO 48
   47 PAXIS(NM+1) = PAXIS(NM+1)+PCINTF
      GO TO 151
   48 PAXIS(NM+1) = PAXIS(NM+1)*(1.+PERPP)
      POINTF = PAXIS(NM+1)-(1.-PERPP)*PAXIS(NM+1)
      PTDEF1 = POINTF/POINTK
  151 WRITE(3,208)POINTF,UT3,PTDEF1,UT1
  208 FORMAT(//T5,'THE POINT DEFL DUE TC',F12.4,1X,A4,' IS',F12.7,1X,
     1A2//)
      L = 1
      DO 37 I = 1,NM
      IF(IPRD.LE.0.AND.IPAXIS.LT.0)FRIC(I) = -FRIC(I)
      IF((PHOLDV.EQ.0..AND.PHOLDH.EQ.C.).AND.IPRD.GT.0.)GO TO 41
      IF(IPRD.GT.0.AND.I.EQ.1)GC TO 36
      IF(IPRD.GT.0)GO TO 38
      IF(IPRD.LE.0)GO TO 39
   36 P(1) = PHOLDV
      P(2) = PHOLDH
   38 P(L+3) = PAXIS(I+1)*XSIN
      P(L+4) = PAXIS(I+1)*XCOS
      GO TO 37
   39 IF(I.GT.1)GO TO 35
      P(1) = PAXIS(NM+1)*XSIN
      P(2) = PAXIS(NM+1)*XCOS
      P(L+3) = P(1)-FRIC(I)*XSIN
      P(L+4) = P(2)-FRIC(I)*XCOS
      GO TO 37
   35 P(L+3) = P(L)-FRIC(I)*XSIN
      P(L+4) = P(L-1)-FRIC(I)*XCOS
      GO TO 37
C     THIS BRANCH TAKES CARE OF PILE WITH MOMENT ONLY
   41 P(L+3) = 0.
      P(L+4) = 0.
   37 L = L+3
C  ******  APPLY P-DELTA EFFECT
      IF(NDELT.LE.0)GO TO 158
      LK=0
      DO 924 KI=1,JTSOIL
      P(LK+3) = -(X(2,1) - X(LK+2,1))*P(1)
  924 LK=LK+3
  158 WRITE(3,906)UT2,UT5
  906 FORMAT(///,T4,'NM',2X,'SLIP, ',A2,2X,'SHEAR, ',A7,6X,'P-MATRIX IN
     13-COLS')
  907 IP1 = 1
      DO 74  I = 1,NM
      IP2 = IP1+1
      IP3 = IP1+2
      WRITE(3,893)I,SLIP(I,1),SS(I),IP1,P(IP1),IP2,P(IP2),IP3,P(IP3)
  893 FORMAT(T3,I3,F10.6,2X,F10.5,3X,I2,F10.3,2X,I2,F10.3,2X,I2,F10.3)
   74 IP1 = IP3+1
      NMP2 = NP-2
      WRITE(3,895)((IM,P(IM)),IM=NMP2,NP)
  895 FORMAT(T31,I2,F10.3,2X,I2,F10.3,2X,I2,F10.3)
      JJJ = JJJ+1
      IF(JJJ.LE.2)GO TO 1120
C     TEST FOR SLIP CONVERGENCE------
      KCOUN = 0
      DO 58  I = 1,NM
      SLIPCH = SLIP(I,1)-SLIP(I,2)
   58 IF(ABS(SLIPCH).LE.0.008)KCCUN = KCCUN+1
      IF(JJJ.GT.7)GO TO 77
 1120 DO 896  J = 1,NM
      X(J,2) = X(J,1)
      SLIP(J,2) = SLIP(J,1)
  896 XMOD(J,2) = XMOD(J,1)
      PTDEF2 = PTDEF1
      IF(JJJ.LE.2.OR.KCOUN.LT.NM)GO TC 2000
      IF(ABS(X(2,1)-XPDEL).GT.0.C20.ANC.NDELT.GT.0)GO TO 2000
   77 II = II+1
      JJJ = JJJ-1
      WRITE(3,170)JJJ
  170 FORMAT('1',///,T5,'THE FINAL CCMPUTED PILE ELEMENT FORCES AND OTHE
     1R DATA AFTER',I2,' ITERATICNS FCLLCWS')
      WRITE(3,59) UT1
   59 FORMAT(//,T5,'MEMNO',5X,'END MOMENTS',10X,'AXIAL FORCE',8X,'SOIL R
```

```
     1EACTIONS',3X,'ELEM. DEFL,',A2,4X,'SOIL SPRINGS')
      GO TO 8000
C     CYCLE TO FIND PILE ELEMENT FORCES AS F = SAT*X
  605 DO 851  I = 1,6
      F(I) = 0.
      DO 851  K = 1,6
      N = NPE(IP,K)
      IF(IP.LT.NM)XPER1 = XL(IP)/(XL(IP)+XL(IP+1))
      IF(IP.GT.1)XPER2 = XL(IP)/(XL(IP-1)+XL(IP))
      XMOD(N,1) = X(N,1)
      IF(I.LE.2)GO TO 851
      IF(I.GE.5)GO TO 851
      IF(K/3*3.EQ.K)GO TO 851
  849 IF(IP.EQ.1.AND.K.LE.3)GO TO 851
      IF(IP.EQ.1.AND.K.GT.3)XMOD(N,1)=X(N,1)*XPER1
      IF(IP.GT.1.AND.K.LT.3)XMOD(N,1) = X(N,1)*XPER2
      IF((IP.GT.1.AND.IP.NE.NM).AND.K.GT.3)XMOD(N,1)=X(N,1)*XPER1
  851 F(I) = F(I) + ESAT(I,K)*XMOD(N,1)
      SUMH = SUMH + (F(5)+F(6))*XSIN
      WRITE(3,62)IP,(F(I),I=1,6),SUMDX(IP),SSPR1,SSPR2
   62 FORMAT(T5,I5,6F10.3,F12.5,2X,2F12.3)
      IF(IP.EQ.NM)GO TO 609
      GO TO 205
  609 WRITE(3,63)SUMH,UT3
   63 FORMAT(//T10,'SUM HOR. SOIL REACTIONS =',F10.2,1X,A4)
      DELV = SUMDX(1)*XSIN*FU1
      DELH=X(2,1)*FU1
      ROT = X(3,1)
      WRITE(3,175)DELV,UT2,DELH,UT2,ROT
  175 FORMAT(//T15,'PILE TOP MOVEMENTS:  VERT =',F10.5,1X,A2/ T36,'HORZ
     1=',F10.5,1X,A2/ T37,'ROT =',F10.5,' RAD')
      POINTF = F(4)/XSIN
      WRITE(3,174)P11B,UT3,P21B,UT3,P31B,UT4,POINTF,UT3
  174 FORMAT(//T5,'****** THE ULTIMATE PILE RESISTANCE =',F12.4,1X,A4,
     1' VERTICAL AND',F9.3,1X,A4,' HORIZ'/ T12,'THE ORIGINAL MOMENT APPL
     2IED =',F12.4,1X,A4/ T15,'POINT FORCE =',F12.3,1X,A4)
      IF(IPRD.LE.0)WRITE(3,241)
  241 FORMAT(T15,'*** PILE RESISTANCE IS COMPUTED')
      JCOUN = JCOUN+1
      IF(JCOUN.GT.NLC)GO TO 7000
      IF(NLC.GT.1)READ(1,102)POINTX
      WRITE(3,7701)JCOUN,POINTX,UT1
 7701 FORMAT(//T10,'CURRENT PROB NO =',I3,3X,'POINTX =',F10.6,1X,A4)
      GO TO 7500
 6000 STOP
      END
```

PROBLEMS

12-1 Use three elements and constant skin friction = 4 ksf for all three elements. Solve parts (*a*) to (*c*) separately, then (*d*). Use a 14BP73 with a length of 32 ft.

(*a*) P(1) = 10 kips (*b*) P(2) = 10 kips
(*c*) P(3) = 10 kips (*d*) P(1) = P(2) = P(3) = 10 kips

12-2 Repeat Prob. 12-1 for $\alpha = 80°$.

12-3 Solve the following problem if $\alpha = 90°$ and the response of these loads is desired (IPRD 1). Use 10 elements, a 16-in-OD × 0.375-in-wall pipe pile. Use metric units.

(*a*) P(1) = −450 kN; pile data (fps): $I = 562.08$ in^4; $A = 18.408$ sq in
(*b*) P(2) = −200 kN; perimeter = 4.18 ft; $L = 50$ ft
(*c*) P(3) = 65 kN-m
(*d*) The $\sum P_i$ of (*a*), (*b*), (*c*)

Use three load-transfer curves with constant values of τ for all slip as follows:

Elements	1–3	4–7	8–10
τ, kN/sq m	12.5	8.7	35

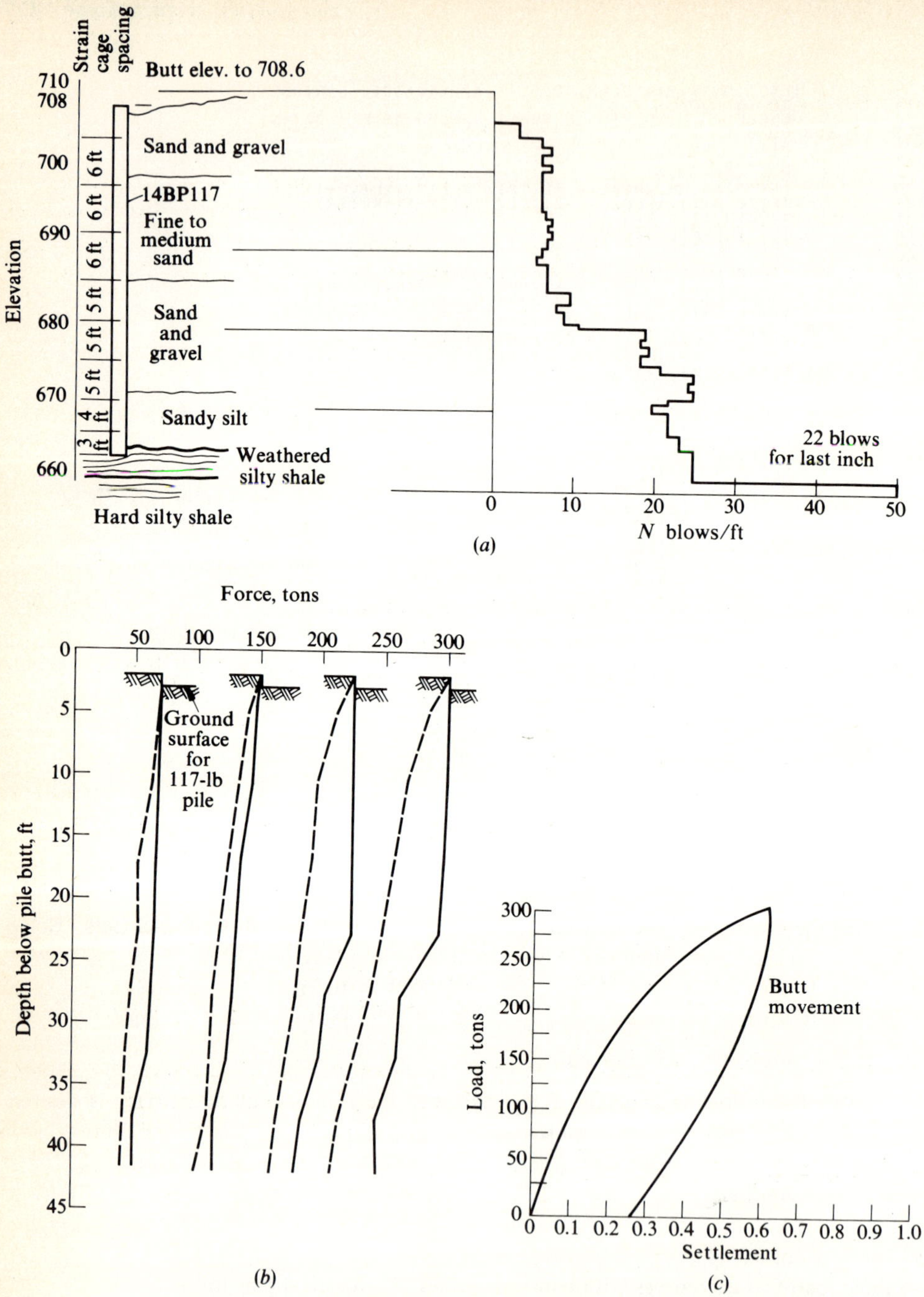

FIGURE 12-8
Point-bearing H-pile tests and data. [*After D'Appolonia and Romualdi* (1963).] (*a*) Soil profile, blow count, and instrumentation. (*b*) Load-transfer data (tons in fps units). (*c*) Load-test data for 14BP89 pile.

12-4 Repeat Prob. 12-3 for $\alpha = 80°$.

12-5 Verify the output of Example 12-4 using Fig. 12-7.

12-6 Repeat Example 12-4 using a stiffer point spring to obtain a better pile-response fit. Subtract the given point deflections (in inches) from the given (or computer listing) element deflections, then add back new estimated point deflections based on the new point spring constant to obtain new deflections. The load-transfer ordinates will not require correction. Other data are unchanged.

12-7 Using the D'Appolonia and Romualdi (1963) data of Fig. 12-8, make a plot of load versus deflection and compare to the field test.

12-8 Use the data of Reese et al. (1969) (Fig. 12-2) and make a load-versus-deflection curve for comparison to field test 1 or 2.

REFERENCES

COYLE, H. M., and L. C. REESE (1966): Load Transfer of Axially Loaded Piles in Clay, *J. Soil Mech. Found. Div., ASCE*, vol. 92, SM2, March, pp. 1–26.

D'APPOLONIA, E. and J. P. ROMUALDI (1963): Load Transfer in End-Bearing Piles, *J. Soil Mech. Found. Div., ASCE*, vol. 89, SM2, pp. 1–25.

MOHAN, D., G. S. JAIN, and V. KUMAR (1963): Load Bearing Capacity of Piles, *Geotech. (Lond.)*, vol. 13, no. 1, March, pp. 76–86.

REESE, L. C., W. R. HUDSON, and V. N. VIJAYVERGIYA (1969): An Investigation of the Interaction between Bored Piles and Soil, *Proc. 7th Int. Conf. Soil Mech. Found. Eng., Mexico City*, vol. 2, pp. 211–215.

SEED, H. B. and L. C. REESE (1957): The Action of Soft Clay along Friction Piles, *Trans. ASCE*, vol. 122, pp. 731–734.

SHERMAN, W. C. (1969): Instrumented Pile Tests in a Stiff Clay, *Proc. 7th Int. Conf. Soil Mech. Found. Eng., Mexico City*, vol. 2, pp. 227–232.

TAVENAS, FRANCOIS A. (1971): Load Test Results on Friction Piles in Sand, *Can. Geotech. J.*, vol. 8, no. 1, February, pp. 7–22.

VESIC, A. S. (1970): Load Transfer in Pile-Soil Systems, *Proc. Conf. Des. Install. Pile Found. Cell. Struct., Lehigh Univ.*, pp. 47–73 (contains large number of references).

13

GENERAL SOLUTION FOR PILE GROUPS

13-1 MATRIX METHOD OF ANALYSIS FOR PILE GROUPS

A matrix approach for obtaining the individual pile forces in a pile group was considered by Hrennikoff (1950). Later Aschenbrenner (1967) and Bowles (1968) presented solutions for pile heads free to rotate. Saul (1968) and Reese et al. (1970) extended the solution to a somewhat more general case. Butterfield and Banerjee (1971) considered a matrix solution based on a pile located in a homogeneous isotropic half-space. This last solution is considered by the author to be too mathematical to be practical and will not be discussed.

Reese and Saul use very similar methods. Saul places one of the pile-head forces in the *XZ* plane and considers the pile force components by using a separate matrix. Reese et al. (1970) place one pile-head force in the *XY* plane and directly consider the contribution of pile-head forces in resisting pile-cap moments. The Reese et al. (1970) solution is essentially the same as the method presented in this chapter. Two of the major differences involved in any matrix method of pile-group analysis are the definition of the direction cosines and the method of defining the pile-head stiffness.

To formulate a three-dimensional matrix solution for a group of piles we will assume that the piles are interconnected at their heads to a rigid pile cap. This implies that relative movement of the pile cap between adjacent pile-head connections is negligible. The pile cap will have six degrees of freedom, including translation in the three coordinate directions as well as freedom to rotate about the three coordinate axes, i.e., translations ΔX, ΔY, ΔZ and rotations αX, αY, αZ. The solution is a relatively simple one using matrix operations, the major part of the effort being that of obtaining direction cosines.

The problem in brief is to solve the usual matrix equations

$$P = AF \tag{a}$$

$$e = A^T X \tag{b}$$

$$F = Se \tag{c}$$

and

$$X = [ASA^T]^{-1}P \tag{d}$$

$$F = SA^T X \tag{e}$$

There are certain dissimilarities in this method, however, in that we must:

1 Develop the A matrix for each pile. The individual-pile A matrix in this case is the ratio of applied group load carried by the pile to satisfy the six static equations of equilibrium.
2 Develop the S matrix for each pile.
3 Recognize that in the equation $e = BX = A^T X$ the reciprocity rule is again valid to obtain $B = A^T$.
4 Sum the individual-pile ASA^T matrices as

$$(ASA^T)_{\text{pile cap}} = \sum_{i=1}^{n} (ASA^T)_i \tag{13-1}$$

to obtain the total pile-cap (and group) matrix to invert.

The sum of the individual-pile forces acting on the foundation cap provides the equilibrium of the system; thus, the signs of the computed pile forces must be interpreted properly; i.e., the direction of the pile forces acting to keep the cap in equilibrium *are opposite in direction to the forces acting on the pile* (Fig. 13-1).

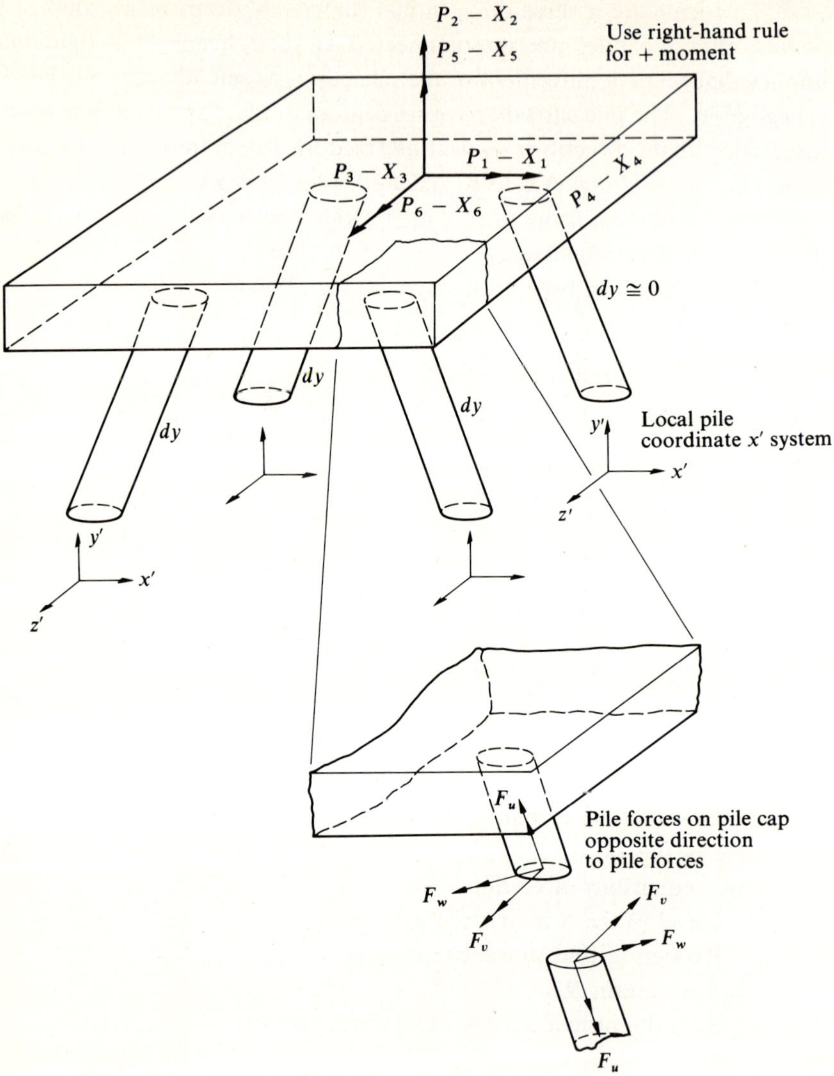

FIGURE 13-1
Pile group with foundation forces (P_i) and individual-pile forces. Positive directions shown.

13-2 THE INDIVIDUAL-PILE *A* MATRIX

Figure 13-2 is any pile in a group with pile-head coordinates (x,y,z) subject to the pile forces F_u, F_v, F_w and moment vectors of M_u, M_v and M_w. A local coordinate system (X', Y', Z' axes) is placed on the pile head such that the pile force F_w is in the $X'Y'$

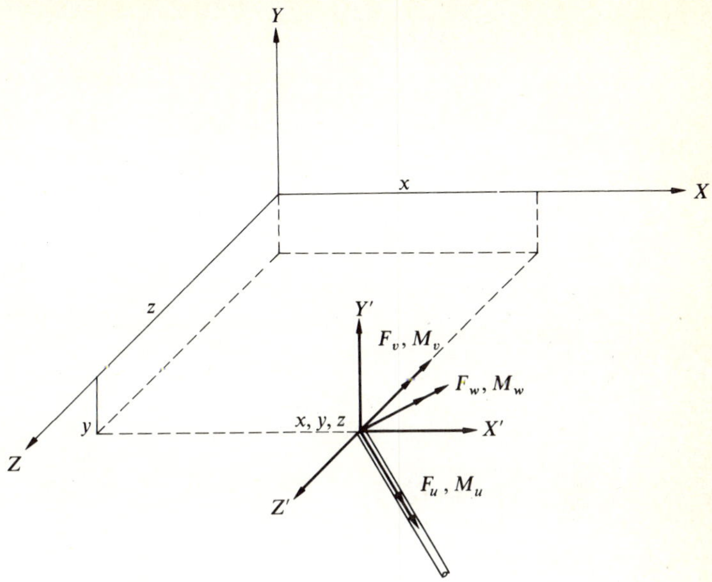

FIGURE 13-2
The pile with the foundation-cap coordinate axis and the local pile coordinate axis. Note that F_w is in the $X'Y'$ plane. The positive axis system is shown. Positive pile forces are as shown.

plane; thus, F_w has no force component along the Z axis. The positive X component of F_w is always in the positive X'-axis direction. Direction cosines are defined as follows:

α_1 = angle of pile projection extension in XZ plane with X axis measured clockwise (obtained graphically or computed)

α_2 = pile slope computed as

$$\alpha_2 = \tan^{-1} \frac{H}{1} \tag{13-2}$$

where H is vertical-to-horizontal pitch taken as 1; thus, 4:1 or 12:1 batter ($H = 4$ and 12 in these cases)

$$\alpha_3 = \tan^{-1} \frac{\sin \alpha_2}{\cos \alpha_1 \cos \alpha_2} \tag{13-3}$$

where the appropriate trigonometric relationships are obtained from Fig. 13-3. Note that α_3 defines the slope of force F_w in the XY plane since

$$\lambda = 90 - \alpha_3$$

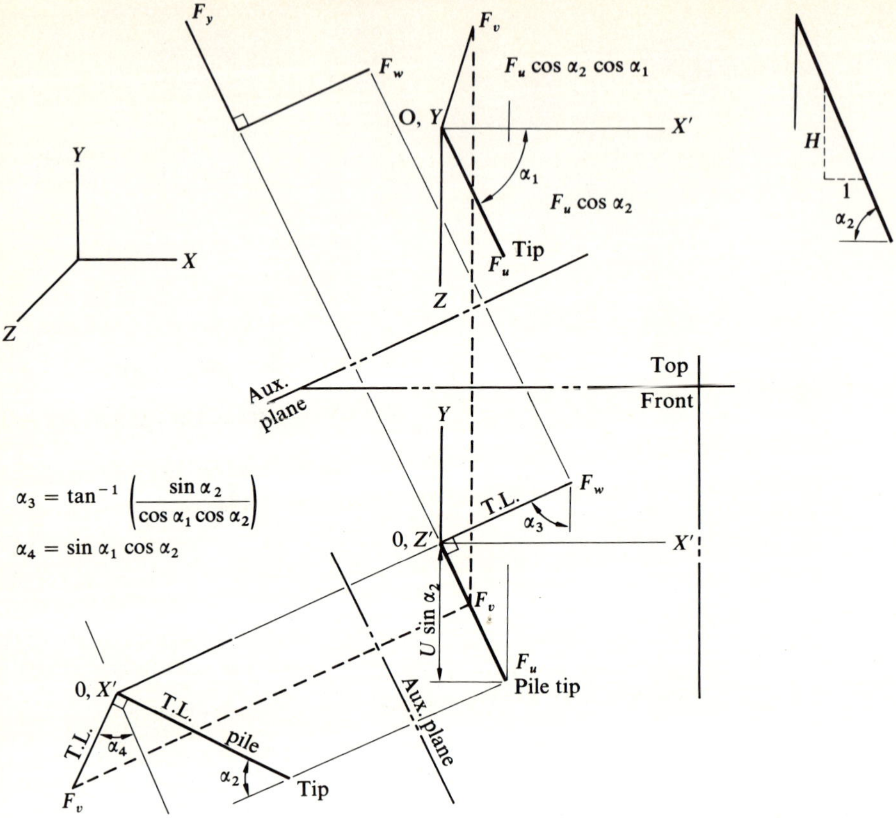

FIGURE 13-3
Method of obtaining the direction cosines for building the A matrix.

From Fig. 13-3 the other angle, α_4, needed to develop the A matrix is

$$\alpha_4 = \sin \alpha_1 \cos \alpha_2 \tag{13-4}$$

That portion of the pile-cap forces P_i' carried by the individual (ith) pile is

$$P'|_i = AF$$

and the following A matrix can be established from this equation. For convenience the familiar static forces $F_x, F_y, \ldots, M_z$ associated with the appropriate P-matrix values are identified.

$$
A_i = \begin{array}{r|llllll}
P \backslash F & F_u = 1 & F_v = 2 & F_w = 3 & M_u = 4 & M_v = 5 & M_w = 6 \\
\hline
P'_1 = F_x = 1 & \cos\alpha_1 \cos\alpha_2 & \cos\alpha_3 \cos\alpha_4 & \sin\alpha_3 & 0 & 0 & 0 \\
P'_2 = F_y = 2 & -\sin\alpha_2 & \sin\alpha_3 \cos\alpha_4 & \cos\alpha_3 & 0 & 0 & 0 \\
P'_3 = F_z = 3 & \sin\alpha_1 \cos\alpha_2 & \sin\alpha_4 & 0 & 0 & 0 & 0 \\
P'_4 = M_x = 4 & -Z(-\sin\alpha_2) + Y(\sin\alpha_1 \cos\alpha_2) & -Z(\sin\alpha_3 \cos\alpha_4) + Y \sin\alpha_4 & -Z\cos\alpha_3 & \cos\alpha_1 \cos\alpha_2 & \cos\alpha_3 \cos\alpha_4 & \sin\alpha_3 \\
P'_5 = M_y = 5 & Z\cos\alpha_1 \cos\alpha_2 - X\sin\alpha_1 \cos\alpha_2 & Z(\cos\alpha_3 \cos\alpha_4) - X\sin\alpha_4 & Z\sin\alpha_3 & -\sin\alpha_2 & \sin\alpha_3 \cos\alpha_4 & \cos\alpha_3 \\
P'_6 = M_z = 6 & -Y(\cos\alpha_1 \cos\alpha_2) + X(-\sin\alpha_2) & -Y(\cos\alpha_3 \cos\alpha_4) + X(\sin\alpha_3 \cos\alpha_4) & -Y\sin\alpha_3 + X\cos\alpha_3 & \sin\alpha_1 \cos\alpha_2 & \sin\alpha_4 & 0
\end{array}
$$

The sum of all the pile-component contributions is

$$\sum^{n} P_i = P_{\text{total found}} = \sum_{i}^{n} AF$$

13-3 THE *S* MATRIX

A relationship can be developed using the lateral-pile solution of Chap. 9 [see also Bowles (1968), p. 551] to obtain the individual pile-head response to rotation and translation.

The torsion resistance of piles is an estimate. In the absence of better data a value can be taken:

$$\text{Torsion constant } E = \psi \frac{GJ}{L'} \qquad FL/\text{rad}$$

where G = shear modulus of pile material

J = polar moment of inertia (Table 13-1) of pile

L' = effective pile length

ψ = coefficient to correct the constant

Since lateral-pile response indicates fixity at approximately $L/3$, a value of $\psi = 2$ to 2.5 does not appear unreasonable.

The compression constant can be computed using the method of Chap. 12 as illustrated in Example 12-2. Alternatively one might use a value of

$$\text{Compression constant } A' = \begin{cases} \dfrac{AE}{L} & \text{for point-bearing piles} \\[2ex] \dfrac{2AE}{L} & \text{for friction piles} \end{cases} \qquad \text{force}/L$$

where A = cross-sectional area of pile

E = modulus of elasticity

The S matrix from the relationship

$$F = Se$$

Table 13-1 H-PILE PROPERTIES INCLUDING TORSION CONSTANT J

			Flange			Elastic properties						
						Axis XX			Axis YY			
Designation	Area A, sq in	Depth d, in	Width b_f, in	Thickness t_f, in	Web thickness t_w, in	I, in^4	S, in^3	r, in	I, in^4	S, in^3	r, in	J,* in^4
HP14X117	34.4	14.23	14.885	0.805	0.805	1,230	173	5.97	443	59.5	3.59	8.106
X102	30.0	14.03	14.784	0.704	0.704	1,050	150	5.93	380	51.3	3.56	5.411
X89	26.2	13.86	14.696	0.616	0.616	910	131	5.89	326	44.4	3.53	3.677
X73	21.5	13.64	14.586	0.506	0.506	734	108	5.85	262	35.9	3.49	2.060
HP12X74	21.8	12.12	12.217	0.607	0.607	566	93.4	5.10	185	30.2	2.91	2.967
X53	15.6	11.78	12.046	0.436	0.436	394	66.9	5.03	127	21.1	2.86	1.124
HP10X57	16.8	10.01	10.224	0.564	0.564	295	58.8	4.19	101	19.7	2.45	1.970
X42	12.4	9.72	10.078	0.418	0.418	211	43.4	4.13	71.4	14.2	2.40	0.830
HP8X36	10.6	8.03	8.158	0.446	0.446	120	29.9	3.36	40.4	9.91	1.95	0.768

* Computed by author using equations from Bethlehem Steel Corp., Torsion Analysis of Rolled Steel Sections, Bethlehem, Pa., 1963.

is obtained for the individual pile as follows:

$$
S = \begin{array}{c|cccccc}
F \backslash e & \Delta u = 1 & \Delta v = 2 & \Delta w = 3 & \alpha_u = 4 & \alpha_v = 5 & \alpha_w = 6 \\
\hline
F_u = 1 & A' & 0 & 0 & 0 & 0 & 0 \\
F_v = 2 & 0 & B_0 & 0 & 0 & 0 & +C \\
F_w = 3 & 0 & 0 & B_1 & 0 & -C & 0 \\
M_u = 4 & 0 & 0 & 0 & E & 0 & 0 \\
M_v = 5 & 0 & 0 & -D & 0 & F_0 & 0 \\
M_w = 6 & 0 & D & 0 & 0 & 0 & F_1
\end{array}
$$

where A' = compression constant using the method of Chap. 12 (approximately AE/L or $2AE/L$) (computer program variable C5)

B_0, B_1 = constants relating head response to cause unit deformation in either v or w direction (translation); for batter piles $B_0 \neq B_1$; however, in absence of better data use $B_0 = B_1$ (computer program variable C1)

C = constant relating effect of pile-head rotation to create a force F_v due to α_w or F_w due to α_v (computer program variable C2)

D = similar to C except these constants cause moment M_v due to translation δ_w and moment M_w due to translation δ_v (computer program variable C3)

E = torsion constant of pile (computer program variable C6)

F_0, F_1 = constants analogous to B relating rotation to moment; for batter piles $F_0 \neq F_1$; however, in absence of better data take $F_0 = F_1$ (computer program variable C4)

The S-matrix constants B and D are obtained by finding the lateral-pile solution of Chap. 9 for translation and specified rotation (which may be zero or an assumed value of rotation, say 0.001, 0.005, 0.01 rad, etc.). This assumed value of rotation will produce a fixed-end-moment effect which can be applied in the P matrix at the appropriate nodes and properly subscripted. The lateral-pile program output will give:

1 Pile-head translation for incremented values of applied head force, say for 5, 10, 15, 20 kips.

2 Pile-head end moment for each of the applied lateral forces of item 1 necessary to maintain the head in an unrotated position. One can now plot:

3 Applied lateral-pile force versus deflection, the slope of which is

$$C1 = B = \frac{P}{\delta_h} \qquad \text{force/length} = FL^{-1}$$

4 Induced pile-end moment versus deflection, the slope of which is

$$C3 = D = \frac{M}{\delta_h} \qquad FLL^{-1}$$

In a similar manner we may use the lateral-pile solution for rotation (but no translation) and apply a series of moments to the pile head. For small translations we can compute fixed-end moments for assumed values of, say, $\Delta = 0.001$, 0.005, and 0.01 ft and apply them appropriately in the P matrix at the two affected nodes. The output will give data to plot curves of:

1 Pile-head rotation θ in radians versus incremented moment, the slope of which is

$$C4 = F = \frac{M}{\theta} \qquad FL/\text{rad}$$

2 Pile-head force P_h necessary to restrain translation when the corresponding moments are applied. This force is plotted versus rotation θ to give

$$C2 = C = \frac{P_h}{\theta} \qquad F/\text{rad}$$

It should be evident that by using these two modified lateral-pile solutions any degree of head fixity can be analyzed. Obviously, if the pile head is pinned to the foundation cap, constants D, E, and F (C3, C4, and C6 of computer program) are zero; if fully fixed, the constants C (C2 in computer program) are zero. Most practical solutions are somewhere between these two extremes, as obtained using all six constants in the S matrix.

These solutions imply that the deflections are small and that the slope of the pile-head response curves is linear over the range of deflections to be considered. If deflections are too large or in the nonlinear range, revised S-matrix values can be used through an iterative process of computing the deflections, obtaining the constants, recomputing the deflections, etc., until reasonable agreement of used versus computed deflection is reached.

13-4 THE GENERAL SOLUTION

With the individual-pile A and S matrices built, one proceeds to build the SA^T and ASA^T matrices for each pile. All the individual-pile matrices are of order 6 × 6; thus, this problem can be solved efficiently on a computer of very modest core capacity.

As the ASA^T is built for each pile, it is added term by term to the foundation ASA^T matrix.

When all piles have contributed their individual ASA^T to the foundation ASA^T matrix, the resulting matrix (also 6 × 6) is inverted and the foundation displacements computed as

$$X = [ASA^T]^{-1}P$$

With the foundation (pile-cap) displacements (X's) known, the individual pile movements are also known since we have assumed a rigid pile cap. The X matrix is at this point related to the components of individual-pile displacements in the six degrees of freedom through the relationship

$$e = A^T X$$

from which the individual-pile movements are computed. Note that the units of displacement will be the units used in the P and S matrices. The individual-pile forces can be computed once the pile displacements are known, using

$$F = Se$$

One can also compute the individual-pile forces directly from the foundation-displacement matrix X and the individual-pile SA^T matrix

$$F = SA^T X$$

This latter computation however, will not yield correct values for pile moments, as the moments computed in this equation also include the effect of pile position (terms in the lower left corner of the A matrix).

13-5 EXAMPLES

Three examples will illustrate the procedure and checking of the solutions. The first problem will use four piles with a simple loading system for ease in checking. The second problem will be more general with nine piles and the maximum possible loads, namely, six. The third problem is a four-pile solution in metric units and with pinned heads (capable of rotation and translation).

EXAMPLE 13.1 Find the individual-pile forces and movements for the conditions given in Fig. E13-1.1. The piles are 14BP73 steel sections approximately 50 ft in length. Soil-modulus variation with depth is shown in Fig. 12-5. The load-transfer data and response are also shown in Fig. 12-5. Assume that the piles are reasonably rigidly attached to the pile cap, and compute the pile forces.

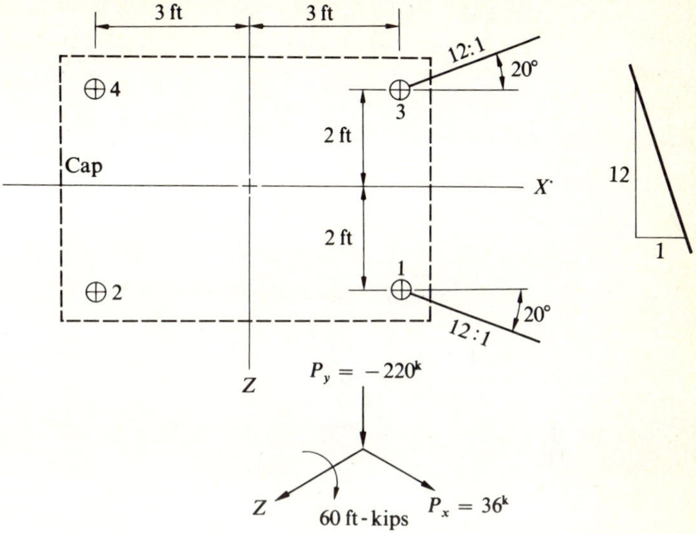

FIGURE E13-1.1
Pile group for Example 13-1.

SOLUTION The lateral-pile solution of Chap. 9, appropriately modified, is used to obtain the pile-head response curves shown in Fig. 13-4. We will assume the S-matrix constant C6 (or E) as (obtain G from Table 13-1 and assume $\psi = 2.0$):

$$C6 = \frac{2.0(12{,}000)(2.060)}{50(144)} = 0.6867 \text{ ft-kips/rad}$$

$$C5 = \frac{(60 - 20)12}{1.87 - 0.29} = 303.8 \text{ kips/ft} \qquad \text{Fig. 12-5}$$

$$C1 = \frac{P_h}{\delta} = \frac{40(12)}{0.977} = 491.3 \text{ kips/ft} \qquad \text{Fig. 13-4}a$$

$$C2 = \frac{P_h}{\phi} = \frac{3.717}{1.289 \times 10^{-3}} = 2{,}883.6 \text{ kips/rad} \qquad \text{Fig. 13-4}b$$

$$C3 = \frac{M}{\delta} = \frac{235.04(12)}{0.977} = 2{,}886.8 \text{ ft-kips/ft} \qquad \text{Fig. 13-4}a$$

$$C4 = \frac{M}{\phi} = \frac{100}{3.222 \times 10^{-3}} = 31{,}036 \text{ ft-kips/rad} \qquad \text{Fig. 13-4}b$$

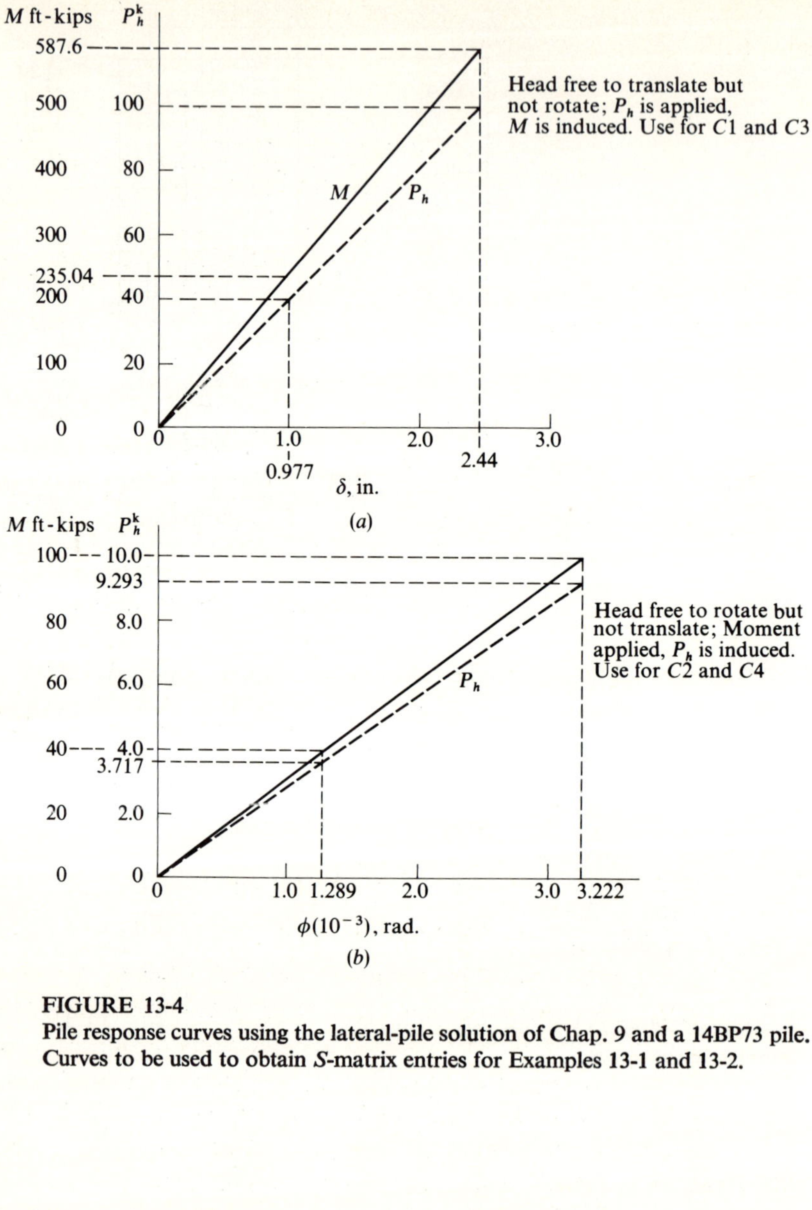

FIGURE 13-4
Pile response curves using the lateral-pile solution of Chap. 9 and a 14BP73 pile. Curves to be used to obtain *S*-matrix entries for Examples 13-1 and 13-2.

Computer input, based on one-load condition (NLC = 1) to obtain a complete listing (LIST = 1) is as follows:

Card	Data
1	TITLE (Note that a UNIT card is not needed)
2	N NP NLC LIST
	4 3 1 1
	C1–C6
3	491.3 2883.6 2886.8 31036.0 303.8 .6867
	X(K) Y(K) Z(K) ALPHA(K) H(K)
4	3.0 0. 2.0 20. 12.
5	Duplicate card 3
6	−3.0 0. −2.0 0. 100.
7	Duplicate card 3
8	3.0 0. −2.0 340. 12.
9	Duplicate card 3
10	−3.0 0. −2.0 0. 100.
11	1 36 P(1) in $+X$ direction Fig. 13-1
12	2 −220. P(2) in $-Y$ direction (downward)
13	6 −60. P(6) moment clockwise about Z axis

These data cards represent the input. The output is shown on Fig. E13-1.2.

CHECK EXAMPLE 13-1 AND COMMENTS

1 Note that $\sum F_x$, $\sum F_y$, and $\sum F_z = 0$ from the last page of computer output (given values are listed in parentheses).

2 The sum of $M_v = 2(-15.8) + 2(27.2) = 22.8$ and the loading system introduces a moment about the Z axis of

$$-2(3)(51.89) + 2(3)(58.11) = 37.4 \text{ ft-kips}$$

$\sum M_v = 22.8 + 37.4 - 60 \cong 0.$

3 Note that the axial loads are not the same in all piles.

4 The M_x, M_y, and M_z components shown in the output include the moments contributed by the pile-head forces and the pile position with respect to the axes of the pile cap. For example, $M_x = 132.03 = 2(58.11) + 16$ for pile 1; $M_y = 2(7.61) + 3(1.02) - 0.028(27.2) + 0.078(15.8) = 18.75$ ft-kips (18.75).

5 The bending moment in pile 1 is $M_v = 27.0$ ft-kips, $M_w = 16$ ft-kips, both values taken directly from the computer output sheet.

6 Additional refinement might have been made in C5 by obtaining the slope across a load increment closer to 58 and 52 kips rather than from 20 to 60 kips.

7 Since this group of piles is symmetrically placed, one would expect the lateral-pressure response of piles 1 and 3 in the Z direction to be equal and opposite in sign (−1.02 and +1.02). ////

```
J E BOWLES   EXAMPLE 13-1 W/4 PILES 14BP73 ;  TWO PILES W/ALPHA AND BATTER

    GENERAL INPUT DATA
PILE NO      X          Y           Z        ALPHA      BATTER
    1      3.00        0.0        2.00      20.00       12.00
    2     -3.00        0.0        2.00       0.0       100.00
    3      3.00        0.0       -2.00     340.00       12.00
    4     -3.00        0.0       -2.00       0.0       100.00
    THE PILE CONSTANTS ARE
PILE NO        C1           C2            C3            C4             C5             C6
    1        491.3        2883.6        2886.8       31036.0         303.8            0.7
    2        491.3        2883.6        2886.8       31036.0         303.8            0.7
    3        491.3        2883.6        2886.8       31036.0         303.8            0.7
    4        491.3        2883.6        2886.8       31036.0         303.8            0.7

    THE A-MATRIX FOR PILE NO  1***                               THE SAT MATRIX--100 FACTORED
  0.0780  0.0022  0.9969  0.0     0.0     0.0          0.2     -3.0      0.1      6.1      0.2     -9.1
 -0.9965 -0.0283  0.0781  0.0     0.0     0.0          0.0     -0.1     -4.9     29.0     17.0     -0.4
  0.0284 -0.9996  0.0     0.0     0.0     0.0          4.9      0.4      0.0     -0.8     10.6     30.0
  1.9931  0.0566 -0.1561  0.0780  0.0022  0.9969       0.0      0.0      0.0      0.0     -0.0      0.0
  0.0709  3.0032  1.9939 -0.9965 -0.0283  0.0781     -28.8     -2.3      0.0      5.2    -66.3   -317.0
 -2.9896 -0.0849  0.2342  0.0284 -0.9996  0.0          0.1     -0.8    -28.9    311.0    110.9     -2.5

    THE A-MATRIX FOR PILE NO  2***                               THE SAT MATRIX--100 FACTORED
  0.0     0.0000  1.0000  0.0     0.0     0.0          0.0     -3.0      0.0      6.1      0.0      9.1
 -1.0000 -0.0000  0.0000  0.0     0.0     0.0          0.0     -0.0     -4.9     28.8    -14.7      0.0
  0.0    -1.0000  0.0     0.0     0.0     0.0          4.9      0.0      0.0     -0.0      9.8     28.8
  2.0000  0.0000 -0.0000  0.0     0.0000  1.0000       0.0      0.0      0.0      0.0     -0.0      0.0
  0.0    -3.0000  2.0000 -1.0000 -0.0000  0.0000     -28.9     -0.0      0.0      0.0    -57.7   -310.4
  3.0000  0.0000 -0.0000  0.0    -1.0000  0.0          0.0     -0.0    -28.9    310.4    -86.6      0.0

    THE A-MATRIX FOR PILE NO  3***                               THE SAT MATRIX--100 FACTORED
  0.0780 -0.0022  0.9969  0.0     0.0     0.0          0.2     -3.0     -0.1     -6.1     -0.2     -9.1
 -0.9965  0.0283  0.0781  0.0     0.0     0.0         -0.0      0.1     -4.9     29.0     17.0      0.4
 -0.0284 -0.9996  0.0     0.0     0.0     0.0          4.9      0.4      0.0      0.8    -10.6     30.0
 -1.9931  0.0566  0.1561  0.0780 -0.0022  0.9969       0.0      0.0      0.0      0.0     -0.0     -0.0
 -0.0709  3.0032 -1.9939 -0.9965  0.0283  0.0781     -28.8     -2.3      0.0     -5.2     66.3   -317.0
 -2.9896  0.0849  0.2342 -0.0284 -0.9996  0.0         -0.1      0.8    -28.9    311.0    110.9      2.5
```

```
     THE A-MATRIX FOR PILE NO   4***                                    THE SAT MATRIX--100 FACTORED
  0.0       0.0000  1.0000  0.0      0.0      0.0           0.0      -3.0      0.0     -6.1      0.0       9.1
 -1.0000 -0.0000  0.0000  0.0      0.0      0.0           0.0      -0.0     -4.9     28.8    -14.7       0.0
  0.0     -1.0000  0.0     0.0      0.0      0.0           4.9       0.0      0.0      0.0     -9.8      28.8
 -2.0000 -0.0000  0.0000  0.0      0.0000   1.0000        0.0       0.0      0.0      0.0     -0.0       0.0
  0.0     -3.0000 -2.0000 -1.0000 -0.0000   0.0000      -28.9      -0.0      0.0     -0.0     57.7    -310.4
  3.0000   0.0000 -0.0000  0.0     -1.0000   0.0           0.0      -0.0    -28.9    310.4    -86.6       0.0


 THE P-MATRIX IS          THE FOUNDATION ASAT MATRIX--1000 FACTORED
  1  PX =     36.00        1      2.0      0.0      0.0      0.0     -0.0     11.6
  2  PY =   -220.00        2      0.0      1.2      0.0      0.0     -0.0      0.5
  3  PZ =      0.0         3      0.0      0.0      2.0    -11.5     -0.5     -0.0
  4  MX =      0.0         4      0.0      0.0    -11.5    129.3      4.7      0.0
  5  MY =      0.0         5     -0.0     -0.0     -0.5      4.7     29.3     -0.0
  6  MZ =    -60.00        6     11.6      0.5      0.0      0.0     -0.0    137.8


       THE FOUNDATION DISPLACEMENTS ARE
 X = 0.039962   Y=-0.180408   Z= 0.000000   ALPHA X= 0.000000   ALPHA Y=-0.000000   ALPHA Z=-0.003205


 THE FOUNDATION DISPLACEMENTS AND PILE FORCES--**NOTE FU,FV,FW,ETC ARE ACTING ON CAP
PILE      DU       DV       DW    ALPHA U ALPHA V ALPHA W        FU        FV        FW        MU        MV        MW
  1   0.1925   0.0055  0.0250 -0.0001  0.0032  0.0000       58.48      2.69      3.05     -0.00     27.24     15.79
  2   0.1708 -0.0000  0.0400  0.0000  0.0032  0.0000       51.89      0.00     10.39      0.00    -15.90      0.00
  3   0.1925 -0.0055  0.0250  0.0001  0.0032  0.0000       58.48     -2.69      3.05      0.00     27.24    -15.79
  4   0.1708 -0.0000  0.0400  0.0000  0.0032  0.0000       51.89      0.00     10.39      0.00    -15.90      0.00


      INDIVIDUAL PILE FORCE COMPONENTS TO CHECK SUM OF FORCES ALONG AXES
      PILE NO        FX          FY          FZ          MX          MY          MZ
         1         7.6078    -58.1128     -1.0251    132.0267     18.7522   -201.5629
         2        10.3922    -51.8871     -0.0000    103.7744     20.7844    171.5627
         3         7.6078    -58.1128      1.0251   -132.0266    -18.7522   -201.5629
         4        10.3922    -51.8871     -0.0000   -103.7742    -20.7845    171.5627
    TOTAL =       36.0000   -219.9998      0.0000      0.0002     -0.0000    -60.0003
             (    36.000)(  -220.000)(     0.0   )(    0.0   )(    0.0   )(  -60.000)
```

FIGURE E13-1.2

I/O (complete listing) for Example 13-1. Note that the last two lines provide the statics check for the problem. Individual piles must resist the axial force F_u (FU) and the bending moment M_v (MV).

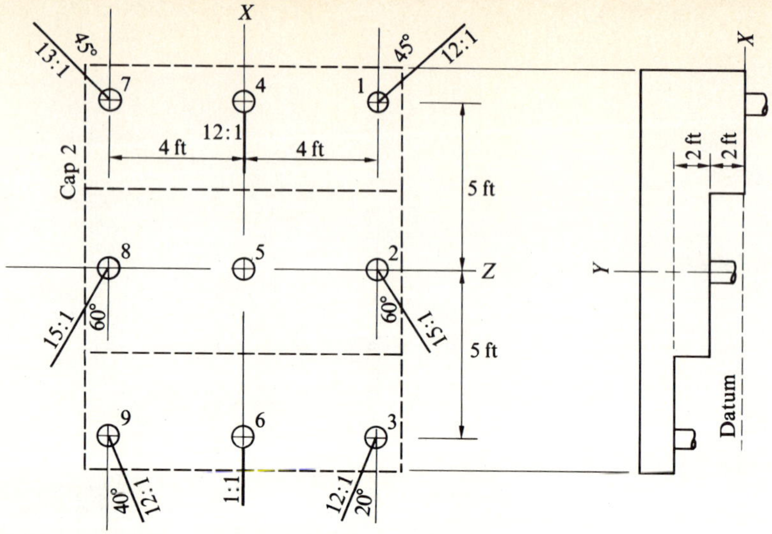

FIGURE E13-2.1
Nine-pile group with pile heads at varying elevations.

EXAMPLE 13.2 A nine-pile group as shown in Fig. E13-2.1 will be analyzed. Piles are 14BP73, and all pile data used in Example 13-1 will be used here. The only change is in using a rather odd pile group to illustrate entering data into the computer program. Note, however, that using the same *S*-matrix entries for this group and that of Example 13-1 is questionable, especially for those piles with heads at $\pm$4.0-ft elevation or for batters at large skew angles.

SOLUTION With data similar to Example 13-1 only the cards containing the pile coordinates will be shown to illustrate entering the α angle.

Card	Data				
1	TITLE				
2	N NP NLC LIST				
	9 6 1 0				
Pile	**X(K)**	**Y(K)**	**Z(K)**	**ALPHA(K)**	**H(K)**
1	5.0	0.	4.	45.	12.
2	0.0	2.	4.	120.	15.
3	−5.0	4.	4.	200.	12.
4	5.0	0.	0.	180.	12.
5	0.0	2.	0.	0.	100.
6	−5.0	4.	0.	180.	1.
7	5.0	0.	−4.	315.	12.
8	0.0	2.	−4.	240.	15.
9	−5.0	4.	−4.	160.	12.
10–15	*P*-matrix entries (each entry on separate card)				
	1 100.0		4 −250.		
	2 −500.		5 −180.		
	3 200.0		6 −120.		

The partial computer input and output sheets follow (Fig. E13-2.2); note again that the problem statics are satisfied. ////

```
J E BOWLES   EXAMPLE 13-2   (9 14BP73 PILES)   USING 6 P-MATRIX ENTRIES

      GENERAL INPUT DATA
PILE NO      X          Y          Z        ALPHA     BATTER
   1        5.00       0.0        4.00      45.00     12.00
   2        0.0        2.00       4.00     120.00     15.00
   3       -5.00       4.00       4.00     200.00     12.00
   4        5.00       0.0        0.0      180.00     12.00
   5        0.0        2.00       0.0        0.0     100.00
   6       -5.00       4.00       0.0      180.00      1.00
   7        5.00       0.0       -4.00     315.00     12.00
   8        0.0        2.00      -4.00     240.00     15.00
   9       -5.00       4.00      -4.00     160.00     12.00
      THE PILE CONSTANTS ARE
PILE NO        C1           C2           C3          C4          C5          C6
   1          491.3       2883.6       2886.8      31036.0       303.8         0.7
   2          491.3       2883.6       2886.8      31036.0       303.8         0.7
   3          491.3       2883.6       2886.8      31036.0       303.8         0.7
   4          491.3       2883.6       2886.8      31036.0       303.8         0.7
   5          491.3       2883.6       2886.8      31036.0       303.8         0.7
   6          491.3       2883.6       2886.8      31036.0       303.8         0.7
   7          491.3       2883.6       2886.8      31036.0       303.8         0.7
   8          491.3       2883.6       2886.8      31036.0       303.8         0.7
   9          491.3       2883.6       2886.8      31036.0       303.8         0.7
THE P-MATRIX IS          THE FOUNDATION ASAT MATRIX--1000 FACTORED
 1  PX =    100.00        1      4.3     -0.1      0.0      0.0     -0.0    -15.6
 2  PY =   -500.00        2     -0.1      2.8      0.0      0.0      0.0      3.3
 3  PZ =    200.00        3      0.0      0.0      4.4     16.7     -3.3     -0.0
 4  MX =   -250.00        4      0.0      0.0     16.7    399.6     51.0      0.0
 5  MY =   -180.00        5     -0.0      0.0     -3.3     51.0    122.8      0.0
 6  MZ =   -120.00        6    -15.6      3.3     -0.0      0.0      0.0    404.6
         THE FOUNDATION DISPLACEMENTS ARE
X = 0.025577   Y=-0.177644   Z= 0.058697   ALPHA X=-0.003268   ALPHA Y= 0.001450   ALPHA Z= 0.002146
THE FOUNDATION DISPLACEMENTS AND PILE FORCES--**NOTE FU,FV,FW,ETC ARE ACTING ON CAP
PILE    DU       DV       DW    ALPHA U  ALPHA V  ALPHA W      FU        FV        FW         MU        MV        MW
 1   0.1582  -0.0422   0.0223  -0.0015  -0.0022  -0.0032     48.05    -29.91     17.40      -0.00   -133.77   -220.53
 2   0.1663   0.0615   0.0326  -0.0012   0.0021  -0.0033     50.52     20.65     10.04      -0.00    -29.88     74.65
 3   0.1714   0.0479   0.0364  -0.0013   0.0022  -0.0034     52.08     13.83     11.60      -0.00    -37.48     33.74
 4   0.1642   0.0514   0.0394  -0.0012   0.0021  -0.0034     49.89     15.54     13.14      -0.00    -47.00     43.69
 5   0.1776  -0.0522   0.0213  -0.0014  -0.0021  -0.0033     53.97    -35.05     16.65      -0.00   -128.05   -252.02
 6   0.1212   0.0529   0.1452   0.0013   0.0021  -0.0033     36.82     16.36     65.16       0.00   -352.61     49.09
 7   0.1775  -0.0620   0.0092  -0.0018  -0.0020  -0.0032     53.93    -39.61     10.40      -0.00    -89.93   -277.54
 8   0.1868   0.0411   0.0218  -0.0015   0.0022  -0.0033     56.74     10.64      4.33      -0.00      5.86     15.83
 9   0.2014   0.0585   0.0269  -0.0011   0.0021  -0.0034     61.18     19.03      7.12      -0.00    -12.09     64.32
      INDIVIDUAL PILE FORCE COMPONENTS TO CHECK SUM OF FORCES ALONG AXES
      PILE NO       FX          FY          FZ          MX          MY          MZ
         1        20.0859    -45.1102     32.6815    -40.1733    -88.1939    -92.0107
         2         8.3125    -51.9366     23.5307    329.4697     32.4844    -46.4556
         3         7.5364    -52.4100     12.3448    292.5769     88.1744    194.4354
         4         8.9566    -50.8076     15.5366     43.5417    -81.3109   -301.0347
         5        16.6454    -53.9683     35.0513   -181.9139      0.0009     94.7558
         6        20.0404    -72.1056     16.3556    100.1302     47.0688    -72.2454
         7        13.6833    -55.4508     36.3775   -498.5479   -258.2166   -187.4813
         8         2.4561    -56.1480      7.3582   -194.0428    -10.0133      0.9417
         9         2.2833    -62.0628     20.7637   -101.0420     90.0065    289.0918
    TOTAL =       99.9999   -499.9995    199.9998   -250.0018   -179.9993   -120.0029
             (   100.000)(  -500.000)(   200.000)(  -250.000)(  -180.000)(  -120.000)
```

FIGURE E13-2.2
Complete input and partial output for Example 13-2. Note again that the statics check in the last two lines.

EXAMPLE 13-3 Solve the pile group shown in Fig. E13-3.1 using metric units. This is the same pile-soil system (14BP73) as in Example 13-1. Assume that the heads are pinned to the pile cap. The loads are $P_x = 222$ kN (50 kips); $P_y = -1{,}780$ kN (400 kips), and $M_y = -108.5$ kN-m (80 ft-kips).

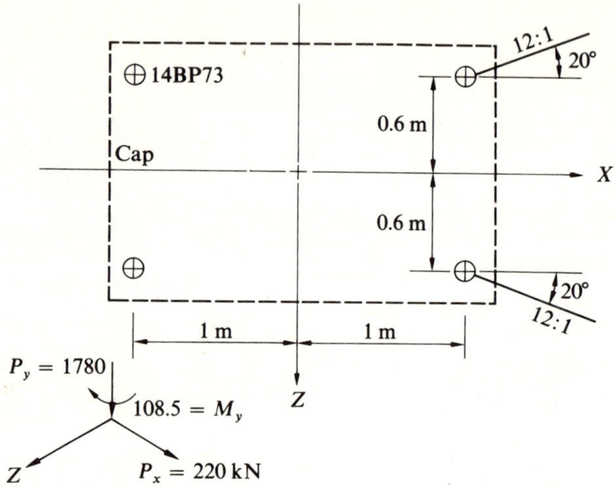

FIGURE E13-3.1
Pile group.

SOLUTION From Fig. E9–3.2 the slope of curve *a* gives

$$C1 = \frac{P}{\Delta} = \frac{100(100)}{3.587} = 2{,}787.8 \text{ kN/m}$$

and from curve *b*

$$C2 = \frac{P}{\theta} = \frac{100}{0.011} = 9{,}090.9 \text{ kN/rad}$$

From Example 13-1 we have

$$C5 = 303.8(14.5914) = 4{,}432.9 \text{ kN/m}$$

Since the pile heads are pinned, C3 = C4 = C6 = 0.0. Data cards are similar:

Card	Data					
1	TITLE					
2	4	3	1	1		
3	2787.8	9090.9	0.	0.	4432.9	0.0
4	1	0	.6	20.	12.	
↓						
11	1	222.2 kN				
12	2	−1780. kN				
13	5	−108.5 kN-m				

The partial output is shown on Fig. E13-3.2 (pages 447 and 448). ////

```
J E BOWLES EXAMPLE 13-3 W/4 PILES 14BP73;   PINNED HEAD & METRIC SOLUTION

     GENERAL INPUT DATA
PILE NO      X           Y            Z          ALPHA       BATTER
    1       1.00        0.0         0.60        20.00        12.00
    2      -1.00        0.0         0.60         0.0        100.00
    3       1.00        0.0        -0.60       340.00        12.00
    4      -1.00        0.0        -0.60         0.0        100.00
     THE PILE CONSTANTS ARE
PILE NO          C1            C2            C3            C4            C5            C6
    1          2787.8        9090.9          0.0           0.0         4432.9           0.0
    2          2787.8        9090.9          0.0           0.0         4432.9           0.0
    3          2787.8        9090.9          0.0           0.0         4432.9           0.0
    4          2787.8        9090.9          0.0           0.0         4432.9           0.0

     THE A-MATRIX FOR PILE NO   1***                            THE SAT MATRIX--100 FACTORED
 0.0780  0.0022  0.9969  0.0     0.0     0.0           3.5     -44.2     1.3    26.5     0.8   -44.2
-0.9965 -0.0283  0.0781  0.0     0.0     0.0           0.1      -0.8   -27.9    91.1    35.0    -0.8
 0.0284 -0.9996  0.0     0.0     0.0     0.0          27.8       2.2     0.0    -1.5    19.2    93.0
 0.5979  0.0170 -0.0468  0.0780  0.0022  0.9969        0.0       0.0     0.0     0.0     0.0     0.0
 0.0184  1.0009  0.5982 -0.9965 -0.0283  0.0781        0.0       0.0     0.0     0.0     0.0     0.0
-0.9965 -0.0283  0.0781  0.0284 -0.9996  0.0           0.0       0.0     0.0     0.0     0.0     0.0

     THE A-MATRIX FOR PILE NO   2***                            THE SAT MATRIX--100 FACTORED
 0.0     0.0000  1.0000  0.0     0.0     0.0           0.0     -44.3     0.0    26.6     0.0    44.3
-1.0000 -0.0000  0.0000  0.0     0.0     0.0           0.0      -0.0   -27.9    90.9   -27.9     0.0
 0.0    -1.0000  0.0     0.0     0.0     0.0          27.9       0.0     0.0    -0.0    16.7    90.9
 0.6000  0.0000 -0.0000  0.0     0.0000  1.0000        0.0       0.0     0.0     0.0     0.0     0.0
 0.0    -1.0000  0.6000 -1.0000 -0.0000  0.0000        0.0       0.0     0.0     0.0     0.0     0.0
 1.0000  0.0000 -0.0000  0.0    -1.0000  0.0           0.0       0.0     0.0     0.0     0.0     0.0

     THE A-MATRIX FOR PILE NO   3***                            THE SAT MATRIX--100 FACTORED
 0.0780 -0.0022  0.9969  0.0     0.0     0.0           3.5     -44.2    -1.3   -26.5    -0.8   -44.2
-0.9965  0.0283  0.0781  0.0     0.0     0.0          -0.1       0.8   -27.9    91.1    35.0     0.8
-0.0284 -0.9996  0.0     0.0     0.0     0.0          27.8       2.2     0.0     1.5   -19.2    93.0
-0.5979  0.0170  0.0468  0.0780 -0.0022  0.9969        0.0       0.0     0.0     0.0     0.0     0.0
-0.0184  1.0009 -0.5982 -0.9965  0.0283  0.0781        0.0       0.0     0.0     0.0     0.0     0.0
-0.9965  0.0283  0.0781 -0.0284 -0.9996  0.0           0.0       0.0     0.0     0.0     0.0     0.0
```

FIGURE E13-3.2
Complete I/O for Example 13-3. Note that forces are in kilonewtons and moments in kilonewton-meters. Displacements are in radians and meters.

```
    THE A-MATRIX FOR PILE NO   4***                                THE SAT MATRIX--100 FACTORED
 0.0      0.0000  1.0000  0.0     0.0      0.0          0.0     -44.3     0.0    -25.6     0.0     44.3
-1.0000 -0.0000  0.0000  0.0     0.0      0.0          0.0     -0.0    -27.9     90.9   -27.9      0.0
 0.0     -1.0000  0.0     0.0     0.0      0.0         27.9      0.0      0.0      0.0   -16.7     90.9
-0.6000 -0.0000  0.0000  0.0     0.0000   1.0000       0.0      0.0      0.0      0.0     0.0      0.0
 0.0     -1.0000 -0.6000 -1.0000 -0.0000  0.0000       0.0      0.0      0.0      0.0     0.0      0.0
 1.0000  0.0000 -0.0000  0.0    -1.0000   0.0          0.0      0.0      0.0      0.0     0.0      0.0


THE P-MATRIX IS       THE FOUNDATION ASAT MATRIX--1000 FACTORED
 1  PX =    222.00     1     11.2     -0.3      0.0      0.0      0.0     36.0
 2  PY =  -1780.00     2     -0.3     17.7     -0.0      0.0     -0.0      1.4
 3  PZ =      0.0      3      0.0     -0.0     11.2    -36.2     -1.4     -0.0
 4  MX =      0.0      4      0.0      0.0      0.1      6.7      0.0      0.0
 5  MY =   -108.50     5     -0.0     -0.0      0.0     -0.0     16.9     -0.0
 6  MZ =      0.0      6     -0.3     -0.0     -0.0     -0.0     -0.0     19.1


            THE FOUNDATION DISPLACEMENTS ARE
X = 0.017217  Y=-0.100274  Z=-0.000684  ALPHA X= 0.000041  ALPHA Y=-0.006422  ALPHA Z= 0.000111


THE FOUNDATION DISPLACEMENTS AND PILE FORCES--**NOTE FU,FV,FW,ETC ARE ACTING ON CAP
PILE   DU      DV      DW    ALPHA U ALPHA V ALPHA W      FU       FV       FW       MU        MV        MW
 1  0.1010 -0.0029  0.0055  0.0064  0.0001 -0.0005    447.93   -12.19    14.69     0.0       0.0       0.0
 2  0.1004  0.0071  0.0134  0.0064 -0.0001  0.0000    445.11    20.18    38.27     0.0       0.0       0.0
 3  0.1013 -0.0086  0.0132  0.0064 -0.0003 -0.0005    448.94   -28.21    39.43     0.0       0.0       0.0
 4  0.1004  0.0071  0.0211  0.0064 -0.0001  0.0000    444.89    20.18    59.75     0.0       0.0       0.0


    INDIVIDUAL PILE FORCE COMPONENTS TO CHECK SUM OF FORCES ALONG AXES
    PILE NO       FX          FY          FZ          MX          MY          MZ
       1        49.5775   -444.8911     24.9051    266.9343      4.8414   -444.8911
       2        38.2668   -445.1072    -20.1783    267.0642      2.7818    445.1072
       3        74.4058   -445.1064     15.4515   -267.0635    -60.0950   -445.1064
       4        59.7498   -444.8911    -20.1783   -266.9346    -56.0282    444.8911
   TOTAL =     221.9999  -1779.9958     -0.0001      0.0005   -108.4999      0.0007
          (    222.000)( -1780.000)(     0.0   )(    0.0   )( -108.500)(     0.0   )
```

FIGURE E13-3.2 *(Continued)*

13-6 COMPUTER PROGRAM FOR PILE GROUPS

This computer program will compute the individual-pile forces and displacements of a pile group of any configuration and any number of piles. The pile-group loading can be general, but the cap must be rigid. The nonzero *S*-matrix entries are obtained separately by the user and read in separately for each pile.

It is not necessary that pile tops be at equal elevation (*X*, *Y*, *Z* coordinates are read for each pile). Piles may be of unequal length. These conditions require that the appropriate *S*-matrix entries be evaluated for the different conditions, and the user is expected to recognize and make the necessary changes. Note the sign convention of the *P*-matrix entries from Fig. 13-1. This program will solve either fps or metric units. The user must insert appropriate units into the input data. Use the TITLE to identify units.

Line	Operation
1–2	Bookkeeping operations
3	READ TITLE (note no UNIT card and no *FU* entries)
7	READ (4I5) N = number of piles in group; NP = number of nonzero values in *P* matrix $0 \leq$ NP ≤ 6; NLC = number of loading conditions so *P* matrix may be changed without reading additional pile data; LIST = IF > 0 will list *A* and SA^T matrices of each pile
19–33	DO loop to read pile data
20	READ (6F10.2) C1(K) – C6(K)
21	READ (5F10.4) X(K), Y(K), Z(K) = pile coordinates; ALPHA(K) = angle of projection of pile with *X* axis; H(K) = pile batter as H:1, read 100 for vertical piles
35	Begins loop to develop individual-pile *A*, SA^T, and pile contribution to foundation ASA^T matrix. I is used as the counter varying from 1 to N (number of piles in group)
35–56	Compute pile-direction cosines
63–90	Zeros and builds individual-pile *A* matrix
91–103	Zeros and builds pile *S* matrix
105–109	Builds individual-pile SA^T matrix
115–118	Builds the foundation ASA^T matrix using the contribution of one pile at a time
121–124	Zeros and read *P* matrix
130	Calls MINV standard single-precision matrix-inversion (IBM) subroutine
131–134	Computes foundation-displacement matrix XF(J) and write values
142–145	Computes individual-pile movements using $e = A^TX$. Note *A* matrix is rebuilt for this operation along with the *S* matrix of next operation by setting I = 1 and INDEX = INDEX + 1
146–149	Computes pile forces using $F = Se$
152–161	Computes pile force and moment components in *X*, *Y*, *Z* directions using $P = AF$ and sums values to check problem statics
170	Checks if NLC is satisfied

```
      C      J E BOWLES 3-DIMENSIONAL PILE GRCUP ANALYSIS--FIXED OR PINNED HEAD
      C      ALL UNITS ARE FT (M), OR KIPS (KN)--OUTPUT = FT,KIPS OR F-K (OR ME
      C      SLOPE BASEC ON H:1 WITH H = VERT AS 4:U; 12:1, ETC (FPS OR METRIC)
      C      SIGNS **** +PX TO RT; +PY IS DCWN; +PZ OUT OF PAPER; +PMX,
      C         +PMY, AND +PMZ ARE POSITIVE M'S ACCORDING TO RT HAND RULE****
      C      SIGNS OF DEFLECTIONS ARE CONSISTENT WITH +AXES
0001         DIMENSION XF(30),ALPHA(30),SLOPE(30),C1(30),C2(30),C3(30),C4(30),
            1C5(30),C6(30),X(30),Y(30),Z(30),A(6,6),SAT(6,6),ASAT(6,6),C(6,6),
            2F(6),G1(30),G2(30),ALPHAR(30),ZZ(6),YY(6),P(6),XPILE(6),BSLOPE(30)
            3,H(30),FCOMP(30,6),TITLE(20)
```

```
      DOUBLE PRECISION ALPHA,SLOPE,H,G1,G2,F1,F2,B1,B2,ALPHAR
 6000 READ(1,1000,END=150)TITLE
 1000 FORMAT(20A4)
      WRITE(3,1001)TITLE
 1001 FORMAT('1',//,T5,20A4,/)
      READ(1,1004)N,NP,NLC,LIST
 1004 FORMAT (4I5)
      WRITE(3,1005)
 1005 FORMAT(T10, 'GENERAL INPUT DATA',/,T5, 'PILE NO',4X, 'X',9X,'Y',9X
     1,'Z', 7X,'ALPHA',5X,'BATTER')
      NPROB = 1
 3300 SUMX = 0.
      SUMY = 0.
      SUMZ = 0.
      SUMMX = 0.
      SUMMY = 0.
      SUMMZ = 0.
      IF(NPROB.GT.1)GO TO 31
      DO 102  K = 1,N
      READ(1,1006)C1(K), C2(K), C3(K), C4(K), C5(K), C6(K)
      READ(1,1006) X(K), Y(K), Z(K), ALPHA(K), H(K)
 1006 FORMAT(6F10.2)
  102 WRITE(3,1007)K,X(K),Y(K),Z(K),ALPHA(K),H(K)
 1007 FORMAT(T8,I2,5(3X,F7.2))
      WRITE(3,1003)
 1003 FORMAT(    T10,'THE PILE CONSTANTS ARE',/,T5, 'PILE NO', 6X,'C1', 1
     A1X,'C2',11X,'C3', 11X,'C4',11X,'C5',11X,'C6')
      DO 103  I = 1,N
  103 WRITE(3,1009)I,C1(I),C2(I),C3(I),C4(I),C5(I),C6(I)
 1009 FORMAT(T8,I2,3X,F11.1,5(2X,F11.1))
      INDEX = 1
      DO 104  I = 1,6
      DO 104  J = 1,6
  104 ASAT(I,J) = 0.
C     FORM A & SAT MATRICES FOR EACH PILE AND SUM THE ASAT MATRIX FOR
C         THE FOUNDATION
      CC = 180./3.1415927
      I = 1
   60 ALPHAR(I) = ALPHA(I)/CC
      B1 = DSIN(ALPHAR(I))
      F1 = DCOS(ALPHAR(I))
      S1 = 1.
      S2 = 1.
      IF(F1)10,11,11
   10 S1 = -1.0
      GO TO 12
   11 S2 = -1.0
   12 IF(H(I).EQ.100.)GO TO 15
      BSLOPE(I) = DSQRT (1.+H(I)**2)
      B2 = H(I)/BSLOPE(I)
      F2 = 1./BSLOPE(I)
      GO TO 16
   15 AFAC = 90./CC
      G1(I) = AFAC
      B2 = 1.
      F2 = 0.
      GO TO 17
   16 G1(I) = DATAN (B2/(F2*F1))
   17 G2(I) = DARCOS (B1*F2)
      B3 = DSIN (G1(I))
      F3 = DCOS (G1(I))
      B4 = DSIN (G2(I))
      F4 = DCOS (G2(I))
      DO 61   II = 1,6
      DO 61   JJ = 1,6
   61 A(II,JJ) = 0.
      A(1,1) = F1*F2
      A(1,2) = F3*F4*S1
      A(1,3) = B3*S1
      A(2,1) = -B2
      A(2,2) = B3*F4*S2
      A(2,3) = F3*S1
   22 A(3,1) = B1*F2
      A(3,2) = B4*S2
      A(3,3) = 0.
      A(4,1) =-Z(I)*A(2,1)  + Y(I)*A(3,1)
      A(4,2) =-Z(I)*A(2,2)  + Y(I)*A(3,2)
      A(4,3) =-Z(I)*A(2,3)
      A(4,4) = A(1,1)
      A(4,5) = A(1,2)
      A(4,6) = A(1,3)
      A(5,1) = Z(I)*A(1,1)  - X(I)*A(3,1)
      A(5,2) = Z(I)*A(1,2)  - X(I)*A(3,2)
      A(5,3) = Z(I)*A(1,3)
      A(5,4) = A(2,1)
      A(5,5) = A(2,2)
      A(5,6) = A(2,3)
      A(6,1) =-Y(I)*A(1,1)  + X(I)*A(2,1)
      A(6,2) =-Y(I)*A(1,2)  + X(I)*A(2,2)
      A(6,3) =-Y(I)*A(1,3)  + X(I)*A(2,3)
      A(6,4) = A(3,1)
      A(6,5) = A(3,2)
      A(6,6) = 0.
   24 DO 62   K = 1,6
      DO 62   J = 1,6
```

```
   62 C(K,J) = 0.
      C(1,1) = C5(I)
      C(2,2) = C1(I)
      C(2,6) = C2(I)
      C(3,3) = C1(I)
      C(3,5) = -C2(I)
      C(4,4) = C6(I)
      C(5,3) = -C3(I)
      C(5,5) = C4(I)
      C(6,2) = C3(I)
      C(6,6) = C4(I)
      IF(INDEX.GT.1)GO TO 65
   26 DO 106  K = 1,6
      DO 106  L = 1,6
      SAT(K,L) = 0.
      DO 106  NN = 1,6
  106 SAT(K,L) = SAT(K,L) + C(K,NN)*A(L,NN)
      IF(LIST.LE.0)GO TO 75
      WRITE(3,1098)I
 1098 FORMAT( //,T10, 'THE A-MATRIX FOR PILE NO',I3,'***', T65, 'THE SAT
     1 MATRIX--100 FACTORED')
 2001 WRITE(3,2020)((A(M,K), K = 1,6),(SAT(M,J), J = 1,6),M=1,6)
 2020 FORMAT(T5,0P 6F8.4,1X,-2P6F9.0)
   75 DO 108  K = 1,6
      DO 108  L = 1,6
      DO 108  M = 1,6
  108 ASAT(K,L) = ASAT(K,L) + A(K,M)*SAT(M,L)
  550 I = I+1
      IF(I.LE.N)GO TO 60
   31 DO 110  L = 1,6
  110 P(L) = 0.
      READ(1,1010)(J,P(J),JJ=1,NP)
 1010 FORMAT(I5, F10.4)
      WRITE(3,968)
  968 FORMAT(    ///,T5, 'THE P-MATRIX IS',T26,'THE FOUNDATION ASAT MATRI
     1X--1000 FACTORED')
      WRITE(3,969)(I,P(I),I,(ASAT(I,J),J=1,6),I=1,6)
  969 FORMAT(T5,I3,2X,'PX =',F10.2,5X,I3,-3P(6F9.1),/,T5,I3,2X,'PY =',0P
     1F10.2,5X,I3,-3P(6F9.1),/,T5,I3,2X,'PZ =', 0PF10.2,5X,I3,-3P(6F9.1)
     2,/,T5,I3,2X,'MX =',0PF10.2,5X,I3,-3P(6F9.1),/,T5, I3,2X,'MY =',0PF
     A10.2,5X,I3, -3P(6F9.1),/,T5,I3,2X,'MZ =',0PF10.2,5X,I3,-3P(6F9.1))
      IF(NPROB.GT.1)GO TO 200
C     **INVERT ASAT MATRIX --USE IBM SUBROUTINE SINCE A ZERO MAY BE ON D
      CALL MINV (ASAT,6,DET,ZZ,YY)
C     COMPUTE FOUNDATION DISPLACEMENTS XF
  200 DO 114  J = 1,6
      XF(J) = 0.
      DO 114  M = 1,6
  114 XF(J) = XF(J) + ASAT(J,M)*P(M)
      WRITE(3,971) (XF(MM), MM=1,6)
  971 FORMAT(//,T15,'THE FOUNDATION DISPLACEMENTS ARE',/, T5, 'X =', F9.
     16,2X,'Y=',F9.6,2X,'Z=',F9.6,2X, 'ALPHA X=',F9.6,2X,
     1'ALPHA Y=', F9.6, 2X, 'ALPHA Z=', F9.6)
      WRITE(3,972)
  972 FORMAT(//,T5, 'THE FOUNDATION DISPLACEMENTS AND PILE FORCES--**NOT
     1E FU,FV,FW,ETC ARE ACTING ON CAP',/,T2, 'PILE',4X,'DU', 6X,'DV',6X
     2, 'DW',3X,'ALPHA U',1X, 'ALPHA V',1X,'ALPHA W',6X,'FU',7X,'FV',7X,
     3'FW',7X,'MU',7X,'MV',7X,'MW')
      INDEX = INDEX + 1
      I = 1
      GO TO 60
C     COMPUTE INDIVIDUAL PILE MOVEMENTS (XPILE) USING E = AT*X
   65 DO 116  K = 1,6
      XPILE(K) = 0.0
      DO 116 J = 1,6
  116 XPILE(K) = XPILE(K) + A(J,K)*XF(J)
C ****   COMPUTE PILE FORCES USING F = C*XPILE(K)  *NOTE COULD USE F = S
      DO 122  L = 1,6
      F(L) = 0.
      DO 122  M = 1,6
  122 F(L) = F(L) + C(L,M)*XPILE(M)
      WRITE(3,1023) I, (XPILE(K), K = 1,6), (F(L), L=1,6)
 1023 FORMAT(T3,I2,2X,F7.4,5(1X,F7.4),2X,F8.1,5(2X,F7.0))
C***********COMPUTE X,Y,Z,COMPONENTS OF INDIVIDUAL PILES USING P = A*F
      DO 118  K = 1,6
      FCOMP(I,K) = 0.
      DO 118  L = 1,6
  118 FCOMP(I,K) = FCOMP(I,K)+A(K,L)*F(L)
      SUMX = SUMX + FCOMP(I,1)
      SUMY = SUMY + FCOMP(I,2)
      SUMZ = SUMZ + FCOMP(I,3)
      SUMMX = SUMMX + FCOMP(I,4)
      SUMMY = SUMMY+FCOMP(I,5)
      SUMMZ = SUMMZ + FCOMP(I,6)
      I = I+1
      IF(I.LE.N)GO TO 60
      WRITE(3,1024)
 1024 FORMAT(//,T10,  'INDIVIDUAL PILE FORCE COMPONENTS TO CHECK SUM OF
     1FORCES ALONG AXES' ,/,T10,'PILE NO',6X,'FX',10X,'FY',10X,'FZ',10X,
     2 'MX',10X,'MY',10X,'MZ')
      WRITE(3,1025)(I,(FCOMP(I,K),K=1,6),I=1,N)
 1025 FORMAT(T10,I3,3X,6F12.4)
      WRITE(3,1026)SUMX,SUMY,SUMZ,SUMMX,SUMMY,SUMMZ,(P(NN),NN=1,6)
 1026 FORMAT(T8, 'TOTAL =',1X,6F12.4,/,T16,'(',F10.3,')(',F10.3,')(',F10
     1.3,')(',F10.3,')(',F10.3,')(',F10.3,')')
      IF(NPROB.EQ.NLC)GO TO 6000
      NPROB = NPROB+1
      GO TO 3300
  150 STOP
      END
```

13-7 GENERAL COMMENTS

While this solution is quite flexible and general for a rigid pile cap, it could be easily extended to flexible pile caps as follows:

1 Revise the finite-element mat program of Chap. 7 to reduce the soil springs to zero.
2 Set up a pile-cap grid system so that pile heads intersect mat nodes as conveniently as possible.
3 Treat cap as rigid and solve for the pile forces on the foundation cap.
4 Treat the foundation cap as a flexible mat and compute the resulting displacements using the forces from step 3.
5 Modify the pile forces with the new displacements for 4 (lines 150 to 153 of this program) and recycle until convergence is achieved.
6 Final mat forces will provide a rational structural design of the pile cap.

PROBLEMS

13-1 Find the individual-pile forces of the group shown. All piles are 14BP74 at 50 ft, full head fixity.

k_s = 200 kcf = const

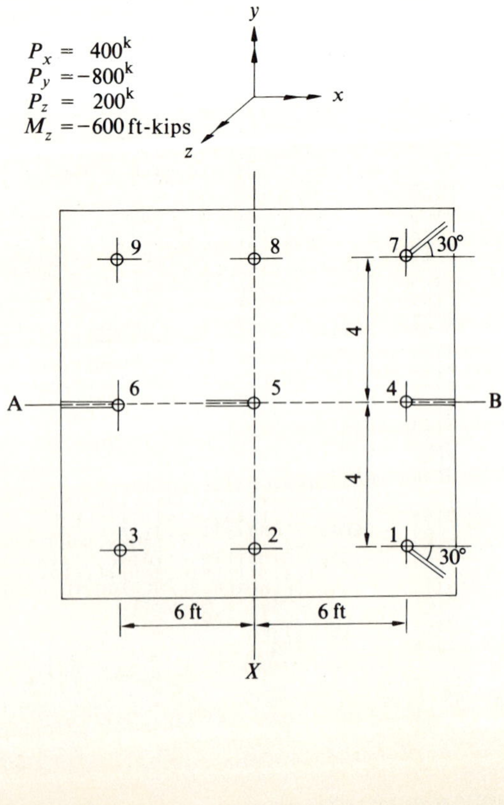

All batters are 6 : 1.

$$E = 30{,}000 \text{ ksi}$$

$$G = 12{,}000 \text{ ksi}$$

$$\psi = 2.0$$

Use Fig. 12-5 for load-transfer data to obtain constant A or C5(K).

13-2 Repeat Prob. 13-1 for 16-in-OD pipe (0.312 wall) piles filled with 3,000-psi concrete. Use $k_s = 10 + 10Z^{0.5}$.

13-3 Verify Examples 13-1 and 13-2.

13-4 Verify Example 13-3.

13-5 Repeat Prob. 13-1 for the forward row (piles 1, 2, 3) cast in a cap stepped along dashed line AB; change in elevation of $+4.0$ ft; repeat with -4.0 ft. Make pile 5 vertical.

13-6 What size H-piles should be used in Example 13-1 to limit $\Delta X \leq 0.10$ in? Value now is 0.16 in.

REFERENCES

ASCHENBRENNER, RUDOLF (1967): Three Dimensional Analysis of Pile Foundations, *J. Struct. Div., ASCE*, vol. 93, ST1, February, pp. 201–219.

BOWLES, J. E. (1968): "Foundation Analysis and Design," chap. 10, McGraw-Hill, New York.

BUTTERFIELD, R., and P. K. BANERJEE (1971): The Problem of Pile Group-Pile Cap Interaction, *Geotech.* (*Lond.*), vol. 21, no. 2, June, pp. 135–142.

HRENNIKOFF, A. (1950): Analysis of Pile Foundations with Batter Piles, *Trans. ASCE*, vol. 115, pp. 351–389.

REESE, L. C., M. W. O'NEILL, and ROBERT E. SMITH (1970): Generalized Analysis of Pile Foundations, *J. Soil Mech. Found. Div., ASCE*, vol. 96, SM1, January, pp. 235–250.

SAUL, WILLIAM E. (1968): Static and Dynamic Analysis of Pile Foundations, *J. Struct. Div., ASCE*, vol. 94, ST5, May, pp. 1077–1100.

14

SLOPE STABILITY

14-1 INTRODUCTION

The stability of earth masses against sliding, or gravity effects, is a serious problem. It must be routinely solved in most earthwork construction involving cut or fill, as in highways, railroads, dams, levees, and stockpiles. Slope stability is also a naturally occurring problem in hilly or mountainous areas (anyone who has ever driven through mountainous regions has seen raw slopes where masses of earth have slipped downhill). Stability failure of an embankment or slope occurs when an outer portion of the mass slides downward and outward with respect to the remainder of the mass, generally along a fairly well-defined slip surface (Fig. 14-1).

Stability analysis consists in (1) analyzing the forces causing and resisting stability failure and (2) determining the soil-strength properties. Methods of analysis include the assumption of the shape of the slip surface; the failure mass may be assumed circular, logarithmic spiral, sliding block, or wedges (Fig. 14-1). Field observations indicate that the failure surface is generally curved unless a definite weaker plane exists in the soil mass which can become a boundary condition. With the observation of curved slip failure surfaces, using either a circular or logarithmic spiral makes a

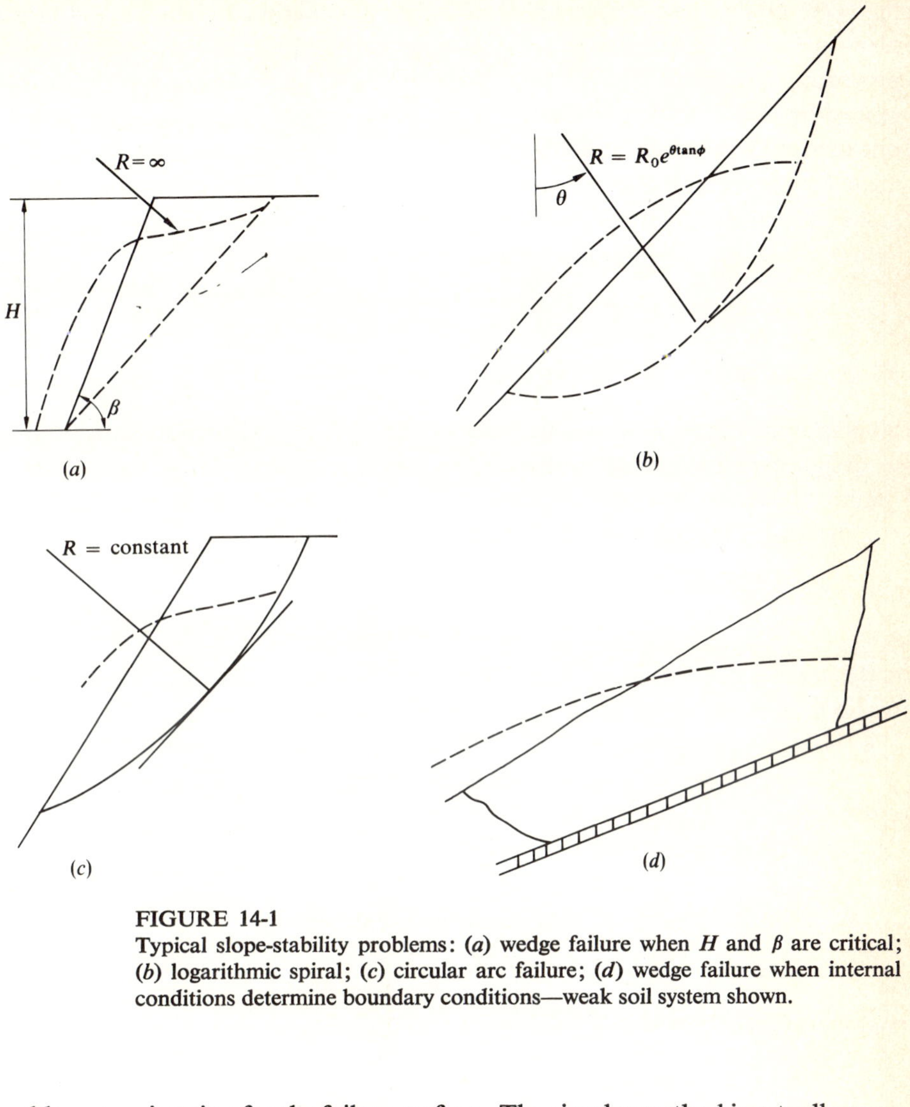

FIGURE 14-1
Typical slope-stability problems: (*a*) wedge failure when H and β are critical; (*b*) logarithmic spiral; (*c*) circular arc failure; (*d*) wedge failure when internal conditions determine boundary conditions—weak soil system shown.

reasonable approximation for the failure surface. The circular method is actually more widely used due to easier computation.

Most analysis methods further subdivide the slip zone into slices (or finite elements) to evaluate the driving and resisting forces as the sum of the slice contributions. The use of slices also provides a convenient means of analysis when the soil properties vary within the slip zone.

The literature subdivides slope-stability analysis into the several methods. The *limit-equilibrium* method evaluates the overall stability of the sliding mass just on the verge of slip, using some or all of the three equations of static equilibrium of a plane problem. The soil stress-strain relationships are not considered. This is the procedure believed to have been first proposed by Fellenius [Peterson (1955)], refined by Bishop (1955), and later used by Morgenstern and Price (1965), and is the one used in this chapter. The *finite-element* method seems in an early stage of development as reported by Whitman and Bailey (1967). A *plasticity-theory* method, suggested by Fang and Hirst (1970), has been shown to provide upper and lower bond values of the limit-equilibrium method.

14-2 SAFETY OF SLOPES

Slope-stability analysis is used to evaluate the [illegible] existing or natural slopes and to design the slopes of embankments so that an adequate safety factor (SF) exists. Figure 14-2*a* illustrates the situation where a cohesionless mass is on a slope of β. An analysis of this mass indicates that

$$\text{SF} = \frac{\tan \phi}{\tan \beta} \tag{a}$$

and is independent of size of mass. If the soil is cohesive or has both angle of internal friction and cohesion and the mass is homogenous and isotropic, the conditions of Fig. 14-2*b* can be considered.

In Fig. 14-2*b* the weight of the sliding wedge of unit width is

$$W = \tfrac{1}{2}\gamma LH \frac{\sin(\beta - \rho)}{\sin \beta}$$

where $\overline{AB} = L$. The shear resistance due to cohesion is

$$C = cL$$

and solving the force polygon of Fig. 14-2*b* for equilibrium gives

$$\frac{\gamma H}{C} = \frac{2 \sin \alpha \cos \phi}{\sin(\alpha - \rho) \sin(\rho - \phi)} \tag{14-1}$$

In the literature [Taylor (1948), Fang and Hirst (1970), among others] the term $\gamma H/C$ (or $C/\gamma H$) is termed a *stability number* N_s.

To obtain the minimum value of N_s or $N_{s,\mathrm{cr}}$, we can differentiate Eq. (14-1) and equate to zero as

$$\frac{dH}{d\rho} = 0$$

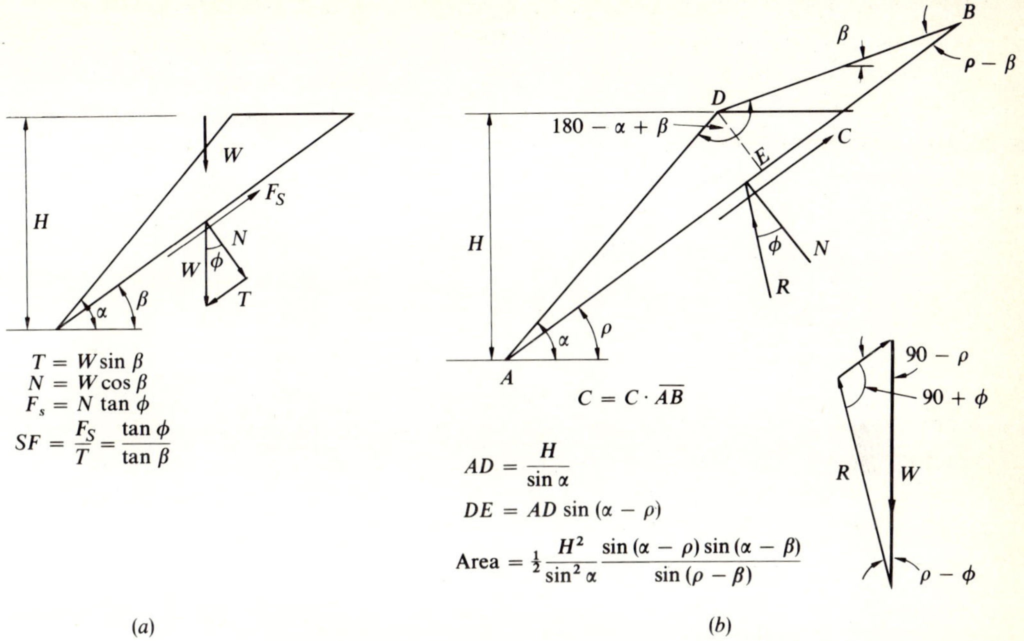

FIGURE 14-2
Slope stability with wedges used as failure zone: (*a*) cohesionless soil; (*b*) cohesive soil.

Solving, we obtain

$$N_s = \frac{4 \sin \beta \cos \phi}{1 - \cos (\beta - \phi)} \tag{b}$$

With $\beta = 90°$, $N_s = 4 \tan(45 + \phi/2)$, and with $\gamma H/C = 4 \tan(45 + \phi/2)$, the critical height of slope is

$$H_c = \frac{4c \tan (45 + \phi/2)}{\gamma} \tag{c}$$

which is the value widely used for the critical height of slope in vertical cuts. The actual height is reduced as $H_c = H_{cr}/\text{SF}$. One may prepare tables or charts for Eq. (14-1).

When the failure surface is taken as in Fig. 14-3, the stability requirement is that the sum of moments about the center of rotation be zero. Let us see how this might be.

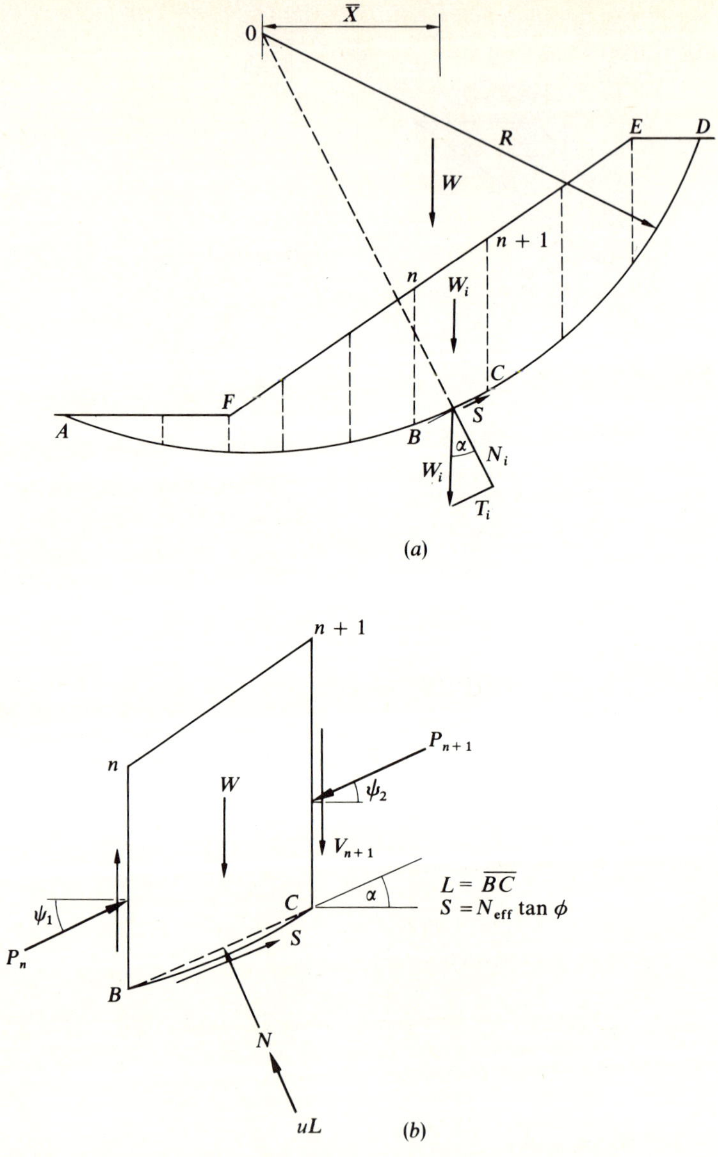

FIGURE 14-3
Resisting and driving forces in the circular slope failure for limit-equilibrium analysis: (*a*) circular sliding mass (mass has been subdivided into slices); (*b*) forces on a typical slice from (*a*).

From the figure it is evident that the cohesion resistance is at a constant distance R from the center of rotation. Thus,

$$C = c \times \text{arc length}$$

$$M_{\text{resist}} = RC \tag{d}$$

The overturning moment can be computed as $W\bar{X}$ or as the product of the radius R and the tangential component of W for the slice

$$\text{SF} = \frac{M_{\text{resist}}}{M_{\text{ot}}} = \frac{CR}{W\bar{X}} \tag{e}$$

When the circular mass is a homogenous purely cohesive soil ($\phi = 0$), it is often easier to cut the failure zone out of cardboard and hang it by a thread at two or more points to find the center of gravity; then compute the arc length of the slip surface and compute the safety factors as

$$\text{SF} = \frac{Rc \times \text{arc length}}{W\bar{X}}$$

For the more general cases of ϕ-c soils, which may be stratified, it is often convenient to use slices and compute the shear and tangential component of the slice weight of each slice. The resisting moment is

$$M_r = \sum(RC + RS)$$

the driving moment is

$$M_d = \sum RT$$

and

$$\text{SF} = \frac{\sum(C + S)}{\sum T} \tag{f}$$

The presence of unbalanced water pressure on the sides of the slice can also be incorporated into Eq. (f) since magnitude, direction, and point of application can be computed. This method is used by the U.S. Corps of Engineers (1960), among others.

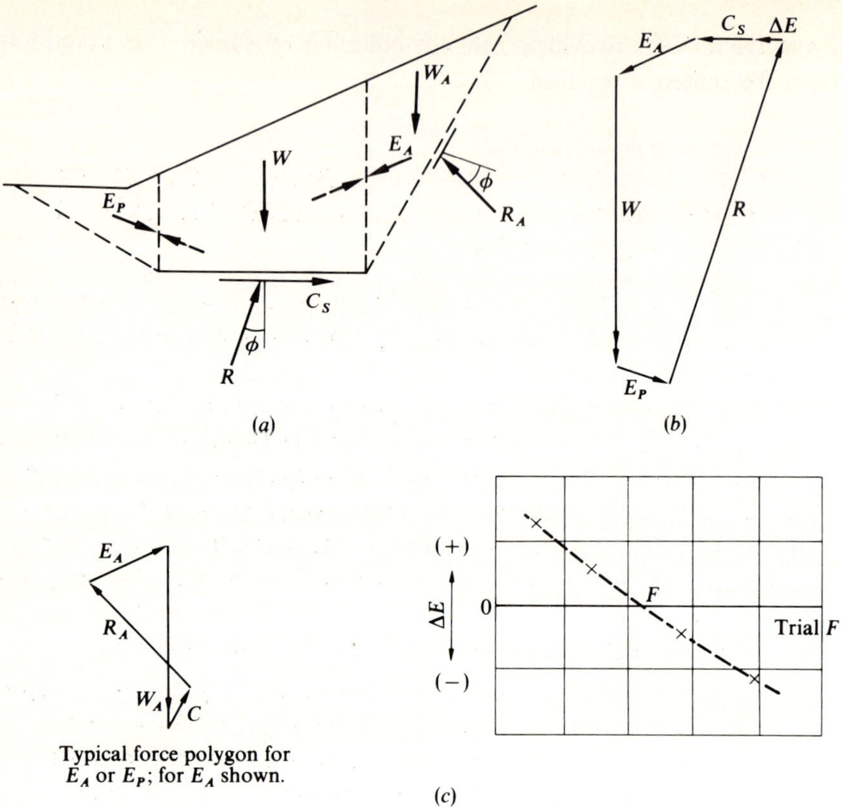

FIGURE 14-4
Trial-wedge solution (includes uplift pressure in forces as appropriate): (*a*) trial wedge; (*b*) force polygon of failure block; (*c*) plot of trial SF versus ΔE to find value at $\Delta E = 0$.

The wedge method (Fig. 14-4) is in equilibrium (SF = 1) if the force polygon of Fig. 14-4*b* closes. The safety factor of the wedge could be defined as

$$\text{SF} = \frac{\sum F_r}{\sum F_d}$$

However, current opinion is that one should apply a safety factor to the soil strength parameters. If we do this and compute the amount of force ΔE needed to close the force polygon (Fig. 14-4*b*), we can plot ΔE versus a series of assumed safety factors. That value obtained graphically at $\Delta E = 0$ is taken as the safety factor. In this procedure it is evident that E_a and E_p, the active and passive pressure wedges acting on the main block, are the resultants of force polygons drawn for those zones and include the effect of pore-water pressure. This method is also used by the U.S. Corps of Engineers (1960).

14-3 STABILITY ANALYSIS USING THE LIMIT-EQUILIBRIUM CONCEPT AND CIRCULAR SLIP SURFACES

The preceding section discussed the safety-factor concept. This section will proceed to the general solution of circular slip surfaces using the method of slices proposed by Bishop (1955).

Figure 14-3*a* is the general situation which occurs with ϕ-c soils. Here the failure surface is taken as circular. One may debate the superiority of a circular or logarithmic spiral or perhaps several segments of circular arcs of varying radii, but with the large number of uncertainties involved, it appears not at all unreasonable to use a circular arc. If we subdivide the failure surface into a number of slices to better describe the soil properties and isolate a slice so that we can determine the body forces, we obtain Fig. 14-3*b*.

For use with Fig. 14-3*b* let us define the terms as follows:

P_n, P_{n+1} = lateral pressures acting on sections n, $n + 1$, respectively, due to adjacent slices inclined at angles ψ_1, ψ_2 as shown
V_n, V_{n+1} = vertical shear forces including ϕ effect and cohesion
W = weight of slice including effect of stratification
N = total normal force acting on base of slice
s = shear force acting on base
h = average height of slice
b = width of slice
L = length of slice over which shear force acts (length BC)
u = pore-water pressure
α = angle between horizontal plane and BC
$\bar{X}$ = horizontal distance of center of slice to center of rotation
SF = safety factor
$\bar{B}$ = factor relating effective stress to total stress

The average normal stress on the slice base is

$$\sigma_n = \frac{N}{L} \tag{g}$$

and the shear strength is

$$s = c + \sigma' \tan \phi \tag{14-2}$$

where c, σ', and ϕ are *effective* shear-strength parameters.

The amount of shear strength for limiting equilibrium, applying the safety factor SF to the strength parameters, is

$$s = \frac{1}{\text{SF}}\left[c + \left(\frac{N}{L} - u\right)\tan\phi\right] \tag{14-2a}$$

Note here that $N/L - u$ is the effective normal pressure perpendicular to BC. In many cases one can obtain the effective pressure by using the submerged unit weight of soil as appropriate (the method is in the included computer program).

The sum of the moments of the soil mass within zone $ABCDEF$ made up of n slices is obtained as follows:

$$S = Ls$$

and

$$\sum W\bar{X} = \sum SR = \sum sLR \tag{h}$$

where the terms W, $\bar{X}$, etc., are for each slice and the summation accounts for the contribution of all the slices to the mass. By inspection the terms $\sum W\bar{X}$ are driving moments; the other terms represent resisting moments.

Substituting Eq. (14-2a) into Eq. (h) and solving for the safety factor gives

$$\text{SF} = \frac{R}{\sum W\bar{X}} \sum [cL + (N - uL)\tan\phi] \tag{i}$$

The normal force on BC (Fig. 14-3b) is obtained as $\sum F_v = 0$ on the slice to obtain

$$N = (W + V_n - V_{n+1})\cos\alpha - (P_n \sin\psi_1 - P_{n+1}\sin\psi_2)\sin\alpha \tag{j}$$

One can now obtain for SF

$$\text{SF} = \frac{R}{\sum W\bar{X}} \sum \{cL + (W\cos\alpha - uL)\tan\phi$$
$$+ [(V_{n+1} - V_n)\cos\alpha + (P_{n+1}\sin\psi_2 - P_n\sin\psi_1)\sin\alpha]\tan\phi\} \tag{14-3}$$

With no external body forces on the slice ($\sum F_v = \sum F_H = 0$)

$$\sum (V_n - V_{n+1} + P_n \sin\psi_1 - P_{n+1}\sin\psi_2) = 0$$

$$\sum (P_n \cos\psi_1 - P_n \cos\psi_2) = 0$$

It is at this point that small differences in the solution occur. Bishop (1955) and Spencer (1967) have shown that neglecting

$$\sum [(V_{n+1} - V_n)\cos\alpha + (P_{n+1}\sin\psi_2 - P_n\sin\psi_1)\sin\alpha]\tan\phi \tag{k}$$

introduces very small to negligible errors. Whitman and Bailey (1967) and Morgenstern and Price (1965) considered assuming $\psi_1 = \psi_2 = \psi$ then further assuming some reasonable variation of ψ from slice to slice including approximations for P_n, P_{n+1}. Reasonableness included observing the resulting line of action of P_n, P_{n+1} such that it remained inside the slip surface. It was considered unreasonable to compute tension in the soil, and so this fact could also be used to adjust ψ. In passing, note that more than one solution is possible with the assumption of ψ values. Whitman and Bailey (1967) showed (Table 14-1), however, that neglecting the P terms was not at all serious.

Other factors are probably much more serious if not correctly evaluated; for example, ϕ and c, stratification, unit weight of soil, and determination of pore pressure. Finally with all the estimation to this point it appears a bit unreasonable to estimate the position of the lateral thrust line to compute a safety factor a few percent higher or lower.

There are several other comments one can make relative to V_n, V_{n+1}, P_n, and P_{n+1}. Bishop (1955) assumed horizontal P values, which made his derivations somewhat easier. In general, one would expect $V = f(P)$ with rather indeterminate values of P and its point of application for stratified soils. The lateral pressures P would be approximately $\frac{1}{2}\gamma H^2 K$, where K is a lateral earth-pressure coefficient (not necessarily the active or passive value) with stratification taken into account. A P value for unbalanced waterhead where both magnitude and point of application can be determined seems to be the only case where P can be found with any reliability. In this case $V = 0$. How serious would neglect of the cases other than water be? As other investigators cited herein have shown, it is not a large amount of error. The discrepancy can be reduced with the computer considerably, as it should be apparent that P_n and P_{n+1} should be equal and collinear if the slice is of a width $dx \to 0$ since the average height would be the same as either the n or $n + 1$ sections. If these forces are equal and collinear, no moment results; also V_n and V_{n+1} would be equal and cancel. Thus one may conclude that thin slices should be used. This requires a large number of slices, but the computer can easily handle them.

The equation form actually used in the computer program is a rearrangement of

Table 14-1 COMPARISON OF SAFETY FACTORS

Example no.	Accurate	Simplified Bishop	Fellenius
1	1.58–1.62	1.61	1.49
2	1.24–1.26	1.33	1.09
3	0.73–0.78	0.70–0.82	0.66
4	2.01–2.03	2.00	1.14

SOURCE: Whitman and Bailey (1967).

Eq. (14-3) as proposed by Bishop (1955) to improve the accuracy of the computations by reducing the effect of the angle α when the variation of α is large from slice to slice.

If N is resolved, and if $L = b \sec \alpha$ and $\bar{X} = R \sin \alpha$ are used, Eq. (14-3) becomes

$$\mathrm{SF} = \frac{1}{\sum W \sin \alpha} \sum \left\{ [cb + W(1 - \bar{B}) \tan \phi] \frac{\sec \alpha}{1 + [(\tan \phi \tan \alpha)/\mathrm{SF}]} \right\} \tag{14-4}$$

This expression omits all the vertical components from P_{n+1}, P_n (and V_{n+1}, V_n) on the slice. The safety factor appears on both sides of Eq. (14-4); thus, the solution becomes an iterative process. Note that $W(1 - \bar{B})$ is simply the effective pressure.

The author considered using a logarithmic spiral to define the failure surface. (The slip surface in the computer program is a circular arc.) According to Terzaghi (1943), a spiral could be of the general form

$$R = R_0 \exp(\theta \tan \phi) \tag{14-5}$$

where R = instant radius at θ from R_0, rad

R_0 = radius at entrance point

exp = e = base of natural logarithms

This raised several points: (1) it is obvious that discontinuities should not exist in stratified soils; (2) what would one do about $\tan \phi$ in these cases? Since the computer program should be able to accommodate two or more different soils in the failure arc, it was the author's opinion that for a dubious increase in computational precision the logarithmic spiral would add an inordinate amount of work and increase the size of the computer program an excessive amount.

14-4 EXAMPLES

The following two examples taken from Bishop (1955) will illustrate the method. To conserve space the computer plots have information added and certain lines made heavier for convenience. Only partial input data will be given except as obtained from the computer output sheets.

EXAMPLE 14-1 This example uses Bishop (1955, fig. 3*a*) because this reference is readily available if additional information is desired. Given data are shown on Fig. E14-1.1. The author has converted the Bishop values to metric data, also shown on the figure.

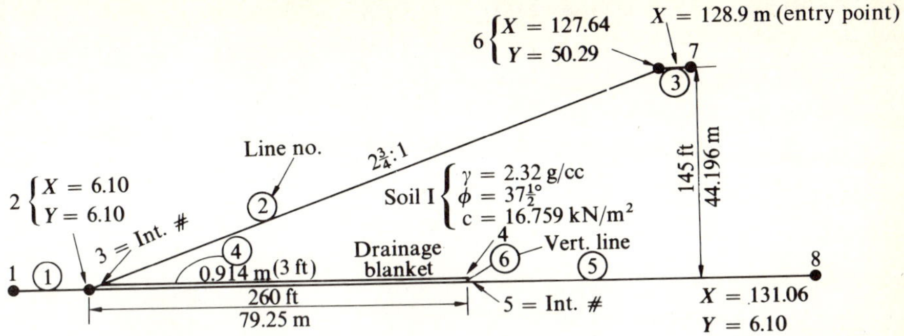

FIGURE E14-1.1
Slope dimensions and line and intersection numbering. Coordinates used in analysis also shown (not started with $Y = 0$).

SOLUTION Due to the large amount of input only typical data cards are given.

Card	Data
1	TITLE
2	UNITS (only M CM FU4 = 9.807)
3	NOL NLIT NOS NOLE ITX ITY PCODE NP DIMEN LIST
	6 8 1 3 4 4 1 4 75 0
4	CX CY ENTX ENTY DELX DELY SWIDTH SCALE
	37.00 127.00 128.93 50.29 2. 2. 1.5 20.
5	2 (Number of line intersections on line 1)
6	1. 2. 0.0 6.10 6.10 6.10 1 2
	C(I,J), J = 1, 6 and line intersect numbers
7	3 (Number of line intersections on line 2)
8	2. 3. 6.10 6.10 127.64 50.29 2 3 6
...	...
17	0.0 6.10 (*X,Y* coordinates of intersection 1)
18	6.10 6.10 (*X,Y* coordinates of intersection 2)
...	...
25	5 22.8 37.5 16.8 1.
	(5 = number of soil lines defining soil 1)
26	2 3 6 (line number; left and right line end intersection numbers)
27	3 6 7
28	4 3 4
29	5 5 8
30	6 4 5

These 30 cards make up the input for Prob. 14-1. The input uses metric units, and Fig. E14-1.2 illustrates computer printout of the input as well as some of the intermediate computation steps and printout checks. The plot of point 1 at a scale of 1 cm = 20 m is on Fig. 14-5*a*. Center coordinates of $X = 37.00$ and $Y = 127.00$ and entrance coordinates of $X = 128.93$ and $Y = 50.29$ m are shown on the computer

```
   J E BOWLES   EXAMPLE 14-1   BISHOP (1955) FIG 3A USING METRIC UNITS

 NO OF LINES =   6        NO OF LINE INTERSECT =   8
 NO OF SOILS =   1        NO OF EXTERNAL SOIL LINES =   3
 NO OF X-INCREMENTS =   1        NO OF Y-INCREMENTS =   1
      INITIAL SLICE WIDTH =   1.5 M

 THE LINE END CCORD MATRIX
LINE NO NO INT     X1          Y1          X2          Y2      SLCPE          LINE INTER NO
  1.     2.       0.0         6.10        6.10        6.10 0.0                 1     2
  2.     3.       6.10        6.10      127.64       50.29 0.363584            2     3     6
  3.     2.     127.64       50.29      129.54       50.29 0.0                 6     7
  4.     2.       8.61        7.01       85.34        7.01 0.0                 3     4
  5.     3.       6.10        6.10      131.06        6.10 0.0                 2     5     8
  6.     2.      85.34        7.01       85.34        6.10 0.100000E 08        4     5

 LINE INTERSECT ARRAY
INT NO         X              Y
  1           0.0            6.10
  2           6.10           6.10
  3           8.61           7.01
  4          85.34           7.01
  5          85.34           6.10
  6         127.64          50.29
  7         129.54          50.29
  8         131.06           6.10

 SOIL DATA ARRAY
SOIL NO   LINE #   LEFT INT    RT. INT    SAT    UNIT WT    PHI     COHESION
  1        2.         3.         6.        1.      22.8      37.5      16.8
  1        3.         6.         7.        1.      22.8      37.5      16.8
  1        4.         3.         4.        1.      22.8      37.5      16.8
  1        5.         5.         8.        1.      22.8      37.5      16.8
  1        6.         4.         5.        1.      22.8      37.5      16.8

 TRIAL CIRCLE NO =   1
 CIRCLE CTR COORDS:   X =      37.00   Y =     127.00
 ENTRANCE PT. CCORDS:   X =     128.93   Y =      50.29
      TRIAL ARC RADIUS =    119.731


 XXX LINE   1 NOT INTERSECTED BY TRIAL CIRCLE

 XXX LINE   4 NOT INTERSECTED BY TRIAL CIRCLE

 XXX LINE   5 NOT INTERSECTED BY TRIAL CIRCLE

 XXX LINE   6 NOT INTERSECTED BY TRIAL CIRCLE

 ARC INTERSECT WITH LINE ARRAY
LINE NO            X               Y
  2.            14.948            9.32
  3.           128.930           50.29

 THE ARRAY WITH ALL INTERSECTIONS FOLLOWS:
 I =  1            0.0         6.100    K =   1    KK =   1
 I =  2            6.100       6.100    K =   1    KK =   2
 I =  3            8.611       7.010    K =   1    KK =   3
 I =  4           14.948       9.317    K =   2    KK =   3
 I =  5           85.344       7.010    K =   2    KK =   4
 I =  6           85.344       6.100    K =   2    KK =   5
 I =  7          127.640      50.290    K =   2    KK =   6
 I =  8          128.930      50.290    K =   2    KK =   6
 I =  9          129.540      50.290    K =   2    KK =   7
 I = 10          131.060       6.100    K =   2    KK =   8

 THE APPLICABLE ARRAY ARCINT FOLLOWS:
 I =  1           14.948       9.317    K =   4    KK =   1
 I =  2          127.640      50.290    K =   7    KK =   2
 I =  3          128.930      50.290    K =   8    KK =   3

 FIND SLICE WIDTH AND NO OF SLICES

 ***** MAXIMUM SLICE WIDTH HAS BEEN INCREMENTED TO 2.00
                FI=     1.00000     FC=      1.38546
                FI=     1.48542     FO=      1.48453

        THE SAFETY FACTOR FOR POINT    1 IS      1.48453
```

FIGURE E14-1.2
I/O for Example 14-1 using metric units.

plot. The safety factor for this set of coordinates was 1.484 (also shown on the computer plot). ////

EXAMPLE 14-2 This example is taken from Bishop (1955, fig. 4). Soil lines, numbering, etc., are shown in Fig. E14-2.1. Note in this example that lines 5 and 6

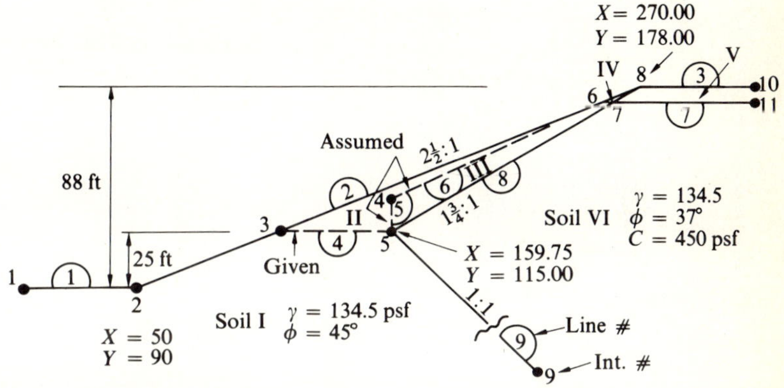

FIGURE E14-2.1
Slope with intersection, soil, and line numbers. Critical dimensions and coordinates given. Refer to Fig. E14-2.2 for remaining coordinates and soil data.

were not given by Bishop. Line 6 was approximated to be that shown by Bishop as the piezometric head. To use effective weights this became a "soil" line. Line 5 was used to close the soil area. Figure E14-2.2 (page 468) is a *partial* computer listing of input. All plots give data for checking. Note that slice lines occur at all line intersections. Figure 14-5*b* (page 470) also shows one of the plots of this problem (point 1, with center coordinates of $X = 150.0$, $Y = 200.0$, entrance coordinates of $X = 275.00$, $Y = 178.00$). The safety factor was 2.072. ////

14-5 THE COMPUTER PROGRAM

This computer program was developed using Bishop's (1955) simplified procedure. With some modification it could be made to include the lateral-pressure forces of Morgenstern and Price (1965) or Whitman and Bailey (1967). The program uses a circular-arc failure surface. It may be possible with some modification to solve a logarithmic spiral, as the major change involved is computing the line intersection for a spiral rather than a circle; however, as the program allows for different soils (ϕ

J E BOWLES EXAMPLE 14-2 SLOPE STABILITY US UNITS FRCM BISHOP (1955) FIG 4

NO OF LINES = 9 NO OF LINE INTERSECT = 11

NC OF SOILS = 6 NO OF EXTERNAL SOIL LINES = 3

NO OF X-INCREMENTS = 1 NO OF Y-INCREMENTS = 1

INITIAL SLICE WICTH = 3.0 FT

THE LINE END COORD MATRIX

LINE NO	NC INT	X1	Y1	X2	Y2	SLCPE	LINE INTER NO			
1.	2.	0.0	90.00	50.00	90.00	0.0	1	2		
2.	4.	50.00	90.00	270.00	178.00	0.4C0000	2	3	6	8
3.	2.	270.00	178.00	310.00	178.00	0.0	8	10		
4.	2.	112.50	115.00	159.75	115.0C	C.0	3	5		
5.	2.	159.75	130.00	159.75	115.00	0.1CCCOOE 08	4	5		
6.	2.	159.75	130.00	252.50	171.00	0.442048	4	6		
7.	3.	252.50	171.00	320.00	171.00	C.0	6	7	11	
8.	3.	159.75	115.00	270.00	178.00	0.571429	5	7	8	
9.	2.	159.75	115.00	275.00	0.0	-0.997831	5	9		

LINE INTERSECT ARRAY

INT NO	X	Y
1	0.0	90.00
2	50.00	90.00
3	112.50	115.00
4	159.75	130.00
5	159.75	115.00
6	252.50	171.00
7	257.75	171.00
8	270.C0	178.C0
9	275.00	0.0
10	310.00	178.00
11	320.00	171.00

SOIL DATA ARRAY

SOIL NO	LINE #	LEFT INT	RT. INT	SAT	UNIT WT	PHI	COHESION
1	1.	1.	2.	1.	134.5	45.0	0.0
1	2.	2.	3.	1.	134.5	45.0	0.0
1	4.	3.	5.	1.	134.5	45.0	0.0
1	5.	5.	5.	1.	134.5	45.0	0.0
1	8.	5.	5.	1.	134.5	45.0	0.0
1	9.	5.	9.	1.	134.5	45.C	0.0
2	2.	3.	6.	0.	118.0	45.0	0.0
2	4.	3.	5.	0.	118.0	45.0	0.0
2	5.	4.	5.	0.	118.0	45.C	0.0
2	6.	4.	6.	0.	118.0	45.0	0.0
2	7.	6.	6.	0.	118.0	45.0	0.0
2	8.	5.	5.	0.	118.0	45.0	0.0
2	9.	5.	5.	0.	118.0	45.0	0.0
3	2.	6.	6.	1.	134.5	45.0	0.0
3	4.	5.	5.	1.	134.5	45.0	0.0
3	5.	4.	5.	1.	134.5	45.C	0.0
3	6.	4.	6.	1.	134.5	45.0	0.0
3	7.	6.	7.	1.	134.5	45.0	0.0
3	8.	5.	7.	1.	134.5	45.C	0.0
3	9.	5.	5.	1.	134.5	45.0	0.0
4	2.	6.	8.	0.	118.0	45.0	0.0
4	3.	8.	8.	0.	118.0	45.0	0.0
4	6.	6.	6.	0.	118.0	45.0	0.0
4	7.	6.	7.	0.	118.0	45.0	0.0
4	8.	7.	8.	0.	118.0	45.0	0.0
5	2.	8.	8.	0.	135.0	37.C	450.0
5	3.	8.	10.	0.	135.0	37.0	450.0
5	7.	7.	11.	0.	135.0	37.0	450.0
5	8.	7.	8.	0.	135.0	37.C	450.0
6	4.	5.	5.	1.	134.5	37.0	450.0
6	7.	7.	11.	1.	134.5	37.0	450.0
6	8.	5.	7.	1.	134.5	37.0	450.0
6	9.	5.	9.	1.	134.5	37.0	450.0

TRIAL CIRCLE NO = 1
CIRCLE CTR COORDS: X = 150.00 Y = 200.00
ENTRANCE PT. COORDS: X = 275.00 Y = 178.00
TRIAL ARC RADIUS = 126.921

FIGURE E14-2.2
Partial I/O for Example 14-2.

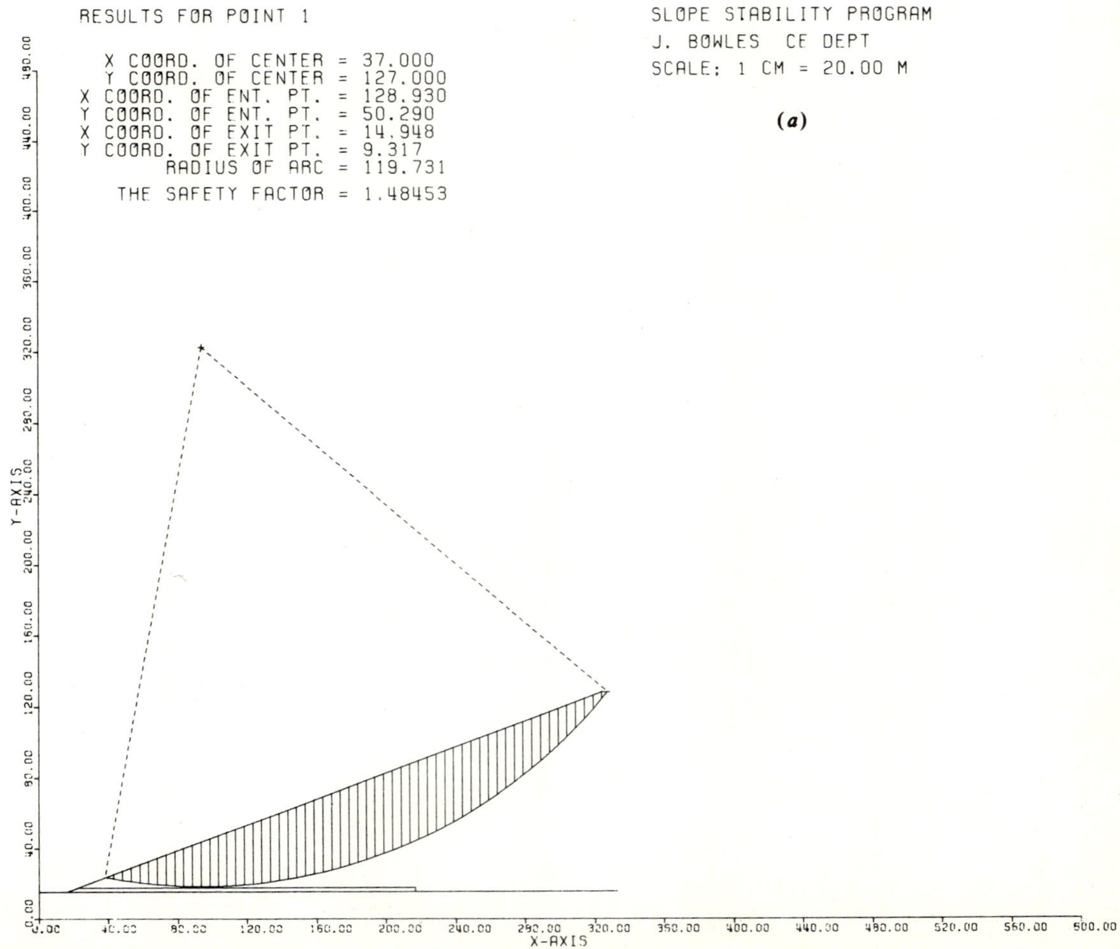

FIGURE 14-5
Plots of trial circles as indicated from Examples 14-1 and 14-2.

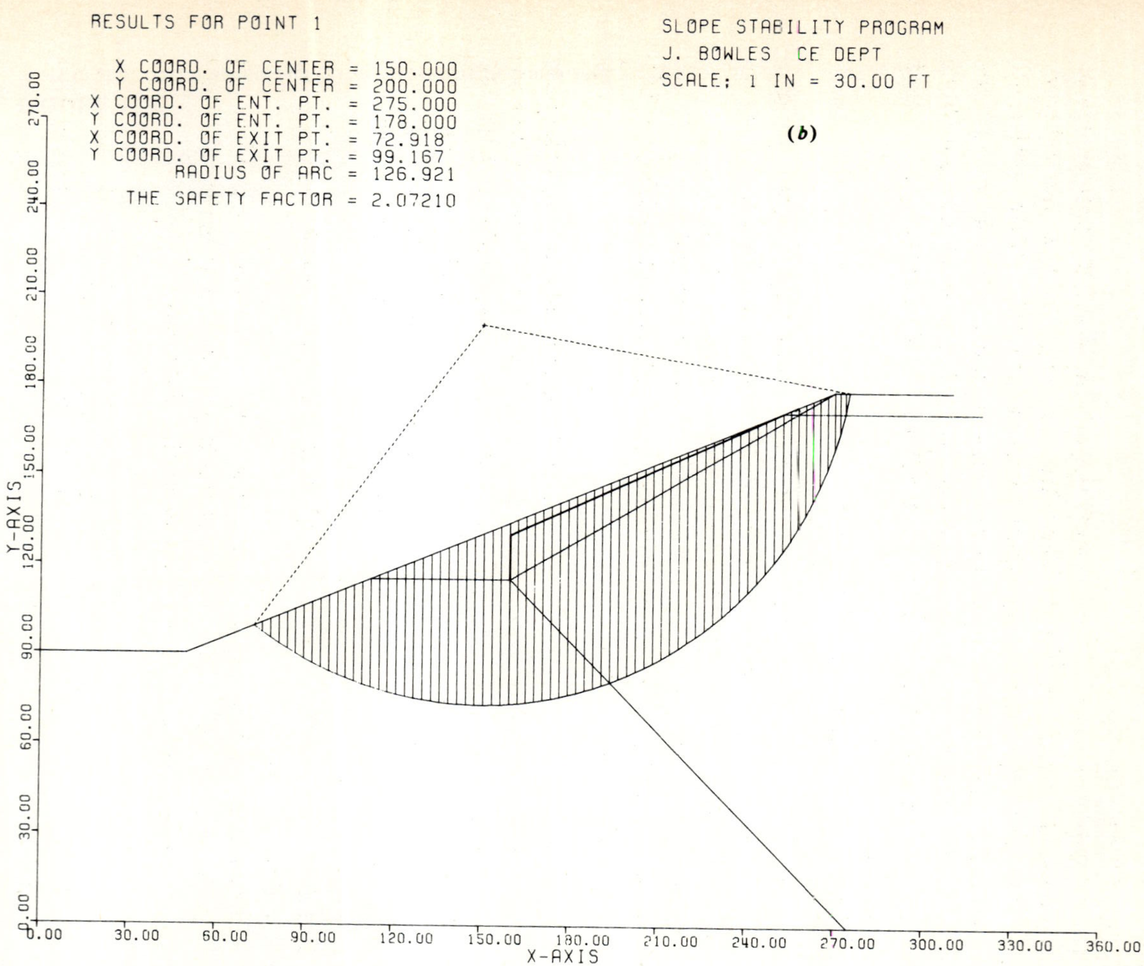

FIGURE 14-5 *(Continued)*

angles) to be intersected by the failure arc, the use of a logarithmic spiral may be too academic. Once the intersection coordinates are found, the areas, angles, etc., use coordinates for the computations.

The program will solve any slope-stability problem where the ground surface can be described by a series of straight lines. Any number of soil lines and soil types can be used (a very large number may require changing the DIMENSION statement). This large number of soils allows for describing the pore pressure by use of saturated-soil unit weights (note this was done in Example 14-2). The program computes both the total and the effective slice weights; thus the $\bar{B}$ used by Bishop (1955) is not used in this program.

It is necessary in using this program to:

1 Number all line intersections in increasing *X* coordinates from left to right.
2 Number the upper external (closest to arc center) soil lines first in order from left to right. Interior soil lines may be numbered in any order.
3 The different soils in the mass may be numbered in any order.
4 Compute all line intersections accurately to 0.005 so that SMLNO = 0.01 will properly test coordinates.

The program locates all line intersections *inside* the trial arc including those lines intersecting the arc. A slice is located on every intersection point. Distances between intersection points are divided according to the slice-width specification (SWIDTH).

This program uses the CALCOMP plotter and standard plot subroutines to plot any or all of the trial circles using the call PCODE and NP on the third data card. For users using other plotters the plot routines/instructions in this program are easily removed as they are all grouped at the end.

The iterative procedure for the safety factor uses the equation proposed by Little and Price (1958).

This program will compute in either metric or fps units. The plot routines will plot the scale on 2-cm ticks. This can be changed (line 355). Use meters and kilonewtons or feet and pounds. The unit card contains three entries:

M	CM	9.807		
FT	IN	62.5	→	H_2O

The user is advised to plot (if that capability exists) the first trial circle and obtain a full listing of input and output if any of the internal soil lines intersect to form odd geometric shapes. The author has run many problems, and they have all been found to work (Example 14-2 contains an unusual intersection pattern); however, it is possible that not all cases have been covered. In the worst possible case it is expected that not over one or two slice weights would be in error, but the user should check. Again especially note step 4 above.

Line	Operation
1–5	Bookkeeping
6	READ TITLE and work units (two cards)
10	READ (15I5) NOL = total number of soil lines; NLIT = total number of line intersections (the end of any line whether or not intersected by another line is a line intersection); NOS = number of soils in mass (same soil submerged is counted twice); NOLE = number of top external soil lines; ITX,ITY = number of circle center points in *X*, *Y* directions to be analyzed for a single entrance point; PCODE = plot subroutine used if >0; NP = plot control counter (if two plots, every other point, three plots, third point, etc.); DIMEN = control number of slices as 75, 80, 90, etc.; LIST = control to obtain extra output (after test run use 0 to conserve paper)
12	READ (8F10.3) CX,CY = initial trial-circle center coordinates; ENTX, ENTY = trial-circle entrance coordinates; DELX,DELY = center *X*, *Y* coordinate increments for each trial; SWIDTH = initial slice width (DIMEN may increase value); SCALE = plotting scale as 20, 30 for 1 in = 20 ft, 1 in = 30 ft, etc.
22–51	READ problem data on line intersections, soils, etc., and forms arrays for later use; also writes data for checking
22	READ NLI = number of line intersections of each line in turn one entry per card
23	READ (C(I,J), (NOLIT(I,N), N = 1,NLI) C(I,J) = line data including line number, number of line intersections for the lines, the *X*, *Y* coordinates of the end points left to right (six entries); NOLIT(I,N) = all the line intersection numbers on the *i*th line including the end values (NLI entries)
25	Gives vertical lines the slope value of BIGNO so computer does not divide by 0.0
33–34	READ INTAR(J,K) INTAR(J,K) = line intersection *X*, *Y* coordinates in increasing intersection numbers
40	READ Soil data on DO loop NSLIN(I) = number of soil lines defining the boundary of the soil. Include lines terminating at a joint. If a line intersects a soil-line boundary between the ends, count the soil boundary line twice; G(I) = unit weight either saturated or wet; PHI(I) = ϕ angle; COHES(I) = cohesion; SAT = 1. if saturated, 0. if wet
43	READ LINSOL = soil line number; INTL,INTR = intersection number on left and right end of line (if a soil line terminates at a joint on a soil boundary, that line is included; the joint number is used for both INTL and INTR)
51	Begins DO to test circles
58–121	Computes trial circle, line intersections; finds lines not used LNU(I), those in circle but not intersected by circle
122–174	Sets up arc-intersection array ARCINT
177–195	Finds slice width and checks total number of slices against DIMEN and increments slice width if necessary
197–251	Finds coordinates of all lines intersecting slices in slice array SLIC(I,J,K). Note that lines not intersected in a slice are given same coordinates as last line intersected and stacked at a point. This is so a routine for both coordinate test and area computations can be used using all lines. SLIC array is sorted for decreasing *Y* coordinates
254–314	Computes areas of slice parts and weights, both effective and total; sums slice weights; finds ϕ and c for soil touching arc surface
315–341	Computes safety factor
342	Tests PCODE and NP for plotting
342–402	PLOT ROUTINE Also note that statement numbers 15, 402, and IBUF(1000) in DIMENSION statement relate to the plot routine

```
C       J E BOWLES--- SLOPE STABILITY PROBLEM BY SIMPLIFIED BISHOP METHOD
C       READ TOP INTERSECTION OF VERTICAL LINE FIRST THEN OTHER END
C       READ LINE INTERSECTION NUMBERS WITH INCREASING X-COORD.
C       NOL = NO OF SOIL LINES--NUMBER EXTERNAL LINES FIRST FROM LT TO RT
C       NLIT = TOTAL NUMBER OF LINE INTERSECTIONS.
C       NOS = NUMBER OF SOILS INCLUCED IN SECTION--USE ANY ORDER
C       NOLE = NUMBER OF EXTERNAL SOIL LINES.
C       DIMEN = CONTROL FOR NUMBER OF SLICES ALLOWED AS 75, 80, 90, ETC
C       C(I,J) = LINE COORD MATRIX--I = LINE NO., J=2 IS NO OF INTERSECT
C          ON LINE, J=3-6 ARE X1,Y1,X2,Y2
C         READ LINE END COORDS X1,Y1,X2,Y2 LEFT TO RIGHT
C       ITX,ITY = # OF CIRCLE CENTER PCINTS IN X, Y DIRECTIONS.
C       LIST IF GREATER THAN 0 WILL LIST ACCITIONAL DATA
C       SWIDTH = INIT. MAX SLICE WIDTH--AUTC CHANGED IF EXCEEDS "DIMEN"
C       SCALE = PLOT SCALE 1" = 10' USE 10.;  1" = 20' USE 20., ETC
C       IF PCODE = 0   THE PLOT RCUTINE IS SKIPPED
C       NP IS A PLOT CONTROL COUNTER; IF NP = 3 EVERY THIRD ONE IS PLOTTED
C       IF A LINE TERMINATES AT A SOIL JOINT, IT IS INCL BUT BOTH INTERSEC
C          ARE THOSE OF THE JOINT
C       SAT = SWITHC FOR SATURATED SOIL IF SAT >0 SOIL SATURATED
C       UT1 = FT OR M FOR SCALE;  UT2 = IN OR CM FOR SCALE;  READ COORDS A
C          UNIT WEIGHT IN US OR METRIC UNITS--USE TITLE TO GIVE UNITS
        DIMENSION C(15,6),NOLIT(15,10),SLCPE(15),INTAR(15,2),EFFWT(100),
       2NSLIN(10),G(10),PHI(10),COHES(1C),SLICX(76),SOIL(8,9,4),SAT(10),
       3ARCINT(20,3),LNU(15),CO(100),ALPHA(100),TITLE(20),SLIC(76,10,2),
       4P(100),B(100),AREA(100),WEIGH(100),ALLINT(30,3),IBUF(1000)
        INTEGER PCODE,DIMEN
        REAL INTAR,METER
        BIGNO = 9999999.
        SMLNO=0.01
   4000 READ(1,1000,END=150)TITLE,UT1,UT2,FU4
   1C00 FORMAT(20A4/A4,6X,A4,6X,F10.4)
        WRITE(3,2000)TITLE
   2000 FORMAT('1',T5,20A4,//)
        READ(1,123)NOL,NLIT,NOS,NOLE,ITX,ITY,PCODE,NP,DIMEN,LIST
    123 FORMAT(15I5)
        READ(1,1010)CX,CY,ENTX,ENTY,DELX,DELY,SWIDTH,SCALE
   1010 FORMAT(8F10.3)
        WHOLD = SWIDTH
        IF(PCODE.NE.0)CALL PLOTS(IBUF,1C00,9)
        NOSP1=NOS+1
        WRITE(3,2001)NOL,NLIT,NOS,NOLE,ITX,ITY,SWIDTH,UT1
   2C01 FORMAT(T5,'NO OF LINES =',I3,5X,'NC CF LINE INTERSECT =',I3,//,T5,
       A'NO OF SOILS =',I3,5X,'NO OF EXTERNAL SOIL LINES =',I3,//,T5,
       A'NO OF X-INCREMENTS =',I3,5X,'NO OF Y-INCREMENTS =',I3,//,T10, 'IN
       2ITIAL SLICE WIDTH =',F5.1,1X,A2//)
        WRITE(3,2003)
   2003 FORMAT(T5,'THE LINE END COORD MATRIX',/,T4,'LINE NC',T12,'NO INT',
       1T21,'X1',8X,'Y1',8X,'X2',8X,'Y2',6X,'SLOPE',5X,'LINE INTER NO')
        DO 333 I=1,NOL
C       NLI = NO OF INTERSECT ON ANY LINE --MUST HAVE AT LEAST 2-VALUES
        READ(1, 123)NLI
        READ(1,1012)(C(I,J),J=1,6),(NCLIT(I,N),N=1,NLI)
   1012 FORMAT(2F5.0,4F10.4,6I5)
        SLOPE(I) = BIGNO
        IF(ABS(C(I,5)-C(I,3)).LE.0.001)GO TO 333
        SLOPE(I)=(C(I,6)-C(I,4))/(C(I,5)-C(I,3))
    333 WRITE(3,2004)(C(I,J),J=1,6),SLOPE(I),(NOLIT(I,KK),KK=1,NLI)
   2C04 FORMAT(T6,F3.0,3X, F3.0,T15,4(2X,F8.2),G13.6,1X,6I5)
C**** READ X AND Y-COORDINATES TC BUILD LINE INTERSECT ARRAY (INTAR)
        WRITE(3,2005)
   2005 FORMAT(//,T5,'LINE INTERSECT ARRAY',/,T4,'INT NO',T16,'X',T28,'Y')
        DO 2 J=1,NLIT
        READ(1,1010)(INTAR(J,K),K=1,2)
      2 WRITE(3,2006)J,(INTAR(J,K),K=1,2)
   2C06 FORMAT(T6,  I3,T12,F10.2,2X,F10.2)
C       BUILD SOIL INTERSECTION ARRAY USING LINE INTERSECT NO. (INTNO)
        WRITE(3,2008)
   2008 FORMAT(//,T5,'SOIL DATA ARRAY',/,T4,'SOIL NO',T13,'LINE #',T21,
       A'LEFT INT',3X,'RT. INT',3X,'SAT',3X,'UNIT WT', 3X,'PHI',3X,'COHESI
       2ON')
C ***   NUMBER SOILS IN ANY ORDER--PREFERABLY FROM LEFT TO RIGHT
        DO 5 I=1,NCS
        READ(1,1016)NSLIN(I),G(I),PHI(I),COHES(I),SAT(I)
   1016 FORMAT(I5,4F10.4)
        NS=NSLIN(I)
        DO 5  K = 1,NS
        READ(1, 123)LINSOL,INTL,INTR
        SOIL(I,K,1)=LINSOL
        SOIL(I,K,2)=INTL

        SOIL(I,K,3)=INTR
        SOIL(I,K,4) = SAT(I)
      5 WRITE(3,2009)I,(SOIL(I,K,MM),MM=1,4),G(I),PHI(I),COHES(I)
   2009 FORMAT(T5,I3,6X,F3.0,7X,F3.0,6X,F3.0,6X,F2.0,5X,F6.1,3X,F4.1,3X,F7
       A.1)
        PCOUN = 0.
C       BEGIN LOOP TO TEST TRIAL CIRCLES
    100 DO 360 IY=1,ITY
        IF(IY.GT.1)CY=CY+DELY
        DO 360 IX=1,ITX
        PCOUN = PCOUN + 1.
        NCOUN = PCOUN
        SWIDTH = WHOLD
```

```
      IF(IX.GT.1)CX=CX+DELX
C**** COMPUTE RADIUS OF TRIAL CIRCLE--TEST EACH LINE FOR INTERSECT WITH
C        TRIAL RADIUS---CX,CY ARE INITIAL TRIAL CENTER OF ARC COORDINATES
      R = SQRT((ENTX-CX)**2+(CY-ENTY)**2)
      WRITE(3,2121)NCOUN,CX,CY,ENTX,ENTY,R
 2121 FORMAT(//,T5,'TRIAL CIRCLE NO =',I3,/,T5,'CIRCLE CTR COORDS:',2X,'
     1X =',F10.2,2X,'Y =',F10.2,/,T5,'ENTRANCE PT. COORDS:',2X,'X =',F10
     2.2,2X,'Y =',F10.2,/,T10,'TRIAL ARC RADIUS =',F10.3,//)
      K1=0
      DO 8 I=1,NOL
      LNU(I)=0
      IF(ABS(SLOPE(I)).LE.0.0001)GO TO 9
      CON=C(I,3)-C(I,4)/SLOPE(I)
      AA=1.0/SLOPE(I)**2+1.0
      BB=2.0*CON/SLOPE(I)-2.0*CX/SLOPE(I)-2.0*CY
      CC = CON**2 - 2.*CX*CON + CX**2 + CY**2 - R**2
      DIFF=BB**2-4.0*AA*CC
      IF(DIFF.LT.0.0)GO TO 20
      YPR=(-BB+SQRT(DIFF))/(2.0*AA)
      YNR=(-BB-SQRT(DIFF))/(2.0*AA)
C     THIS PART COMPUTES X-COORDS IF CIRCLE INTERSECTS ANY BUT HORIZONTA
C        LINES (SLOPE(I)>0.0001).
      XPR=YPR/SLOPE(I)+CON
      XNR=YNR/SLOPE(I)+CON
      GO TO 10
C     FOLLOWING STEPS USED FOR HORIZONTAL LINES
    9 DIFF = R**2 - (CY-C(I,4))**2
      IF(DIFF.LT.0.)GO TO 20
      XPR = CX + SQRT(DIFF)
      XNR = CX - SQRT(DIFF)
      YPR=C(I,4)
      YNR=C(I,4)
   10 J1=0
      J2=0
      IF(ABS(SLOPE(I)).GE.BIGNO)GO TO 11
      IF(XPR.GE.C(I,3).AND.XPR.LE.C(I,5))J1=1
      IF(XNR.GE.C(I,3).AND.XNR.LE.C(I,5))J2=1
      GO TO 12
   11 IF(SLOPE(I))66,66,666
   66 IF(YPR.GE.C(I,6).AND.YPR.LE.C(I,4))J1=1
      IF(YNR.GE.C(I,6).AND.YNR.LE.C(I,4))J2=1
      GO TO 12
  666 IF(YPR.GE.C(I,4).AND.YPR.LE.C(I,6))J1=1
      IF(YNR.GE.C(I,4).AND.YNR.LE.C(I,6))J2=1
   12 IF(J2.EQ.0)GO TO 13
      K1=K1+1
      ARCINT(K1,1)=I
      ARCINT(K1,2)=XNR
      ARCINT(K1,3)=YNR
   13 IF(J1.EQ.0)GO TO 7
      K1=K1+1
      ARCINT(K1,1)=I
      ARCINT(K1,2)=XPR
      ARCINT(K1,3)=YPR
         GO TO 8
    7 IF(J1.NE.0.OR.J2.NE.0)GO TO 8
C     END OF LOOP FOR ARC & LINE INTERSECTIONS WITH ARC INT. ARRAY BUILT
   20 LNU(I)=I
      WRITE(3,2101)I
 2101 FORMAT(//,T5,'XXX LINE',I3,' NOT INTERSECTED BY TRIAL CIRCLE')
    8 CONTINUE
      DO 400 I=1,NOL
      IF(LNU(I).EQ.0)GO TO 400
      R1=SQRT((CX-C(I,3))**2+(CY-C(I,4))**2)
      R2=SQRT((CX-C(I,5))**2+(CY-C(I,6))**2)
      IF(R.LT.R1.AND.R.LT.R2)GO TO 400
      LNU(I) = 0
      IF(SLOPE(I).EQ.BIGNO)LNU(I)=I
      IF(SLOPE(I).EQ.BIGNO)WRITE(3,403)LNU(I)
  403 FORMAT(//,'**** LINE',I3,' IS IN ARC BUT VERT. AND NOT USED')
      WRITE(3,401)LNU(I)
  401 FORMAT(//,T5,'*** LINE',I3,' IS NOT INTERSECTED BUT IS IN ARC')
  400 CONTINUE
C     FIND WIDTH OF SECTIONS BETWEEN LINE INTERSECTIONS SO SLICE ARRAY
C        CAN BE SET UP.
C     FIND LOCATION OF ARC EXIT AND NEAREST ADJACENT INTERSECTION.
C     NOTE K1 = TOTAL ENTRIES IN CIRCLE-LINE INTERSECTION ARRAY(ARCINT).
C     SORT ARC INTERSECTIONS FROM LEAST TO GREATEST X-COORDINATES.
      K1M=K1-1
   24 DO 26 KY=1,K1M
      IF(ARCINT(KY,2).LE.ARCINT(KY+1,2))GO TO 26
      DO 25 KX=1,3
      SAVE=ARCINT(KY,KX)
      ARCINT(KY,KX)=ARCINT(KY+1,KX)
      ARCINT(KY+1,KX)=SAVE
   25 CONTINUE
      GO TO 24
   26 CONTINUE
      WRITE(3,2112)
 2112 FORMAT(//,T5,'ARC INTERSECT WITH LINE ARRAY',/,T4,'LINE NO',T19,
     A'X',T32,'Y')
      WRITE(3,2114)((ARCINT(KZ,JJ),JJ=1,3),KZ=1,K1)
 2114 FORMAT(T5,F3.0,T13,F10.3,2X,F10.2)
      LINE1=ARCINT(1,1)
      S1=ARCINT(1,2)
      S2=ARCINT(1,3)
```

```
      WRITE(3,8053)
 8053 FORMAT(//,T5,'THE ARRAY WITH ALL INTERSECTIONS FOLLOWS:')
      ICOUN = 0
      K=1
      KK=0
      LL=NLIT+K1
C******COMBINE ARCINT AND INTAR IN ORDER OF INCREASING X-VALUES
      DO 70 I=1,LL
      KK=KK+1
      DO 75 J=1,2
   75 ALLINT(I,J)=INTAR(KK,J)
      IF(I.NE.1.AND.ALLINT(I-1,1).EQ.INTAR(KK,1))GO TO 70
      IF(ARCINT(K,2).GE.INTAR(KK,1))GO TO 70
      IF(ICOUN.GT.0)GO TO 70
   72 DO 73 L=1,2
   73 ALLINT(I,L)=ARCINT(K,L+1)
      IF(K.EQ.K1)ICOUN = 1
      K=K+1
      IF(K.GT.K1)K = K1
      KK=KK-1
   70 WRITE(3,8051)I,(ALLINT(I,J),J=1,2),K,KK
 8051 FORMAT(T5,'I =',I3,2X,2F12.3, 2X,' K =',I3,2X,'KK =',I3)
      WRITE(3,8052)
 8052 FORMAT(//,T5,'THE APPLICABLE ARRAY ARCINT FOLLOWS:')
      LAL=0
      DO 77 I=1,LL
      R2=SQRT((CX-ALLINT(I,1))**2+(CY-ALLINT(I,2))**2)
      IF(R2.GT.(R+SMLNO))GO TO 77
      LAL=LAL+1
      DO 78 K=1,2
   78 ARCINT(LAL,K)=ALLINT(I,K)
      WRITE(3,8051)LAL,(ARCINT(LAL,J),J=1,2),I,LAL
   77 CONTINUE
C     THE ARRAY ARCINT HAS BEEN REDEFINED.  IT NOW CONTAINS ALL THE
C     APPLICABLE INTERSECTION POINTS IN INCREASING X VALUE.
C     CALCULATE THE WIDTH BETWEEN EACH INTERSECTION POINT.
   33 SLICX(1)=S1
      SLIC(1,1,1)=S2
      SLIC(1,1,2)=LINE1
C     DIVIDE WIDTH(L) INTO SLICES OF SIZE NOT GREATER THAN SWIDTH
C         SET SLICE ARRAR SIZE NOSLIC X NO OF LINES & X,Y COORDS
      WRITE(3,8057)
 8057 FORMAT(//,T5,'FIND SLICE WIDTH AND NO OF SLICES')
   98 N=1
      NOSLIC=1
      KM = 1
      K=LAL-1
      DO 45 L=1,K
      MM=1
      IF((ARCINT(L+1,1)-ARCINT(L,1)).LE.SWIDTH)GO TO 46
      DO 47 MM=1,100
      AM=MM
      WIDTH=(ARCINT(L+1,1)-ARCINT(L,1))/AM
      IF(WIDTH.LE.SWIDTH)GO TO 49
   47 CONTINUE
   46 WIDTH=ARCINT(L+1,1)-ARCINT(L,1)
   49 NOSLIC=NOSLIC+MM
C*******CONTROL FOR DIMENSION OF 75 SPACES IN SLICE ARRAY
      IF(NOSLIC.LT.DIMEN)GO TO 99
      SWIDTH=SWIDTH+0.5
      WRITE(3,9999)SWIDTH
 9999 FORMAT('0',T5,'***** MAXIMUM SLICE WIDTH HAS BEEN INCREMENTED TO',
     1F5.2,1X,A2)
      GO TO 98
   99 NSM1 = NOSLIC-1
      DO 51 I=N,NOLE
      IF(LNU(I).EQ.I)GO TO 51
  101 DO 52 JJ=KM,NSM1
      SLICX(JJ+1)=SLICX(JJ)+WIDTH
      SLIC(JJ+1,1,1) = SLIC(JJ,1,1)+WIDTH*SLOPE(I)
   52 SLIC(JJ+1,1,2)=I
      DIFF=SLICX(JJ+1)-C(I,5)
      IF(ABS(DIFF)-.010)50,50,48
   50 N = I+1
   48 KM = NOSLIC
      GO TO 45
   51 CONTINUE
   45 CONTINUE
C     COMPLETE SLICE ARRAY--CHECK INTERNAL SOIL LINES---SLICE ARRAY SORT
C  FOR DECREASING Y-COORDS.   *LINES NOT INTERSECTED GIVEN SAME COORDS
      NOLP1=NOL+1
      NOLEP1 = NOLE+1
      DO 60 I=1,NOSLIC
      N=2
      ARCY=CY-SQRT(R**2-(CX-SLICX(I))**2)
      DO 59  J = NOLEP1,NOL
      IF(LNU(J).EQ.J)GO TO 59
      SLIC(I,N,2)=J
      SLIC(I,N,1) = C(J,4) + (SLICX(I)    -C(J,3))*SLOPE(J)
      IF(SLICX(I).LT.(C(J,3)-SMLNO).OR.SLICX(I).GT.(C(J,5)+SMLNO))SLIC(I
     1,N,1) = -10.
      IF(SLIC(I,N,1).GT.(SLIC(I,1,1)+SMLNO).OR.SLIC(I,N,1).LT.(ARCY-
     ASMLNO))SLIC(I,N,1)=-10.0
   57 N = N+1
   59 CONTINUE
      SLIC(I,N,1)=ARCY
```

```
   60 SLIC(I,N,2) = NOLP1
C     SORT SLICE ARRAY IN DECREASING ORDER.
      IF(LIST.NE.0)WRITE(3,2116)
 2116 FORMAT(T6,'SLICE #',2X,'X-COORD',4(2X,'LINE NO',2X,'Y-COORD'))
      MCOUN=N
      N=MCOUN-1
      DO 81 KZ=1,NOSLIC
   84 NUM = 1
      DO 85  KY = 1,N
      IF(SLIC(KZ,KY,1).LE.-9.50)GO TO 82
      IF((SLIC(KZ,KY,1)+SMLNO).GE.SLIC(KZ,KY+1,1))GO TO 85
      SAVE=SLIC(KZ,KY,1)
      SLIC(KZ,KY,1)=SLIC(KZ,KY+1,1)
      SLIC(KZ,KY+1,1)=SAVE
      SAVE=SLIC(KZ,KY,2)
      SLIC(KZ,KY,2)=SLIC(KZ,KY+1,2)
      SLIC(KZ,KY+1,2)=SAVE
      GO TO 85
   82 SLIC(KZ,KY,1)=SLIC(KZ,KY-1,1)
      SLIC(KZ,KY,2)=SLIC(KZ,KY-1,2)
   85 IF(KY.NE.1.AND.SLIC(KZ,KY,1)-SMLNO.LE.SLIC(KZ,KY-1,1))NUM=NUM+1
      IF(NUM.NE.N)GO TO 84
   81 IF(LIST.NE.0)WRITE(3,3)KZ,SLICX(KZ),((SLIC(KZ,KY,2),SLIC(KZ,KY,1),
     1KY=1,MCOUN))
    3 FORMAT(T6,I5,9F9.2,/,T20,8F9.2,/,T20,8F9.2,/,T20,8F9.2)
      DO 306 I=1,NSM1
      SAREA=0.0
      WEIGHT=0.0
      EFWT = 0.
      ISOIL=0
      NN=MCOUN-1
      IF(LIST.NE.0)WRITE(3,350) I
  350 FORMAT(/,T5,'SLICE LINE NUMBER',I4)
      DO 303 J=1,NN
      DA=(SLIC(I,J,1)+SLIC(I+1,J,1)-SLIC(I,J+1,1)-SLIC(I+1,J+1,1))*
     A(SLICX(I+1)-SLICX(I))/2.0
      IF(DA.LE.SMLNO)GO TO 303
      DO 305 II=1,NOSP1
      IF(II.EQ.NOSP1)GO TO 308
      IF(ISOIL.EQ.II)GO TO 305
      N=NSLIN(II)
      ICOUNT=0
      JCOUNT=0
  311 DO 304 JJ=1,N
      IF(JCOUNT.EQ.2)GO TO 305
      INTL=SOIL(II,JJ,2)
      INTR=SOIL(II,JJ,3)
      IF(ICOUNT.EQ.1) GO TO 310
      IF(SLIC(I,J,2).NE.SOIL(II,JJ,1))GO TO 304
      ICOUNT=1
      JSOIL=II
      IF((SLICX(I)+SMLNO).GE.INTAR(INTL,1).AND.(SLICX(I)-SMLNO).LE.
     AINTAR(INTR,1))GO TO 310
      ICOUNT=0
      GO TO 304
  310 IF(SLIC(I+1,J,2).NE.SOIL(II,JJ,1))GO TO 304
      IF((SLICX(I+1)+SMLNO).LT.INTAR(INTL,1).OR.(SLICX(I+1)-SMLNO).GT.
     AINTAR(INTR,1))GO TO 304
      ICOUNT=2
      IF(JSOIL.NE.II)GO TO 305
  302 ISOIL=II
  308 SAREA=SAREA+DA
      GSUB = G(ISOIL)
      IF(SAT(ISOIL).GT.0.1)GSUB=G(ISOIL) - FU4
      EFWT = EFWT + DA*GSUB
      WEIGHT=WEIGHT+DA*G(ISOIL)
      IF(LIST.EQ.0)GO TO 9092
      WRITE(3,351)J,I,DA
  351 FORMAT(T10,'DSLICE NO',I2,' OF SLICE -',I3,' WITH DA OF',F10.3)
      WRITE(3,352)ISOIL,J,SAREA,WEIGHT,EFWT
  352 FORMAT(T15,'SOIL -',I3,' LIES IN DSLICE -',I3,/,T15,'TOTAL AREA ='
     A,F10.3,3X,'TOTAL WEIGHT =',G10.3,5X,'EFFECT. WT =',G10.3)
 9092 GO TO 303
  304 CONTINUE
      IF(ICOUNT.EQ.1)JCOUNT=JCOUNT+1
      IF(ICOUNT.EQ.1)GO TO 311
  305 CONTINUE
  303 CONTINUE
      ALPHA(I)=ARSIN(ABS(CX-((SLICX(I+1)-SLICX(I))/2.0+SLICX(I)))/R)
      AREA(I)=SAREA
      EFFWT(I) = EFWT
      WEIGH(I)=WEIGHT
      CO(I)=COHES(ISOIL)
  306 P(I)=PHI(ISOIL)
      IF(LIST.EQ.0)GO TO 365
      WRITE(3,354)
  354 FORMAT(T5,'SLICE #',3X,'AREA',3X,'WEIGHT',4X,'COHESION',3X,'PHI',
     A3X,'ALPHA')
      DO 307 I=1,NSM1
      IF(AREA(I).LE.SMLNO)GO TO 362
      WRITE(3,353)I,AREA(I),WEIGH(I),CO(I),P(I),ALPHA(I)
      GO TO 307
  362 WRITE(3,353)I,AREA(I),WEIGH(I)
  353 FORMAT(T7,I2,2X,3F10.3,F6.2,F9.4)
  307 CONTINUE
C     SUM OVERTURN. MOMENT AND COMPUTE PART OF RESIST. TERM
  365 DO 367  I = 1,NSM1
```

```
      IF(AREA(I).LE.SMLNO)P(I) = 0.00
      IF(AREA(I).LE.SMLNO)CO(I) = 0.00
  367 P(I) = P(I)/57.2958
      FI=1.0
  383 ZUM=0.0
      TF=0.0
      DO 382 K=1,NSM1
      CENTR=(SLICX(K+1)-SLICX(K))/2.0+SLICX(K)
      IF((CENTR-CX).LT.0.0)T=-WEIGH(K)*SIN(ALPHA(K))
      IF((CENTR-CX).EQ.0.0)T=0.0
      IF((CENTR-CX).GT.0.0)T=WEIGH(K)*SIN(ALPHA(K))
  394 TF=TF+T
      B(K) = (SLICX(K+1)-SLICX(K))/COS(ALPHA(K))
      Z =(CO(K)*B(K)+EFFWT(K)*TAN(P(K)))/(COS(ALPHA(K))+TAN(P(K))*SIN(AL
     1PHA(K))/FI)
  382 ZUM=ZUM+Z
      FO=ZUM/TF
      WRITE(3,116)FI,FO
  116 FORMAT(20X,'FI= ',F10.5,3X,'FO= ',F10.5)
      DEN=0.0
      ZUM=0.0
      DO 384 K=1,NSM1
      F=FO
      Z = (CO(K)*B(K)+EFFWT(K)*TAN(P(K)))/(F*COS(ALPHA(K))+TAN(P(K))*SIN
     1(ALPHA(K)))
      ZUM=ZUM+Z
      D = (CO(K)*B(K)+EFFWT(K)*TAN(P(K)))*TAN(P(K))*SIN(ALPHA(K))/(F*COS
     1(ALPHA(K)) + TAN(P(K))*SIN(ALPHA(K)))**2
  384 DEN=DEN+D
      FI= FO*(1.0-(TF-ZUM)/(TF-DEN))
  386 IF(ABS( FI-FO)-.001)385,385,383
  385 WRITE(3,108)NCOUN,FO
  108 FORMAT('0',15X,'THE SAFETY FACTOR FOR POINT ',I3,' IS ',F10.5)
      IF(PCODE.EQ.0.OR.R.GE.CY)GO TO 360
      IF(NCOUN.GT.1)GO TO 405
      XAXIS=12.
      YAXIS=9.
      PSCALE=SCALE
      DATA METER/'M'/
      IF(UT1.NE.METER)GO TO 405
      XAXIS=15.
      YAXIS=12.
      PSCALE=SCALE*2.
  405 IF(NCOUN.NE.1.AND.(NCOUN/NP*NP.NE.NCOUN).AND.NCOUN.NE.ITX*ITY)GO T
     1O 360
C   PLOT   PLOT   PLOT   PLOT   PLCT   PLCT   PLOT   PLOT   PLOT   PLOT
      CALL PLOT (5.0,-11.0,-3)
      CALL PLOT(0.0,0.5,-3)
      IF(UT1.EQ.METER)CALL FACTOR(.7874)
      CALL AXIS(0.0,0.0,'X-AXIS',-6,XAXIS,0.0,0.0,PSCALE)
      CALL AXIS(0.0,0.0,'Y-AXIS',6,YAXIS,90.0,0.0,PSCALE)
      CALL FACTOR(1.0/SCALE)
      DO 387 I=1,NOL
      CALL PLOT(C(I,3),C(I,4),3)
  387 CALL PLOT(C(I,5),C(I,6),2)
      DO 388 I=1,NOSLIC
      CALL PLOT(SLICX(I),SLIC(I,1,1),3)
  388 CALL PLOT(SLICX(I),SLIC(I,MCOUN,1),2)
      CALL PLOT(SLICX(1),SLIC(1,MCCUN,1),3)
      DO 389 I=1,NOSLIC
  389 CALL PLOT(SLICX(I),SLIC(I,MCOUN,1),2)
      CALL SYMBOL(CX,CY,1.4,3,0.0,-1)
      CALL DASHPT(ENTX,ENTY,1.0)
      CALL PLOT(CX,CY,3)
      CALL DASHPT(SLICX(1),SLIC(1,1,1),1.0)
      CALL FACTOR(1.0)
      CALL SYMBOL(7.0,10.0,0.14,23HSLOPE STABILITY PROGRAM,0.0,23)
      CALL SYMBOL(7.0,9.7,0.14,18HJ. BOWLES  CE DEPT,0.0,18)
      CALL SYMBOL(7.0,9.4,0.14,'SCALE; 1 ',0.0,9)
      CALL SYMBOL(999.,999.,0.14,UT2,C.0,2)
      CALL SYMBOL(999.,999.,0.14,' = ',C.C,3)
      CALL NUMBER(999.,999.,0.14,SCALE,0.0,2)
      CALL SYMBOL(999.,999.,0.14,1H ,0.0,1)
      CALL SYMBOL(999.,999.,0.14,LT1,C.C,2)
      CALL SYMBOL(0.5,10.0,.14,18HRESULTS FOR POINT ,0.0,18)
      CALL NUMBER(999.,999.,.14,PCCUN,0.0,-1)
      CALL SYMBOL(0.5,9.50,.14,23H  X CCORD. OF CENTER = ,0.0,23)
      CALL NUMBER(999.,999.,.14,CX,0.C,3)
      CALL SYMBOL(0.5,9.300,.14,23H  Y COORD. OF CENTER = ,0.0,23)
      CALL NUMBER(999.,999.,.14,CY,0.0,3)
      CALL SYMBOL(0.5,9.10,.14,23HX CCORD. OF ENT. PT. = ,0.0,23)
      CALL NUMBER(999.,999.,.14,ENTX,0.0,3)
      CALL SYMBOL(0.5,8.900,.14,23HY CCCRD. OF ENT. PT. = ,0.0,23)
      CALL NUMBER(999.,999.,.14,ENTY,C.C,3)
      CALL SYMBOL(0.5,8.70,.14,23HX CCORD. OF EXIT PT. = ,0.0,23)
      CALL NUMBER(999.,999.,.14,SLICX(1),0.0,3)
      CALL SYMBOL(0.5,8.500,.14,23HY CCCRD. OF EXIT PT. = ,0.0,23)
      CALL NUMBER(999.,999.,.14,SLIC(1,1,1),0.0,3)
      CALL SYMBCL(0.5,8.30,.14,23H       RADIUS OF ARC = ,0.0,23)
      CALL NUMBER(999.,999.,.14,R,0.0,3)
      CALL SYMBOL(0.5,8.00,.14,23H   THE SAFETY FACTOR = ,0.0,23)
      CALL NUMBER(999.,999.,.14,FC,0.0,5)
      CALL PLOT(15.0,0.0,-3)
C   PLOT   PLOT   PLOT   PLOT   PLOT   PLOT   PLOT   PLOT   PLOT   PLOT
  360 IF(IX.EQ.ITX)CX=CX-(IX-1)*DELX
      GO TO 4000
  150 IF(PCODE.NE.0)CALL PLOT(5.0,0.0,999)
      END
```

PROBLEMS

14-1 Verify the output of Example 14-1. Use ITX = ITY = 4, DELX = DELY = 10. and search for a lower SF.

14-2 Repeat Prob. 14-1 for Example 14-2 using DELX = DELY = 5.0.

14-3 Change the entrance coordinates to any point on the slope of Example 14-1 and see if a lower SF can be found. Use same initial center coordinates as in Example 14-1.

14-4 Repeat Prob. 14-3 for Example 14-2.

REFERENCES

BISHOP, A. W. (1955): The Use of the Slip Circle in the Stability Analysis of Slopes, *Geotech. (Lond.)*, vol. 5, no. 1, March, pp. 7–17.

———, and N. R. MORGENSTERN (1960): Stability Coefficients for Earth Slopes, *Geotech. (Lond.)*, vol. 10, no. 4, December, pp. 129–150.

FANG, H. Y., and T. J. HIRST (1970): Application of Plasticity Theory to Slope Stability Problems, *Highw. Res. Rec.* 323, HRB, Washington, pp. 26–38.

JANBU, N. (1957): Earth Pressures and Bearing Capacity Calculations by Generalized Procedure of Slices, *Proc. 4th Int. Conf. Soil Mech. Found. Eng., Lond.*, vol. 2, pp. 207–212.

LITTLE, A. L., and V. E. PRICE (1958): The Use of an Electronic Computer for Slope Stability Analysis, *Geotech. (Lond.)*, vol. 8, pp. 113–120.

LOWE, JOHN, III (1967): Stability Analysis of Embankments, *J. Soil Mech. Found. Div., ASCE*, vol. 93, SM4, July, pp. 1–33.

MORGENSTERN, N. R., and V. E. PRICE (1965): The Analysis of the Stability of General Slip Surfaces, *Geotech. (Lond.)*, vol. 15, no. 1, March, pp. 79–93.

PETERSON, K. E. (1955): The Early History of Circular Sliding Surfaces, *Geotech. (Lond.)*, vol. 5, no. 4, December, pp. 275–296 (see also Obituary, ibid., December 1957, pp. 198–200).

SPENCER, E. (1967): A Method of Analysis of the Stability of Embankments Assuming Parallel Inter-Slice Forces, *Geotech. (Lond.)*, vol. 17, no. 1, March, pp. 11–26.

TAYLOR, D. W. (1948): "Fundamentals of Soil Mechanics," chap. 16, Wiley, New York.

TERZAGHI, K. (1943): "Theoretical Soil Mechanics," Wiley, New York.

U.S. CORPS OF ENGINEERS (1960): Stability of Earth and Rockfall Dams, *Eng. Des. Man.* EM1110-2-1902, Washington (including change 1 of 1962).

WHITMAN, R. V., and W. A. BAILEY (1967): Use of Computers for Slope Stability Analysis, *J. Soil Mech. Found. Div., ASCE*, vol. 93, SM4, July, pp. 475–498.

APPENDIX A

DATA ON PILE-DRIVING EQUIPMENT

Table A-1 REPRESENTATIVE PILE-HAMMER (PILE-DRIVER) DATA[a]

Manufacturer	Total weight of hammer, lb	Weight of ram, lb	Length of stroke, in	Strokes per minute	Energy per blow (manufacturer's rating), ft-lb
			Single-acting hammers		
McKiernan-Terry Corp.					
S5	12,460	5,000	39	60	16,250
S8	18,300	8,000	39	55	26,000
S10	22,380	10,000	39	55	32,500
S14	31,700	14,000	32	60	37,500
S20	38,650	20,000	36	60	60,000
Vulcan Iron Works[b]					
2	6,700	3,000	29	70	7,260
1	9,700	5,000	36	60	15,000
06	11,200	6,500	36	60	19,500
08	16,750	8,000	39	50	26,000
010	18,750	10,000	39	50	32,500
014	27,500	14,000	36	60	42,000
016	30,250	16,250	36	60	48,750
020	41,670	20,000	36	60	60,000
Raymond Concrete Pile Division, Raymond International, Inc.					
1	11,000	5,000	36	60	15,000
1S	12,500	6,500	36	58	19,500
0	16,100	7,500	39	52	24,375
2/0	18,550	10,000	39	50	32,500
3/0	21,225	12,500	39	48	40,625
4/0	23,800	15,000	39	46	48,758
5/0	26,450	17,500	39	44	56,875
22X	31,750	22,050	31	58	56,900
30X	52,000	30,000	30	70	75,000
8/0	34,000	25,000	39	40	81,250
40X	62,000	40,000	39	64	100,000

[a] Consult manufacturer's catalogs for additional data and models later than 1973.
[b] All with standard bases.

Table A-1 REPRESENTATIVE PILE-HAMMER (PILE-DRIVER) DATA (*Continued*)

Manufacturer	Total weight of hammer, lb	Weight of ram, lb	Length of stroke, in	Strokes per minute	Energy per blow (manufacturer's rating), ft-lb
Single-acting hammers					
British Steel Piling Co., Ltd.					
4b	4,595	3,360	54[c]		[d]
5b	5,820	4,480	54		
6b	7,060	5,600	54		
7b	7,840	6,720	54		
9b	10,080	8,960	54		
9c	10,870	8,960	72		
10b	13,100	11,200	54		
10c	13,220	11,200	72		
12b	15,456	13,440	54		
12c	15,904	13,440	72		
Double-acting hammers					
McKiernan-Terry Corp.					
No. 6	2,900	400	$8\frac{3}{4}$	275	2,500
No. 7	5,000	800	$9\frac{1}{2}$	225	4,150
9B3	7,000	1,600	17	145	8,750
10B3	10,850	3,000	19	105	13,100
11B3	14,000	5,000	19	95	19,150
C5	11,880	5,000	18	100–110	16,000
C826	17,750	8,000	18	85–95	24,000
Vulcan Iron Works, Inc.					
18C	4,139	1,800	10.5	150	3,600
30C	7,036	3,000	12.5	133	7,260
50C	11,782	5,000	15.5	120	15,100
65C	14,886	6,500	15.5	117	19,200
80C	17,885	8,000	16.5	111	24,450
140C	27,984	14,000	15.5	103	36,000
200C	39,050	20,000	15.5	98	50,200
400C	83,000	40,000	16.5	100	113,488
Union Iron Works of New Jersey, Inc.					
00	21,000	6,000	36	85	54,900
0A	17,000	5,000	21	90	22,050
1	10,500	1,850	21	130	13,100
1A	10,500	1,600	18	120	10,020
$1\frac{1}{2}$A	9,200	1,500	18	125	8,680
2	6,600	1,025	16	145	5,755
3	5,200	820	$13\frac{1}{2}$	150	4,390
3A	4,700	700	14	160	3,660
4	2,800	370	12	200	2,100
5	1,625	210	9	250	1,010

[c] Maximum strokes given; operator can control stroke.
[d] Not rated.

Table A-1 REPRESENTATIVE PILE-HAMMER (PILE-DRIVER) DATA (*Continued*)

Manufacturer	Total weight of hammer, lb	Weight of ram, lb	Length of stroke, in	Strokes per minute	Energy per blow (manufacturer's rating), ft-lb	
Double-acting hammers						
Raymond Concrete Pile Division, Raymond International, Inc.						
65C	14,675	6,500	16	100	19,500	
65CH	14,615	6,500	16	130	19,500	
80CH	17,782	8,000	16½	130	24,450	
150C	32,500	15,000	18	100	48,750	
Diesel hammers						
McKiernan-Terry Corp.						
DE10	3,100	1,100	108[e]	48	6,600[f]	9,900[g]
DE20	5,375	2,000	113	48	12,000	18,800
DE30	8,125	2,800	129	48	16,800	30,100
DE40	9,825	4,000	129	48	24,000	43,000
IDH-J22[h]	10,800	4,850	120[i]	48	39,100	48,500
Link-Belt						
105	3,885	1,445	35.23	90–98	6,500	
180	4,550	1,725	37.60	90–95	8,100	
312	10,375	3,857	30.89	100–105	15,000	
520	12,545	5,070	43.17	80–84	26,300	
Delmag Maschinenfabrik (The Foundation Equipment Corp.)						
D5	2,401	1,100	[j]	42–60	9,100	
D12	5,440	2,750		42–60	22,600	
D22	10,054	4,850		42–60	39,800	
D30	12,320	6,600		39–60	23,870–54,200	
D44	19,842	8,819		40–60	72,300	

[e] Maximum stroke; stroke increases with increased driving resistance.
[f] Based on 6-ft stroke.
[g] Maximum striking energy.
[h] Manufactured by Ishikawajima-Harima Heavy Industries, Tokyo, Japan (McKiernan-Terry is United States distributor).
[i] Maximum.
[j] Not given.

Table A-2 WEIGHT OF DRIVE-CAP PACKAGE*

Hammer stock no.	For model	Helmet nominal size, in	Helmet guide insert no.† Slot F	Slot D	Hammer jaw no.	Drive-cap weight, lb
H-51	D-5	12 × 12	101732	(1755)	30290	991
H-52		14 × 14	101732		30290	1,150
H-53		16 × 16		(1740)	30290	1,630
H-57		12 × 12	101748	(2425)	33901	991
H-58		14 × 14	101748		33901	1,200
H-59		16 × 16	101732	(1746)	33901	1,600
H-54		18 × 18	101732		33901	1,820
H-121‡	D-12	12 × 12	101732	(1755)	5122	991
H-122		14 × 14	101732		5122	1,150
H-123		16 × 16		(1740)	5122	1,630
H-128		12 × 12	101748	(2425)	6502	991
H-129		14 × 14	101748		6502	1,200
H-120		16 × 16	101732	(1746)	6502	1,600
H-124		18 × 18	101732		6502	1,820
H-221	D-22 or	12 × 12	101748	(2425)	1787	991
H-222	D-30	14 × 14	101748		1787	1,270
H-223‡		16 × 16	101732	(1746)	1787	1,670
H-224		18 × 18	101732		1787	1,820
H-227		16 × 16	101748		32348	1,680
H-228		18 × 18	101748		32348	1,870
H-229		22 × 22	101732		32348	2,340
H-226		24 × 24	101732		32348	2,500

SOURCE: The Foundation Equipment Corp.

* The drive-cap package consists of helmet, helmet guide, two each split pins, wood cushion block, and anvil plate. The drive-cap package for U leads includes helmet-guide channel in lieu of helmet-guide insert.

† Slots F and D refer to designation impressed into helmet casting to differentiate between the slots for helmet inserts.

‡ Normally ordered with this model.

Table A-3 TYPICAL WEIGHTS OF ANVIL BLOCKS (PILE CAPS)

Hammer no.	Weight of anvil block, lb	Pile type
S5	1,575	H and small sheet
	1,660	Z-sheet
S8	1,780	H and small sheet
	1,765	Z-sheet
	2,350	18–24-in pipe
	2,890	18–24-in pipe
S14, S20	3,415	12–14-in H
	4,500	20–36-in pipe
9B3	1,360	All H and sheet
	1,130	All Z-sheet
	1,060	14–18-in pipe
10B3	1,765	All H
	1,710	14–18-in pipe
11B3	1,550	All H
	2,110	12–24-in pipe
C5, C826	1,340	All H
	1,470	12–20-in pipe
	1,258	Z-sheet

SOURCE: McKiernan-Terry Corp.

APPENDIX B

DESIGN DATA FOR H-PILES, PRECAST-CONCRETE PILES, AND STEEL SHEET PILING (REPRESENTATIVE)

Table B-1 U.S. STANDARD STEEL H-PILES WITH STANDARD AISI DESIGNATION*

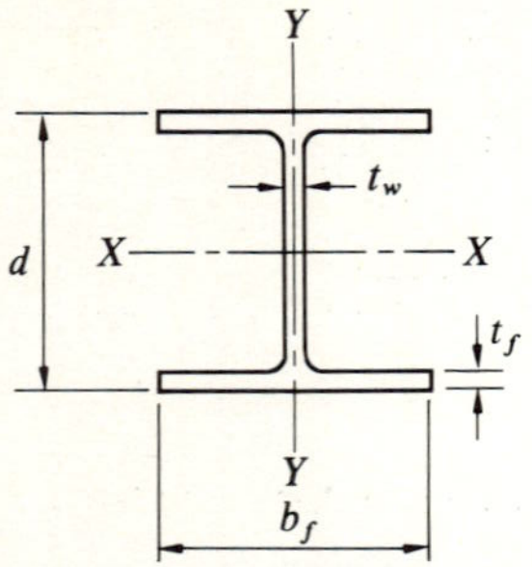

Designation	Area A, sq in	Depth d, in	Flange, in: Width b_f	Flange, in: Thickness t_f	Section properties: Web thickness t_w, in	Axis XX: I, in^4	Axis XX: S, in^3	Axis XX: r, in	Axis YY: I, in^4	Axis YY: S, in^3	Axis YY: r, in	Nominal weight per ft, lb
HP14 × 117	34.4	14.23	14.885	0.805	0.805	1,230	173	5.97	443	59.5	3.59	117
HP14 × 102	30.0	14.03	14.784	0.704	0.704	1,050	150	5.93	380	51.3	3.56	102
HP14 × 89	26.2	13.86	14.696	0.616	0.616	910	131	5.89	326	44.4	3.53	89
HP14 × 73	21.5	13.64	14.586	0.506	0.506	734	108	5.85	262	35.9	3.49	73
HP12 × 74	21.8	12.12	12.217	0.607	0.607	566	93.4	5.10	185	30.2	2.91	74
HP12 × 53	15.6	11.78	12.046	0.436	0.436	394	66.9	5.03	127	21.1	2.86	53
HP10 × 57	16.8	10.01	10.224	0.564	0.564	295	58.8	4.19	101	19.7	2.45	57
HP10 × 42	12.4	9.72	10.078	0.418	0.418	211	43.4	4.13	71.4	14.2	2.40	42
HP8 × 36	10.6	8.03	8.158	0.446	0.446	120	29.9	3.36	40.4	9.91	1.95	36

* These pile dimensions represent standard British steel H-pile sections. Normal material specifications: ASTM A36, ASTM A572 grades 42 through 60 (HP 14 × 117 not available in grade 60).

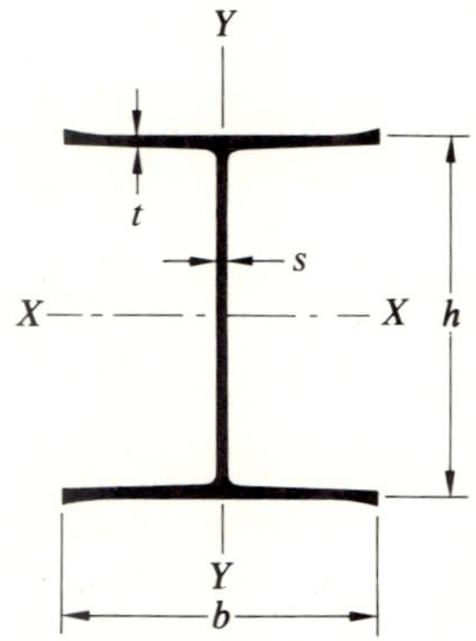

Table B-2 REPRESENTATIVE EUROPEAN (GERMAN) H-PILES

Profilbezeichnung (section)	*Abmessungen* (dimensions), mm				*Gewicht* (weight), kg/m	*Umfang* (perimeter), cm		*Querschnitt* (area), sq cm		*Trägheitsmoment* (moment of inertia), cm^4		*Widerstandsmoment* (section modulus), cm^3		*Kleinster Trägheitshalbmesser* (minimum radius of gyration) i_{min}, cm
	h	b	s	t		*Abwicklung* (metal)	*Umriss* (outline)	*Stahl* (steel)	*Umriss* (outline) $b \times h$	J_x	J_y	W_x	W_y	
300S	304	322	9	13.0	96	190	134	122	980	22,410	8,720	1,470	545	8.5
350L	350	382	9	12.5	111	224	154	142	1,340	35,040	14,290	2,000	748	10.0
350S	354	383	10	14.5	126	226	156	161	1,360	40,350	16,380	2,280	855	10.0
400L	400	378	10	13.5	122	232	164	155	1,510	48,750	14,410	2,440	762	9.6
400S	404	380	12	15.5	140	234	166	179	1,540	56,290	16,650	2,790	876	9.6
500L	500	379	11	14.5	140	252	184	179	1,900	84,230	14,940	3,370	788	9.2
500S	504	380	12	16.5	156	254	186	198	1,920	95,970	16,970	3,780	893	9.3
600L	600	380	12	14.5	153	272	204	195	2,280	127,500	15,140	4,250	797	8.8
600S	604	380	12	16.5	165	274	206	210	2,300	141,400	16,970	4,680	893	9.0
700S	700	380	14	17.7	192	292	224	245	2,660	211,320	17,410	6,040	917	8.4
900S	900	322	14.5	21.3	215	304	251	273	2,900	363,800	11,360	8,080	705	6.5

SOURCE: Stahlwerke Peine-Salzgitter AG.

Table B-3 STANDARD U.S. ROLLED-STEEL PIPE PILES

Size OD, in	Wall thickness, in	Weight per foot, lb	Moment of inertia, in^4	Radius of gyration, in	Section modulus, cu in	Area of external surface, sq ft/lin ft	Area of steel in cross section, sq in	Inside cross-sectional area, sq/in	Concrete per foot, cu yd	Approx. collapsing pressure, psi
10	0.115	12.14	43.54	3.495	8.70	2.61	3.571	74.96	0.0192	104
	0.172	18.05	63.97	3.475	12.79	2.61	5.306	73.23	0.0188	338
	0.179	18.77	66.59	3.472	13.31	2.61	5.531	73.00	0.0187	382
	0.188	19.70	69.46	3.469	13.89	2.61	5.780	72.75	0.0187	435
	0.250	26.03	90.89	3.448	18.17	2.61	7.657	70.88	0.0182	1,005
$10\frac{3}{4}$	0.115	13.06	54.23	3.760	10.08	2.81	3.842	86.92	0.0223	87
	0.172	19.42	79.76	3.740	14.84	2.81	5.711	85.05	0.0218	273
	0.179	20.24	83.04	3.737	15.44	2.81	5.954	84.80	0.0218	309
	0.188	21.15	86.64	3.734	16.11	2.81	6.221	84.54	0.0217	352
	0.250	28.04	113.51	3.713	21.11	2.81	8.246	82.51	0.0212	814
12	0.155	14.59	75.68	4.202	12.61	3.14	4.293	108.80	0.0279	68
	0.172	21.71	111.51	4.182	18.58	3.14	6.386	106.71	0.0274	198
	0.179	22.60	116.11	4.179	19.35	3.14	6.658	106.43	0.0273	224
	0.188	23.72	121.17	4.176	20.19	3.14	6.958	106.13	0.0272	256
	0.250	31.37	159.05	4.155	26.50	3.14	9.228	103.86	0.0267	591
$12\frac{3}{4}$	0.115	15.51	90.93	4.467	14.26	3.33	4.564	123.11	0.0316	60
	0.172	23.09	134.09	4.447	21.03	3.33	6.791	120.88	0.0310	166
	0.179	24.07	139.64	4.444	21.90	3.33	7.080	120.59	0.0310	188
	0.188	25.16	145.75	4.442	22.86	3.33	7.399	120.27	0.0309	214
	0.250	33.38	191.48	4.420	30.03	3.33	9.817	117.85	0.0303	496
14	0.115	17.05	120.68	4.909	17.24	3.66	5.016	148.92	0.0383	48
	0.172	25.38	178.17	4.889	25.45	3.66	7.466	146.47	0.0376	126
	0.179	26.42	185.57	4.886	26.51	3.66	7.785	146.15	0.0375	143
	0.188	27.66	193.72	4.883	27.67	3.66	8.136	145.80	0.0375	163
	0.250	36.71	254.84	4.862	36.40	3.66	10.799	143.13	0.0368	377

16	0.115	19.50	180.70	5.616	22.58	4.18	5.738	195.32	0.0502	36
	0.125	21.19	196.04	5.612	24.50	4.18	6.234	194.82	0.0501	43
	0.134	22.79	210.57	5.609	26.32	4.18	6.703	194.35	0.0499	51
	0.141	23.82	219.90	5.607	27.48	4.18	7.006	194.05	0.0499	56
	0.156	26.40	243.24	5.601	30.40	4.18	7.764	193.29	0.0497	71
	0.164	27.74	255.94	5.599	31.99	4.18	8.178	192.88	0.0496	79
	0.172	29.06	267.20	5.596	33.40	4.18	8.546	192.51	0.0495	88
	0.179	30.30	278.35	5.593	34.79	4.18	8.911	192.15	0.0494	96
	0.188	31.66	290.63	5.590	36.32	4.18	9.314	191.74	0.0493	110
	0.203	34.25	313.74	5.585	39.21	4.18	10.074	190.98	0.0491	139
	0.219	36.87	337.08	5.580	42.13	4.18	10.845	190.21	0.0489	173
	0.230	38.70	353.67	5.576	44.20	4.18	11.394	189.66	0.0487	200
	0.250	42.05	382.98	5.569	47.87	4.18	12.370	188.69	0.0485	256
	0.281	47.22	428.32	5.558	53.54	4.18	13.888	187.17	0.0481	360
	0.312	52.36	473.11	5.547	59.13	4.18	15.401	185.66	0.0477	490
	0.344	57.48	517.37	5.536	64.67	4.18	16.907	184.15	0.0473	667
	0.375	62.48	562.08	5.526	70.26	4.18	18.408	182.65	0.0470	830

SOURCE: ARMCO Steel Corporation, Metal Products Division. All calculations based on wall thicknesses per manufacturer's standard gage tables. Sizes listed here are standard.

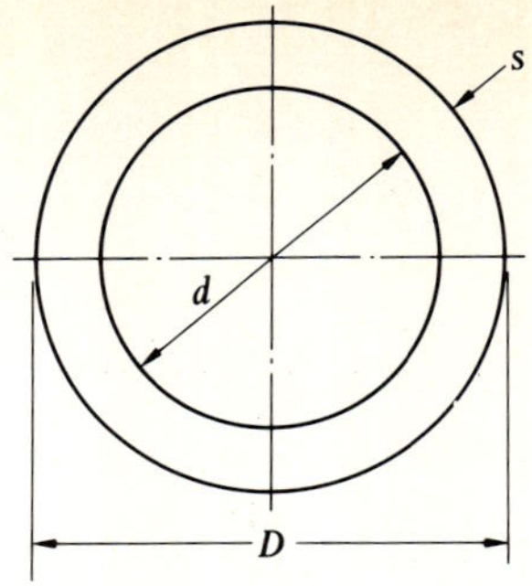

Table B-4 SPIRAL WELDED STEEL PIPE PILES (GERMAN)

Aussendurchmesser (outside diameter) *D*		*Wanddicke* (wall thickness) *s*, mm	*Gewicht* (weight) *G*, kg/m	*Querschnitt* (area)		*Umfang* (perimeter) *U*, m	*Trägheitsmoment* (moment of inertia) J_x, cm^4	*Widerstandsmoment* (section modulus) W_x, cu cm	*Trägheitshalbmesser* (radius of gyration) i_x, cm
mm	in			*Stahl* (steel) *f*, sq cm	*Rohr* (pipe) *F*, sq m				
323.8	$12\frac{3}{4}$	10.0	77.4	99	0.082	1.02	12,150	750	11.1
		12.7	97.4	124			15,040	929	11.0
355.6	14	10.0	85.2	109	0.099	1.12	16,220	912	12.2
		12.7	107.4	137			20,140	1,130	12.1
406.4	16	10.0	97,8	125	0.130	1.28	24,480	1,210	14.0
		12.7	123.3	157			30,470	1,500	13.9
457.2	18	10.0	110.3	140	0.164	1.44	35,140	1,540	15.8
		12.7	139.2	177			43,840	1,920	15.7
508.0	20	10.0	122.8	156	0.203	1.60	48,520	1,910	17.6
		12.7	155.1	198			60,640	2,390	17.5

609.6	24	10.0	147.9	188	0.292	1.92	84,680	2,790	21.2
		12.7	187.0	238			106,110	3,480	21.1
762.0	30	10	185.4	236	0.456	2.39	167,030	4,380	26.6
		13	240.1	306			214,580	5,630	26.5
		16	294.4	375			260,990	6,890	26.4
914.4	36	10	223.0	284	0.657	2.87	290,530	6,360	32.0
		13	289.0	368			373,980	8,180	31.9
		16	354.5	452			455,770	9,970	31.8
1,066.8	42	10	260.6	332	0.894	3.35	463,530	8,690	37.4
		13	337.9	430			597,510	11,200	37.3
		17	440.1	561			772,580	14,480	37.1
1,168.4	46	10	285.7	364	1.072	3.67	610,480	10,450	41.0
		14	398.6	508			845,900	14,480	40.8
		18	510.7	651			1,076,430	18,430	40.7
1,219.2	48	10	298.2	380	1.167	3.83	694,360	11,390	42.8
		14	416.1	530			962,550	15,790	42.6
		18	533.2	679			1,225,390	20,100	42.5
1,320.8	52	12	387.3	493	1.370	4.15	1,056,620	16,000	46.3
		15	483.0	615			1,311,780	19,860	46.2
		18	578.3	737			1,563,410	23,670	46.1
1,524.0	60	12	447.5	570	1.824	4.79	1,629,100	21,380	53.5
		15	558.2	711			2,024,350	26,570	53.4
		18	668.5	852			2,414,860	31,690	53.3
1,625.6	64	12	477.5	608	2.075	5.11	1,980,050	24,360	57.1
		15	595.8	759			2,461,360	30,280	56.9
		18	713.6	909			2,937,270	36,140	56.8

SOURCE: Stahlwerke Peine-Salzgitter AG.

Table B-5 UNION MONOTUBE PILES

Type	Length, ft	Diameter, in		Theoretical weight of steel, lb					Est. concrete vol, cu yd
		Point	Top	11 gage	9 gage	7 gage	5 gage	3 gage	
F, taper, 0.14 in/ft	25	8½	12	338	421	502	591	711	0.43
	30	8	12	388	484	579	681	820	0.55
	40	8½	14	595	748	900	1,059	1,275	0.95
	60	8	16		1,213	1,465	1,733	2,093	1.68
	75	8	18			1,962	2,312	2,792	2.59
J, taper, 0.25 in/ft	17	8	12	225	279	332	390	468	0.32
	25	8	14	364	457	549	645	777	0.58
	33	8	16		653	786	924	1,112	0.95
	40	8	18			1,038	1,221	1,469	1.37
Y, taper, 0.40 in/ft	10	8	12	139	171	202	239	285	0.18
	15	8½	14	229	288	345	404	484	0.34
	20	8½	16		412	494	579	696	0.56
	25	8½	18			663	778	934	0.86
N12	20	12	12	317	398	478	558	668	0.51
	25			394	495	593	694	831	0.64
	30			471	589	708	829	993	0.77
	35			547	687	825	967	1,158	0.89
	40			625	784	942	1,100	1,317	1.02
N14	20	14	14	392	490	587	689	823	0.70
	25			485	606	731	858	1,025	0.87
	30			581	727	877	1,023	1,222	1.05
	35			679	849	1,018	1,194	1,427	1.22
	40			773	967	1,166	1,368	1,634	1.40
N16	20	16	16		555	666	781	933	0.90
	25				687	829	971	1,161	1.13
	30				824	988	1,158	1,384	1.35
	35				957	1,153	1,352	1,615	1.58
	40				1,095	1,320	1,539	1,840	1.80
N18	20	18	18			755	880	1,052	1.16
	25					934	1,095	1,308	1.45
	30					1,119	1,311	1,566	1.75
	35					1,305	1,522	1,819	2.04
	40					1,486	1,741	2,081	2.33

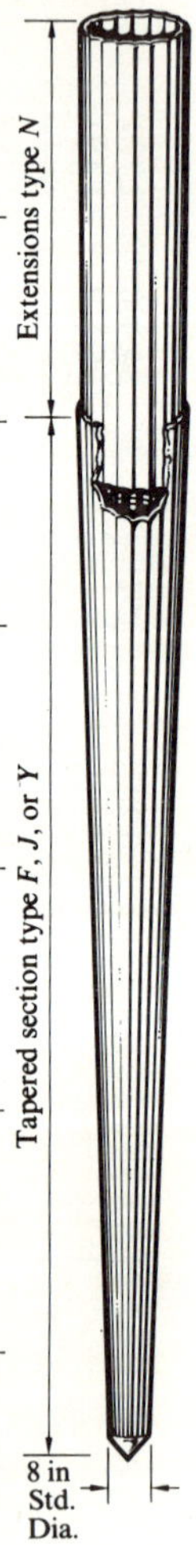

SOURCE: The Union Metal Manufacturing Co., Canton, Ohio.

Table B-6*a* RAYMOND STEP-TAPER CORES AND STEP-TAPER SHELLS

Step-taper cores				Raymond step-taper shells					
						Nominal surface area,[b] sq ft			
Section no.	Nominal OD, in	Average diam, in	Approximate weight, lb/lin ft	Section no.	OD,[a] in	8-ft steps[c]	12-ft steps[c]	Approximate weight, lb/lin ft[d]	Nominal cross-sectional area, sq in[e]
000	8	$7\frac{7}{16}$	75	000	$8\frac{3}{8}$	18	27	8.5	58
00	9	$8\frac{5}{16}$	90	00	$9\frac{1}{2}$	20	30	9	71
0	10	$9\frac{3}{16}$	105	0	$10\frac{3}{8}$	22	33	10	85
1	11	10	120–200	1	$11\frac{1}{4}$	24	36	11	100
2	12	$11\frac{1}{8}$	120–220	2	$12\frac{1}{4}$	26	39	12	118
3	13	$12\frac{1}{8}$	125–200	3	$13\frac{1}{4}$	28	42	13	138
4	14	$13\frac{1}{8}$	135–165	4	$14\frac{1}{4}$	30	45	14	160
5	15	$14\frac{1}{8}$	150–165	5	$15\frac{1}{4}$	32	48	15	182
6	16	$15\frac{1}{8}$	160	6	$16\frac{1}{4}$	34	51	16	208
7	17	16	200	7	$17\frac{1}{4}$	36	54	17	234

[a] Diameters shown are at bottom of shell section. Top diameter of shell section is $\frac{1}{8}$ in larger.

[b] Circumscribed area only; does not include full contact surface of corrugated section or projected end area due to taper.

[c] Sections are available also in lengths of 4, 16, and 24 ft; check with manufacturer.

[d] Gages range from no. 12 to 20. Manufacturer assumes responsibility for selection of proper gages.

[e] Total area.

SOURCE: Raymond Concrete Pile Division, Raymond International, Inc.

Table B-6*b* **RAYMOND STEP-TAPER PILE AND STEP-TAPER CORE ASSEMBLIES (FOR DRIVING)***

Step-taper core and shell makeup

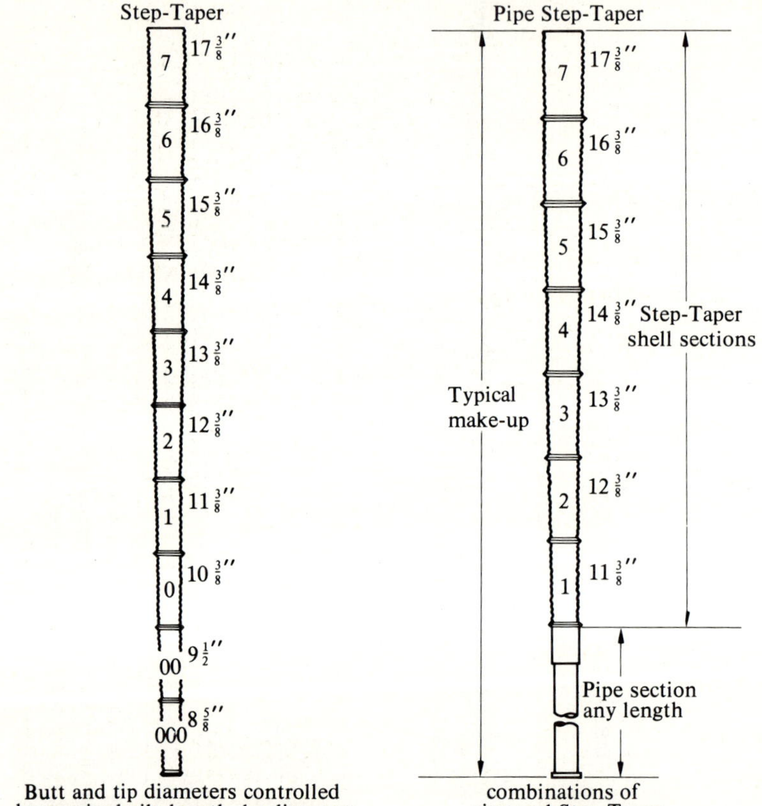

Butt and tip diameters controlled by required pile length, loading conditions and shell section lengths.
Sections made of various gages and in lengths of 4. 8 12 and 16 feet.

combinations of pipe and Step-Taper shell sizes available.

* Also shown is the standard uniform taper pile available in 8-in size (shown) and 10.8 in.
† Weights include head, point, and pins but not shell.

Weights† of step-taper core

Point size	Top size	8-ft steps Length, ft	8-ft steps Weight, lb	12-ft steps Length, ft	12-ft steps Weight, lb	16-ft steps Length, ft	16-ft steps Weight, lb
000	3	48	10,500	72	14,500	96	18,000
	4	56	12,000	84	16,500	112	21,000
	5	64	13,500	96	19,000	128	24,500
	6	72	15,000	108	22,000	144	28,500
	7	80	18,500	120	26,000	160	34,000
00	3	40	9,500	60	13,000	80	16,000
	4	48	11,000	72	15,000	96	19,000
	5	56	12,500	84	17,500	112	22,500
	6	64	14,000	96	20,000	128	26,500
	7	72	17,500	108	24,500	144	32,000
0	3	32	8,500	48	11,000	64	14,000
	4	40	10,000	60	13,500	80	17,000
	5	48	11,500	72	15,500	96	20,000
	6	56	13,000	84	18,500	112	24,000
	7	64	16,500	96	23,000	128	29,500
1	3	24	7,000	36	9,500	48	11,500
	4	32	9,000	48	11,500	64	4,500
	5	40	10,500	60	14,000	80	17,500
	6	48	12,000	72	17,000	96	21,500
	7	56	15,500	84	21,000	112	27,000
2	3	16	5,500	24	7,000	32	8,500
	4	24	7,000	36	9,500	48	11,500
	5	32	8,500	48	11,500	64	14,500
	6	40	10,500	60	14,500	80	18,500
	7	48	14,000	72	19,000	96	24,000
3	3	8	4,000	12	5,000	16	5,000
	4	16	5,500	24	7,000	32	8,500
	5	24	7,000	36	9,500	48	11,500
	6	32	8,500	48	12,500	64	15,500
	7	40	12,000	60	16,500	80	21,000

Table B-7 PRESTRESSED CONCRETE PIPE PILES

Size				
OD, in	ID, in	Wall thickness, in	Area, sq in	I, in^4
24	16	4	251	13,070
34	26	4	377	43,170
36‡	28	4	402	52,280
	27	4½	445	56,360
	26	5	487	60,000
48	39	4½	615	147,000
	38	5	675	158,200
	36	6	792	178,100
54‡	45	4½	700	216,100
	44	5	770	233,400
	42	6	904	264,600
66	56	5	958	448,700
	54	6	1,131	514,000
72	62	5	1,052	593,800
	60	6	1,244	683,000
78	67	5½	1,253	827,800
	65	6½	1,460	940,700
84	72	6	1,470	1,124,700
	70	7	1,693	1,265,300
90	78	6	1,583	1,403,600
	76	7	1,825	1,582,900

SOURCE: Raymond Concrete Pile Division, Raymond International, Inc.

* Unit weight of concrete = 150 lb/cu ft.

† Cable = 12 each 0.192-in-diam wire stress-relieved strands (140,000 lb/sq in stress).

‡ Standard sizes; other sizes available only in quantity.

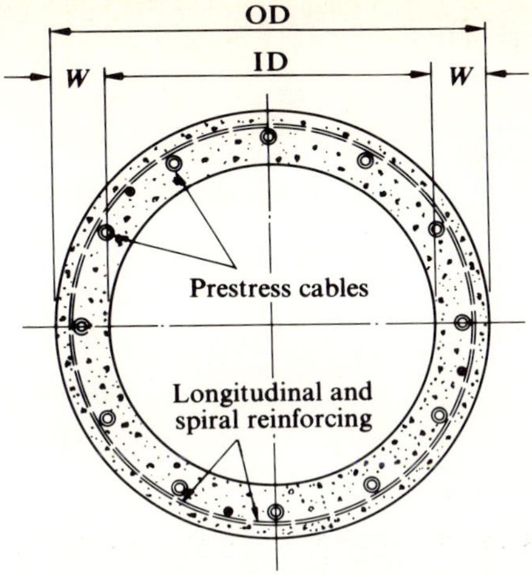

S, in³	Circumference, in	Point area, sq ft	Weight per foot, lb*	Concrete design stress, per cable, lb/sq in†
1,089	6.28	3.14	261	193.8
2,540	8.90	6.30	393	129.0
2,900	9.43	7.07	419	121.0
3,130	9.43	7.07	464	109.3
3,330	9.43	7.07	507	99.9
6,130	12.57	12.57	641	79.1
6,590	12.57	12.57	703	72.0
7,420	12.57	12.57	825	61.4
8,000	14.14	15.90	729	69.5
8,640	14.14	15.90	802	63.2
9,800	14.14	15.90	942	53.7
13,600	17.28	23.76	998	50.8
15,580	17.28	23.76	1,178	43.0
16,500	18.85	28.27	1,096	46.2
18,970	18.85	28.27	1,296	39.1
21,230	20.42	33.18	1,305	38.8
24,120	20.42	33.18	1,521	33.3
26,780	21.99	38.48	1,531	33.1
30,130	21.99	38.48	1,764	28.7
31,190	23.56	44.18	1,649	30.7
35,180	23.56	44.18	1,901	26.6

Table B-8 PRECAST-CONCRETE PIPE PILES

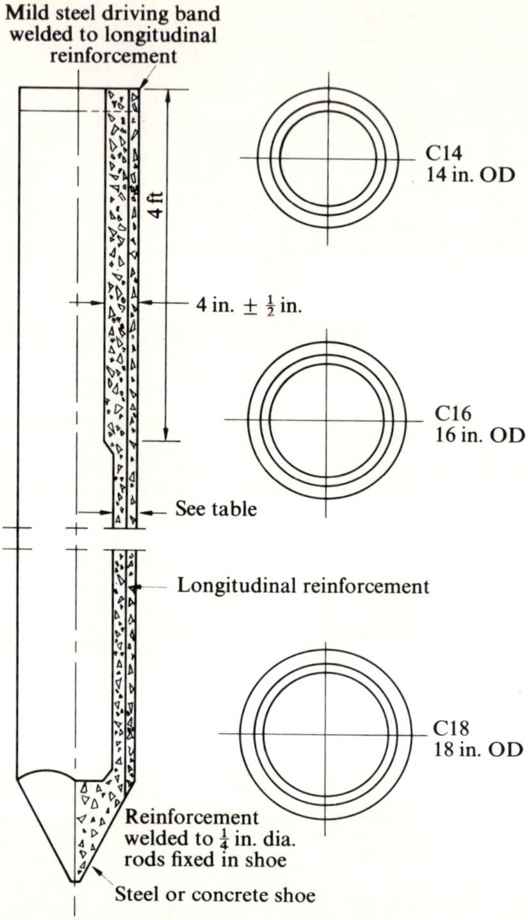

SOURCE: ROCLA Concrete Piles, Ltd., Brisbane, Australia.

NOTES ON DRIVING: Always use helmet at least ½ in thick with ID at least ½ in larger than OD of pile with a 1-in-thick mild-steel diaphragm plate. Use a timber cap block above plate and at least 1-in pine board below diaphragm and in contact with pile (spring cushion). For piles longer than 30 ft the pile hammer should weigh more than 4,500 lb.

Dimensions and loads

Type	Pile OD, in	Thickness of wall of barrel, in	Current maximum length, ft	Maximum working load, tons
C14	14	2	60	50
		2½	60	60
C16	16	2	60	60
		2½	60	70
		3	60	80
C18	18	2	40	70
		2½	40	80
		3	40	90

Approximate weight per pile, lb

Pile length, ft	Outside diameter / Wall thickness, in: 14 in, 2	14 in, 2½	16 in, 2	16 in, 2½	16 in, 3	18 in, 2	18 in, 2½	18 in, 3
20	2,080	2,400	2,415	2,815	3,120	2,830	3,210	3,610
25	2,540	2,960	2,940	3,465	3,865	3,445	3,945	4,470
30	3,000	3,520	3,465	4,115	4,610	4,060	4,680	5,330
35	3,460	4,080	3,990	4,765	5,355	4,675	5,415	6,190
40	3,920	4,640	4,515	5,415	6,100	5,290	6,150	7,050
45	4,380	5,200	5,040	6,065	6,845	5,905	6,885	7,910
50	4,840	5,760	5,565	6,715	7,580	6,520	7,620	8,770
55	5,300	6,320	6,090	7,365	8,335	7,135	8,355	9,630
60	5,760	6,880	6,615	8,015	9,080	7,750	9,090	10,490

Table B-9 STEEL SHEET PILING PRODUCED IN THE UNITED STATES*

Designation	Profile†	Weight, lb		Area A, sq in	Driving width, in	Surface area, sq ft/ft		Section properties: Axis XX						Axis YY		Producer§
								Single section			Per lineal foot of wall			Single section		
		Per foot	Per sq ft of wall			Including interlock	Coating area‡	I, in^4	S, cu in	r, in	I, in^4	S, cu in	r, in	I, in^4	S, cu in	
PZ38	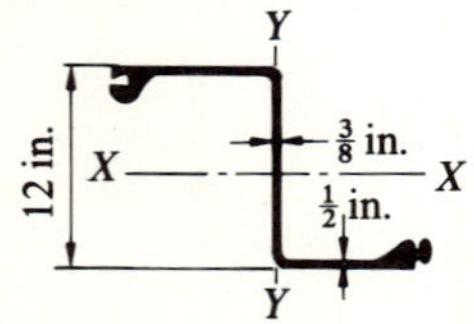	57.0	38.0	16.77	18	5.52	5.06	421.2	70.2	5.01	280.8	46.8	5.01	471.0	49.6	B,U
PZ32	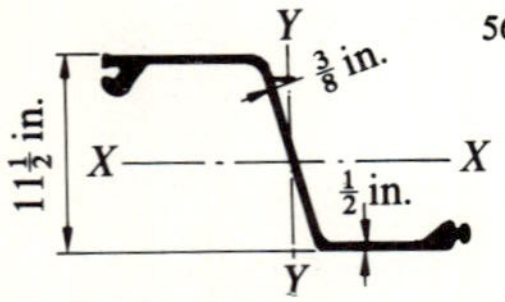	56.0	32.0	16.47	21	5.52	5.06	385.7	67.0	4.84	220.4	38.3	4.84	705.0	63.9	B,U
PZ27	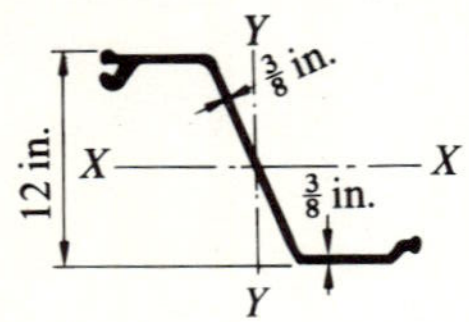	40.5	27.0	11.91	18	4.94	4.48	276.3	45.3	4.82	184.2	30.2	4.82	340.0	36.0	B,U
PDA27	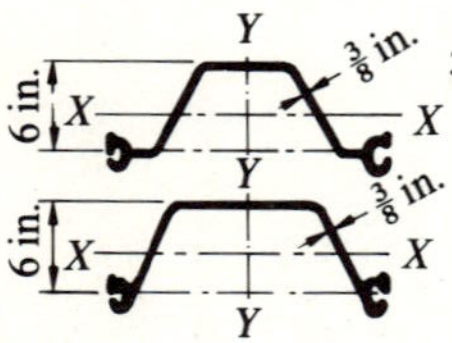	36.0	27.0	10.59	16	4.52	3.86	53.0	14.3	2.24	39.8	10.7	2.24	327.0	39.1	U B

PMA22	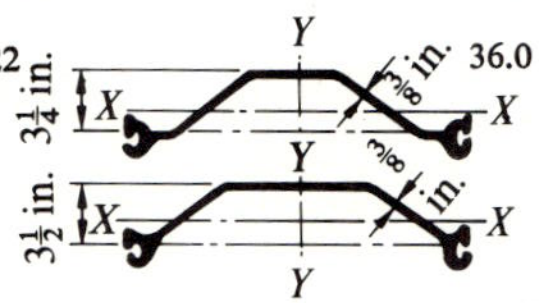	36.0	22.0	10.59	19$\frac{5}{8}$	4.54	3.88	22.4 26.0	8.8	1.57	13.7 15.9	5.4	1.57	486.0	47.9	U B
PSA28¶	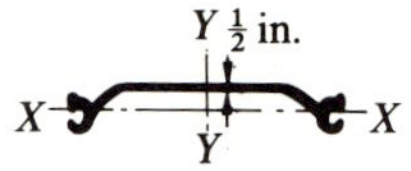	37.3	28.0	10.98	16	3.74	3.06	6.0	3.3	0.74	4.5	2.5	0.74	332.0	39.8	B,U,W
PSA23¶	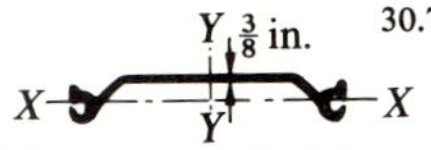	30.7	23.0	8.99	16	3.76	3.08	5.5	3.2	0.78	4.1	2.4	0.78	288.0	34.5	B,U,W
PS32¶	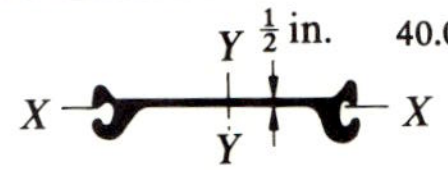	40.0	32.0	11.76	15	3.66	2.82	3.6–4.6	2.4–3.0	0.55–0.63	2.9–3.7	1.9–2.4	0.55–0.63			B,U,W
PS28¶	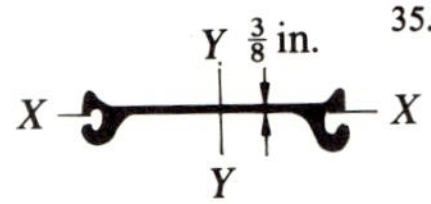	35.0	28.0	10.29	15	3.70	2.86	3.5–4.6	2.4–3.0	0.58–0.67	2.8–3.7	1.9–2.4	0.58–0.67			B,U,W
PWZ27	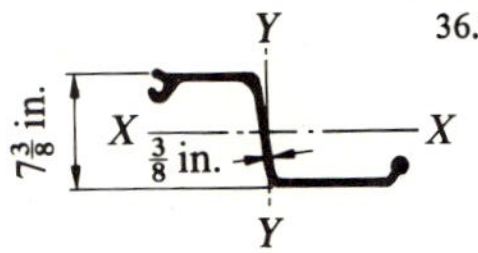	36.0	27.0	10.59	16	4.33	3.90	92.1	25.3	2.97	70.1	19.0	2.57			W
PWZ22	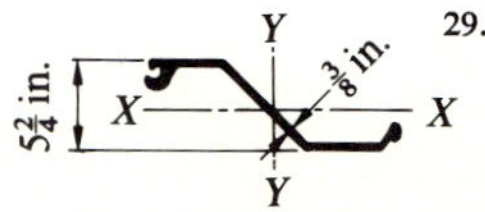	29.3	22.0	8.62	16	3.73	3.30	42.4	14.8	2.22	31.7	11.0	1.97			W

Table B-9 *(Continued)*

Designation	Profile†	Weight, lb: Per foot	Weight, lb: Per sq ft of wall	Area A, sq in	Driving width, in	Surface area, sq ft/ft: Including interlock	Surface area, sq ft/ft: Coating area‡	Section properties, Axis XX, Single section: I, in^4	Axis XX, Single section: S, cu in	Axis XX, Single section: r, in	Axis XX, Per lineal foot of wall: I, in^4	Axis XX, Per lineal foot of wall: S, cu in	Axis XX, Per lineal foot of wall: r, in	Axis YY, Single section: I, in^4	Axis YY, Single section: S, cu in	Producer§
PSX35¶	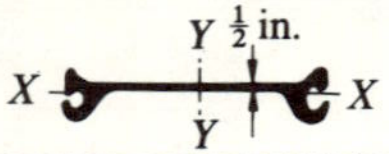 Y ½ in.	44.5	35.0	13.09	15¼	3.74	2.84	5.2	3.3	0.63	4.1	2.6	0.63			B
PSX32¶	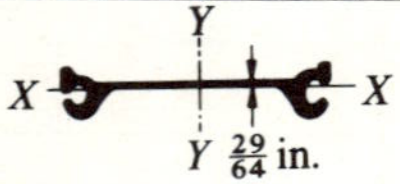 Y 29/64 in.	44.0	32.0	12.94	16½	3.88	3.10	5.1	3.3	0.63	3.7	2.4	0.63			U

SOURCE: AISI.

* Normal materal specifications: ASTM A328, ASTM A572 grades 42 through 55.

† Sections produced by different manufacturers may not interlock properly. Consult the manufacturer.

‡ Excludes bowl and ball of interlock. (Divide value by 2 for area on one side of pile.)

¶ These sections generally used in applications involving interlock strength rather than section modulus. Section properties shown for information purposes only. PSX designation refers to piling with high strength interlocks.

§ B, Bethlehem Steel Corporation; U, United States Steel Corporation; W, Weirton Steel Division of National Steel Corporation.

Table B-10 STEEL SHEET PILING PRODUCED IN EUROPE

Regular	Profile (section)	b, mm	h, mm	d,* mm	*Umfang*† (wall area) two sides, cm/m	*Stahlquer-schnitt*† (steel area), sq cm/m	*Gewicht* (weight), kg Per pile	Per sq m†	*Wider-stands-moment*† (section modulus) S, cm³	*Trägheits-moment*† (moment of inertia) I, cm⁴	*Trägeitshalb-messer* (radius of gyration) r, cm
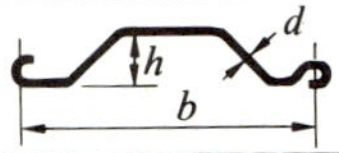	K45L	550	90	4.5	235	57.3	24.8	45	155	695	3.5
	K55L	550	90	5.5	235	70.0	30.2	55	190	850	3.5
	K62L	550	90	6.0	235	79.9	34.1	62	213	960	3.5
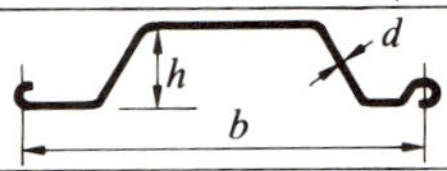	K60S	700	150	5.5	275	76.4	42.0	60	375	2,810	6.0
	K71S	700	150	6.5	275	90.4	50.0	71	444	3,320	6.0
	K76S	700	150	7.0	275	96.7	53.0	76	478	3,585	6.0
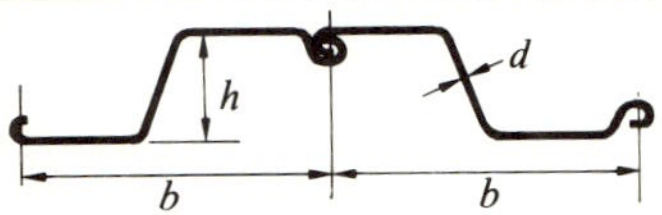	K78Z	550	200	6.5	265	99.3	42.9	78	680	6,800	8.26
	K95Z	550	200	8.0	265	121.0	52.3	95	835	8,350	8.26
	K107Z	550	200	9.0	265	136.4	58.9	107	940	9,400	8.26
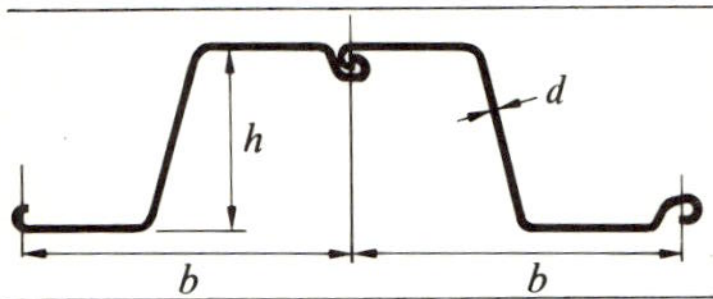	K121Z	600	340	9.0	295	154.0	72.6	121	1,800	30,600	14,08
	K134Z	600	340	10.0	295	171.0	80.4	134	2,000	34,000	14.08
	K148Z	600	340	11.0	295	188.0	88.7	148	2,200	37,400	14.08
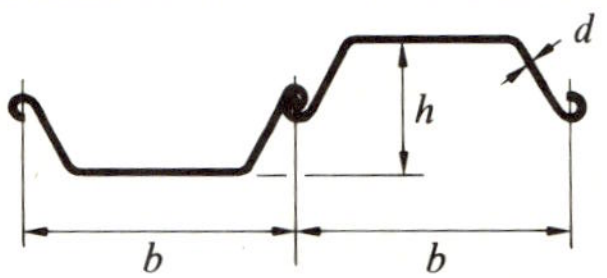	K79U	500	240	6.5	250	98.7	39.5	79	600	7,200	8.4
	K95U	500	240	8.0	250	121.0	47.5	95	714	8,570	8.4
	K107U	500	240	9.0	250	136.2	53.5	107	804	9,640	8.4
	K119U	500	240	10.0	250	151.8	59.6	119	894	10,710	8.4
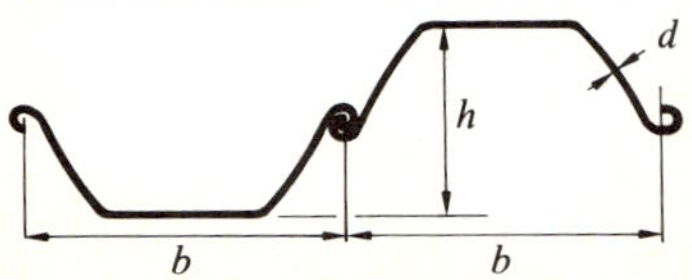	K110U	575	360	9.0	265	140.0	63.3	110	1,150	20,340	12.05
	K116U	575	360	9.5	265	148.0	66.8	116	1,200	21,600	12.05
	K122U	575	360	10.0	265	155.0	70.2	122	1,250	22,500	12.05

SOURCE: Fried. Krupp Huttenwerke, AG, 4140 Rheinhauser.

* Sections are constant thickness.

† For 1 m of wall width per meter of length.

NAME INDEX

NAME INDEX

SUBJECT INDEX

SUBJECT INDEX